CHORDATE DEVELOPMENT

A Practical Textbook With Atlases and Techniques for Experimental and Descriptive Embryology

THIRD EDITION

Text and Illustrations
by
H. Eugene Lehman

University of North Carolina
Chapel Hill

Hunter Textbooks Inc.

ISBN 0-88725-083-1

Library of Congress Catalog Card Number 77-83249

823 Reynolda Road
Winston-Salem, North Carolina 27104

ACKNOWLEDGEMENTS

This book is written with gratitude to my teachers for their generous gifts to a richly interwoven story of descriptive and experimental embryology that has been of endless interest and fascination to me. In an attempt to repay them, I gratefully acknowledge my indebtedness to Lyle L. Williams, Maryville College; Donald Paul Costello, The University of North Carolina, Chapel Hill; Victor Chandler Twitty, Stanford University; and Friedrich Baltzer, University of Bern, Switzerland.

It is a pleasure to express profound appreciation to my wife, Lillian; to former students, laboratory assistants, and colleagues who over the years have contributed ideas, and made suggestions or criticisms; and to Donald Briggs who has given me encouragement to bring this work to completion.

This, the third edition of *Chordate Development,* has benefited greatly from suggestions of users of earlier editions in classroom situations. The incorporation of corrections in fact and mechanics should substantially enhance the value of the present work. It is a pleasure to express my appreciation to all those who have helped in this way and it is my hope that they and other users of the present edition will bring additional corrections and improvements to my attention so future printings can be improved.

There have been no major changes, additions, or rearrangements of the text and illustrations in this revision. There is the same generous use of color throughout the text. It is my hope that the principles of development and directions for experimentation and observation presented in this handbook will assist in giving a new generation of students sustained interests in embryology and a desire to contribute to the analysis of development.

H.E. Lehman
Chapel Hill, 1987

TABLE OF CONTENTS

LIST OF TABLES

INTRODUCTION FOR STUDENTS

CHAPTER I

CHAPTER III

CHAPTER IV

CHAPTER V

CHAPTER VI

CHAPTER VII

CHAPTER VIII

CHAPTER IX

LIST OF ILLUSTRATIONS

INTRODUCTION FOR STUDENTS

CHAPTER I

CHAPTER II

CHAPTER III

CHAPTER IV

CHAPTER V

CHAPTER VI

CHAPTER VII

CHAPTER VIII

CHAPTER IX

CHAPTER X

FOREWORD TO INSTRUCTORS

For several decades the study of embryology has increasingly become an extension of general studies in cellular and molecular biology as applied to developing systems. Major advances are being made in chemical and ultrastructural descriptions of cell differentiation. However, embryology is more than the cellular and molecular biology of growth and differentiation. In the face of the very great sophistication in modern experimental techniques, one sometimes loses sight of the fact that there are basic principles of embryology which elevate it to the status of a major biological discipline. This discipline has its own primary principles, concepts, and rich history of monumental fact. These perspectives have over the years provided the foundation for the rational analysis of developmental processes.

It is the purpose of this work to review some of these major principles and facts as they relate in particular to the development and organization of the vertebrate body. The materials selected for laboratory study include only a very small fraction of the wealth of materials available. However, among them are some of the best direct evidence available for illustrating many basic concepts of descriptive, comparative, and experimental embryology. By providing, at the very beginning, an understanding of representative patterns and causes underlying vertebrate structure, it is hoped that a solid foundation will be laid for those who continue in the study of development.

Each chapter has a major theme or principle stated in its title. Materials appropriate to each title have been assembled for laboratory study or demonstration. Many are the traditional ones for undergraduate courses in embryology. It is hoped that their incorporation as illustrations of primary principles will enhance their intrinsic worth in the study of development.

The selection of concepts and materials included is far from complete with regard both to basic principles and representative types of organisms and to techniques. In making choices for inclusion, I have been governed by the following practical considerations.

Permanent supplies in the form of charts, models, films, and prepared slides can be obtained from commercial suppliers, some of whom are given in Appendix III. Although materials may be initially expensive and must be obtained gradually, their useful life is very long.

Where traditional species and developmental stages long used in courses in embryology continue to serve adequately for present purposes, they have been retained. Thus, there is no radical departure that entails abandoning familiar laboratory materials that have already been acquired.

Where innovations have been made, they are optional additives designed to enrich the laboratory experience. Among these descriptive materials are included the study of the Ammocetes larva of cyclostomes and the salamander larva (see Chapters 1 and 5), both of which have such high intrinsic value that they should, in my opinion, replace adult Amphioxus and the late frog tadpole respectively as preferred materials for embryology courses.

Where innovations in the use of living materials have been made, these have been chosen from species that are generally common in nature and/or can be obtained at reasonable cost from biological supply houses. Wherever possible, forms have been recommended that are hardy under laboratory conditions and that respond reliably to the treatments suggested.

The experiments on living materials have all been greatly simplified. They are designed primarily for use in circumstances where cost is a major consideration and where instructors may not be professionally trained in the arts and crafts of experimental embryology. The experiments can all be done with the improvisation of instruments and culture dishes readily available from standard laboratory stocks of supplies and materials. These methods can be greatly improved by substitutions in the event that instructors have expanded resources in expertise and equipment.

Sophisticated experiments requiring expensive chemicals, analytical electronics, or special temperature equipment have been omitted. Preference in experimental methods has been given to a representative body of strictly embryological techniques involved in the obtaining, handling, and culture of eggs and embryos under a variety of normal and experimental conditions. Most of these techniques are not included in other laboratory courses and accordingly do not suffer from redundancy with the standard methodology of physiology, genetics, or cell and molecular biology.

Each chapter includes optional slides and living materials for study with suggested supplements from models and films (Appendix II). It is recognized that notwithstanding the many omissions, this laboratory guide still has included too much material for all of it to be included in a standard, one semester course in embryology. The excess will be justified when one considers some of the special problems encountered in organizing embryology courses, namely:

a. Most animal breeding cycles are seasonal. A laboratory guide that aspires to be useful in the fall, winter, spring, or summer must take into account that alternative living materials must be suggested for use at different times of the year and the description of substitutes must be provided. Materials available in life in one season must be omitted or provided in slides, models, and films in another part of the year.

b. Fixed specific developmental time schedules are difficult to coordinate with specific class hours. Everyone who has made the attempt to include living developmental materials in scheduled laboratory periods knows the great problems and efforts that go into preparations only to find that often the embryos are too young or too old to be used at a particular time on a specific day of the week. Since failures do occur in spite of one's best efforts, there is a real need for alternative stand-by materials in the form of slides, models or films; these back-up reserves must also be described in the directions.

c. Slides, models, and films are often prohibitively expensive. It is unlikely that any single course at the beginning will be equipped with all materials suggested for study in this guide. Most slide and demonstration materials have a long laboratory life and the teaching collection can be built up with nominal annual expense so that within a few years a reasonably comprehensive set of course materials will be on hand for routine use. In recognizing these facts from the beginning, an attempt has been made to provide a reasonably complete set of descriptions that will serve as substitutes for primary living or preserved materials which for one reason or another may not be available at a given time. In large university classes it is becoming more and more common to make the laboratory aspects of a course optional. For students who take a "lecture only course" in embryology, this guide could serve as a marginal substitute for direct laboratory contact with primary materials.

d. Detailed drawings and diagrams beyond the immediate needs of an introductory course have been included. The justification for these illustrations rests on several grounds. Although drawing is a valuable learning technique for students to acquire, excessive amounts of required drawing will greatly reduce the amount and diversity of laboratory experience a student can have. To encourage more laboratory experience on the part of students, complete sets of drawings of slide materials have been provided. For the general student, detailed illustrations serve to direct attention to refinements of information otherwise easily overlooked. It is also hoped that the guide will provide a permanent reference source to persons who go beyond the introductory level and need occasional morphological information on developmental stages in the design of advanced experimental work.

e. Explanatory information is given on the significance of factual data. The inclusion of theoretical implications and interpretations relating to embryology, evolution, and experimental design is justified in that it helps the student to fit the laboratory work into a conceptual framework of total development. On the practical side, the inclusion of these ideas also relieves the necessity of having supplementary reference materials in the laboratory where a limitation in desk space is often most critical. It is not, however, to be implied that the brief account given here is any substitute for reading a comprehensive text on general embryology and development.

f. There is redundancy in reviewing patterns of embryonic development in representative chordate groups. The degree of repetition is directly proportional to the degree to which different forms reveal the basic developmental common plan for the phylum. This kind of repetition is unavoidable if a comparative approach is given to the subject. Repetition has the educational advantage of reinforcing general principles but, more importantly, it provides the student with an opportunity independently to differentiate between the general characters and the uniquely specific ones that are encountered among organisms. Thus, by examining corresponding stages, the student learns to recognize homologies. Expressed in another way, experience is given in distinguishing between significant and trivial details. These and other factors have all contributed to the length and, hopefully, to the greater usefulness of this set of laboratory instructions.

The interpretation of evolutionary and embryological similarities and causes has for the most part followed traditional explanations from the older literature. In a short work such as this, inclusion of all known variations and interpretations would render the main theme and generalizations obscure to the beginning student. It is hoped that specialists will be tolerant of omissions of their own work and of current views which may differ from some of the oversimplified and occasionally categorical statements that are made without qualification. In some cases this is surely due to my ignorance, but in many more the simplifications have been deliberate in order to keep the size of this book within reasonable bounds.

Traditional terminology has been retained in all cases where it continues to serve a useful purpose. However, in certain cases I have taken liberties in the creation of new terms where old ones seem inadequate or misleading. An attempt has been made to select among synonyms for homologous structures the term that is most widely used in mammalian and human embryology. This is a concession to the fact that most undergraduate students in introductory embryology courses have professional goals in medicine or health related science. The use of mammalian terms is of long term value to them and increases the relevance of the course to their professional goals.

A word of explanation is in order for the usage of italics and capitalization in association with the names of organisms in this work. In keeping with standard practice in the technical literature, italics are used only with complete scientific names (i.e., when both generic and specific names are combined as with *Rana pipiens* or *R. pipiens,* etc.). Nude generic names are not italicized; however, generic names and names of higher systematic categories are capitalized when given in taxonomic rather than Anglicized form (i.e., Amphioxus, Ascaris, Amphibia, Mammalia, but amphioxids, ascarids, amphibians, mammals, etc.).

In all cases the descriptions of embryological materials have been based on direct observation and study. Likewise, the methods given for the handling of living and preserved materials have been tested in laboratory situations over a number of years at the University of North Carolina-Chapel Hill and at the Bermuda Biological Station. All illustrations are original. Aside from the obviously schematic and diagrammatic text figures (some of which are modified from acknowledged sources, and others from *Laboratory Studies in General Zoology,* H.E. Lehman, 1981, Hunter Publ. Co.), all drawings have been made directly from life or from preserved or sectioned materials. The lack of professionalism in draftsmanship is explained on grounds that professional assistance on such a large scale would have increased the cost of publication beyond that which an average student could afford. I have therefore elected to do my own artwork, such as it is.

Criticism can be leveled at the amount of space devoted to descriptive and comparative morphogenesis of vertebrate organ systems. I will hasten to recommend that, wherever it is at all possible, one should elect to use living rather than preserved materials. It is a regrettable fact, however, that as university classes become larger and multisection laboratories become the rule, the efforts and expense in providing living laboratory materials expand logarithmically. One simply cannot provide the same freedom for individual study to classes with several hundred students that is accorded one or two dozen if the same laboratory space and equipment is available for both. In large classes, therefore, one is often pressured into a reliance on models, films, and preserved materials that can be counted on to meet rigid scheduling demands for time, space, and equipment.

As to criticisms that may be leveled at this work for providing detailed information and removing from the student the joys of independent discovery, I will simply reply by quoting from the old but admirable *Laboratory Manual for Comparative Anatomy* by Libby H. Hyman who said in reply to a similar criticism, "It is my opinion that human beings in general see chiefly that which is pointed out to them; this has been proved over and over again in the history of biology. The large number and complexity of the anatomical facts to be acquired, the limited time allowed for their acquisition, and the large size of the classes, and the limited number of laboratory assistants available seem to me to necessitate that detailed and specific laboratory directions be provided. If the directions are not given in the manual, then the laboratory instructors are compelled to provide them verbally" (page vii, University of Chicago Press, 1922). The present text is designed to instruct in the belief that the larger the body of factual information acquired by individuals, the more rational and profound will be the questions that are raised by the inquiring mind. I have not included so-called "thought questions" and leave these to the ingenuity and personal predilections of the instructor who may choose to use this book as an aid to teaching. In all candor, I must add that when I have used laboratory aids with such questions, they have usually been so inane or abstruse as to be virtually useless as a teaching method. Students will almost always ask the assistant for the answers and assistants will check out their reply with the instructors. I see little virtue in going around this "Adam's barn" if, in the end, one is to give the answers to students anyway.

Finally a word can be said concerning the overall organization of subject matter. Embryology, probably more than any other general biological discipline, has an intrinsic organization that can hardly be improved upon. It is reasonable to begin with gamete formation and proceed to fertilization, cleavage, gastrulation, and organ formation in the same sequence in which the embryo unfolds the phenotypic expression of its final body plan. There should be compelling reasons for departing from this schedule as I have done with the materials presented in Chapter 1. Chapter 1 deals with the concept of the vertebrate body plan as seen in larvae of tunicates, Amphioxus, and cyclostomes. This chapter originally was written to enter the sequence between the chapters presently numbered 4 and 5. That is to say, between a consideration of amphibian neurulation in Chapter 4 and amphibian larval development in Chapter 5. That is still a reasonable place for the material of Chapter 1 to be reinserted should one so desire. The reasons for moving this subject matter to an introductory position rest on practical considerations. A great number of my students come to embryology after having had a course either in general vertebrate biology or in vertebrate comparative morphology and evolution. Therefore most of them are familiar with the vertebrate common plan. This review at the outset provides a valuable link between familiar ideas from former courses and the special subject matter of embryology. A second consideration rests on the fact that not all students have had this vertebrate background. For them it is particularly desirable to have a brief overview of concepts on the vertebrate body and its primary organization. The material in Chapter 1, therefore, serves to equalize the disparity and to provide a framework of common knowledge upon which the remaining evolutionary and morphogenetic material in the course can rationally be arranged. The materials described in Chapter 1 are rarely included in an embryology course. Instructors may choose to begin the actual laboratory experience with Chapter 2. In that case Chapter 1 will be supplemental to the general information on vertebrate structure reviewed in the Introduction.

Last, one will note two major omissions from this laboratory guide, namely, the 96 hour chick and the development of fish. With regard to the former, I feel that, if one is to choose between the 96 hour chick and the 10 mm pig, there can be no doubt that the pig is more instructive of late amniote development. From the point of view of student interest in mammalian forms, the study of the pig generates far more spontaneous interest than does another chick stage. As for the fish, it was omitted entirely for practical reasons of economizing on space. Should there be an indication of sufficient interest in this subject for experimental reasons, a section will be added in subsequent printings of this work.

In conclusion, it is hoped that the instructions given in the following chapters are sufficiently detailed and clear that the subject of vertebrate embryology will essentially teach itself to those who elect to follow this program of study. Embryology is inherently vested with more natural interest than any other biological discipline. Coupled with ideas of vertebrate evolution and experimental morphogenesis, it is an unbeatable combination. It is hoped that this study will enhance that interest through increased comprehension of patterns and mechanisms that are responsible for the growth and development of the vertebrate body.

SUGGESTED LABORATORY SCHEDULES FOR THE STUDY OF EMBRYOLOGY

LABORATORY SUBJECT MATTER	DEMONSTRATIONS, MODELS, & FILMS	DESCRIPTIVE SLIDE STUDY	LIVING & EXPERIMENTAL METHODS	ALL DESCRIPTIVE	ALL EXPERIMENTAL	MIXED FALL PLAN	MIXED SPRING PLAN	INSTRUCTOR'S OPTION
CHAPTER I:								
Tunicate larvae and metamorphosis	+	+		+				
Amphioxus larvae and metamorphosis	+	+		+				
Ammocetes larva, whole mount & sections	+	+		+				
CHAPTER II:								
Gametogenesis in the sea urchin ovary		+		+		+	+	
Gametogenesis in the frog ovary		+		+		+	+	
Gametogenesis in the mammal ovary		+		+		+	+	
Gametogenesis in the mammal testis		+		+		+	+	
Maturation in the Ascaris uterine eggs	+	+		+		+	+	
Fertilization of living echinoderms	+		+		+	+	+	
Polytene chromosomes of Chironomous			+		+	+	+	
CHAPTER III:								
Spiral determinate cleavage in Ilyanassa			+		+		+	
Early development of the starfish	+	+	+	+	+	+	+	
Early development of the sea urchin	+	+	+	+	+	+	+	
Early development of tunicates	+							
Early development of amphioxus	+							
Early development of frogs to neurulation	+	+	+	+	+	+	+	
Fertilization of living frog eggs	+		+		+		+	
Ovulation and egg transport in living frogs	+		+		+	+	+	
CHAPTER IV:								
The 4 to 5 mm frog tailbud embryo	+	+	+	+	+	+	+	
The 6 to 7 mm frog hatching larva	+	+	+	+	+	+	+	
Induction experiments with inverted frog eggs			+		+		+	
Pressure experiments on frog blastopore lips			+		+		+	
CHAPTER V:								
Salamander larvae, whole mount & sections		+		+		+	+	
Operations on frog or salamander eggs:								
by extirpation of primordia	+		+		+		+	
by transplantation of primordia	+		+		+		+	
by isolating primordia in tissue culture	+		+		+		+	
CHAPTER VI:								
The 18 to 24 hour chick embryo	+	+	+	+	+	+	+	
The 33 hour chick, whole mount & sections	+	+	+	+	+	+	+	
Camera lucida reconstruction of the 33 hour chick		+		+	+	+	+	
The use of vital dyes on living embryos			+		+	+	+	
CHAPTER VII:								
The 48 hour chick, whole mount & sections	+	+	+	+	+	+	+	
Circulatory injection & embryo dissection	+	+	+	+	+	+	+	
Fixing, staining & mounting whole embryos			+		+	+	+	
CHAPTER VIII:								
The 72 hour chick, whole mount & sections	+	+	+	+	+	+	+	
Injection of chemicals into living eggs			+		+	+	+	
Primordium removal by heat cautery			+		+	+	+	
Organ transplants on the chorionic membrane			+		+	+	+	
CHAPTER IX:								
The 10 mm pig, serial sections		+		+		+	+	
Living mouse fetus and placenta			+		+	+	+	
CHAPTER X:								
Microdissections of advanced embryos			+		+	+	+	
In toto skeletal staining of embryos			+		+	+	+	
Silver nitrate thick section preparations			+		+	+	+	

INTRODUCTION FOR STUDENTS

TABLE OF CONTENTS

TABLES AND ILLUSTRATIONS

INTRODUCTION FOR STUDENTS

This book has been written as an aid to students in their study of vertebrate embryology. It is designed to be used with the microscopic study of slides and living embryos, and with the experimental study of mechanisms responsible for the unfolding patterns of development in different organisms.

A number of special features are included in the preparation of the book to make it easier to learn the facts and principles of embryology. **Bold face type** is used to highlight technical terms that should be learned, and to direct attention to structures that should be identified in text figures and/or on embryos being studied. A smaller type is used for the legends and explanations of text figures of whole mounts and serial sections of embryos. In many cases, the legends are prepared as "checklists" to be used during study of slides so that all important features will be seen during the laboratory study. Occasionally, items will be included as part of a checklist that are not labelled in some of the drawings. An attempt has been made to be consistent in the use of the same abbreviations for homologous parts in embryos of different age, or species. This coordination of abbreviations will assist greatly in the recognition of similarities in the comparative study of development in different vertebrate groups.

The **glossary** at the end of the book gives a short definition for the more important embryological terms that are introduced. Abbreviations accompany the terms in the glossary for which common symbols are used in figures throughout most of the accompanying text.

The incorporation of theoretical and explanatory information into this text, which is primarily a practical guide to the study of vertebrate embryos, is designed to provide a connecting bridge between laboratory study and the information obtained from reading a more conventional textbook on embryology or developmental biology.

Embryology arose in the mid-nineteenth century as a distinct biological discipline under two parallel stimuli. The first and foremost of these was the work of Ernst von Baer, the so-called father of embryology. He wrote the first general textbook of comparative embryology and first clearly set down the primary principles of development. These, now known as von Baer's Laws, will be summarized in Chapters 8 and 9. The second stimulus was an outgrowth of the overwhelming impact of evolutionary theory following the publication of Darwin's great work on the origin of species. Darwin himself made little use of embryological evidences in his theory of natural selection. However, Ernst Haeckel advanced his recapitulation theory with great vigor; some of the evidence for his theory is reviewed in Chapter 5. Since the recapitulation theory proposed that embryonic stages bear a close correspondence to evolutionary stages in the origin of species, this stimulated a tremendous incentive for the study of embryology in all forms of animal life. By the end of the past century detailed information of great value had accumulated for virtually all taxonomic levels of the animal kingdom. This was the golden age of descriptive and comparative embryology. With the turn of the century, research trends in embryology became analytical and more concerned with causes than with relationships.

It is pertinent at the beginning of this study to review the general concepts of biology which are fundamental to embryology. Most of these ideas have been part of the experience of students in introductory biology courses. It is a saving of time to review them, nonetheless, in the event that these ideas and terms have been omitted or forgotten.

A. THE GENERAL NATURE OF EMBRYONIC DEVELOPMENT

The study of embryology concerns the initial period in the life history of an individual. It extends from the origins of gametes in parents through the development of a zygote into a free-living being possessing the major characters of its species. The central problems of embryology concern the origin of complexity in structure and an understanding of the causes that produce these changes. The factual body of embryology falls into three major categories, namely:

- **a.** The **description** of normal developmental processes in different species.
- **b.** The **comparison** of developmental histories for the purpose of establishing homologies and evolutionary relationships among species.
- **c.** The **experimental analysis** of developmental processes with the aim of explaining and discovering the causes underlying developmental change.

Embryology is one of the general biological disciplines (along with genetics, physiology, cytology, taxonomy, etc.) and, as such, has application across the entire breadth of the animal kingdom. More specifically, it has value in providing answers and information on problems relating to such diverse subjects as mechanisms for reproduction, development and growth; for validity of classification based on homology; for the activation of genetic and enzymatic synthetic systems during ontogeny; and for the rational understanding of anomalies in human development which are of special interest to the medical sciences.

The embryonic period in the life history of most metazoans is arbitrarily divided into the following stages, which are, however, continuous and generally overlap one another:

MAJOR PERIODS IN DEVELOPMENT

I. GAMETOGENESIS. The origin, growth, and specialization of eggs and sperm in parent bodies.
II. FERTILIZATION. The union of gametes with the activation of the egg or zygote to undergo total development.
III. CLEAVAGE. The early mitotic period which generates a multicellular system essential for further differentiation and morphogenesis.
IV. GASTRULATION. The formation of the primitive gut, the separation of primary germ layers, and the establishment of axial symmetry.
V. ORGANOGENESIS. The molding of organ rudiments from germ layer derivatives.
VI. INTEGRATION. The final functional coordination of structure and function that results in a unified organism.

Each chapter that follows will deal with one or more of these basic stages and their inherent problems. Materials have been selected which show to special advantage the various facets of the large and complex field of embryonic development. Emphasis is clearly on the vertebrate story of development. However, wherever invertebrate materials are advantageous, they are included to throw additional light on the problems at hand. A special attempt has been made to employ descriptive, comparative, and experimental approaches to the problems of vertebrate development.

B. GENERAL LABORATORY INFORMATION ABOUT MATERIALS AND PROCEDURES

With the approval of the instructor, the following general laboratory procedures are suggested as routine regulations that will contribute toward an efficient use of laboratory time.

1. Laboratory equipment and attendance. It is intended that the laboratory directions given in the following chapters will be used directly in conjunction with slides, models, and living materials provided in the laboratory. Although these directions and illustrations are complete, some oral announcements will always be given at the beginning of each laboratory period. Promptness in attendance is essential at all laboratory periods. In laboratories in which living materials are scheduled, lateness cannot be tolerated.

a. Personal equipment. Each person should bring a copy of the laboratory directions, a dissecting kit (containing a small scissors, two dissecting needles, two pairs of fine stainless steel forceps, and a small scalpel), a drawing kit (containing eraser, ruler, double weight paper, Manila envelope, and black, blue, red, orange, and green colored pencils), and a small towel or wash cloth. In general, lecture notes and textbook will not be needed but probably would be useful if not burdensome.

b. Course equipment. Each student will have a compound microscope and a dissecting microscope for personal use, and a set of laboratory slides of materials to be studied during the course. A key will be issued for laboratory lockers in which materials can be kept with security assured.

c. Demonstration materials. Charts, models, and special demonstration slides and whole mount preparations will be available for study at appropriate periods. These are an integral part of the study. Allow time for appropriate study of these important supplements.

2. Purpose and methods for laboratory drawings. Making laboratory drawings forces one to observe details and relationships among parts carefully. Drawings, when accurate and fully labelled, provide a valuable record that can be relied on for future reference.

Although the laboratory directions are accompanied with reasonably complete sets of drawings, each slide preparation is of unique value in its own right. Original drawings are therefore corroborative and additive to the body of scientific facts available for reference information. The following printed directions include very few specific drawing requirements. However, from time to time drawings will be suggested. They should always vary significantly in details from any drawings in the laboratory directions, and they should be made with the following instructions in mind.

a. Before starting any drawing, you should thoroughly study and understand the material. Then select the best and most instructive material to record, and decide upon the best magnification and arrangement of the drawing or drawings on the page.

b. Use a double weight drawing paper with a hard surface that will permit erasure. Use a sharp 3-H or 4-H drawing pencil. First lightly sketch the outline of the material and the major topographic features to be included in the drawing. Make constant reference to the original material. Gradually correct the original sketch until its proportions are optically correct in the two-dimensional plane of the paper.

c. Erase all unnecessary guide lines and then make clear sharp outlines of all important features to be shown. Omit unessential or uninformative details. Drawings should always be accurate representations of actual laboratory materials, but it is not desirable blindly to include miscellaneous trivia or dramatic shading that obscures rather than reveals details.

d. Draw on only one side of the page and, in general, plan an aesthetically pleasing arrangement of labels which start at even margins on the right and left side of the page. Label lines should not cross one another, and they should end in, rather than point at, designated structures. Print labels in a clear block style that can be easily read. Label all structures that are drawn.

e. After the drawing is complete and labelled in black, colors may be used to indicate the germ layer origin of special structures. This is an optional additive that should be used sparingly and only for special graphic value. In no case use color merely to mimic the stains on a slide.

f. Last, title the drawing at the bottom, giving the species, developmental stage, plane of view, or section. In the upper right hand corner give your name, laboratory seat, section number, and the date turned in.

g. All required drawings should be turned in on the day of the practical laboratory examination on the material covered by the drawing.

h. Additional optional sketches and drawings are encouraged as experience in illustrated note taking. These should be made whenever individual specimens depart significantly from the record in the laboratory guide. They will serve as valuable reminders of the range in variation that can be expected in the study of individual specimens of similar age.

3. Laboratory evaluation and grades. Laboratory work may be graded on three types of performance.

a. Required drawings prepared in the manner described above will be graded for accuracy, completeness, and originality.

b. Laboratory reports on experiments will be prepared in two to three page write-ups using the following organization unless otherwise directed: I. Statement of Problem, II. Description of Materials and Methods, III. Observations and Results, IV. Discussion and Conclusions, and V. Illustrations including drawings or tables of data as appropriate to the work in question.

c. Laboratory practical examinations will be given in the identification of collateral materials either by projection or individual location of structures on slide and demonstration materials.

The relative value or weight given to each category will be found in the laboratory schedule provided separately at the first laboratory meeting of the course.

4. Care and use of the microscope. (Fig. Int.-1). The microscope is a precision instrument which must be handled with understanding and care if it is to be kept in efficient working order. In general, only low and medium power magnifications will be necessary for the study of most materials. Particular care should be given to whole mount preparations which are exceptionally thick. They usually do not permit the use of short focal length (high power) objectives. To prevent damage to slides and to the microscope, always observe the following rules.

Rules for the Use of the Microscope

a. Always pick up the microscope by the arm; place it on the table about 4 inches from the edge with mirror away from you. Clean the lenses with lens paper provided for that use. Do not tilt the microscope body and stage. This is particularly important when you are using fresh preparations with liquids.

b. Revolve the nosepiece to bring the low power objective (the shortest) into line with the body tube. Correct alignment is indicated by a slight click of the spring catch which holds the

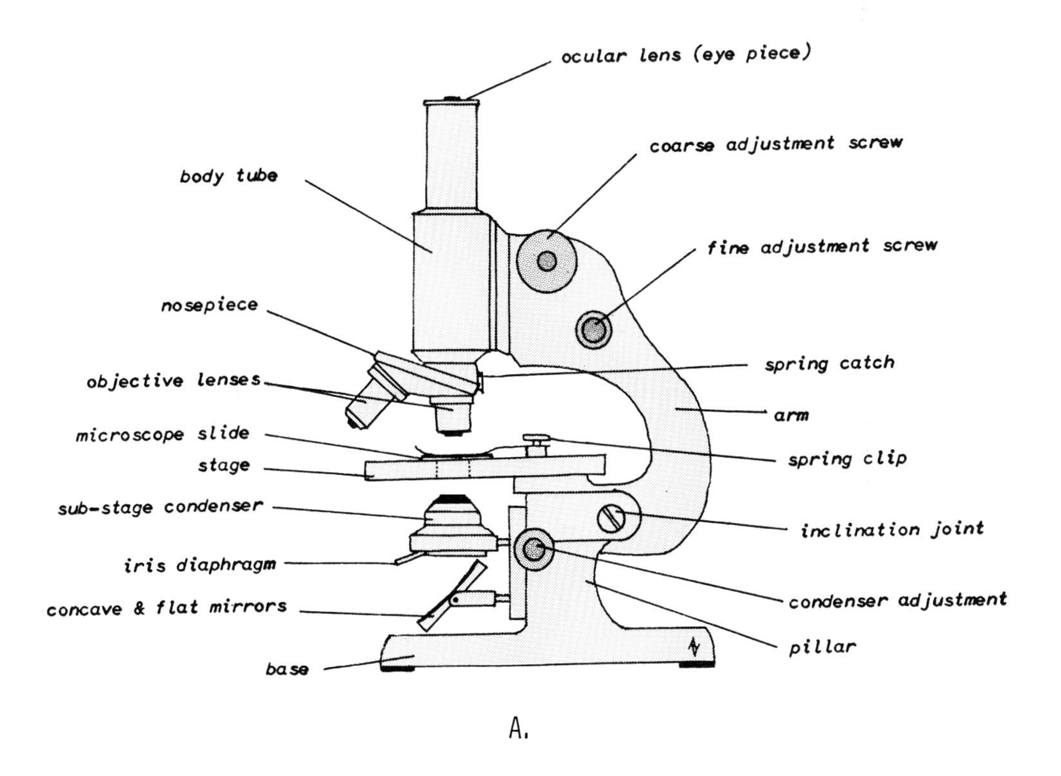

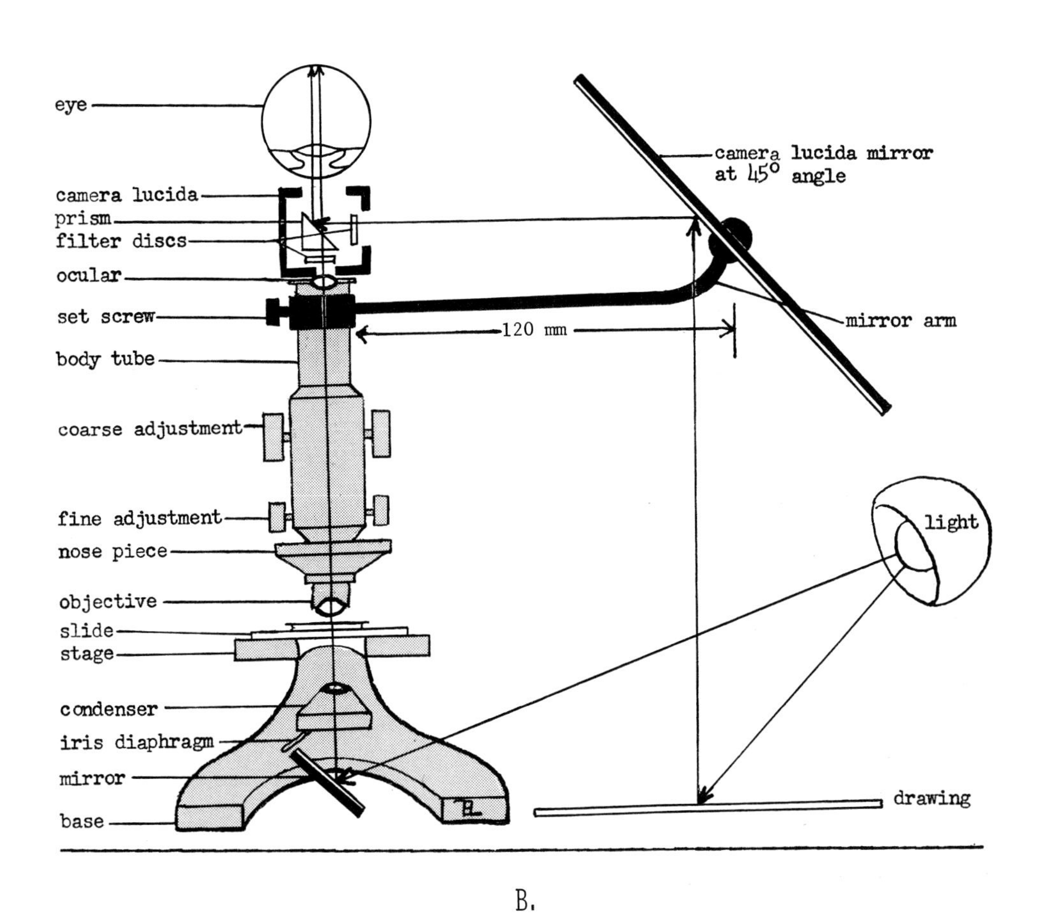

Figure Int.-1 Diagrams of microscopes and camera lucida. 1 A. Diagrammatic side view of a standard student microscope. 1 B. Diagrammatic front view of a microscope with a camera lucida attached.

nosepiece in centered position. Look through the ocular and **adjust the light** by moving the mirror and iris lever until uniform bright illumination is obtained over the entire field. Use the plane surface of the mirror when using a condenser lens.

c. Place a prepared slide on the stage with the coverslip up. Anchor the slide under the spring clips. Then, watching from the side, move the slide so that the object to be examined lies in the center of the stage aperture.

d. Looking from the side, lower the body tube by means of the coarse adjustment screw so that the tip of the low power objective is about 1/4th inch from the cover glass surface.

e. Look through the ocular and slowly focus upward with the coarse adjustment until a clear image comes into view; correct the focus with the fine adjustment screw.

f. Move the slide so that the object lies in the center of the field. Readjust the light to get uniform illumination and most favorable brilliance for the subject matter in question.

g. For use of medium or high power objectives, **always begin with the low power and get the object centered with maximal illumination.** Then, while watching from the side, turn the nosepiece slowly until the medium power objective comes into line. Do not raise the body tube unless you see that the end of the objective is in danger of touching the cover glass as may happen in some of the thick "whole mount" preparations. Look through the ocular and correct the focus with the fine adjustment. Center the object, and adjust the illumination as described above. To get to high power, repeat the procedure outlined above for making the transition from medium to high power. **Never under any circumstances try to find an object under medium or high power without first finding it under low power.** Also, **never focus down with the coarse adjustment while looking through the ocular.** This is the most common single cause for damaged slides and chipped lenses.

h. While examining an object, continually use the fine adjustment to bring into focus details at different levels of the object; this is particularly necessary on whole mount preparations. **Use the lowest magnification which will show you the details with which you are particularly concerned.**

Microscope magnification. The magnification of the compound microscope is achieved by combining the powers of magnification of an **ocular (eye piece) lens** and an **objective lens** (located on the nosepiece). The microscope is usually supplied with two different powers of oculars. The longer ocular is marked 5× or 6× to indicate that it has a magnifying power of 5 or 6. The shorter ocular is 10×; this is the one that will be used for most laboratory materials.

There are three objective lenses. The shortest is the lowest power and has a magnifying power of 4×. The medium power objective magnifies 10×, and the longest objective is the highest power and magnifies 44×. Note that with an increase in magnifying power, the size of the lens diameter decreases, permitting less light to enter the body tube. The focal length of these lenses (i.e., the distance of the lens surface from the material viewed on the microscope stage) is also stamped on the side of each lens mount. The 4× objective has a focal length of approximately 32 mm; the 10× objective has a focal length of 10 mm, and that of the 44× lens is about 4 mm.

The magnification obtained in any given ocular and objective lens combination is the product of the individual magnifying powers of the separate lenses. That is, a 10× ocular combined with a 10× objective will give a magnifying power of 100×. Table Int.-2 gives the magnifications obtained with the above combinations of objective and ocular lenses.

A practical method for estimating the size of objects seen and drawn under the microscope. Estimating the real size of objects viewed under the microscope can be approximated by use of the following instructions and the information given in Table Int.-1.

a. First determine the optical diameter of the objective lens being used.

Low power (4× objective & 10× ocular) = field diameter of about 3.0 mm.

Medium power (10× objective & 10× ocular) = field diameter of about 1.0 mm.

High power (44× objective & 10× ocular) = field diameter of about 0.3 mm.

b. Next estimate the relative size of the object in the microscopic field and calculate its actual size as seen in the microscope (See Table Int.-1).

For example, assume that an object is one-quarter the total field width, then:

At low power it would really be 0.75 mm long (i.e., 25% × 3 mm = 0.75 mm).

TABLE INT.-1
REFERENCE TABLE FOR ESTIMATING THE ACTUAL SIZE OF OBJECTS BASED ON THEIR APPARENT SIZE WHEN VIEWED IN THE MICROSCOPE

OPTICAL FIELD DIAMETER IN MM		ACTUAL SIZE IF THE OBJECT IS OF THE FOLLOWING APPARENT SIZE:			
		25% of field width	50% of field width	75% of field width	100% of field width
Low power	(3.0 mm)	0.75 mm	1.5 mm	2.25 mm	3.0 mm
Medium power	(1.0 mm)	0.25 mm	0.5 mm	0.75 mm	1.0 mm
High power	(0.3 mm)	0.075 mm	0.015 mm	0.225 mm	0.3 mm

TABLE INT.-2
MICROSCOPE AND CAMERA LUCIDA MAGNIFICATIONS

A. TABLE OF MICROSCOPE MAGNIFICATIONS

OCULARS	OBJECTIVES 4× (32 mm)	10× (15 mm)	45× (4 mm)
5×	20×	50×	225×
10×	40×	100×	450×

B. CALCULATE CAMERA LUCIDA MAGNIFICATIONS WITH MIRROR ARM SET AT 120 mm*

OCULARS	OBJECTIVES 4×	10×	45×
5×			
10×			

* See instructions for use of the camera lucida in text.

At medium power it would be 0.25 mm long (i.e., 25% × 1 mm = 0.25 mm).
At high power it would be 0.075 mm long (i.e., 25% × 0.3 mm = 0.075).

c. If you are also making a drawing of the object under the microscope, or if you wish to determine the magnification of a textbook illustration, you can easily do so.

First, measure the length of the drawing with a millimeter rule.

Next divide the measured length of the drawing by the estimated real length of the object as calculated (by the method given in Part b, above) for the observed subject under the microscope at a particular magnification (See Table Int.-1). The value obtained will approximate the real magnification of the drawing.

For example, assume that the apparent size of an object in the microscope is one-quarter of the diameter of the optical field, and, if the drawing in each case is 60 mm long, then:

At low power the drawing magnification would be 80× (i.e., 60 mm / 0.75 mm = 80×).

At medium power the drawing would be 240× (i.e., 60 mm / 0.25 mm = 240×).

At high power the drawing would be 800× (i.e., 60 mm / 0.075 mm = 800×).

It takes very little time to estimate magnifications of objects and drawings. A record of these values is of interest and enhances the scientific value of drawings.

5. Use of the Camera Lucida. The camera lucida is a microscope attachment consisting essentially of a mirror and prism which enables you to see a microscopic subject and trace the material at the same time. The image of your hand and pencil are superimposed on the microscopic field in such a manner that you can record complicated details of microscopic structure with ease and extreme accuracy without distortion. Use the following directions for the attachment and use of the camera lucida.

Rules for the Use of the Camera Lucida

a. The microscope stage must be horizontal.

b. Place a piece of drawing paper and a pencil on the desk close beside the microscope. Move the desk lamp so that it will illuminate the paper brightly without actually being over the paper. Then, locate and center an object under low power according to the previous instructions and get **maximal** illumination through the microscope.

c. Insert the camera lucida mirror arm into the prism assembly and fasten it firmly with the setscrew. The mirror arm should be locked at 120 mm at each subsequent use in order that the same scale of magnification can be duplicated.

d. Remove the ocular from the microscope and slip the camera lucida attachment ring over the body tube. Replace the ocular and lower the prism assembly into position over the ocular. Raise the camera lucida attachment ring just high enough to permit the prism assembly to rest flat, 1 mm above the ocular. Tighten the setscrew on the attachment ring so that it is rigidly fastened to the microscope body tube.

e. Raise the camera lucida prism assembly and set the filter discs on the side and bottom **at open.** Tilt the camera lucida mirror to 45° as indicated by the scale on the lever arm.

f. Lower the camera lucida prism assembly over the ocular and look through the prism aperture and microscope. Center the prism over the ocular lens by means of the lateral adjustment screw located at the back of the camera lucida attachment ring. The prism is centered when a circular field of vision can be seen through the microscope.

g. Place a pencil so that the point rests in the center of the area for your drawing. While looking through the prism assembly and microscope, move the drawing paper until the image of the pencil tip rests on the center of the object you intend to draw.

h. Adjust the illumination so that you can see the microscopic object and the pencil at maximal illumination. To do this, first move the lamp either closer to the paper or to the microscope mirror, depending upon which is dimmest. If you cannot get a good image of the pencil or the microscope object by this means, rotate the filter disc rings to reduce the light on the brightest field (filter discs on the side of the prism assembly control light from the drawing; filter discs on the underside of the prism assembly control light coming through the microscope).

The camera lucida magnifies an object from 3 to 5 times over that obtained by the microscope. In using the camera lucida, **it is generally necessary to go to one power less than that used in studying the material** in question, otherwise the drawing will be too large for the page size. The magnification of the camera lucida can be determined by comparing the "apparent length" of an object as seen or drawn with the camera lucida with its actual length. The quotient obtained when the "apparent length" is divided by the actual length is the magnification. **Calibrate your camera lucida** by placing a millimeter rule on the microscope stage and drawing the "apparent length" of 1 millimeter on the paper. Use each of the lens combinations indicated in Part B of Table Int.-2. The length of your drawings of "apparent lengths" in millimeters is the magnification obtained in each lens combination. Record the camera lucida magnifications.

C. GENERAL TERMINOLOGY RELATING TO THE ORGANIZATION OF STRUCTURE IN EMBRYOS

1. Concepts of polarity, axes, and planes of symmetry (Fig. Int.-2). These terms all refer to spatial and geometric distributions of substances or structures in eggs, embryos or adult organisms. Eggs of vertebrates all begin development as radially symmetrical systems. Gradually they acquire bilateral symmetry during the course of embryonic differentiation. The process of acquiring bilateral symmetry is called **axiation.** Structures in the body which are aligned with the various body axes are said to be axial organs. Among the major axial organs are the nervous, sensory, muscle, skeleton, locomotor appendages, excretory, and circulatory systems. In perfectly bilateral animals, all organs are axiate with regard to one or more of the body axes and planes of symmetry.

The following categories deal first with poles, then axes, and finally planes and body plans of symmetry in the major kinds of organization in animal bodies.

a. Poles of symmetry are the various ends or surfaces of the body.

DORSAL = The back surface of the body
DORSAD = Toward the back

VENTRAL = The belly surface of the body
VENTRAD = Toward the belly

ANTERIOR = The front end of the body
ANTERIAD = Toward the head

POSTERIOR = The tail end of the body
POSTERIAD = Toward the tail

MEDIAN = The middle of the body
MESIAL = Toward the midline

LATERAL = The right and left sides
LATERAD = Toward the sides of the body

b. Axes of symmetry are imaginary one-dimensional lines that connect opposite poles of symmetry.

DORSOVENTRAL OR DV AXIS = The axis from back to belly

ANTEROPOSTERIOR OR AP AXIS = The axis from head to tail

MEDIOLATERAL OR ML AXIS = The axis from right to left

c. Planes of symmetry are imaginary, two-dimensional partitions which are usually described as slices or sections through the body.

TRANSVERSE PLANE = The cross sectional plane; it is at right angles to the AP axis and parallel with the DV and ML axes.

FRONTAL PLANE = The horizontal sectional plane; it is at right angles to the DV axis and parallel with the AP and ML axes.

MEDIAN SAGITTAL PLANE = The plane of bilateral symmetry; it is the only section that effectively divides the vertebrate body into right and left mirror halves. Parasagittal sections are all other sections in the sagittal plane; they are at right angles to the ML axis and parallel with the AP and DV axes.

d. Body plans of symmetry refer to the overall geometric distributions of body parts in terms of poles, axes, and planes of symmetry.

ANAXONIC, ASYMMETRIC, OR CHAOTIC SYMMETRY = Symmetry with no axes or planes that can divide the body into equal mirror halves. Example: a potato.

MONAXONIC, POLYPLANAR, OR RADIAL SYMMETRY = Symmetry with a distinct AP axis and many planes that can divide it longitudinally into mirror halves. Example: a sunflower.

DIAXONIC, BIPLANAR, OR BIRADIAL SYMMETRY = Symmetry with an AP axis and 2 ML axes which give only two planes that can separate the body into mirror halves. Example: a canoe.

TRIAXONIC, MONOPLANAR, OR BILATERAL SYMMETRY = Symmetry with single AP, DV, and ML axes and with only one plane that can separate the body into two mirror halves. Example: the adult vertebrate body.

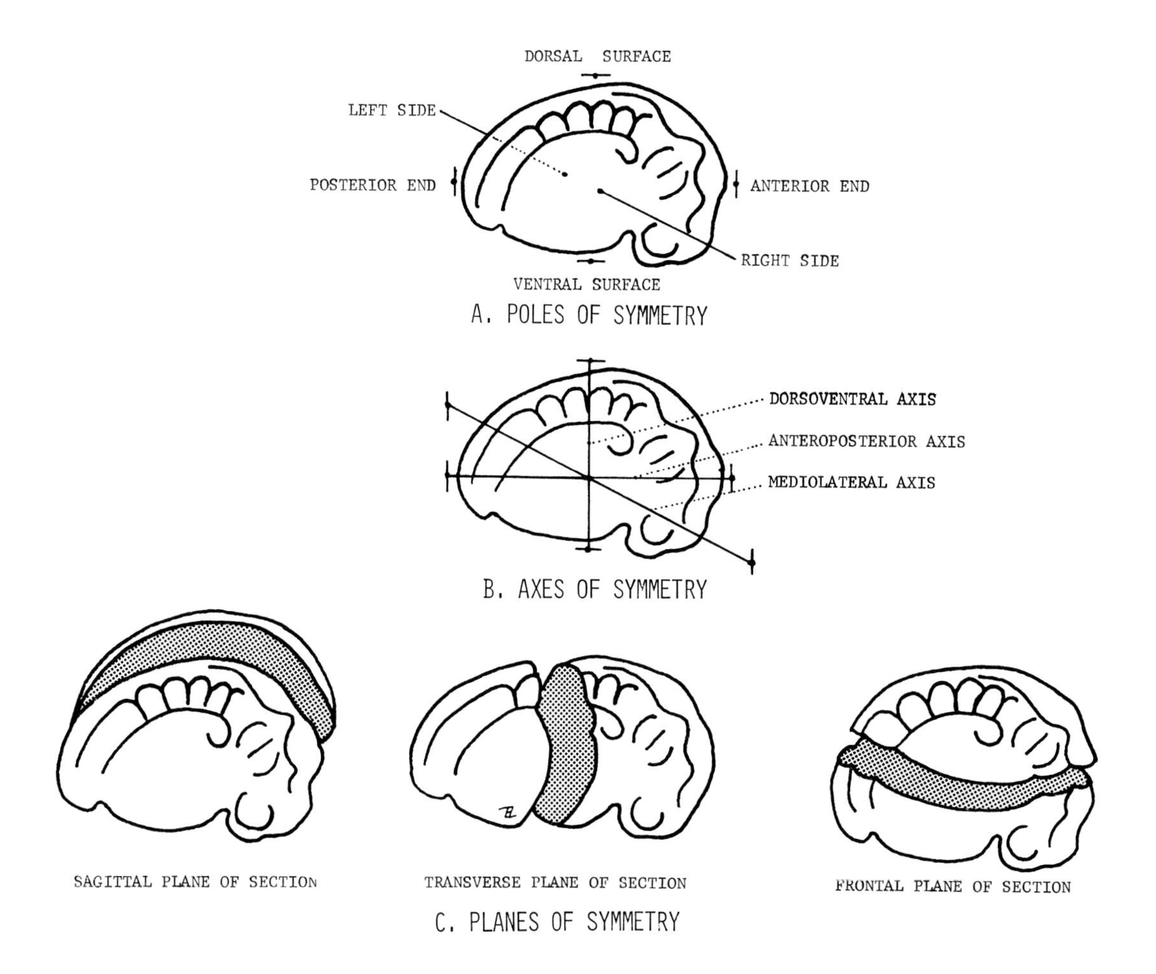

Figure Int.-2. Diagram of poles, axes, and planes of symmetry.

2. Concepts of biological cycles and relationships. These all relate in one form or another to reproductive and/or evolutionary associations among organisms and parts of organisms.

a. Primary biological cycles entail three aspects of reproduction and development in the biological world.

THE CELL CYCLE: This is the life history of cells and normally is divided into a growth and DNA replicating period called interphase, and a period of cell reproduction or division called mitosis.

THE ONTOGENIC CYCLE: This is the life history of individual organisms. In sexual reproduction it begins with fertilization of an egg to produce a zygote which undergoes embryonic development, then larval or juvenile stages, and finally becomes a mature sexually reproductive adult. Senescence and death end the ontogenic cycle.

THE PHYLOGENETIC CYCLE: This is the life history of species and higher taxonomic groups; it is the evolutionary history of races, from their origin to extinction.

b. Concepts of stability and change in ontogenic cycles. Embryonic development is itself affected by evolutionary change. Indeed, it is by change in embryonic histories that adult evolutionary advances are achieved. In a course in embryology concerned, in part, with comparative and evolutionary aspects of development, it is necessary to distinguish clearly between aspects of development in organisms that have been subject to evolutionary modification and those that have been resistant to such change. The following terms have been specially coined from Greek roots to identify special aspects of evolutionary change as they relate to ontogenic cycles in organisms.

PALEOMORPHIC CHARACTERS (*paleo* = old, *morphosis* = to mold or shape). These are aspects of embryonic development that have resisted evolutionary change and represent ancient ancestral ways for the molding and shaping of organs during embryonic development. These are characters of primary value in determining homologies and relationships among species. Example: The formation of gill slits and notochord in mammalian embryos.

CAENOMORPHIC CHARACTERS (*caeno* = new). These are aspects of development which have been modified by evolutionary change and do not exhibit the ancestral, ancient way of forming organs. Usually caenomorphic characters are also adaptive characters which equip embryos with some special ability for survival as embryos or adults. Caenomorphic characters show the way in which evolutionary advances are added to life cycles. They permit divergence of different types from common ancestral beginnings. An example is the development of a chorionic placenta by mammalian embryos to adapt them for intrauterine life. It can be seen that, after a caenomorphic character arises and is included into the embryonic pattern of development, it will become a paleomorphic character for all of its descendants. An example of a caenomorphic character becoming a paleomorphic one is the evolution of an egg with amnion and yolk sac in early stem reptiles, and, thereafter, the retention of these characters in their mammalian descendants.

TACHYMORPHIC CHARACTERS (*tachy* = rapid). These are changes in ontogenic cycles which tend to eliminate paleomorphic or primitive ways of forming organs. The embryo goes directly to the formation of evolutionarily advanced structures without repeating the ancestral steps. As an example we can cite tropical tree frogs which complete their embryonic development and hatch as juvenile froglets without passing through a tadpole stage.

PAEDOMORPHIC CHARACTERS (*paedo* = child). These are aspects of development in which characters that are typically associated with early embryonic or juvenile stages are retained beyond their normal period. Paedomorphic characters can be kept into the adult stages as vestigial reminders of an earlier evolutionary level when they were functional and normal. Atavistic characteristics and many anomalies in adult development are paedomorphic reminders of normal steps in embryonic development that have persisted. Examples include the normal retention of a very short face and a large cranial region in the adult head of man, or the abnormal occurrence of a cleft lip and cleft palate in newborn children.

c. Concepts of similarity in animal structure and function. Fundamental similarities among organisms are those that are derived by genetic, embryological and evolutionary means from the same ancestral source. Convergent similarities are superficial resemblances that are unrelated to each other in basic structure of manner of origin. The terms generally used to describe these different kinds of similarities are as follows:

HOMOLOGOUS STRUCTURES. These are structures in different organisms that are derived from the same embryonic primordia; they have the same fundamental adult morphological plan, and are derived paleontologically from the same structure of a common ancestral stage. It is irrelevant whether or not they share similar functions. Examples: Forelimbs of a whale, frog, bird, bat, horse or man.

ANALOGOUS STRUCTURES. These are structures that share a common function, and they may be convergent adaptations, or homologous ones. Examples of structures that are both homologous and analogous are wings of birds and bats. Examples of structures that are analogous but not homologous are wings of birds and wings of insects.

METAMERIC STRUCTURES. These are structures in the same individual that are structurally similar and repeated at periodic intervals along one or more axes of the body. Examples: spinal nerves or ganglia, or bones in the vertebral column, or aortic arches, etc.

SERIALLY HOMOLOGOUS STRUCTURES. These structures are structurally and developmentally similar in their origin; they come from the same germ layers but from different segmental regions of the body. Example: a cervical vertebra is serially homologous with all other vertebrae in the axial skeleton of the same individual.

D. SUMMARY OF ANIMAL CLASSIFICATION AND THE TAXONOMIC ORGANIZATION OF CHORDATES

In using a comparative approach to the study of any biological system, the relative positions of organisms in the evolutionary and taxonomic (classification) organization of the animal kingdom must constantly be kept in mind. Table Int.-3 provides a brief outline of the living kingdoms and their major subdivisions. The major animal phyla are grouped in several supra-phyletic clusters that are widely recognized as related groups in terms of body organization and probable evolutionary relationship. Emphasis is given in Table Int.-4 to chordate relationships and taxonomic groups. Fluent familiarity with this level of classification will be expected in the remaining chapters of this work. In the long run it will be a great saving of time and avoid confusion if the chordate-taxonomic categories and their minimum definitions are memorized at the beginning. The hemichordates are generally given separate phylum status (see Table Int.-3) but may be included as a subphylum of the chordates (see Table Int.-4).

TABLE INT.-3
SURVEY OF THE LIVING WORLD WITH SPECIAL REFERENCE TO VERTEBRATE RELATIONSHIPS AND CLASSIFICATION

Kingdom PROKARYOTA: Acellular and anucleate forms of very low organization.
Divisions: Viralia (viruses), Schizophyta (bacteria), Cyanophyta (bluegreen algae).

Kingdom MYCOTA: Nucleate and largely syncytial non-photosynthetic plants.
Divisions: Myxomycota (slime molds), Eumycota (true fungi).

Kingdom PROTISTA: Unicellular, colonial and largely uninucleate primitive forms.
Divisions: Protophyta or Thallophyta (nucleate algae), Phylum Protozoa (non-photosynthetic protists).

Kingdom METAPHYTA: Tissue-grade nucleate green plants of higher organization.
Divisions: Bryophyta (mosses), Pteridophyta (ferns), Spermatophyta (seed plants).

***Kingdom METAZOA:** Tissue-grade nucleate animals above the colonial level of organization.

1. Sub-kingdom HYPOMETAZOA: Metazoans lacking nerve cells and extracellular digestion.
   Phyla: Mesozoa (dicyemids and orthonectids), Porifera (sponges).

2. Sub-kingdom RADIATA: Metazoans with radial symmetry and 2 germ layers.
   Phyla: Cnidaria — Coelenterata (polyps and medusae), Ctenophora (comb jellies).

3. Sub-kingdom ACOELOMATA: Metazoans with bilateral symmetry and 3 germ layers, but no mesodermal body cavity beyond simple tissue spaces.
   Phyla: Platyhelminthes (flatworms), Nemertea (ribbon worms).

4. Sub-kingdom BLASTOCOELOMATA — PSEUDOCOELOMATA: Metazoans with bilateral symmetry, 3 germ layers, and with a persistent blastocoele as a body cavity incompletely lined by mesoderm.
   Phyla: Aschelminthes (roundworms, hair worms, rotifers, etc.), Acanthocephala (spiny-headed worms), Entoprocta (entoproct bryozoa).

5. ***Sub-kingdom COELOMATA:** Higher triploblastic metazoans with a true mesodermal body cavity lined entirely with peritoneum, i.e., a coelomic cavity.

   Super-phylum LOPHOPHORA: Coelomates with a ring of filter-feeding hollow tentacles surrounding the mouth but excluding the anus, and pores of nephridial kidneys.
   Phyla: Phoronida (phoronids), Bryozoa (moss animals), Brachyopoda (lamp and tongue shell molluscoids).

   Super-phylum PROTOSTOMIA — SCHIZOCOELA: Coelomates with the embryonic blastopore marking the position of the adult mouth, and the coelom formed in primitive members of major phyla by splitting of paired mesoderm bands formed by mitotic division from primordial mesodermal teloblasts.
   Phyla: Sipunculida (peanut and invert worms), Echiurida (proboscis worms), Mollusca (chitons, clams, snails, squids, etc.), Annelida (segmented worms), Arthropoda (crustacea, arachnids, insects, etc.), and lesser groups.

   ***Super-phylum DEUTEROSTOMIA — ENTEROCOELA:** Coelomates in which the embryonic blastopore persists as, or marks the position of, the adult anus, and in which the coelom in primitive members of major groups is formed by evaginations from the wall of the archenteron.
   Phyla: Echinodermata (starfish, sea urchin, sea cucumber, etc.), Chaetognatha (arrow worms), Hemichordata (acorn worms and pterobranchs), Pogonophora (beard worms), ***Chordata (protochordates and vertebrates).**

* Taxonomic categories to which vertebrates belong are starred.

TABLE INT.-4
SURVEY OF THE PHYLUM CHORDATA

Phylum CHORDATA: Deuterostomes with notochord, gill slits, and tubular dorsal nerve cord.

Sub-phylum HEMICHORDATA: Worm-like chordates with the homologue of the notochord restricted to the prosoma (proboscis), the tubular part of the nerve cord restricted to the mesosoma (collar), and gill slits restricted to the metasoma (trunk or body); now the group is generally given separate phylum status.
Classes: Enteropneusta (acorn worms) and Pterobranchia (pterobranchs).

Sub-phylum TUNICATA — UROCHORDATA: Chordates with the notochord restricted to the tail, and the tubular part of the dorsal nerve cord restricted to the larval cerebral vesicle.
Classes: Appendicularia (larvaceans), Thaliaceans (salps), and Ascidia (sea squirts).

Sub-phylum CEPHALOCHORDATA: Chordates with the notochord and tubular dorsal nerve cord extending throughout the entire length of the body, and lacking a distinct brain and special sense organs.
Class Leptocardia (amphioxids).

Sub-phylum VERTEBRATA — CRANIATA: Chordates with notochord extending from the level of the midbrain to the tip of the tail; with dorsal tubular nerve cord differentiated into a forebrain, midbrain, hindbrain and spinal cord; usually with paired nose, eyes, and ears on head; with segmental axial skeleton and muscles; and heart always differentiated into a sinus venosus, atrium, ventricle, and usually a conus arteriosus (or their derivatives).

Super-class PISCES: Fish-like vertebrates with gill respiration, fins, a simple tubular heart, and mesonephric type of kidney in the adult.

Class AGNATHA: Jawless and limbless fish with seven or more gill slits; with a single unpaired nasal organ, and lacking a conus arteriosus in the heart.
Orders: Ostracodermi (extinct armored agnaths), and Cyclostomata (living smooth skinned lampreys and hagfish).

Class PLACODERMI: Extinct ancestral jawed fish with from 1 to 7 pairs of spiny fins.
Orders: Acanthodia, Arthrodira, Antiarcha, and others.

Class CHONDRICHTHYES: Cartilaginous jawed fishes with a pair of pectoral and pelvic fins; with a spiracle and 5 pairs of gill slits that open separately from the pharynx to the exterior, and lacking a swim bladder.
Sub-classes: Elasmobranchii (sharks and rays), Holostii (chimaeras).

Class OSTEICHTHYES: Bony fishes with jaws, paired pectoral and pelvic fins; 3 pairs of gill slits that are covered by an operculum, and with a swim bladder.
Sub-classes: Actinopterygii (ray finned fishes) and Choanichthyes — Sarcopterygii (lung fish and lobe-finned fishes).

Super-class TETRAPODA: Terrestrial jawed vertebrates with paired pectoral and pelvic walking legs; with lungs for respiration, and with the heart always partly or completely divided into right (pulmonary) and left (systemic) circulatory systems.

Class AMPHIBIA: Cold-blooded tetrapods with naked moist skin; with heart having a partly divided conus and completely divided atrium; adults with a mesonephric type of kidney; with a single ear bone, and the egg lacking an amnion and therefore generally requiring an aquatic environment for development.
Sub-classes: Urodela (salamanders), Anura (frogs), and Gymnophiona (Caecilians).

Class REPTILIA: Cold-blooded tetrapods with scaly skin; heart with divided conus and atria; adult with metanephric kidney; with single ear bone and the egg with an amnion and other characteristic extraembryonic membranes.
Orders: Chelonia (turtles), Crocodilia (alligators), Rhynchocephalia (sphenodon), Squamata (lizards and snakes), and many extinct orders.

Class AVES: Warm-blooded tetrapods with feathers; with heart completely divided; with metanephric kidney, single ear bone, and the egg with characteristic reptilian shell and extraembryonic membranes.
Sub-classes: Archaeornithes (extinct toothed birds) and Neornithes (modern birds).

Class MAMMALIA: Warm-blooded tetrapods with hair, mammary glands, heart completely divided, metanephric kidney, amniote egg, and ear with three ear bones.
Sub-classes: Prototheria — Monotremata (egg-laying mammals), Metatheria — Marsupiala (pouched mammals) and Eutheria — Placentalia (placental mammals), & extinct orders.

E. THE CONCEPT OF UNIFORMITY IN GERM LAYER ORIGIN OF HOMOLOGOUS TISSUES AND ORGANS IN VERTEBRATES

Von Baer was the first to describe early embryos from all vertebrate classes. In these studies he pointed out the remarkable fact that, notwithstanding the great differences exhibited by adults, the homologous organs and tissues of fish, amphibians, reptiles, birds, and mammals all could be traced back to the same source in the primary germ layers of the gastrula and neurula. On this ground he proposed that the manner of embryonic origin of parts (i.e., the germ layer origin) should be a major consideration in the determination of homologies of adult structures. This generalization is universally accepted today as one of the broadest generalizations in all biology. The fact that changes in expected cell fate can be brought about by experimental means in no way weakens the generalization as it was originally proposed. It was stated to be, and is, a valid principle of normal development. The experimental modifications that can be produced, which divert normal cell fates, are merely evidence of labile determination of primordia at the time of experimental manipulation.

On pages 15, 16, and 17 are three tables which give the germ layer genealogy for the origin of most adult structures. Table Int.-5 gives ectodermal derivatives. Table Int.-6 gives mesodermal derivatives, and Table Int.-7 gives endodermal derivatives. The encyclopaedic information in these germ layer tables will be of constant reference value throughout the remainder of this study. In a sense these tables provide a brief review of the entire pattern of vertebrate development. Students will gradually absorb the information in each of these three tables as the germ layer derivatives are studied in progressively older embryos from different vertebrate classes.

Following the germ layer tables is Figure Int.-3 which shows a generalized vertebrate embryo early in organogenesis. Figure Int.-3 A is a whole mount reconstruction showing the major organ primordia. Figure Int.-3 B is a diagrammatic cross section through the midgut region at the level of the heart. These are provided for future reference as an overview of major germ layer derivations in vertebrate development. Note in Figure Int.-3 B that the same structures are labelled at the right and left in different ways. At the right the embryonic germ layers themselves are named; at the left their primary definitive contributions are named.

Following a very old tradition in embryology, colored illustrations in the remainder of this text will follow the convention for using **blue** for ectodermal derivatives, **red** for mesodermal derivatives, and **yellow** for endodermal derivatives. Exceptions to this conventional use of colors will be clearly labelled in each case where there is a departure from this plan.

TABLE INT.-5
DERIVATIVES AND FATE OF THE EMBRYONIC ECTODERM

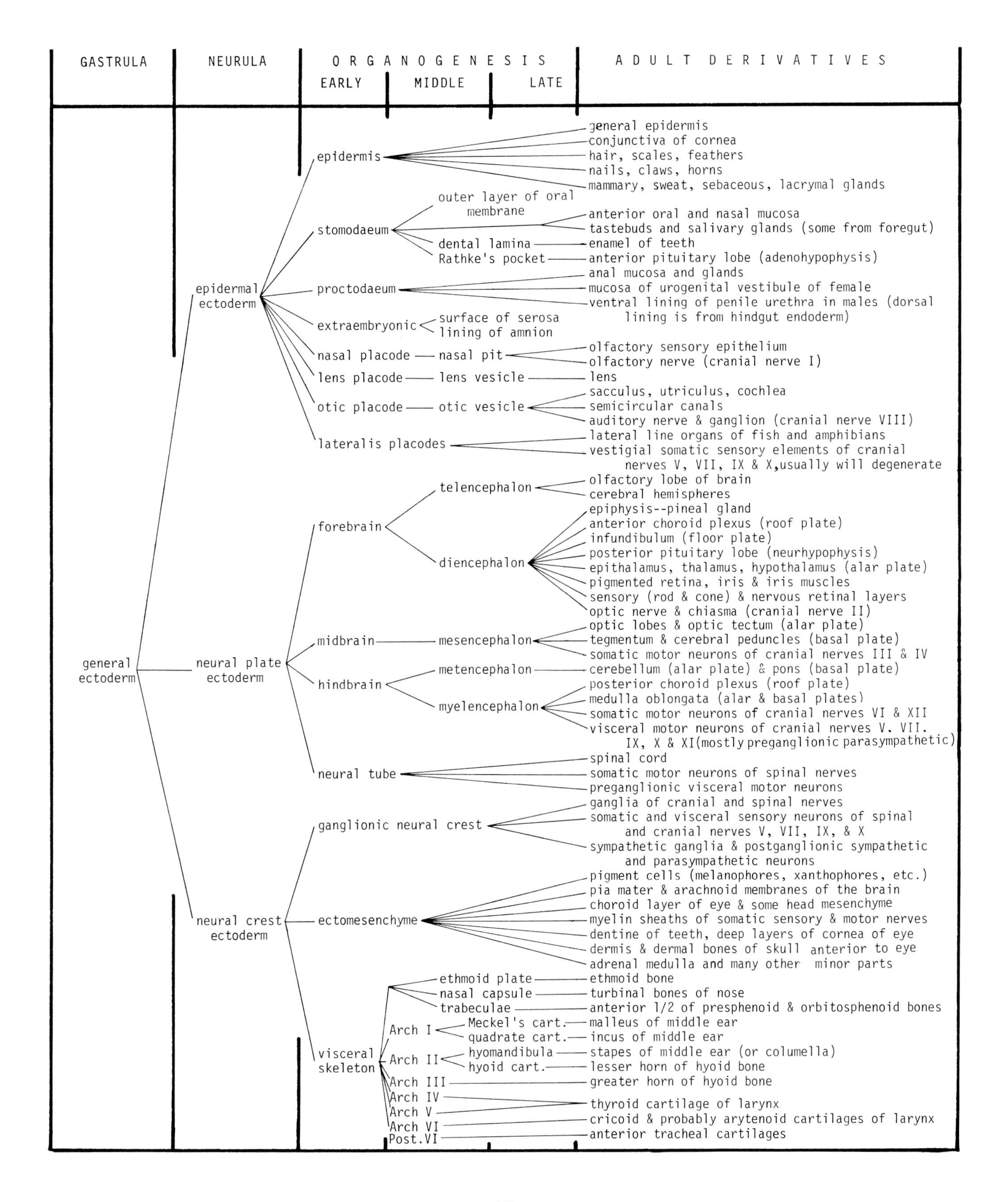

TABLE INT.-6
DERIVATIVES AND FATE OF THE EMBRYONIC MESODERM

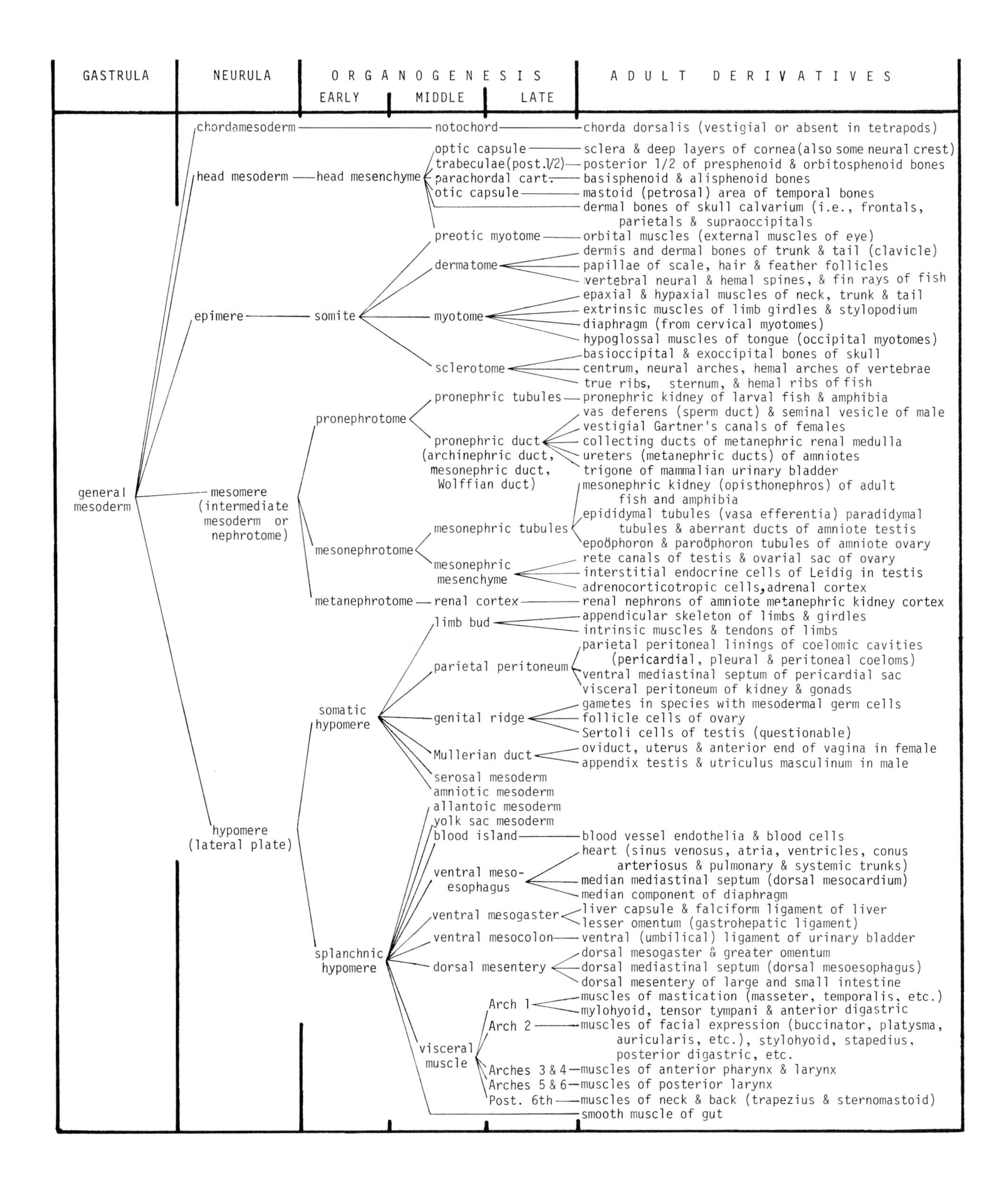

TABLE INT.-7
DERIVATIVES AND FATE OF THE EMBRYONIC ENDODERM

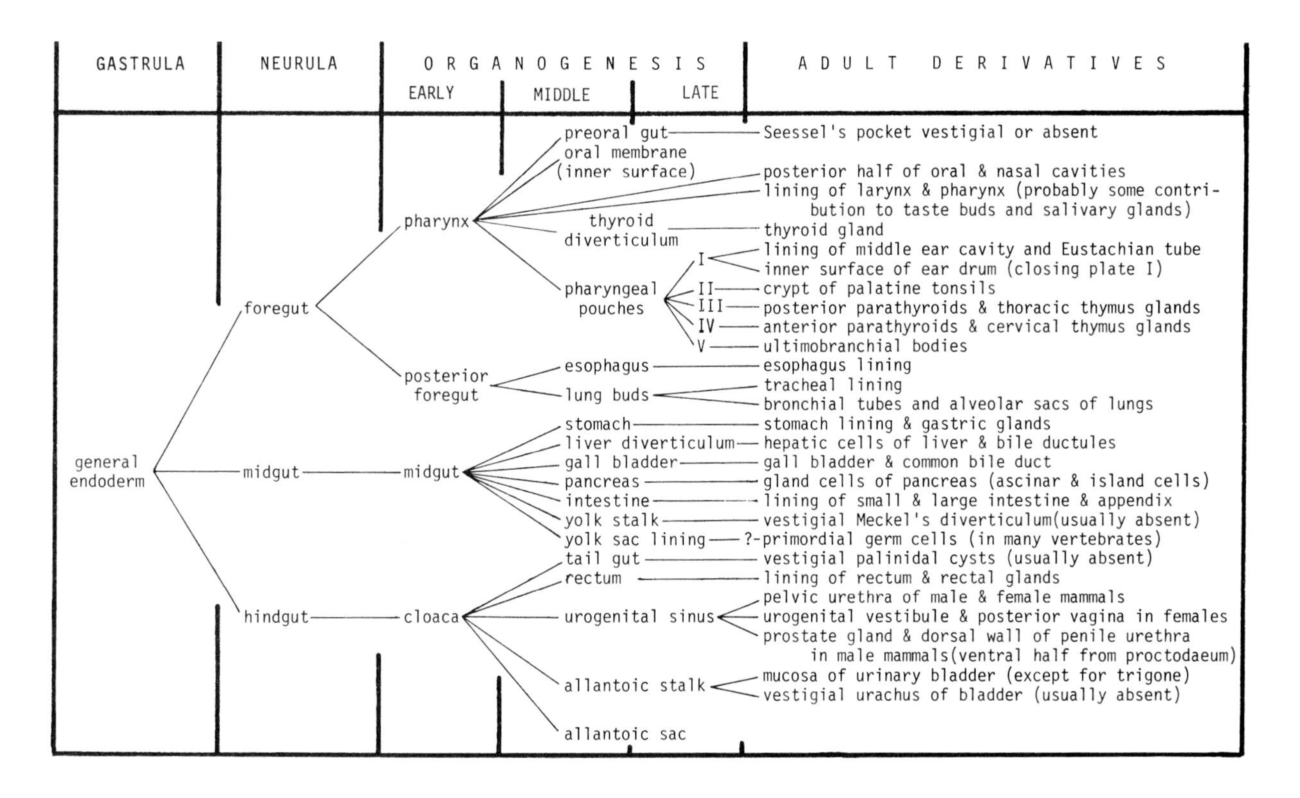

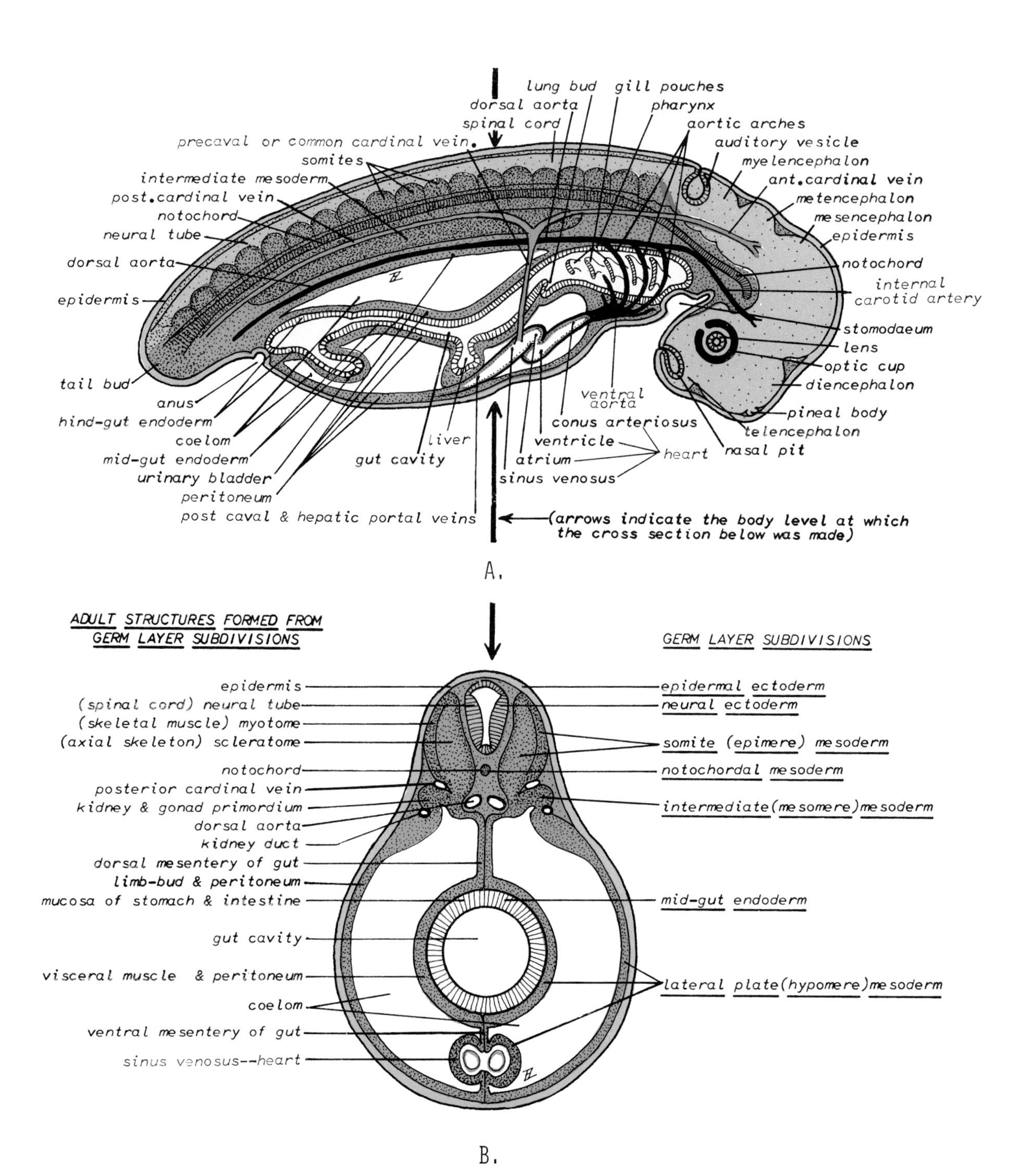

Figure Int.-3. Generalized vertebrate embryo early in organogenesis. A. Generalized early salamander embryo as a model for early vertebrate development. B. Diagrammatic cross section through the body at the level of the midgut and heart, as indicated by heavy arrows. Color code: blue — ectoderm, red — mesoderm, yellow — endoderm. (From H.E. Lehman, *Laboratory Studies in General Zoology,* 1981, Hunter Publishing Company).

CHAPTER I

THE CONCEPT OF THE VERTEBRATE DEVELOPMENTAL COMMON PLAN

Illustrated by Larval Stages of Tunicates, Amphioxus, and Cyclostomes

TABLE OF CONTENTS

TABLES AND ILLUSTRATIONS

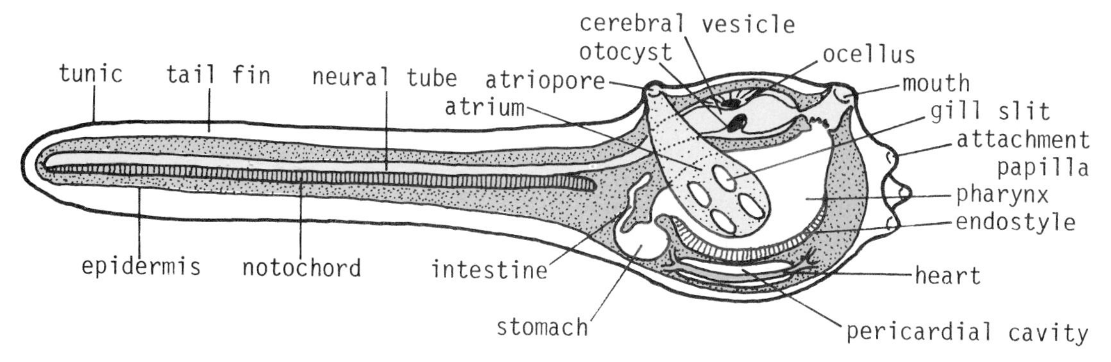

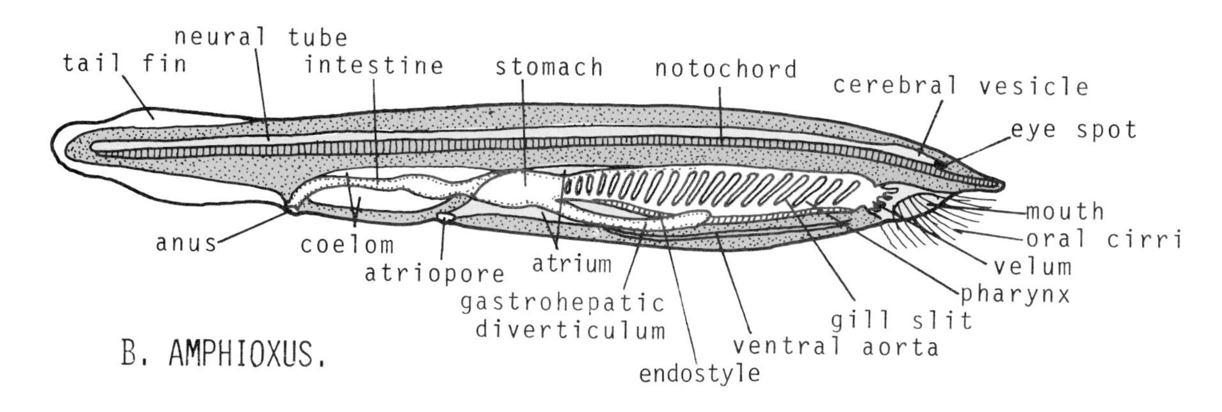

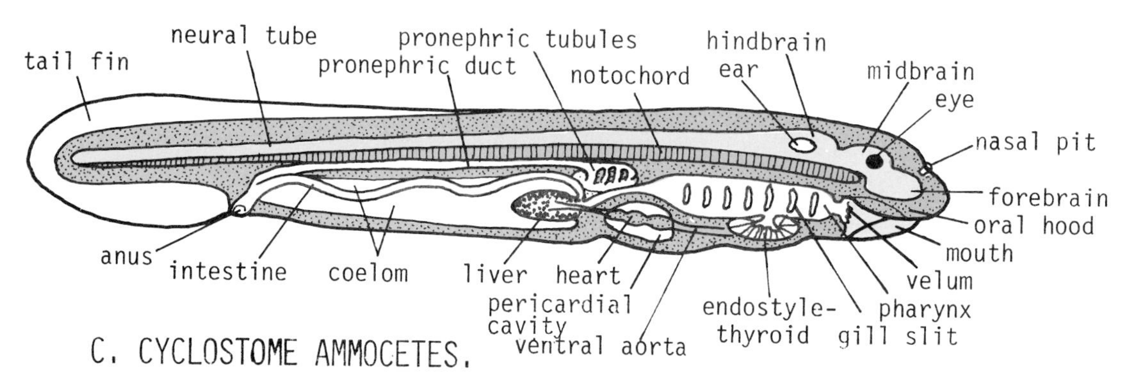

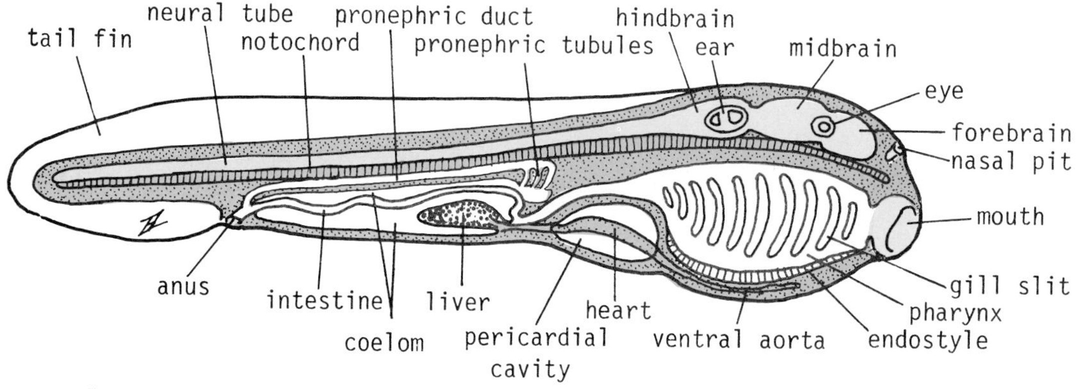

Figure 1-1. Examples of the basic chordate body plan. 1-1 A. Diagrammatic tunicate tadpole. 1-1 B. Diagrammatic young Amphioxus larva. 1-1 C. Diagrammatic early Ammocetes larva of cyclostomes. 1-1 D. Hypothetical chordate archaetype including some qualities of the other three larval types. Color code: blue — ectoderm, red — mesoderm, yellow — endoderm.

CHAPTER I

THE CONCEPT OF THE VERTEBRATE DEVELOPMENTAL COMMON PLAN

Illustrated by Larval Stages of Tunicates, Amphioxus, and Cyclostomes

In strict usage, a larval stage is a phase in the life cycle during which the individual is an independent, free-living organism that is fundamentally different from the adult in some major morphological ways, or in its adaptations for life in a conspicuously different ecological habitat. The larval stage is characteristically terminated by a cataclysmic transformation called **metamorphosis,** during which the larval characters are remodeled or lost, and the definitive juvenile-adult characters are acquired.

The three chordate groups named in the title of this chapter easily qualify as having true larval stages. Schematic diagrams of the larval tunicate, amphioxid, and cyclostome are given in Figure 1-1. When we tabulate the characters shared by such a representative group in a phylum, it is possible to extract the general traits common to virtually all members of the phylum at some comparable stage during their embryonic or larval development. With these common characters we are then able to reconstruct a generalized embryo as a model of early development for the group with the diagnostic characters of the phylum. Such a generalized plan, devoid of all special characters that distinguish particular groups from one another, is called a **developmental common plan.** Examples of developmental common plans are given in Figures Int.-3 and 1-2. The characters on which a developmental plan is based are provided in Table I-1. The diagnostic characters for defining the Phylum Chordata and Subphylum Vertebrata are given along with a checklist for the appearance of these characters in embryonic or larval stages of representative protochordates and vertebrates.

One must be impressed by the truly remarkable and comprehensive range of characters shown in Table I-1 that are shared across the breadth of the phylum by individuals early in their embryonic development. Such homologies are generally explained on the basis of descent from a common ancestor in the remote geologic past.

The concept of an ancestral species or **archaetype** from which all members of a phylum, class, or other taxonomic level arose was a concept of great importance to the development of ideas about the cause of homology. Although the concepts of developmental common plans and archaetypes are related, they are, in fact, fundamentally quite different, as can be seen by comparing Figures 1-1D and 1-2. The concept of the archaetype involves hypothetical extrapolations from observable homologies to an ancestral extinct species of adult animal in some specific imagined habitat. For chordates it is usually a pelagic filter feeder as shown in Figure 1-1D, but these speculations go beyond provable fact. The concept of the developmental plan generalizes in the opposite direction. Instead of supposing details of form and function, it strips away special details and attempts to lay bare the essential common features that are actually present during some part of the life cycles of most living members of a particular taxonomic group. Thus the concept of the common plan is highly conservative in the face of an abundance of fact, whereas the concept of the archaetype is often carried to extremes of fantastic nonsense. For example, in the past serious consideration has been given to marvelously elaborate and ingenious schemes for deriving chordates from such diverse groups of invertebrates as the cephalopod molluscs, trilobites, annelids, arachnids, brachyopods, phoronids, hemichordates, echninoderms, nemerteans, and even sea anemones. It is clear that, at the very most, only one of these theories could be right.

The concept of the archaetype is of only transient interest for the present study which will emphasize the concept of the developmental common plan in vertebrate life cycles. The materials to be studied review the major factual bases for establishing this concept. It can be mentioned that the embryological and larval record for the chordates is unusually rich and complete in comparison with those of many invertebrate phyla in which the ontogenic story is clouded by gaps, or by many alternative pathways for producing similar adult forms.

TABLE I-1

MAJOR DISTINCTIVE CHORDATE AND VERTEBRATE CHARACTERS WITH THEIR DISTRIBUTIONS IN REPRESENTATIVE EMBRYOS AND LARVAE

	DISTINCTIVE CHARACTERS	TUNICATE TADPOLE	AMPHIOXUS LARVA	CYCLOSTOME AMMOCETES	AMPHIBIAN TADPOLE	72 HOUR CHICK EMBRYO	6 mm HYMAN EMBRYO
	GENERAL CHARACTERS: Chordates are triploblastic coelomate deuterostomian metazoan animals.	+	+	+	+	+	+
	DIAGNOSTIC CHORDATE CHARACTERS:						
1.	Dorsal tubular nerve cord	+	+	+	+	+	+
2.	Notochord or chorda dorsalis	+	+	+	+	+	+
3.	Foregut or pharynx with gill slits	+	+	+	+	+	+
	DIAGNOSTIC VERTEBRATE CHARACTERS:						
4.	Body with post-anal tail	+	+	+	+	+	+
5.	Brain with forebrain, midbrain, and hindbrain regions	-	-	+	+	+	+
6.	Head with sensory nose, eyes, and ears	-[1]	-[1]	+	+	+	+
7.	Mesodermal somites derived from archenteron (gut) wall	-	+	+	+	-	-
8.	Body voluntary muscles mainly from mesodermal somites	-	+	+	+	+	+
9.	Notochord is primary axial skeletal support	+	+	+	+	+	+
10.	Major definitive axial skeleton (vertebrae) from somites	-	-	-+	+	+	+
11.	Enterocoelic coelom (derived from archenteron lumen)	-	-+[2]	-+[2]	-	-	-
12.	Coelom is formed by splitting of lateral mesoderm	-	-+[2]	-+[2]	+	+	+
13.	Pericardial coelom surrounding heart as a separate cavity	+[4]	-	+	+	+	+
14.	Primary kidney tubules open into coelom by nephrostomes	-	-	+	+	+[3]	+[3]
15.	Closed circulatory system lined by endothelium	-+[5]	+	+	+	+	+
16.	Heart derived from foregut ventral mesentery	-[4]	-+[6]	+[7]	+	+	+
17.	Arterial common plan with ventral aorta, aortic arches, dorsal aorta, and its branches	-[5]	+	+	+	+	+
18.	Venous common plan with pre- and postcardinal branches of common cardinal veins, and hepatic portal vein	-[5]	+	+	+	+	+
19.	Ectodermal stomodaeum and proctodaeum form the anterior and posterior ends of the gut tube	-+[8]	-+[8]	+	+	+	+
20.	Endodermal foregut with gill pouches and midventral endostyle --thyroid diverticulum	+	+	+	+	+	+
21.	Endodermal midgut with hepatopancreas digestive glands	+	+	+	+	+	+
22.	Endodermal hindgut or cloaca that receives the terminal openings of gut, kidney, and gonads	-	-	+	+	+	+

1. With only simple (mainly unicellular) sensory receptor cells.
2. Only the anterior myocoeles form as enterocoelous pouches from the archenteron; posterior parts of the mesodermal coelomic cavities are formed by splitting of lateral plate mesoderm.
3. Nephrostomes are vestigial and never function in excretion.
4. Tunicate heart and pericardial coelom are derived from general mesoderm; the pericardial cavity is the only coelom of tunicates.
5. The tunicate heart can reverse its beat at will and the circulatory system is open with no capillary walls; homologies of arteries and veins are therefore difficult to establish.
6. Amphioxus ventral aorta is contractile and functions as a heart in the ventral mesentery; however, there is no recognizable development of a distinct heart or pericardial cavity.
7. The cyclostome heart has a sinus venosus, atrium, and ventricle but no conus arteriosus.
8. The tunicate anus opens into the ectodermal atrial sac and not into a proctodaeum. In amphioxus there is no proctodaeum; however, both amphioxus & tunicates have a stomodaeum.

A. THE TUNICATE TADPOLE AS AN EXAMPLE OF A VERY PRIMITIVE CHORDATE DEVELOPMENTAL PATTERN

Chapter 3 will review patterns of early cleavage, gastrulation, and tailbud stages in tunicate development. The following description of larval structure and metamorphosis is a part of that story. The overall summary of this section is illustrated by diagrams in Figure 1-3.

The materials to be studied consist of whole mount preparations of several tunicate tadpoles which provide a range in variation that permits one to grasp rather completely the basic chordate characters of tunicates. Unquestionably they are the most primitive group that can be assigned unequivocally to the Phylum Chordata. It is important to note that until tunicate embryology and larval metamorphosis were studied, tunicates had been assigned to the Phylum Mollusca as shell-less bivalves. Only the presence of a larval tubular nerve cord, notochord, and the manner of gill slit formation permitted their true relationships to be recognized by Kowalevski and Balfour in the 1880's.

With the aid of the diagrams, identify the following general structures in all representative materials available for study. Special features of interest will be summarized below.

1. General morphology of tunicate tadpoles (Fig. 1-3 A–D). Note the two conspicuous body regions, namely, the **head** and very long **postanal tail.** First identify the dorsal side of the tadpole by noting the position of the jet black **ocellus** or eyespot which will appear somewhat larger and more flattened than the spherical black **otocyst** or balance organ. Both are located in the wall of the **cerebral vesicle** or brain on the dorsal side of the head. At the anterior end of the head, note the presence of three or four sucker-like **attachment papillae** which are often large and may equal up to one-third the size of the whole head. Here the **tunic,** which surrounds the entire tadpole, will be very conspicuous. The tissue components that make the internal structure will be seen to extend out from the so-called **internal zooid** and expand into the sucker-like ends of the attachment papillae.

On the dorsal surface of the head, identify the **mouth,** or incurrent siphon, which is located between the cerebral vesicle and the attachment papillae. Trace the mouth into the zooid. The first space will be the ectodermally lined **stomodaeal cavity.** A small duct called the **anterior neuropore** will open from the stomodaeum into the cerebral vesicle, but it is rarely visible in whole mounts. Attempts have been made to homologize the anterior neuropore with the nasal pits and/or hypophyseal duct of vertebrates. The internal end of the stomodaeum opens into the endodermal **foregut,** or **pharynx.** The latter is a large, bulb-like expansion of the gut that nearly fills the internal spaces of the head. Its midventral border is often conspicuously thickened as an **endostyle;** the latter is strongly ciliated and may function as a food concentrating groove that consolidates bacteria trapped in mucous strands with trapped bacteria from the wall of the pharynx before passing it on into the **stomach.** The stomach lies posterior to the pharynx and ends in a short, J-shaped intestine that bends sharply towards the right dorsal side where it ends blindly, without an anal opening. The anus will first be formed at metamorphosis by an opening into an ectodermal invagination called the **atrium.** Identify the **atriopore** or external siphon; it lies just posterior to the cerebral vesicle and opens at the dorsal surface of the body. The atrium in many tadpoles first appears as paired right and left ectodermal invaginations that grow in and fuse with the lateral walls of the pharyngeal endoderm. From one to four openings called **primary gill slits** perforate the walls of the **pharyngeal** and **atrial cavities.** The two atrial cavities usually fuse and one of the atriopores (usually the left) degenerates so that there is usually only one remaining in fully formed tadpoles. It is clear from their manner of origin that the atrial sacs are entirely homologous with the ectodermal visceral grooves which will unite with pharyngeal pouches to form gill slits in higher vertebrates (See Fig. 5-6).

Now examine the tail after determining its dorsal side by reference to the location of the ocellus or otocyst in the cerebral vesicle of the head. Use the cross section diagram of a tunicate tadpole tail (Fig. 1-3C) as an aid in identifying the following parts. The **notochord** runs the entire length of the tail but does not enter the head. It occupies a conspicuous median position in the tail axis. Often the individual cells of the notochord can be distinguished by their appearance as a stack of coins or thick discs. Dorsal to the notochord is a very small **tubular dorsal nerve cord** which is less than one-quarter the diameter of the notochord. Its presence is usually revealed by the presence of a discrete row of closely and uniformly spaced spherical nuclei which represent the neurons of

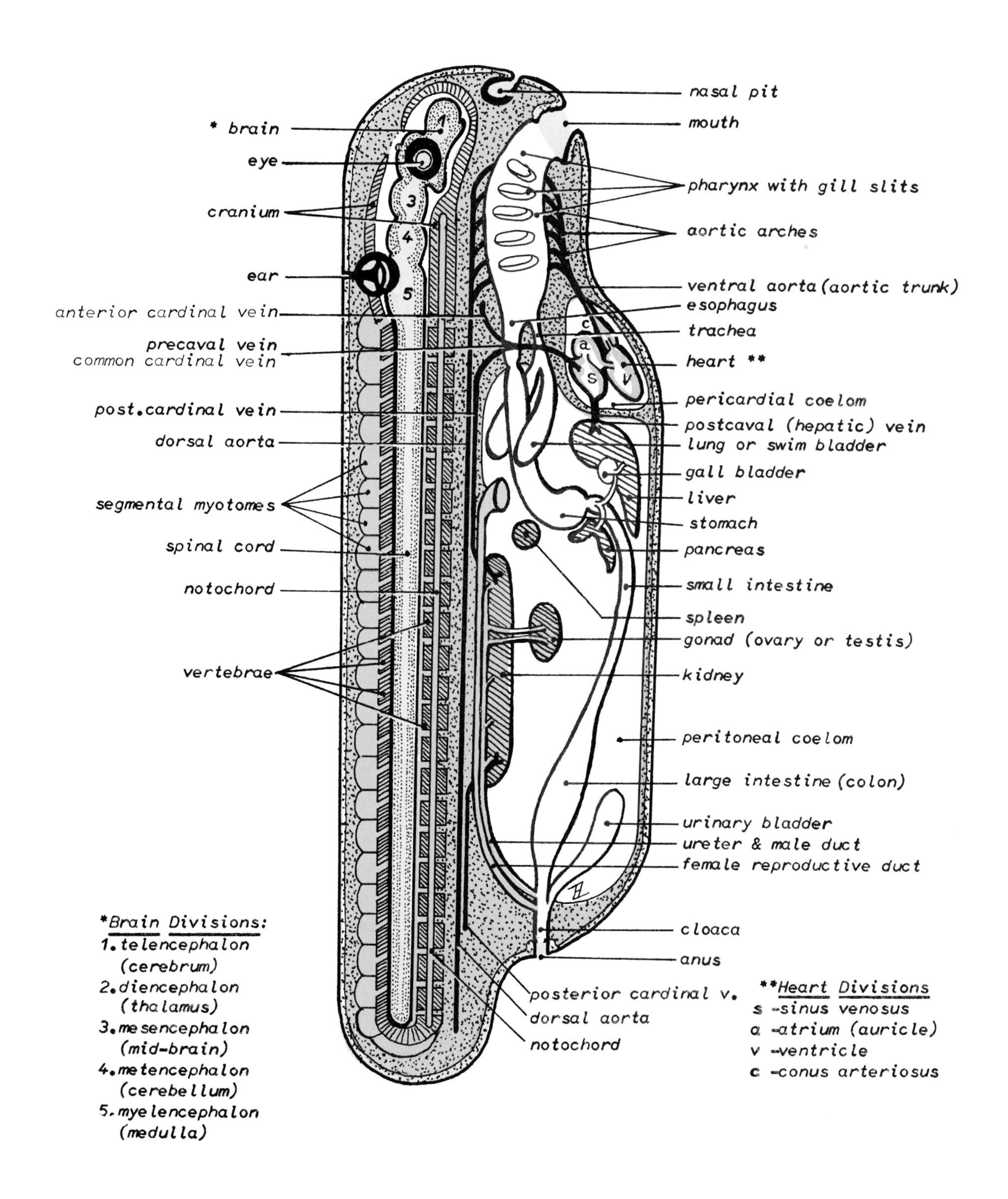

Figure 1-2. A vertebrate developmental common plan. From H.E. Lehman, *Laboratory Studies in General Zoology,* 1981, Hunter Publishing Company. Color code: blue — ectoderm, red — mesoderm, yellow — endoderm.

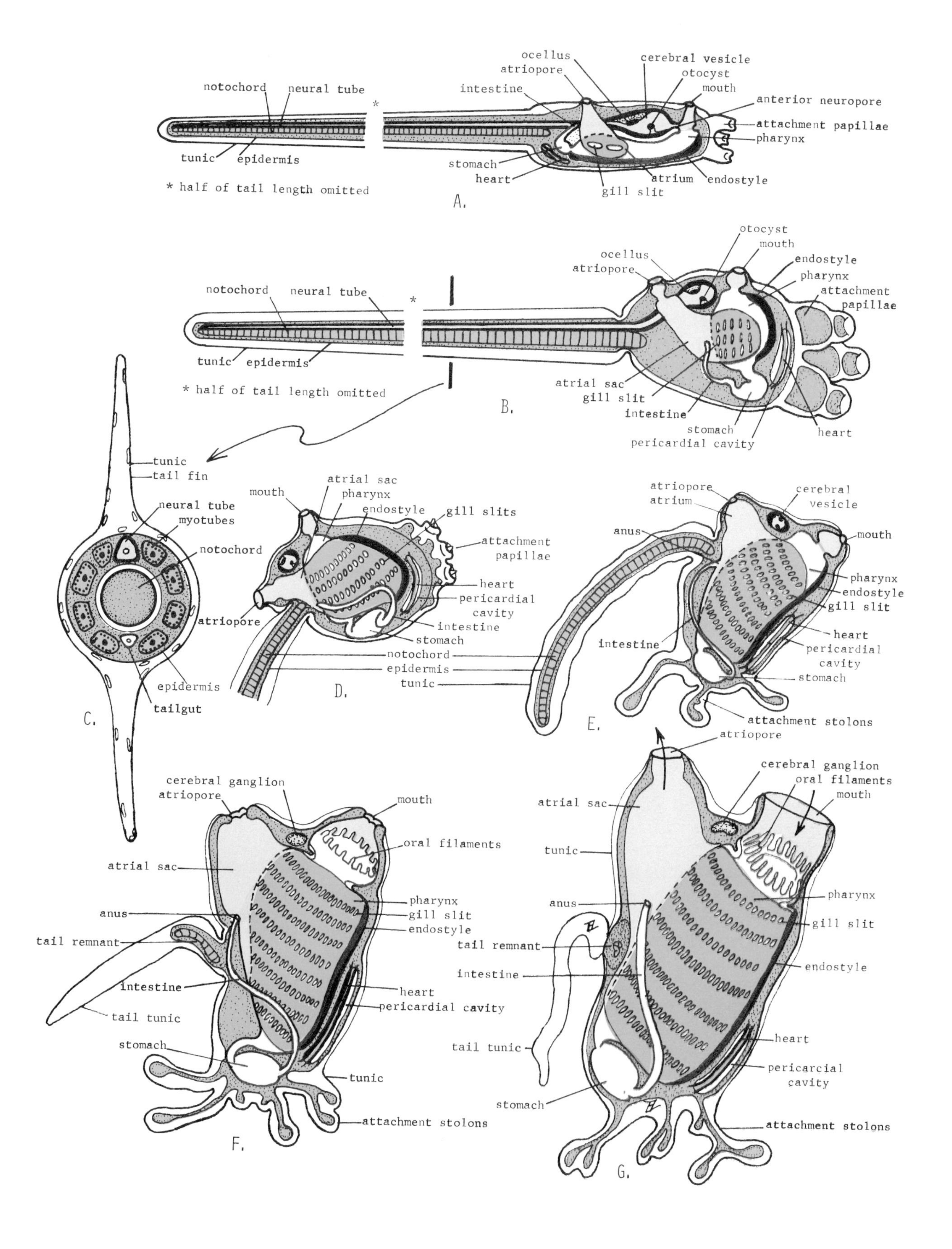

Figure 1-3. Tunicate larvae and metamorphosis. 1-3 A. Larvae of Ascidia. 1-3 B. Larva of Amaroecium. 1-3 C. Diagrammatic section through a tunicate tadpole tail (Modified from Grasse, *Traite de Zoologie,* Vol. 11) 1-3 D-G. Metamorphosis of a tadpole of Ecteinascidia. Color code: blue — ectoderm, red — mesoderm, yellow — endoderm, green — atrium covering pharynx.

the neural tube. On the ventral side of the notochord is a very thin strand of **tail endoderm;** it has no known function and usually cannot be distinguished in whole mount preparations. **Tail muscles** form so-called **pseudo-segments,** each of which consists of from 3 to 8 spindle-shaped muscle cells arranged one above the other on each side of the notochord as shown in Figure 1-3C. The tail is usually composed of about 20 pairs of pseudo-segments. The individual muscle cells, called **myotubes,** are exceedingly primitive. In them the myofibrils for contraction form a ring under the cell membrane. The muscle cell nucleus occupies a central position in clear cytoplasm lacking fibrils. The muscles of the tunicate tail cannot be strictly homologous with the myotomes of amphioxids or vertebrates since there is no somite formation, and hence no segmental myotome formation in tunicates. This primitive myotube type of muscle persists in vertebrates but in higher chordates is restricted to the muscles of the iris which control the size of the pupil of the eye.

2. Metamorphosis and adult structure in tunicates. (Fig. 1-3 D–G). Metamorphosis and adult structure will be shown in whole mounts of *Ecteinascidia turbinata,* a moderately advanced colonial ascidian that shows the basic structure of the subphylum to optimal advantage. Metamorphosis results in the loss of the entire tail with its tubular dorsal nerve cord and notochord. There are far-reaching reorganizations within the head, including the loss of the ocellus, otocyst, and cerebral vesicle. Additions include the formation of an anal opening from the intestine into the atrium, and the differentiation of a heart and pericardial cavity from the general mesoderm of the head. The heart lies ventral to the endostyle of the pharynx and will extend for a considerable distance along the length of the pharynx. True capillaries do not form; there is an open haemocoel with relatively discrete channels that carry blood to and from the pharynx, visceral organs, and tunic, but it is not possible to homologize these channels with the primary arteries and veins of vertebrates. After a long period of juvenile growth, individuals will reach hermaphroditic sexual maturity. The male and female elements of the general **ovotestis** will form between the loop of the intestine and the stomach. **Gonadal lobules** often become intertwined with the large digestive gland called the **hepatopancreas** although it is not certain at all that it is homologous with comparable organs in vertebrates. These visceral organs lie in a thin matrix of mesenchymal cells. The **heart** lies ventral to the pharynx in a pericardial cavity. This is the only body coelom. Colonial tunicates often have an **epicardiac diverticulum,** or **stolon,** which arises from the floor of the pharynx and functions as an asexual reproductive organ. In colonial species it grows out and branches with species specific patterns. Asexual buds will form and can generate colonies of hundreds or even thousands of individuals indistinguishable from those that develop from eggs.

It is appropriate to insert a brief word here concerning the evolution and primary function of the chordate notochord, since this is the type-character for which the phylum is named. It is clear, in the tunicate tadpole, that the function of the stiff elastic notochord is to keep the tail from shortening when muscles undergo contraction. If it were not for the rigid elastic notochord, the tail would merely shorten much as an earthworm shortens when peristaltic waves pass down its body. In the tunicate tail, muscle contractions alternate on opposite sides, and the elastic notochord permits the lateral whip-like movements that are characteristic of tail movements in aquatic vertebrates. Only in mammals such as porpoises, whales, seals, etc., which have secondarily returned to the water, is tail movement vertical instead of horizontal or notochordal in type. The aquatic mammals are clearly carrying back to the water the imprint of the ancestral locomotion of terrestrial mammals and their aquatic adaptations are, therefore, analogous but not homologous with the primitive lateral swimming movements of protochordates and lower vertebrates.

B. THE BASIC CHORDATE DEVELOPMENTAL COMMON PLAN ILLUSTRATED BY LATE EMBRYONIC AND LARVAL STAGES OF AMPHIOXUS

The material for study will include whole mounts of late Amphioxus embryos and whole mounts with representative cross sections of larval and early adult stages. Use models, charts and the diagrams provided in Figures 1-4 and 1-5 as aids in identifying the following parts. It should be noted that there are several genera in the Subphylum Cephalochordata all of which are referred to non-technically as amphioxids. The most common species is *Branchistoma lanceolatum.* The name Amphioxus is an older generic name that can also refer non-specifically to the entire group.

There is general agreement among zoologists

that Amphioxus embryology is closer to the vertebrate archaetype than any other living chordate. We will not take issue with that point of view in so far as cleavage, gastrulation, and coelom formation are concerned (see Chapter 3). However, we will find some peculiarities in Amphioxus larvae that cannot be considered altogether primitive and simple.

1. Late embryonic and pre-larval stages in Amphioxus (Fig. 1-4 A–B). With the aid of the figures and the knowledge gained from tunicates, identify the **anterior neuropore, neurenteric canal, neural plate ectoderm, general epidermis, and archenteron** with a roof of **notochordal mesoderm** and a floor of **endoderm.** Focus high on the embryos and then count the **mesodermal somites** which will vary in number from as low as 4 to as many as 12. Somites are added progressively during development and will often exceed 60 or 70 in adults. Hatching embryos have 2 somite pairs; by the time the anus perforates and feeding begins, there will be about two dozen pairs.

The late embryo is of particular importance for developmental and evolutionary reasons because this stage shows the basic **enterocoelous origin of the mesodermal somites** from the dorsolateral walls of the archenteron. If sections of early embryos are not available for study, refer to models and to the diagrams in Figures 1-4 A–D which will be of value in understanding the following account of coelom formation in Amphioxus. The lumen of a somite is the **myocoele.** Collectively myocoeles are primordia of the general body **coelom.** The latter is formed by an expansion and fusion of the myocoels. Fusion begins at the anterior end with the second pair of somites and progresses posteriorly back to the level of the anus. Only about 6 to 8 pairs of myocoeles are actually formed enterocoelously by evagination from the wall of the archenteron; the remaining myocoeles are formed by secondary cavitation of solid somite blocks as is the case in somites of vertebrates. During the fusion of myocoeles to form the expanding coelomic cavity, the mesial half of the somite nearest the notochord and dorsal nerve cord remains thick; these cells will form the **segmental myotomes,** and the axial muscles are derived from them. The cells originally in the dorsolateral walls of the somites become thin and will form the **somatic** and **visceral layers of lateral mesoderm** that line the coelomic cavity. The visceral mesoderm will surround the gut and will suspend it in the coelomic cavity. The visceral mesoderm will surround the gut and will suspend it in the coelomic cavity by a **dorsal mesentery** in a manner also typical of vertebrates. These steps are shown in Figure 1-4 B.

2. Mid-larval stages of Amphioxus. (Fig. 1-4 C, D). Larval stages available for study are in highly variable developmental ages; accordingly, it is difficult to make brief general descriptions that will be suitable for all specimens. Larval stages are primarily distinguished by the development of the mouth, anus, gill slits, and chevron or V-shaped myotomes. Since there are some problems associated with each of these, we will begin with the identification of general structures already described. Identify first the **notochord** that is now differentiated much as in the tunicate tail; however, note that it runs the entire length of the body from the anterior to posterior end. Above the notochord identify the small and compact **dorsal nerve cord.** At its anterior end identify the small expansion called the **cerebral vesicle,** in the wall of which may be seen one or more densely **pigmented cells** that may have a sensory function. Just short of the posterior extremity of the body, identify the **anus** which opens to the exterior on the ventral surface of the tail. The **anus** forms at the site of the closure of the blastopore and, therefore, amphioxids qualify clearly as deuterostomian coelomates (See Table Int.-3).

The peculiarities in larval development involve a number of seemingly pointless asymmetries in development which will later be corrected to form a perfectly bilateral, late larva. The mouth first forms in a completely normal manner by invagination of stomodaeal ectoderm that meets and fuses with the anterior end of the foregut. However, almost immediately, the mouth grows asymmetrically to the left side (Fig. 1-4 D), where it will form a very large, lens-shaped opening in the side of the pharynx.

At about this time, the first pair of myocoeles begin to expand unequally. The first right myocoele becomes disproportionately large and forms what is called the **head coelom.** The left myocoele retains a very thick wall which sends out a duct that opens to the exterior; the opening is called **Hatschek's pit.** Attempts have been made to homologize the pit and its underlying left first myocoele with the hypophyseal duct and hypophysis of the pituitary gland of vertebrates, even though the pit of Amphioxus is of mesodermal origin and the vertebrate hypophysis is ectodermal.

The grossly asymmetrical development of left and right gill pouches occurs simultaneously with

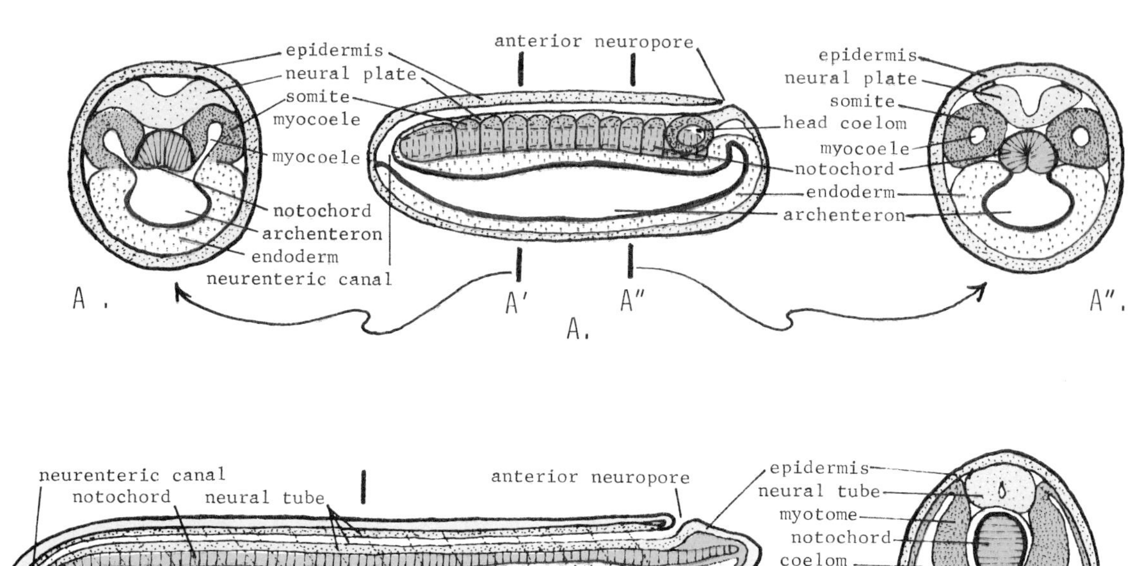

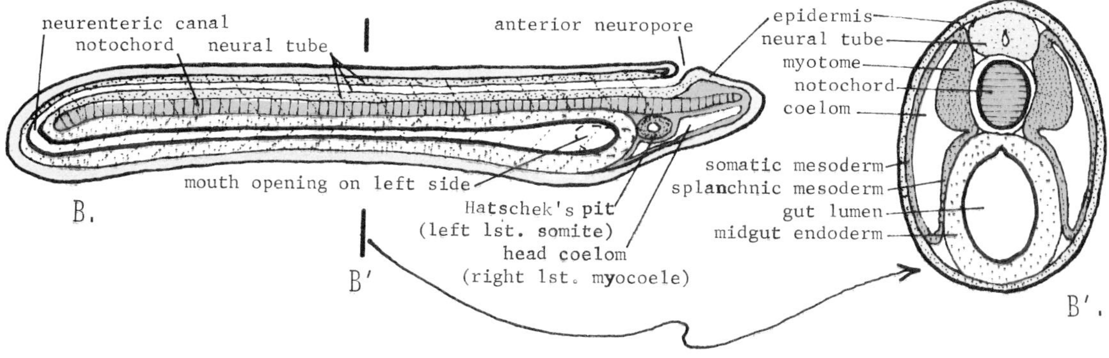

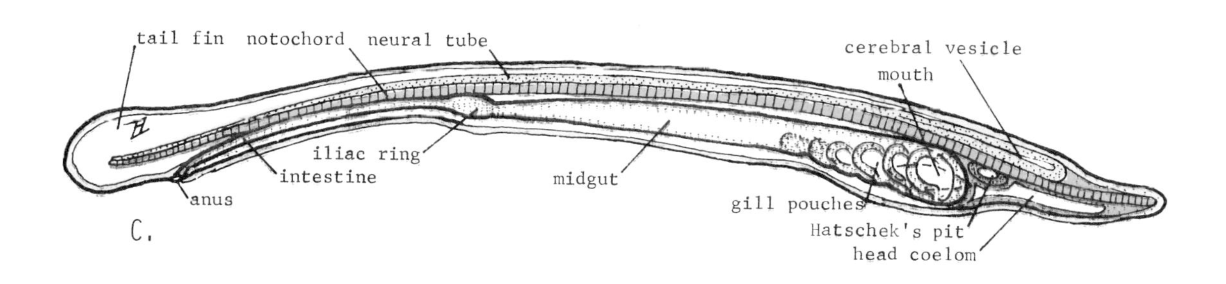

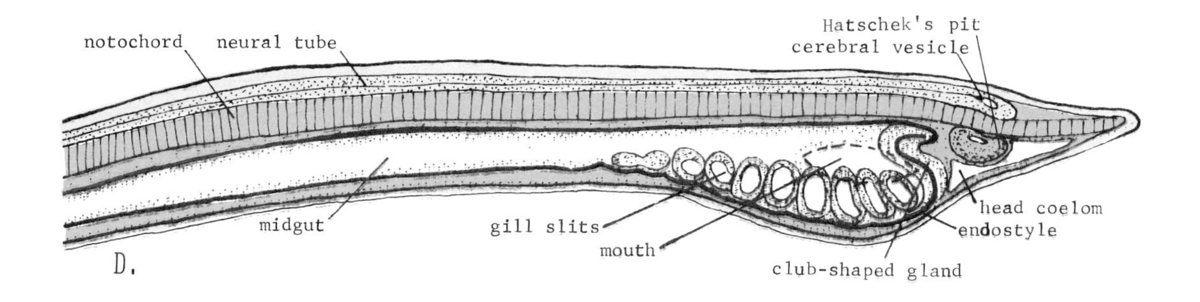

Figure 1-4. Amphioxus late embryos and early larvae. 1-4 A. Optical section through a 10 to 12 somite embryo. 1-4 A'. Cross section through 1-4 A at the level marked A' showing enterocoelous myocoele formation. 1-4 A". Cross section through 1-4 A at the level marked A" and showing separation of chordamesoderm and somite mesoderm from the archenteron wall. 1-4 B. Optical section of a 16 to 20 somite embryo. 1-4 B'. Cross section through 1-4 B at the level marked B' showing expansion of the myocoele to form the coelom. 1-4 C. Optical reconstruction of a 4-gill slit larva. 1-4 D. Optical reconstruction of an 8-gill slit larva. Color code: blue — ectoderm, red — mesoderm, yellow — endoderm.

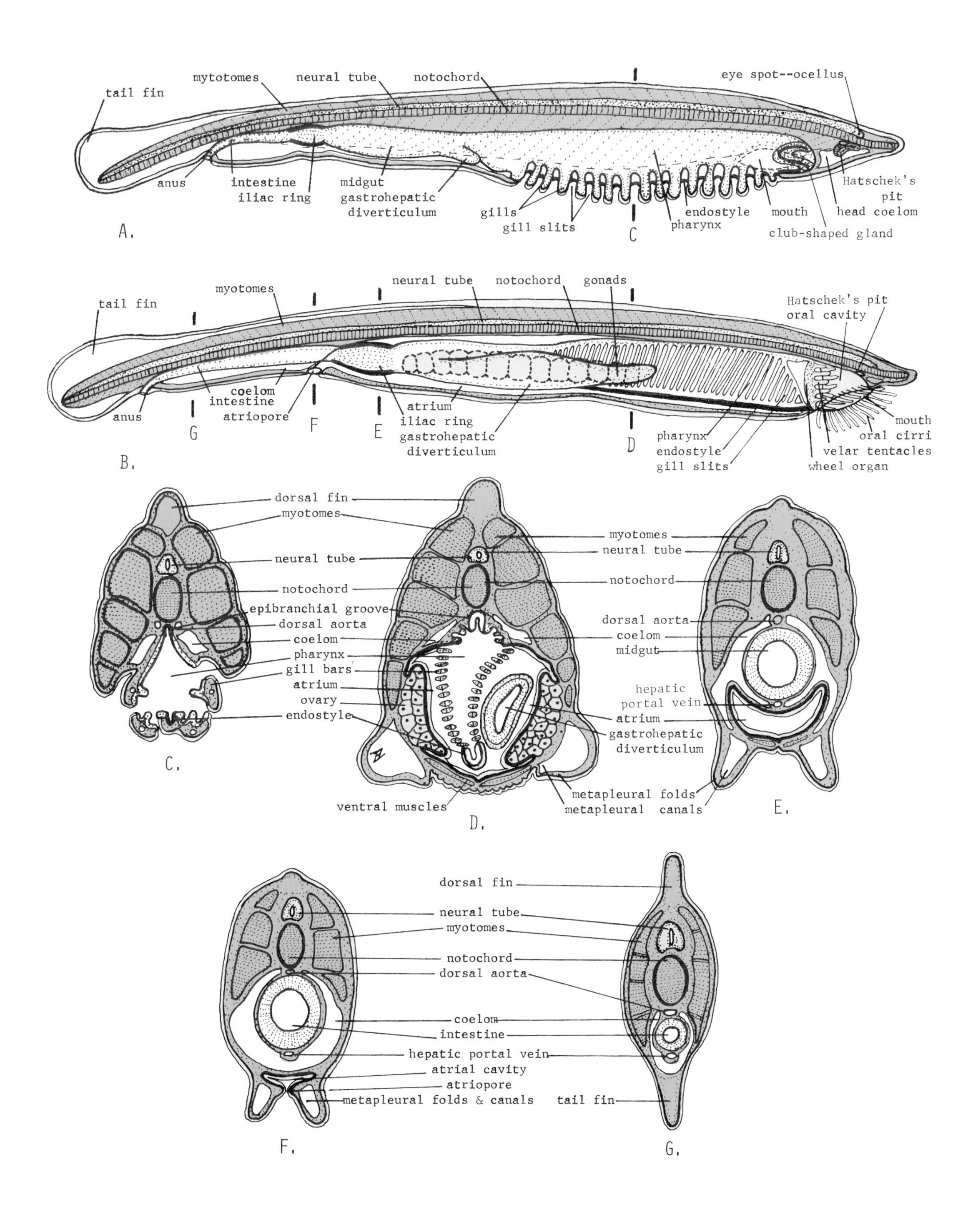

Figure 1-5. Metamorphosis and early adult structure in Amphioxus. 1-5 A. Late larva with 16 to 20 gill slits before the atrium has formed. 1-5 B. Young adult Amphioxus. 1-5 C. Cross section through 1-5 A at the level marked C. 1-5 D–D. Cross sections through 1-5 B at levels marked C through G. Color code: blue — ectoderm, red — mesoderm, yellow — endoderm, green — gonad.

the above events. The left series of pouches first begin to appear. These pouches develop low, near the ventral midline instead of on the side. In all, fourteen left pouches form; afterwards the last six degenerate and the remaining eight belatedly migrate back into typical positions on the left side of the pharynx. During this final migration, the more conservative eight right pouches develop symmetrically with those that have finally taken appropriate positions on the left side.

The last transitional alteration is a return of the mouth to the typical subterminal ventral surface of the head. In the return of the mouth, the buccal cavity enlarges so as to include Hatschek's pit in the roof of the mouth, where it will remain into juvenile and adult stages. The final result of these aimless perambulations of parts is perfect bilateral symmetry; had it been left to me, I would have done it more simply and directly.

Preparations available for study may be at any one of the transitional periods described in the foregoing account. Most preparations will be at a stage when the mouth is still on the left side and when from 4 to 8 of the left gill slits are forming in the ventral midline of the pharynx. Hatschek's pit and vesicle will almost always be very conspicuous just anterior to the oval mouth opening and immediately ventral in lateral view to the anterior notochord and cerebral vesicle.

3. Metamorphosis and juvenile stages of Amphioxus. (Fig. 1-5 A). Larval stages characteristically have up to 20 pairs of gill slits. Metamorphosis is anticipated most clearly by an increase in the number of slits, by the addition of new ones, and by the splitting of old ones until the number of gill slits at metamorphosis reaches about two dozen. Metamorphosis entails changes that transform a free-living planktonic larva into one adapted to assume a burrowing habit in fine silt and sand. The transformations primarily entail protective measures for the mouth and gill structures which, up to this time, have been unadorned and naked.

The mouth and buccal cavity are changed by the addition of oral cirri, a wheel organ, and velum with tentacles which will be described in association with the adult stage (see part 4 which follows).

The most conspicuous change, however, is the development of a protective cover for the pharyngeal gills. This is accomplished by the formation of a pair of lateral folds in the body wall called **metapleural folds.** They begin to form posterior to the pharynx and grow laterally and then ventrally so as to overgrow the gills. In the ventral midline the metapleural folds meet and finally fuse progressively in a zipper-like action from posterior to anterior. In this way, an ectodermally lined cavity called the **atrial sac** comes to surround the ventral and lateral wall of the body, and the gills open into it. The atrial pore is located slightly to the right side at the posterior end of the pharyngeal level of the body. It is clear that the atrial sac of Amphioxus is not homologous with the atrial sacs of tunicates. In the latter, we pointed out that they are homologous with vertebrate visceral grooves. The atrial sac of Amphioxus is much more closely related to opercular folds that cover the gills of many fish and larval amphibians, but in all likelihood these are all convergent adaptations that have evolved independently several times during chordate evolution.

During metamorphosis, the endostyle in the ventral floor of the pharynx becomes very distinctly thickened. In addition, the divisions of the gut take form posterior to the pharynx and three regions become clearly recognizable. The part immediately behind the pharynx is called by various names, of which **midgut** or **stomach** are acceptable if one does not demand strict identity in enzymatic or physiological function with vertebrate organs. From the midgut-stomach region is a ventral evagination, the **gastrohepatic diverticulum,** which grows out anteriorly on the right side and extends for a considerable distance forward into the level of the pharynx. Posterior to the diverticulum is a region that develops a very thick wall with intense staining characteristics; this is the so-called **iliac ring.** Back of it is the hindgut or colon, which opens directly to the exterior by the **anus** without an ectodermally lined proctodaeal component which is characteristic of all higher chordates.

By the time these characters appear, metamorphosis can be said to be complete and almost all adult characters have already been acquired. The larva, after metamorphosis, settles to the bottom and assumes a burrowing habit although it retains the ability to swim with surprising speed and agility.

4. Adult Amphioxus Fig. 1-5 B–G). A full consideration of the adult anatomy of Amphioxus is beyond the scope of the present study. Most adult diagnostic features have been mentioned in association with the treatment of embryonic, larval, and metamorphic changes. The present account will refer to Figure 1-5 B for general features. It should be pointed out that the **mouth cavity** is protected from being clogged with dense particles of sand and silt by a ring of **oral cirri.** The **wheel**

organ at the back of the buccal cavity is the primary organ for generating the ciliary water currents that bring food and water into the pharynx. The **velum,** with its small tentacles, is a small membrane that acts as a secondary line of defense against contamination of the pharynx with coarse particles. Adult animals have a very conspicuous **pharyngeal region** which may have over 200 pairs of **gill slits.** Note also the chevron-shaped **myotomes,** of which there may be as many as 70 or more pairs. It is of interest that the right and left pairs of myotomes are offset so that they alternate in position with one another. It is for this reason that cross sections of Amphioxus are only rarely symmetrical in appearance with regard to the myotomes. In mature animals, 20 to 35 pairs of block-like **gonads** are to be seen in the ventrolateral walls of the metapleural folds at pharyngeal levels of the body. Sexes are separate; gametes are shed into the atrial cavity, and are then carried out of the atriopore by excurrent water flow.

The general structure of nerve cord, notochord, gut components, etc., described for metamorphosed juveniles in the preceding section, apply with equal appropriateness to the adults. It is not possible, in whole mount specimens, to trace the circulatory path. It is sufficient here to note that the arterial plan, including ventral aorta, aortic arches, dorsal aorta, and segmental arteries, conforms with the basic vertebrate plan. Likewise the venous plan is essentially vertebrate in nature. That is, there are paired anterior and posterior cardinal veins and common cardinal veins; there is a single subintestinal or hepatic portal vein; and they all empty into a contractile sinus venosus that lies ventral to the posterior end of the pharynx. The major exceptions in the circulatory plan are to be found in the absence of a pericardial cavity or of a complete heart. It appears that the ventral aorta joins directly with the sinus venosus. Both of them are contractile as also are parts of the aortic arches, dorsal aorta, and subintestinal vein. In Amphioxus the function of the heart is distributed rather diffusely throughout the arterial and venous systems. This may well be a primitive chordate character.

The most peculiar aspect of Amphioxus structure is the total absence of a nephridial excretory system with nephrostomes draining the coelom. Instead, each gill possesses clusters of up to several hundred ciliated flame cells or **protonephridia.** These structures are found in no other chordates, nor in any other deuterostomes. Very similar structures are found in certain flatworms, annelids, phoronids, echiuroids, and other exceedingly remote groups. The presence of these cells is a most puzzling occurrence and, at present, lacks an adequate basis for explanation on common grounds of homology. It is customary to consider the protonephridia of amphioxids to be one of the most remarkable examples of convergent evolution in the animal kingdom, since it is difficult to imagine a direct line of genetic continuity between amphioxids and the lower protostomian invertebrate groups.

In concluding our review of Amphioxus development, we can say that, if tunicates barely achieved the level of specialization to qualify as *bona fide* chordates, then amphioxids have barely failed to qualify as being unequivocal vertebrates. Their position remains secure as being very close to the stem from which the vertebrates arose. The 8 gill slit larval stage is the most characteristic chordate period in their life cycle. For those who maintain that adult Amphioxus is a degenerate form adapted to burrowing, we can only say that the larval stage gives no evidence of there ever having been a larger brain or more highly developed sense organs than those possessed by the adult at present. The position can be defended that amphioxids represent a primary primitive state in chordate evolution.

C. THE PRIMITIVE VERTEBRATE BODY PLAN REPRESENTED BY THE AMMOCETES LARVA OF CYCLOSTOMES

Adult cyclostomes are the most primitive living vertebrates. They belong to the Class Agnatha; the name refers to their jawless condition. In addition to having many other primitive characters, they also lack the paired appendages which are typical of all higher vertebrates. As adults, cyclostomes exhibit a number of specializations adaptive to special habitats; they are ectoparasitic in lampreys, scavenging in hag fish, and detritus, feeding in brook lampreys. The latter most resemble the larval mode of existence which, as in Amphioxus, is burrowing in aquatic sediments. The details of adult structure are beyond our present interest in the embryology and evolution of primitive vertebrate form. For details of structure and function in cyclostomes, refer to standard texts in comparative vertebrate biology.

The present account will be limited to a review of the basic homologies detectable in the pattern of embryonic development and larval form in lampreys, as demonstrated by the **Ammocetes larva.** This name is carried over from a time when the very long-lived larval stage was believed to be an independent species for which the generic name, *Ammocetes,* was assigned. Embryonic and larval stages occupy approximately half the entire life cycle. Larvae live for two or three years in the sediments of rapidly moving freshwater streams. After metamorphosis they become active, predatory parasites on bony fish for which most modifications of metamorphosis are adaptive. Without question, the condition of the larva in habitat and structure is much closer to the unmodified developmental common plan than any other adult or larval living vertebrate.

The universal and profound interest that man has in his origins easily justifies the few hours required to examine and understand this, the most primitive of vertebrate forms.

1. Cleavage, gastrulation and early morphogenesis in the lamprey. (Fig. 1-6 A–J). Laboratory materials will not be available for the study of early embryonic development in lampreys. The present section is included to give a very brief account of early development which is essential as background for understanding the larval stage which follows.

Note in Figure 1-6 A–G that the early cleavage, blastula, and gastrula stages in the lamprey are remarkable for their similarity to those described in Chapter 3 for the amphibian (Fig. 3-9 A–J). It is probably safe, therefore, to say that both are good examples of the early vertebrate plan for cleavage and germ layer formation. In gastrulae of both cyclostomes and amphibians there is a surface ectoderm, and an internal endoderm that forms only the floor of the spacious archenteric cavity. The thin roof of the archenteron will form the general mesoderm, including its notochord, somite, and lateral plate derivatives.

The tailbud stages shown in Figure 1-6 I, J, coincide with hatching; however, the embryo will continue growing without feeding for a considerable time until its large midgut yolk reserves are exhausted. During this pre-larval stage, considerable organogenesis takes place which can be summarized briefly as follows.

In the early tailbud stage, (Fig. 1-6 I', J'), **somite** and **myocoele** formation in the anterior 2 or 3 segments is of special interest in that the myocoele (as in Amphioxus) is formed synchronously with somite separation. That is, here is true enterocoelous formation of the coelomic primordia from the archenteron cavity. In no other true vertebrates does this primitive condition persist. Posterior myocoeles form by the splitting of a solid mesodermal cord; this is also the case for myocoele formation in posterior somites of Amphioxus and for all myocoeles in higher vertebrates. The fate of mesoderm is, in general, like that in amphibians in that the general archenteron roof, after separating from the endoderm, becomes subdivided into a middorsal strand of **notochordal cells.** Progressively more lateral regions will form **somite, mesomere,** and **lateral plate mesoderm.** Initially a cavity forms that is continuous through all three lateral regions of the mesodermal cord. Later they become constricted into discrete **myocoeles, mesocoeles,** and **coelom** with interconnecting narrow channels. Even later, as is characteristic of higher vertebrates, the myocoeles of lampreys degenerate without a trace.

The last element of interest in the strictly embryonic stage concerns the manner in which the central nervous system in lampreys is formed. It is very similar to that found in advanced bony fishes. No open neural plate is formed; instead, a solid cord of ectodermal cells grows directly to the interior from the dorsal midline. This is called the **neural keel.** The surface cells of the keel retain their original epidermal character. At an appreciably later time, the deep cells of the keel will secondarily hollow out and form a characteristic **dorsal tubular nerve cord.** This manner of neural tube formation is not considered primitive. Open neural plates such as those to be described for amphibians are encountered with minor modifications in tunicates, amphioxids, sharks, lung fish, primitive bony fish, amphibians, reptiles, birds, and mammals. Therefore, neural keel formation, wherever it occurs, is believed to be a secondary, or caenomorphic, character in neural tube formation.

2. The late pre-larva and Ammocetes larval stage in lamprey development. (Fig. 1-7 A–K). Whole mount stages and representative serial cross sections will be available for study. Since the developmental age of whole mount preparations will vary considerably, general descriptions will be made of common landmarks even though specific descriptions may not conform with individual preparations. For convenience in making cross-reference between whole mount and cross section slides, the discussion of Ammocetes will deal in sequence with the derivatives of ectoderm, endoderm, and lastly mesoderm.

a. Ectodermal derivatives include the **epidermis** and **neural tube.** The epidermis forms the covering of the **caudal fin** which is continuous with a low **dorsal fin** that extends forward along most of the length of the trunk. It also forms the **otic vesicle** which can be seen as a moderately transparent oval vesicle on the side of the head at the boundary between the midbrain and hindbrain. At the anterior tip of the head, note the very small invagination from the dorsal epidermis that extends under the anterior end of the forebrain. This is the **naso-hypophyseal pit** and, as the name indicates, it will form the olfactory organ and hypophyseal part of the pituitary gland. The remaining epidermal derivative of note is the **stomodaeum.** It will be discussed in association with the development of the anterior gut. The tubular dorsal nerve cord is well developed and forms clear **forebrain, midbrain,** and **hindbrain** vesicles. It is believed that they evolved in association with their nervous connections to primary head sensory organs, that is, the nose, eyes, and ears, respectively. Note the close association of the forebrain with the naso-hypophyseal pit where neuronal connections will ultimately be made between the pit and the anterior forebrain wall. Evaginations project from the lateral wall of the forebrain; these are the **optic vesicles** or **eyes.** In whole mounts they can readily be identified as round black spots at mid-level in the brain at the boundary between the forebrain and midbrain. Throughout larval stages, the eye remains deeply embedded inside the muscles of the head and does not function in vision. At metamorphosis, eye development is resumed and a typical vertebrate eye with cornea, lens, iris, and orbital muscles will develop. At that time neuronal connections from the retina will be made via the optic chiasma to the roof of the **midbrain** in the region called the optic tectum. Note again the location of the otic vesicle near the boundary between the midbrain and hindbrain. The otic vesicle will undergo little further development during larval stages. At metamorphosis, it will form the inner ear with a pair of semicircular canals. At this time neuronal connections will be made with the inner ear and cerebellar element of the anterior hindbrain.

b. Endodermal derivatives are associated with the gut. It is characteristic of vertebrates that an epidermal invagination, called the stomodaeum, will form the anterior end of the gut. The stomodaeum forms the **oral cavity** and, in Ammocetes, the mouth cavity is extended forward by a large upper lip called the **oral hood.** Inside the mouth are the **oral tentacles** which serve as prefilters to the mouth and pharyngeal cavity. As an identifying landmark for the tentacles, note that they are in a dorsoventral line with the naso hypophyseal duct on the roof of the head. During embryonic development the stomodaeum fuses with the anterior foregut and forms an oral membrane which breaks down when the mouth opening is completed. At approximately the level of the former oral membrane, a second valve-like partition is secondarily formed as paired flaps from the lateral walls of the mouth. This valve is called the **velum.** The landmark that characteristically locates the velum is that it is in a dorsoventral line with the pigmented optic vesicle. The very wide part of the gut posterior to the velum is the **pharynx.** There are 7 pairs of feathery gill arches, each of which is followed by a gill slit that opens into the pharynx. The numbering of arches and pouches begins at the anterior end. The landmark for the **first,** or **mandibular, arch** is its dorsoventral alignment with the otic vesicle. Note that it alone among the arches has secondary **gill filaments** only on its posterior side, that is, on the side facing the first pharyngeal pouch. The remaining arches have pouches both in front and behind and, therefore, have gill filaments on both sides of the arches.

Note the ovoid evagination from the floor of the pharynx at the level of arches 2 to 5; this is the **endostyle-thyroid diverticulum.** It is a homologue of the endostyles in Amphioxus and tunicates, and of the thyroid gland in higher vertebrates. In larval Ammocetes its secretory function is primarily that of a mucous gland; that is, it is involved in food capture or consolidation. In this respect it functions much like the endostyle of amphioxids. At metamorphosis the endostyle produces a hormonal factor from different cells; the hormone stimulates metamorphosis. This is a characteristic thyroid function in anamniote vertebrates.

Before proceeding to the post-pharyngeal levels, a brief digression will be made to indicate the manner that will be used to identify body regions. This will be done by reference to somite numbers along the anteroposterior axis. Since anterior somites are very difficult to count, the reference number will be the somite that marks the level at the posterior margin of the pharynx; this one is **Number 16** (± 1) in actual myotome count.

The gut posterior to the pharynx is greatly reduced in diameter. The first region of the post-pharyngeal gut is the **esophagus** located between somite levels 16–18. The pronephros lies above it and the heart beneath it. At somite levels 19–21, the esophagus disappears into the mass of the **liver** or **hepatopancreas,** which lies just posterior to the heart near the ventral midline. There is no distinctive enlargement of the esophagus into a stomach before it enters the somewhat larger **duodenal intestine.** Although it will not be visible in whole mount preparations, it is at this level that the common bile duct from the hepatopancreas passes from the large **gall bladder** into the intestine. Posterior to somite level 21–22, the gut passes straight back to the posterior end of the trunk to about somite level 85–90; this can conveniently be called the **midgut** or **intestine.** At its posterior end the hindgut turns abruptly ventrad through a short ectodermal canal, the **proctodaeum,** that opens to the exterior ventrally by way of the anus near the posterior end of the body.

c. Mesodermal derivatives include the notochord and products of the somites, mesomere, and hypomere. Note that the **notochord** extends from the level of the midbrain under the dorsal nerve cord back to the tip of the tail. It has differentiated into cells almost identical with those previously seen in Amphioxus and tunicate tadpoles. The most conspicuous somite derivatives are the **myotomes;** their usefulness in identifying specific body levels has already been indicated. In Ammocetes the posterior myotomes are conspicuous short block-like structures that are most easily recognized in the dorsal third of the body; however, they do extend down into the lateral parts of the body wall. The mesomere or kidney primordium first differentiates at its anterior end into about 5 pairs of **pronephric tubules** that open by **nephrostomes** into the dorsolateral wall of the **peritoneal coelom.** They can be most readily seen at somite levels 16–18, dorsal to the esophagus. In older Ammocetes there is a tendency for pigment cells to develop around the pronephric tubules. They often can be identified by their dark color, but details of structure will be obscured. The pronephric tubules on each side enter a **pronephric duct** which then grows back to the posterior end of the trunk. Pronephric ducts enter the dorsal wall of the hindgut at about somite level 82–86. In older Ammocetes the **mesonephros** may have begun to develop in the midtrunk region in the roof of the peritoneal cavity. Mesonephric tubules begin to form at about somite level 25 and are added progressively in a posterior direction. They will utilize the pronephric duct that passes dorsal and lateral to the newly forming mesonephric tubules. In Ammocetes the mesonephric tubules form nephrostomes of an unusual kind. They capture a small vesicle from the coelom into which an arterial capillary knot called a glomerulus will project. Thus the nephrostome and coelomic vesicle are converted directly into a Bowman's capsule. Ammocetes, thus, throws very interesting light on the evolutionary steps probably involved in the early development of higher vertebrate kidneys.

Last, we will consider the circulatory system. It is altogether of a vertebrate type. There is a ventral heart that develops from the ventral mesentery of the esophagus; accordingly, we will find the heart ventral to the esophagus at somite levels 16–18. The heart has a well formed **sinus venosus, atrium** and **ventricle,** but seems to be lacking a distinct conus arteriosus. The heart connects anteriorly with a ventral aorta which gives off 7 **aortic arches** that pass through the visceral arches and converge dorsally into a **dorsal aorta.** The venous system includes a pair of **anterior** and **posterior cardinals,** and **common cardinals.** Blood from the gut returns to the sinus by the way of a **subintestinal-hepatic portal vein.** Thus, all major components of the primary vertebrate circulatory plan are present in a very simple condition in Ammocetes. Aside from the heart, details of the circulatory plan cannot be seen in whole mount preparations. However, the positions of the major vessels are indicated in the representative serial cross sections that have been provided in Figure 1-7 C–K.

In conclusion, we have reviewed the major sources of evidence that have led to the widely accepted and most useful concept of the developmental common plan as a basis for generalizing on the basic structure and mode of origin of the vertebrate body. In the next three chapters we will not be concerned so much with the idea of general vertebrate structure as with pre-gastrula events that set in motion the cause and effect sequence of events that will lead to the origin of this plan. We shall, however, return again to the concept of the developmental common plan in Chapter 5 and thereafter, with varying degrees of emphasis, the concept of the common plan will be a recurring theme to the end of this study.

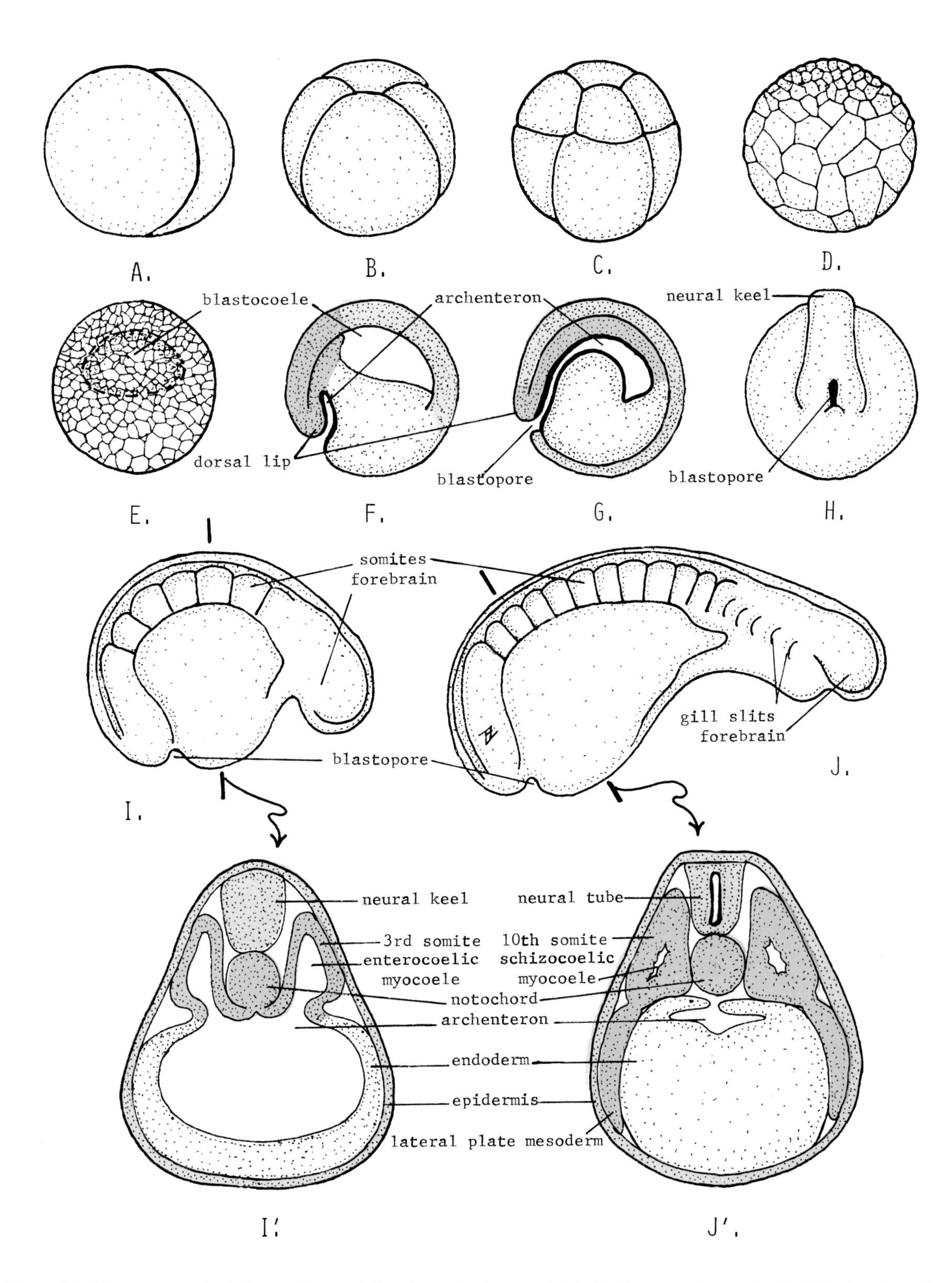

Figure 1-6. Cleavage, gastrulation and neurulation in cyclostomes. 1-6 A–D. Cleavage stages. 1-6 E. Optical section of a blastula. 1-6 F & G. Median sagittal sections of early and mid-gastrulae. 1-6 H. Posterior blastoporal view of a late gastrula with a blastopore and high neural keel. 1-6 I Early tailbud stage. 1-6 I'. Cross section through 1-6 I at the level marked I' and showing enterocoelous formation of myocoele in one of the anterior 3 pairs of somites. 1-6 J. Late tailbud stage. 1-6 J'. Cross section through 1-6 J at the level marked J' and showing schizocoelic myocoele formation in posterior somites. (Fig. 1-6 F and G based on Pflugfelder, *Lehrbuch der Entwicklungsgeschichte und Entwicklungsphysiologie der Tiere.*, 1962; 1-6 I and J adapted from Grasse's, *Traite de Zoologie,* Vol. 13, 1958; 1-6 I' and J' modified from Parker and Haswell, *Textbook of Zoology,* Vol. 2, 1930). Color code: blue — ectoderm, red — mesoderm, yellow — endoderm.

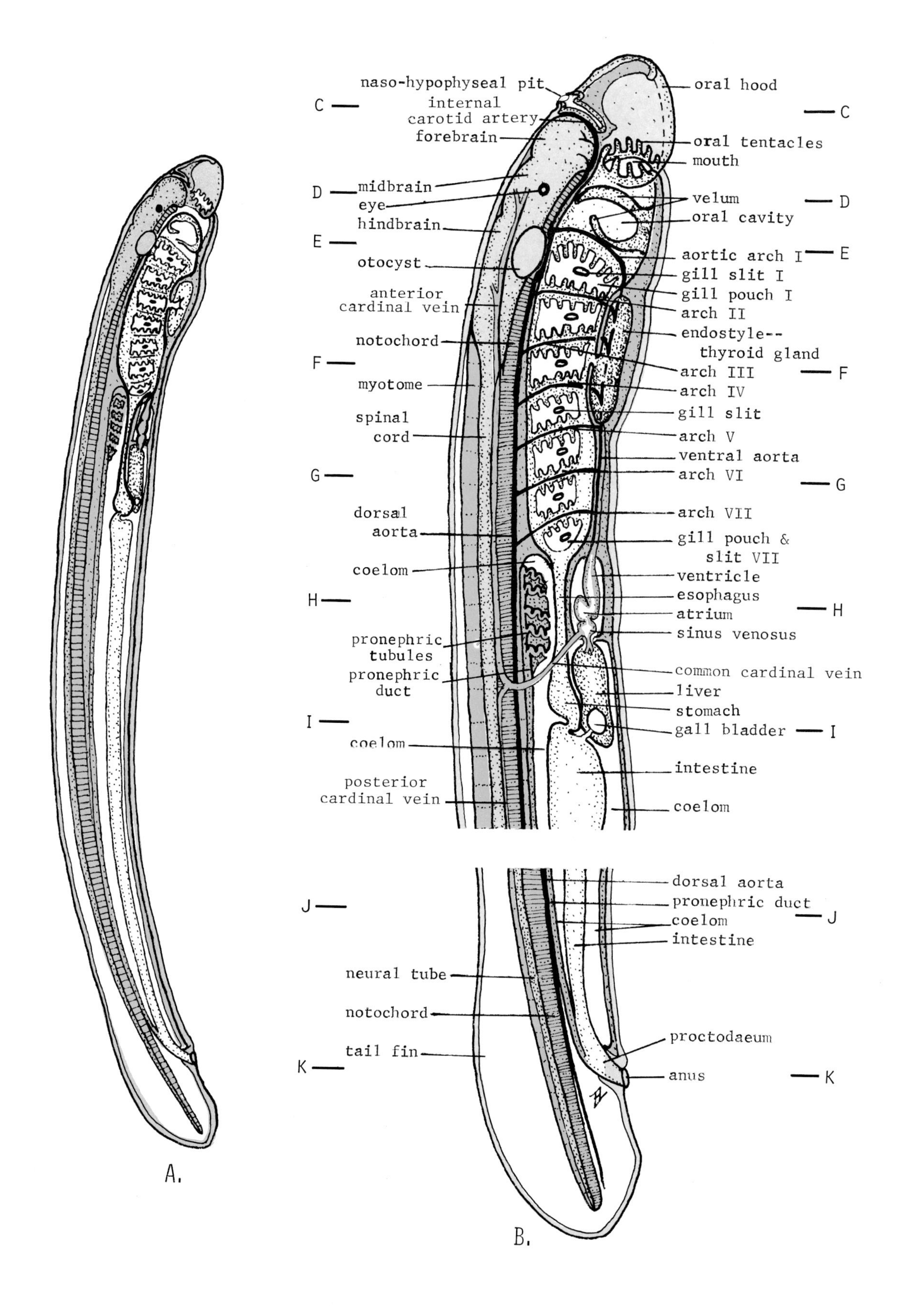

Figure 1-7. Structure of the young Ammocetes larva of cyclostomes. 1-7 A and B. Optical reconstructions of whole mounts of 12 mm Ammocetes larva. Letters C through K at the right and left of 1-7 B correspond with cross section levels for Figures 1-7 C–K. Color code: blue — ectoderm, red — mesoderm, yellow — endoderm.

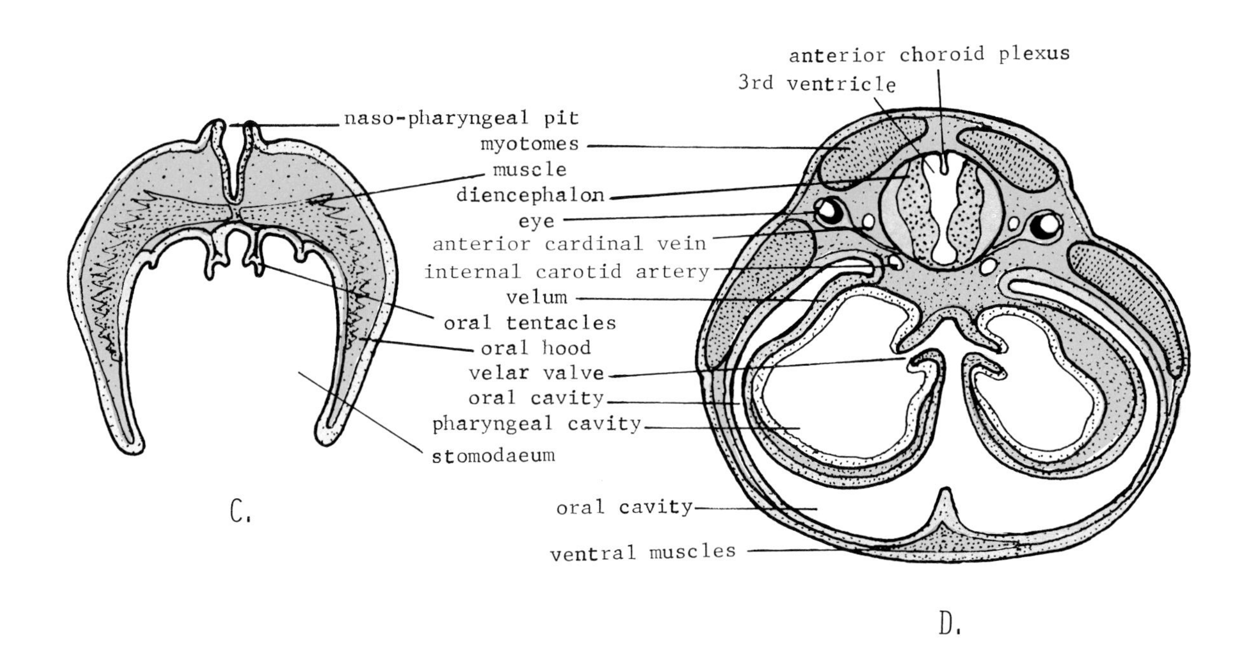

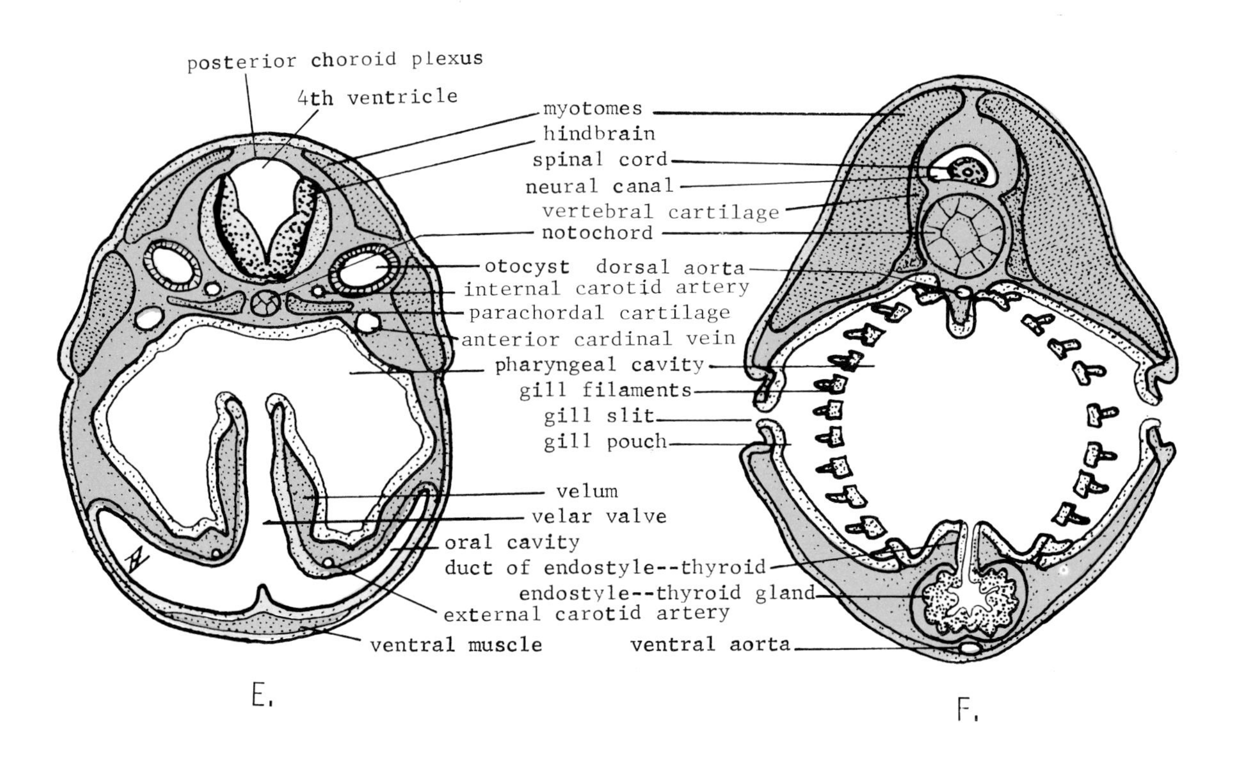

Figure 1-7 C–F. Representative cross sections through an Ammocetes larva at levels corresponding to C–F in Figure 1-7 B. Color code: blue — ectoderm, red — mesoderm, yellow — endoderm.

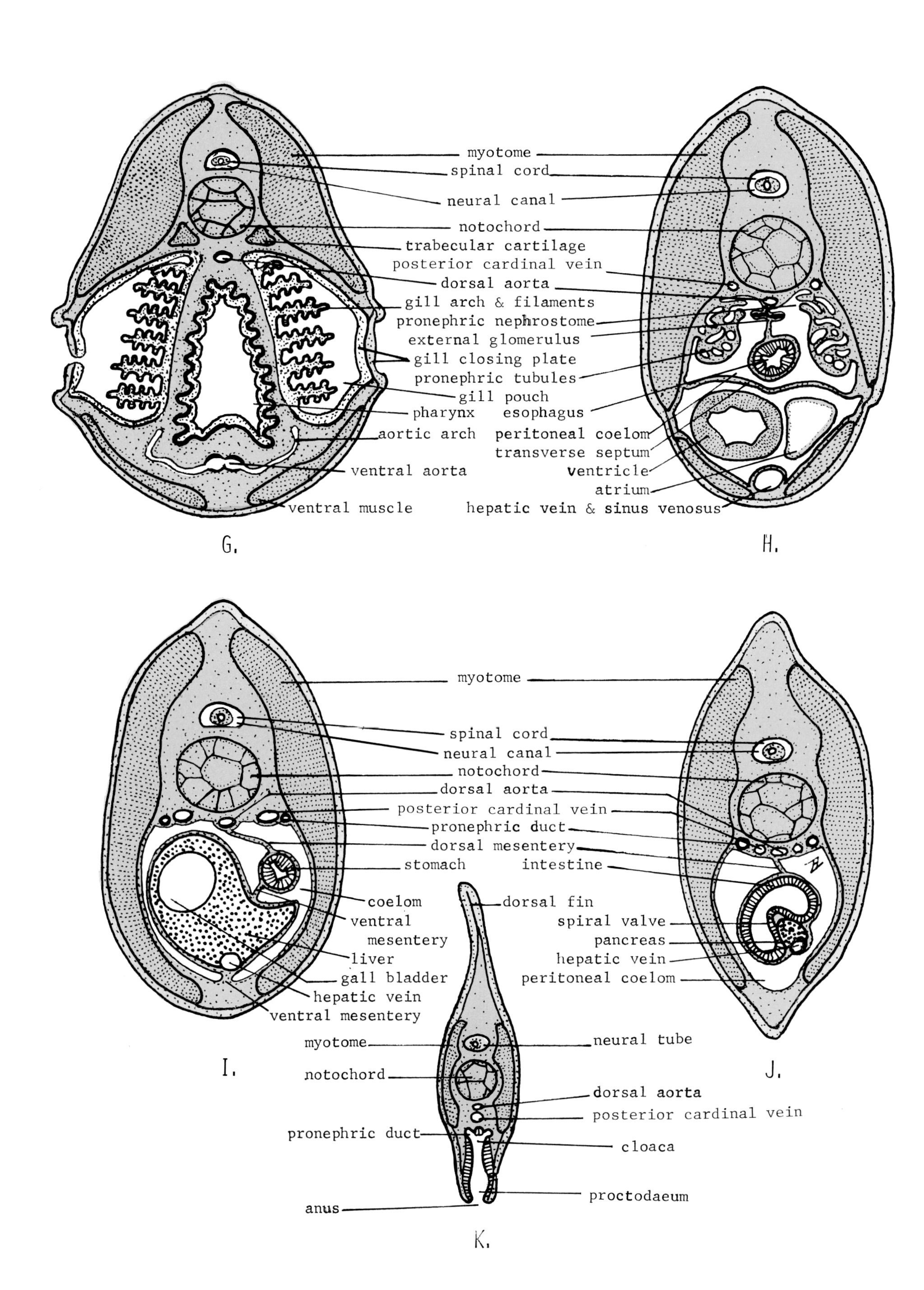

Figure 1-7 G–K. Representative cross sections through an Ammocetes larva at levels corresponding to G–K in Figure 1-7 B. Color code: blue — ectoderm, red — mesoderm, yellow — endoderm.

CHAPTER II

THE STORAGE AND ACTIVATION OF DEVELOPMENTAL INFORMATION

Illustrated by Gamete Formation, Meiosis and Fertilization in Echinoderms, Ascaris, Amphibians, Mammals, and Chironomous

TABLE OF CONTENTS

TABLES AND ILLUSTRATIONS

DIAGRAM OF MITOSIS AND MEIOSIS

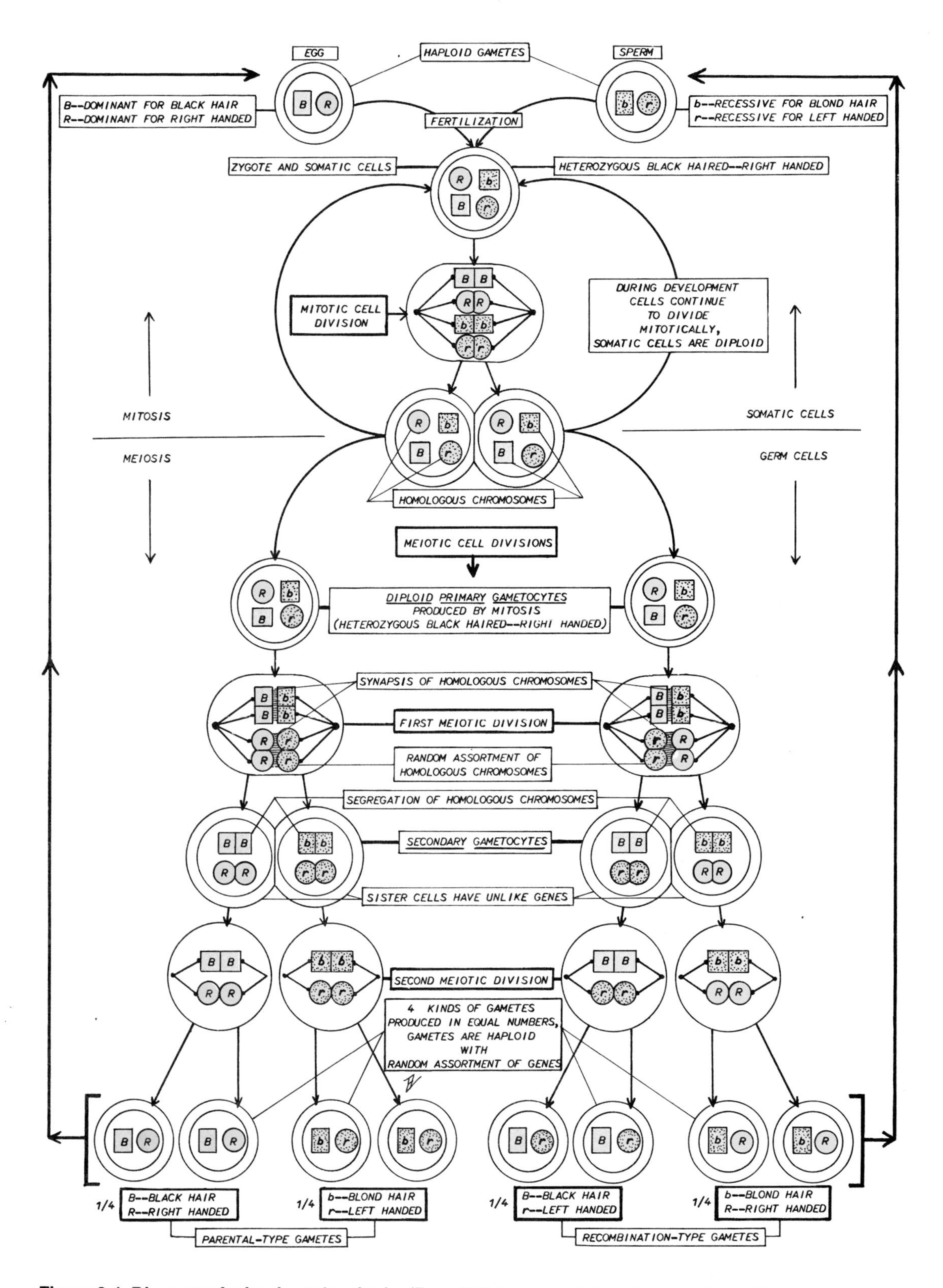

Figure 2-1. Diagram of mitosis and meiosis. (From H.E. Lehman, *Laboratory Studies in General Zoology,* Hunter Publishing Company, 1981).

CHAPTER II

THE STORAGE AND ACTIVATION OF DEVELOPMENTAL INFORMATION

Illustrated by Gamete Formation, Meiosis and Fertilization in Echinoderms, Ascaris, Amphibians, Mammals, and Chironomous

Although sexual reproduction in some very primitive green algae entails the fusion of two identical sex cells called **isogametes,** all metazoan animals have the sex cells of males and females differentiated as **heterogametes;** that is, the gametes are differentiated into nutrient-rich sessile **eggs** and motile **sperm.** Functionally ripe sex cells of both kinds are collectively called **gametes.**

All chemical information used in the programming of embryonic development is encoded in the nucleus and cytoplasm of gametes. It long has been established that the major material of heredity is contained in the chromosomes of the cell nucleus, and that the macromolecules that perform this function are the nucleic acids of DNA (deoxyribonucleic acid). It is also known that small amounts of DNA are found in cytoplasmic organelles such as mitochondria, plastids, and possibly the basal granules of flagella or centrioles. These organelles possess their own self-replicating mechanisms, but they are not of any special importance to the unfolding patterns of embryonic development to be followed in this and following chapters. Hereditary controls over development are overwhelmingly located in nuclear chromosomes. Accordingly, the behavior of these nuclear organelles becomes a matter of paramount concern in the transfer of developmental information from male and female parents to their offspring.

Genes do not control events of development directly. Gene products in the form of RNA (ribonucleic acids), or RNA products in the form of proteins, or protein products in the form of enzyme syntheses are the actual working loci for cell differentiation. Therefore, gene controlled products in the cytoplasm of gametes also contribute substantially to the cellular legacy for development provided by parental gametes to offspring. However, the cytoplasmic contribution is almost exclusively a maternal one, and not an equally shared biparental aspect of storage and transfer of developmental information.

The present chapter will deal with some of the most basic aspects of development. In particular we refer to the essential nature of sexual reproduction, namely, the production of the **fertilized egg** or **zygote.** The subjects will be dealt with under three related but distinctly separate phases in the initiation of embryonic development. These three phases are:

A. Gametogenesis: The formation of ripe eggs and sperm.

B. Maturation: The meiotic reduction of the diploid somatic chromosome number to the haploid number characteristic of mature gametes.

C. Physiological activation: The fertilization events that set in motion capacities for total normal development that lie dormant in ripe eggs and sperm.

Figure 2-1 diagrammatically shows the interrelatedness of gamete fusion or fertilization, mitotic replication, and meiotic reduction in so far as the basic behavior of chromosomes is concerned. Chromosomal markers in the forms of specific allelic gene pairs are used to clarify the genetic consequences of these processes.

A. GAMETOGENESIS: THE FORMATION OF RIPE GAMETES ILLUSTRATED BY OVARIAN DEVELOPMENT IN ECHINODERMS, AMPHIBIANS, AND MAMMALS

Since the cycle of development of male germ cells is inseparably linked with processes of meiotic maturation, general aspects of spermatogenesis will be deferred until the consideration of the second major section in this chapter in which meiotic mechanisms will be given special attention.

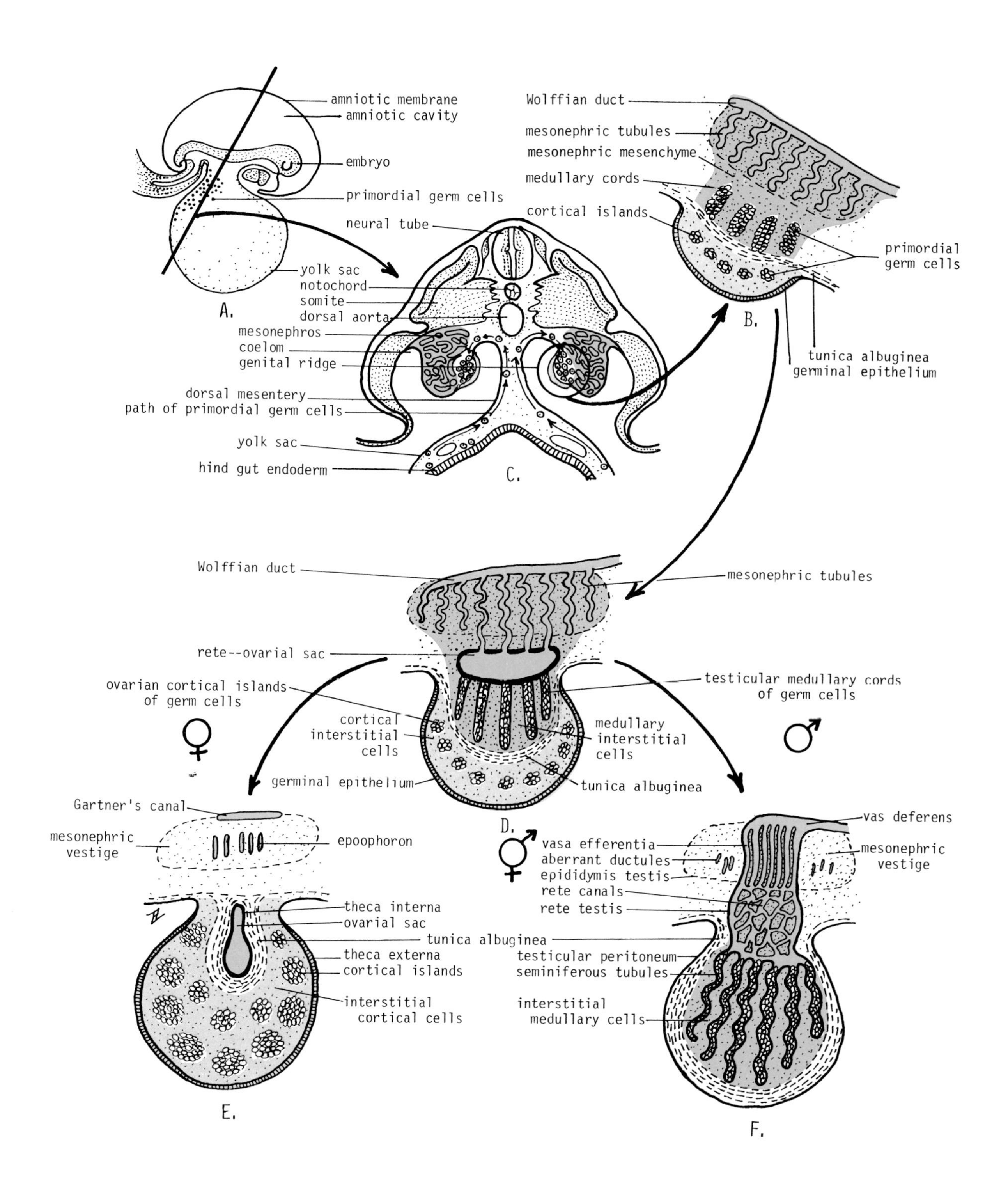

Figure 2-2. Diagrams of primordial germ cell and gonad development in vertebrate embryos. 2-2 A. Diagram of a young tailbud embryo. 2-2 B. Major elements in the embryonic genital ridge. 2-2 C. Cross section of the embryo at the level of the guideline shows the path of primordial germ cell migration. 2-2 D. Intermediate period in the development of the ovotestis with cortical (ovarian) and medullary (testicular) elements. 2-2 E. Ovary after regression of medullary and mesonephric elements. 2-2 F. Testis after regression of cortical elements. Color code: blue — peritoneal derivatives of the germinal epithelium, red — mesonephric derivatives, yellow — primordial germ cells derived from the yolk sac.

1. The origin of vertebrate germ cells (Fig. 2-2). The origin of primordial germ cells during early vertebrate embryonic development will not be studied in any laboratory material directly. However, it is of sufficient importance to an understanding of all following aspects of gametogenesis, meiosis, and fertilization that it will be synoptically reviewed with the aid of diagrams in Figure 2-2. This plate schematically outlines the early aspects of vertebrate germ cell origin during embryonic development in a generalized vertebrate. Although many exceptions exist, it is nevertheless the general rule among vertebrates that the **primordial germ cells** are first identifiable in the region of the midgut or hindgut of the very early embryo (Fig. 2-2 A). The primordial germ cells then leave the gut and migrate through the dorsal mesentery of the gut to a position between the parietal peritoneum and the developing mesonephric kidney (Fig. 2-2 B). If, for natural or experimental reasons, the primordial germ cells fail to reach this location, then the gonad will develop into an ovary or testis (depending upon the genetic sex: XX = female, XY = male in most mammals). However, the gonad will be sterile and possess no reproductive cells. The early **gonad** or **sex gland** initially undergoes identical steps in development, irrespective of the final male or female sex of the individual. Both ovarian and testicular structures will form; the gonad at this hermaphroditic stage is called an **ovotestis** (Fig. 2-2 C, D). The definitive ovarian components will develop from the outer shell, or cortex, of the ovotestis. The functional testicular elements will be derived from the central or medullary parts. Highly schematic, definitive vertebrate gonads are shown in Figures 2-2 E, F.

In the developing ovary (Fig. 2-2 E), note that the primordial germ cells make the **cortical islands** of **oogonial cells;** by mitotic replication these cells will give rise to all future eggs of the ovary. Note the secondary or accessory ovarian structures. These include the **theca externa,** which is merely the covering layer of peritoneum that surrounds the ovary. The **ovarial sac** inside the ovary is lined by the **theca interna;** this is derived from the lining of the reteovarial sac of the ovotestis. The center of the ovary contains many blood vessels and connective tissue elements which collectively make up the **ovarian stroma** or **medulla.** This represents the testicular part of the ovotestis in which all male germ cell functions have degenerated. Much of the ovarian stroma is derived from the **tunica albuginea** of the ovotestis which originally provided a landmark between primitive cortical and medullary regions.

In the developing testis (Fig. 2-2 F), note that it is the medullary components of the ovotestis inside the tunica albuginea that persist. The **primary sex cords,** which are at first solid, will later hollow out and form the **seminiferous tubules,** which are lined primarily by **spermatogonial cells.** Mitotic replication of these cells will give rise to all mature spermatozoa when maturity is reached. Note that the seminiferous tubules unite with the reteovarial sac; the sac then becomes divided into a ramifying series of channels called the **rete canals** (collectively the rete testis). The rete canals then fuse with some of the mesonephric tubules. The latter then serve as secondary conducting ducts for spermatozoa; these tubes are the **vasa efferentia** (collectively the epididymus testis). They retain their primitive connection with the Wolffian duct which in adult males serves as the **sperm duct** or **vas deferens.** It will conduct ripe sperm away from the testis to whatever copulatory accessory structures may be required.

An examination of the plan of the definitive female gonad will reveal remnants of the Wolffian duct as a persistent Gartner's canal. Vestiges of mesonephric tubules also persist as the epoophoron and paroophoron in the mesovarium which support the ovary in the peritoneal cavity. In the male, it is very rare for female cortical elements to persist and, therefore, the tunica albuginea lies in contact with the theca externa or testicular peritoneum.

Exceptions to the above account of gonadal development in males and females are numerous in different species and in abnormal cases of true hermaphroditism. In the later case there is a paedomorphic persistence of the early normal ovotestis in which cortical and medullary elements persist as functional ovaries and testes in the same individual. In some lower vertebrates, the capacity for sex reversal exists throughout life, and the same individual, at different times in its life cycle, can function as a male with medullary elements forming a testis; then at a later time the testis regresses and gonadal cortex develops as a functional ovary. The bipotentiality of sex is present in all individuals as part of the normal embryonic cycle; it is the more remarkable that adult hermaphroditism is so exceedingly rare among adult vertebrates.

2. Oogenesis illustrated by ovarian development in echinoderms, amphibians and mammals (Fig. 2-3. 2-4, and 2-5). There are three phases in the gonadal development of eggs; these are:

The gonial phase, a period of mitotic multiplication of germ cells. Among most vertebrates this occurs at early juvenile stages. In amphibians it occurs a few weeks following metamorphosis. In mammals it takes place during the later stages of fetal development. By gonial replication, the original primordial cell number is usually amplified many thousandfold.

The primary oocyte phase is a period of prolonged cell growth without cell division. During this period the nucleus characteristically becomes greatly enlarged and is called a **germinal vesicle.** The cytoplasm becomes packed with yolk and other macromolecules to be utilized in embryonic development. Eggs will be from 50 times to 50-million times as large as other somatic cells of the ovary depending upon the amount of stored materials in them. A significant part of the storage during the primary oocyte stage is in the form of RNAs which are suspected of being a part of the molecular information utilized in releasing developmental capacities in the embryo.

The ovulation and meiotic reduction phase is the release of fully developed eggs from their investing ovarian follicles, and the reduction in chromosome number in gametes from diploid to haploid. In vertebrates and many invertebrates, ovulation is initiated by hormonal stimuli. Pituitary luteinizing hormone induces ovulation in vertebrates and also activates meiotic reduction divisions that will result in the haploid chromosome number in ootids. Meiotic division (but not ovulation) is activated by progesterone of the corpus luteum.

a. Oogenesis in echinoderms, (Fig. 2-3 A) can be demonstrated equally well with cross sections of ovaries of the starfish, Asterias, or the sea urchin, Lytechinus. In both cases, the ovary is a highly branched, tubular organ constructed somewhat like a stalk of bananas with many branched tubular lobes. In cross sections these lobes will appear as circles or ovals of various sizes. The outer wall of the ovary is a specialized part of the parietal peritoneum called the **germinal epithelium** or **theca externa.** From this layer, germ cells are produced by gonial cell multiplication. In the ovary these very small cells, or **oogonia,** will appear as very densely stained cells that are only about four to six times the diameter of cells in the germinal epithelium itself. **Primary oocytes,** at various stages in growth and yolk accumulation, can be identified readily by the large **nucleus** or **germinal vesicle** each of which will contain a large, intensely stained **nucleolus.** When oocyte growth is complete, the oocytes separate from their connecting point on the ovarian wall (which usually coincides with the animal pole of the egg). This release of eggs to the interior of the ovarian lobules is called ovulation. Note that there is no theca interna lining the ovarian lumen. Within the ovarian lumen the eggs will undergo meiotic reduction and, following this, their haploid nuclei will be exceedingly small as compared with that in primary oocytes.

Trace development in a favorable ovarian lobe which shows the transitional stages in the formation of **oogonia.** These are small cells attached to the ovarian wall. They contain dense cytoplasm and a small granular nucleus. The **primary oocytes** are the largest attached cells; they contain a large vesicular nucleus with a conspicuous single nucleolus. The cavity of the ovarian lobes and the spaces surrounding them contain many **secondary oocytes** and **ootids** (mature ova). It is not possible to distinguish between them on the basis of nuclear structure; both possess a small, vesicular nucleus. Sea urchin eggs are only fertilizable in the fully mature ootid stage, whereas starfish eggs can be fertilized at any time following ovulation through all meiotic stages.

b. Oogenesis in amphibians, as represented by the grass frog, *Rana pipiens,* is shown in Figure 2-4. The primitive bilaminate character of the amphibian ovary almost ideally illustrates the fundamental structure of the vertebrate ovary. Study prepared slides and with the aid of the diagrams, identify the outer peritoneal layer of the ovary, the **theca externa;** from it the secondary generations of oogonia are proliferated by mitotic multiplication of germ cells. Most eggs in the sections are at approximately the same stage of primary **oocyte** growth. Each possesses a very large and lightly stained **germinal vesicle** containing a large number of **nucleoli** just inside the **nuclear membrane.** Note that each egg is covered by a layer of **follicle cells;** these cells are also the cells of the **theca interna** which line the central ovarial sac. It is important to realize that each developing oocyte is growing in the space between the theca interna and theca externa. It is also through the space between these two layers that blood vessels bring nutrients to the egg during its growth phase. Look for a graded series of growing oocytes. The smallest and youngest oocytes, along with oogonial cells, will be seen just under the theca externa. At various locations throughout the ovary of a mature frog, that has passed through at least one reproductive season, will be found large dark bodies

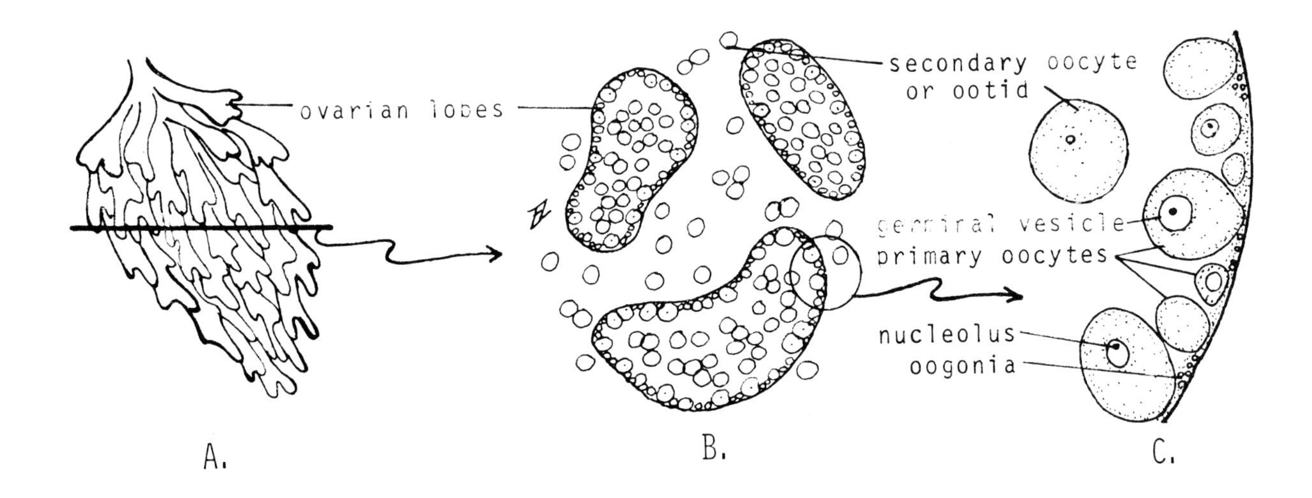

Figure 2-3. Diagrams of echinoderm ovarian structures. 2-3 A. General structure of the ovary. 2-3 B. Low power appearance of a section through the ovary. 2-3 C. Oocytes in the wall of the ovary.

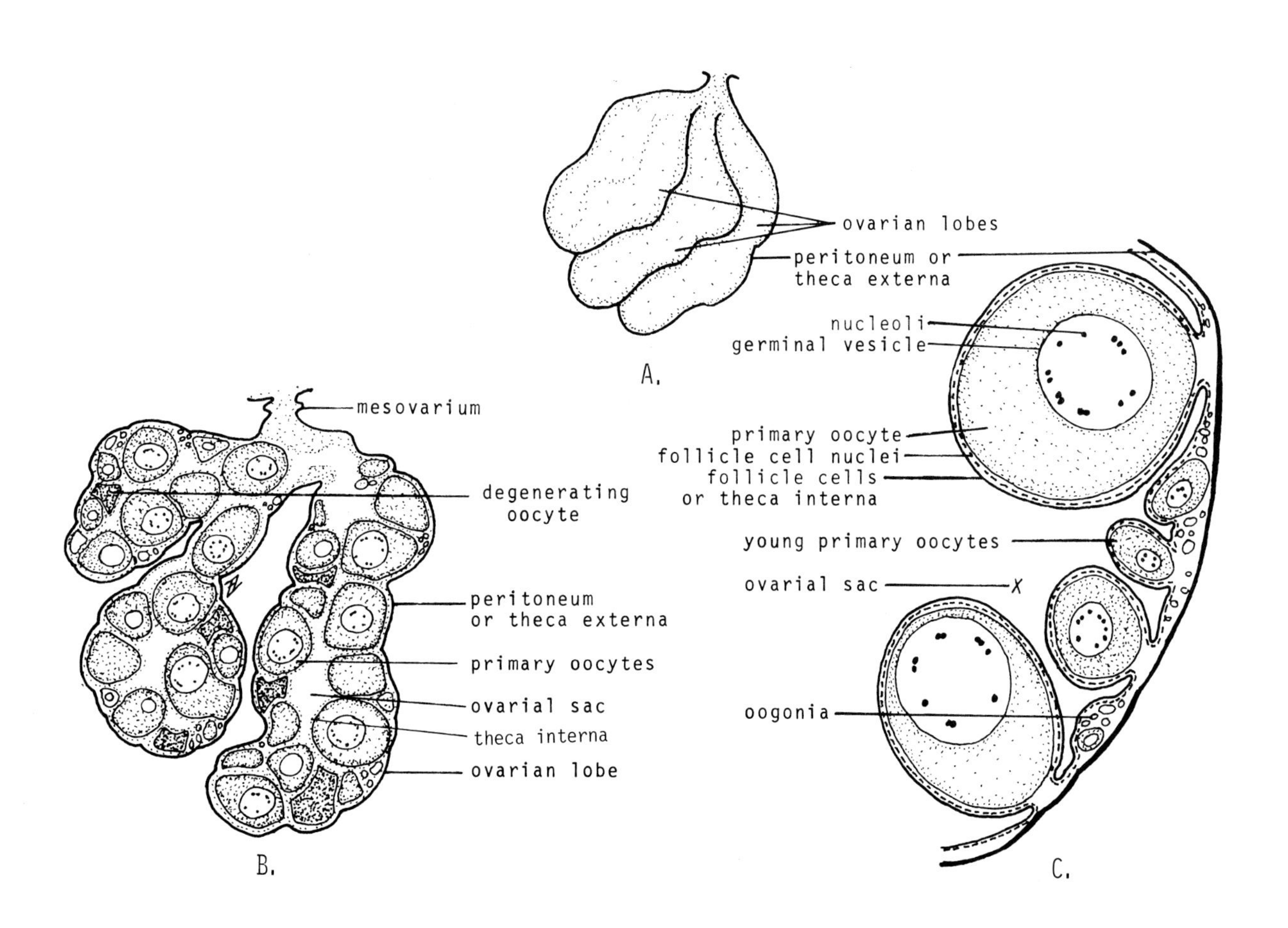

Figure 2-4. Diagrams of amphibian ovarian structure. 2-4 A. General structure of the ovary. 2-4 B. Low power appearance of a section through the ovary. 2-4 C. Oocytes in the wall of the ovary.

with very poorly staining cytoplasm. These are degenerating eggs in advanced stages of resorption and generally represent eggs that were not shed at the preceding breeding season.

If living frogs are available for fertilization experiments (see Chapter 3), examine the living ovary in which ovulation is induced by pituitary and progesterone injection. Note that the ovary in life is thrown into many loose folds. Snip off a segment of one of these lobes and transfer it to a dish of isotonic (0.7%) sodium chloride solution. Examine the fragment at low magnification and observe the distribution of eggs in the living ovarian wall.

With a fine insect pin dissecting needle prick one of the eggs while observing it under low magnification. The yolk will immediately begin to flow out. As it does so, the large germinal vesicle carried with it will appear as a water-clear sphere. With a micropipette, made by secondarily drawing out a Pasteur bacterial pipette, pick up the nucleus and transfer it to a fresh drop of isotonic NaCl in a clean dish. Wash the nucleus in the drop and then transfer the nucleus to a clean glass slide containing clean isotonic NaCl. Add a very small drop of 0.1% Toluidine Blue and add a coverslip. With a filter paper wick, draw off all excess water from the preparation so that the nucleus is flattened and finally ruptured by pressure. Observe the nuclear contents under high power and attempt to identify the **nucleoli** and **lampbrush chromosomes** of the germinal vesicle. The latter are chromosomes with extensive lateral loops that create the impression of fuzzy threads. The lateral chromosomal loops are believed to represent local areas of gene activity associated with the metabolic syntheses accompanying oocyte growth. Because of their high RNA content resulting from gene activity, they will stain intensely with Toluidine Blue, a dye with a strong affinity for nucleic acids.

c. Oogenesis in mammals can be shown equally well in cross sections of ovaries of the mouse, rat, cat, or rabbit. A generalized mammalian ovary with diagnostic stages in oogenesis is shown in Figure 2-5. In this study, emphasis will be placed on follicle development. Identify first the very indefinite boundary between the **cortical** and **medullary parts** of the ovary. The latter will consist of blood vessels and connective tissue cells at the center of the ovary. The cortex forms a peripheral belt around the ovary underlying the surface coating of peritoneum or **theca externa.** An ovarial sac is usually absent, or it may be present as a small vesicle at the center of the medullary mass of blood vessels. All stages in oogenesis will be mixed at random in the cortical region. Identify each of the following stages in oogenesis. (1) **Oogonia** are large cells located just inside the covering peritoneum. These cells are frequently very numerous and make a distinct layer from two to five cells thick. The nuclei of oogonia are much larger than those of surrounding cells that make up the interstitial cells of the ovary. Note that oogonia are not surrounded by any follicle cells. (2) **Young primary oocytes in primary follicles** are in the primary growth phase but their nuclei and cytoplasm will appear to be from 3 to 5 times larger in diameter than are nuclei of interstitial cells. They possess lightly stained cytoplasm and nuclei which have a distinct spherical nucleolus. They are surrounded by a **single layer of cuboidal follicle cells.** Primary follicles lie somewhat deeper in the ovarian cortex than oogonial cells. (3) **Midprimary oocytes in secondary follicles** are characterized by relatively conspicuous but lightly stained cytoplasmic yolk granules, by a round germinal vesicle and conspicuous nucleolus, and by two to ten layers of follicle cells forming a solid mass of cells surrounding the primary oocyte. A cavity has not yet developed in the follicle cell layer. (4) **Late primary oocyte and tertiary follicles** can be identified by the presence of two or three isolated cavities called **antra** (singular, antrum) that have formed by splitting within the follicle cell layers. (5) **Mature primary oocyte and Graafian follicles** form the most conspicuous structures in the ovarian wall. The antrum may be truly enormous and readily visible to the naked eye. The primary oocyte can be seen attached to the side of the Graafian follicle by a small pillar of cells called the **cumulus oophorus.** Follicle cells, however, continue to surround the oocyte and a heavy egg membrane, the **zona pellucida,** will separate them from the surface of the fully developed egg.

The ovarian cycle in mammals is under complex periodic feedback, hormonal controls involving products of the pituitary gland and ovarian follicle cells. The physiology of female reproduction is beyond the scope of this study. However, in very broad outlines, the control system is as follows: (1) First **FSH, the follicle stimulating hormone** of the anterior pituitary causes follicles and eggs to begin development. (2) Then the follicle cells acquire endocrine competence and produce the hormone **estrone** or **female sex hormone,** which acts back on the anterior pituitary and stimulates a repression of FSH and initiates the production of **LH,** the **luteinizing hormone,** by the anterior pituitary. (3) LH acts upon the follicles and causes

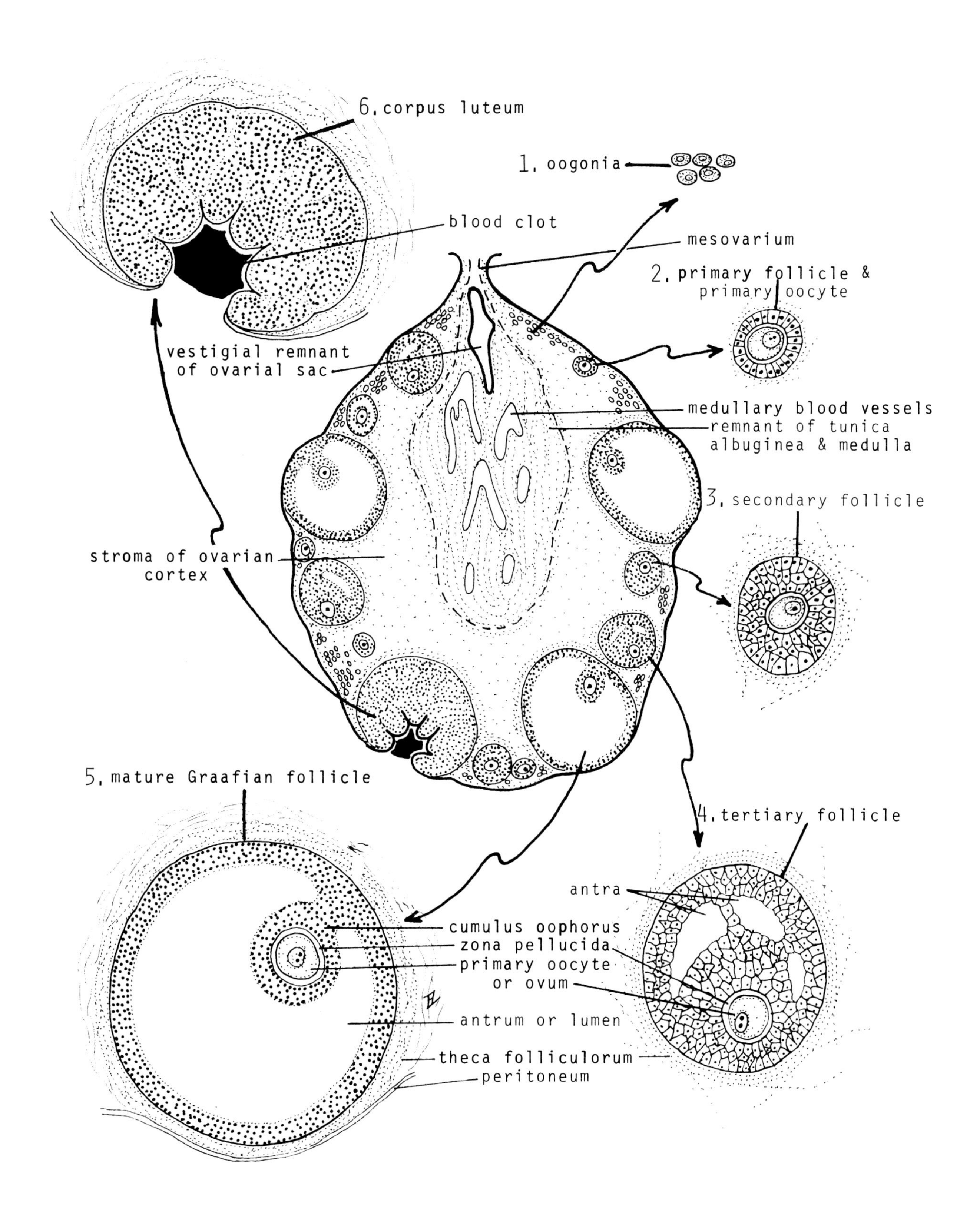

Figure 2-5. Diagram of ovarian structure and follicle development in mammals. The central figure shows the appearance of an ovary semidiagrammatically. Beginning at the upper right are a series of figures at higher magnification that show the major progressive steps in the development of ovarian follicles and of the corpus luteum from follicle cells.

ovulation of the egg and the conversion of the follicle cells into **corpus luteum.** (4) The corpus luteum of the ovary then produces the hormone **progesterone,** which stimulates the uterine wall into final readiness for implantation of the embryo and for placenta formation. It was mentioned earlier that this hormone is also an activator of meiotic reduction in ovulated eggs. In the event that fertilization and implantation fail to occur, the corpus luteum degenerates and the cycle is started anew. (5) However, when fertilization and pregnancy, with implantation of the embryo in the uterus, occur, the placental hormone, **chorionic gonadotropin,** is produced. It takes over the function of pituitary luteinizing hormone and maintains the corpus luteum through pregnancy. Thus, the fetus assures that the physiological conditions necessary for its own growth will be maintained until full term, when the secretion of chorionic gonadotropin ceases and birth occurs.

With the preceding background on follicle cell function, try to identify a **corpus luteum** in the section. One may not be present if the ovary is from an immature or non-pregnant female. Luteal cells are usually quite large and roughly cuboidal, with light staining properties. Generally the antrum of the original follicle is completely replaced by ingrowth of luteal cells, or by an invasion of the region by connective tissue from the ovarian stroma, or by the presence of a massive blood clot that may have occurred at the time of ovulation. In some ovaries, in particular the rabbit, corpora lutea may exceed in volume all other ovarian tissues in a gestating female.

B. MATURATION: THE MEIOTIC REDUCTION OF GERM CELLS TO HAPLOID ILLUSTRATED BY MAMMALIAN SPERMATOGENESIS AND MEIOSIS IN ASCARIS

Refer to the schematic presentation of meiosis in Figures 2-1 and 2-6 for an overview of the events that lead to chromosomal reduction to haploid in gametes, and for a comparison of similarities and differences in this process as they relate to the formation of sperm and eggs.

Maturation is the formation of functional reproductive cells (i.e., sperm and eggs). In the development of both types of gametes, four periods are generally recognized, two of which have already been discussed with regard to egg development. The four periods are: (1) The **gonial stage** is a period during which germ cells undergo little cell specialization but participate in active mitotic replication. (2) The **primary gonocyte stage** is a period of no cell division and great metabolic activity. In a very real sense it can be considered an enormously prolonged prophase during which the diploid chromosome number has replicated and homologous chromosomes have (for a time at least) undergone synapsis. The latter is a form of chromosomal pairing of homologues during which crossing over of segments of chromatids often occurs. Synapsed, or chiasmata-linked, homologous chromosomes are called **bivalents.** Since the homologues of each bivalent consist of two chromatids, the synapsed pair will form a cluster of four chromatids. These are called **tetrads.** The term **pseudoreduction** is applied to the synapsed bivalents or tetrads because there is an apparent reduction in the chromosome number when homologues are tightly synapsed. This, however, has nothing to do with actual reduction which will require two cell divisions. (3) The **maturation stage** is the period of meiosis, or true reduction, during which the primary gametocytes, with diploid diplotene chromosomes (i.e., 4 haploid DNA doses per homologous pair), are reduced to haploid haplotene chromosomes in gametes (i.e., 1 haploid DNA dose per homologous chromosome pair). This will require two cell divisions. The **first meiotic** or **maturation division** is reductional and separates maternal and paternal diplotene (dyad) homologous chromosomes from one another. That is, maternal and paternal chromatids are separated from one another in all regions except those involving crossovers. This is called **genetic segregation.** At the same time, **random recombination** of maternal and paternal characters in different sets of homologous chromosomes also occurs, as shown in Figure 2-1. The cells resulting from the first meiotic division are **secondary gonocytes.** They possess the haploid chromosome number but the chromosomes are diplotene; i.e., they have a 2X DNA dose as compared with haploid values. The second meiotic division is essentially mitotic in character and is said to be equational even though crossover regions are not genetically equivalent. (4) The **gonotid stage** is the final period in meiosis. These cells are the products of the second meiotic division. They will possess fully reduced cells which are both haploid and haplotene; that is, they possess the basic haploid DNA and chromosome value for the species.

The maturation stages described above and diagrammed in Figure 2-6 will be illustrated for

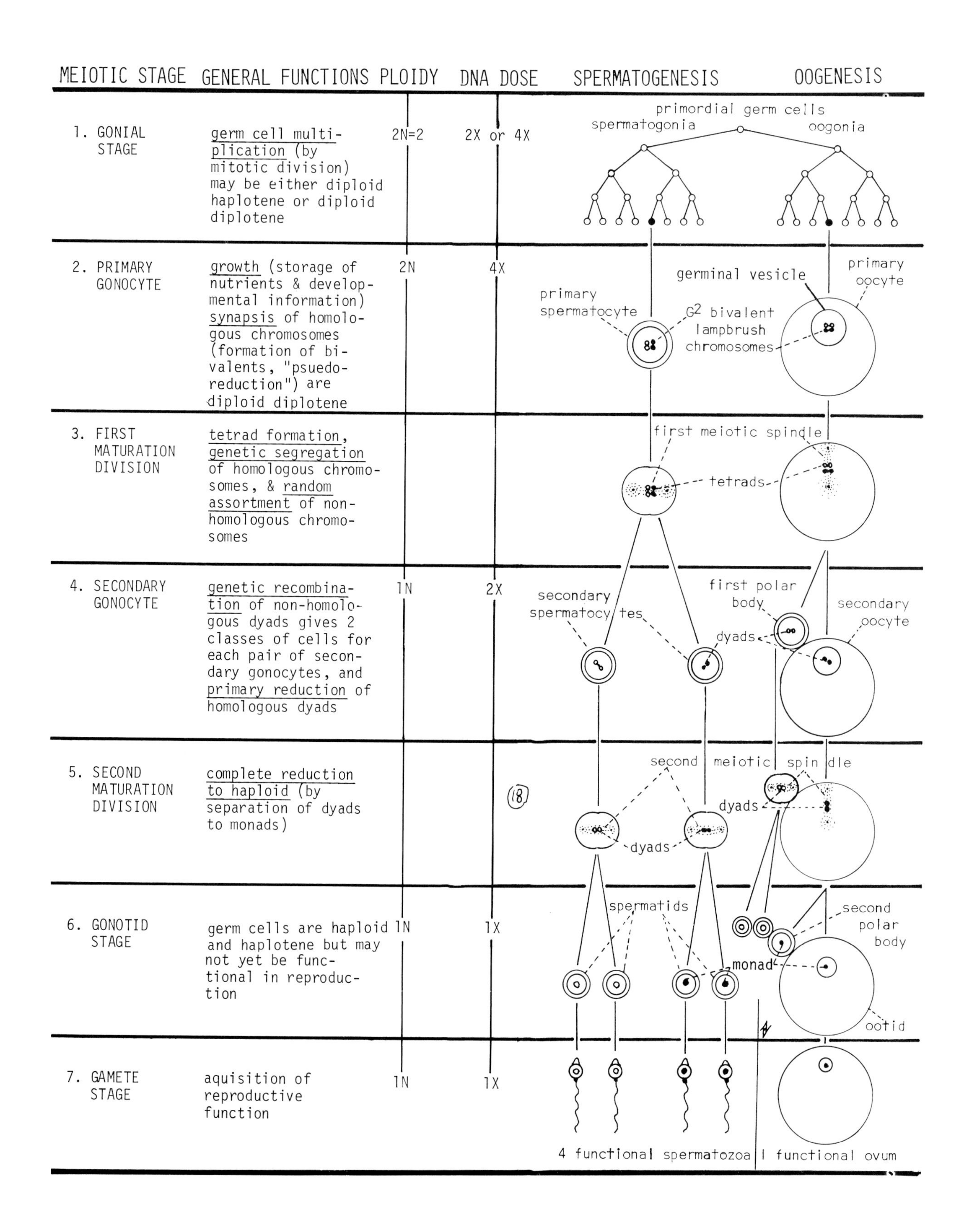

Figure 2-6. Comparison of major aspects of meiosis in spermatogenesis and oogenesis during maturation.

developing sperm in the seminiferous tubules of the mouse, and for the egg in the roundworm, *Ascaris megalocephala bivalens.*

1. Spermatogenesis and meiotic reduction in the mouse testis (Fig. 2-7). The general plan for gametogenesis is illustrated with fewer modifications in the development of sperm than is the case with the development of eggs. Each spermatogonial cell develops into a primary spermatocyte which undergoes two maturation divisions that result in the formation of four functional spermatozoa. The primary deviation from the standard plan is associated with functional specialization of the sperm for motility. This involves the loss of most of the spermatid cytoplasm and the development of a flagellum. The transformation of a spermatid into a spermatozoon is called **spermiogenesis.** During this transformation spermatids become firmly imbedded in nurse cells of the seminiferous tubule. The nurse cells nourish the developing sperm until they have achieved full development. Nurse cells are called **sustentacular cells of Sertoli;** one Sertoli cell usually contains four spermatids (presumably all from the same parent primary spermatocyte). A Sertoli cell can host more than four spermatids at one time. Mature spermatozoa are released from the Sertoli cell and pass into the lumen of the seminiferous tubule under the stimulus of the pituitary **leuteinizing hormone.** Therefore, it is appropriate to consider sperm release from Sertoli cells to be functionally analogous with ovulation of eggs. Both types of gamete release are under the control of the same pituitary hormone, namely, LH.

The slides for study have been stained with a single nuclear dye. Cytoplasmic and cell boundaries are very indistinct. Therefore, the following identifications of cell types and meiotic stages will be based on distinctive nuclear characteristics.

First note that the cross section of the mouse testis is packed with seminiferous tubules cut at random in all conceivable angles; however, most tubules will appear as circles or ovals. Keep in mind that meiotic waves pass along the length of individual tubules. Accordingly, it is unlikely that all meiotic stages will be seen in a single tubule. Identify the following meiotic stages by the study of appropriate tubules. (1) **Spermatogonial nuclei** are located just inside the outer wall of seminiferous tubules. When present, they form a single layer of small, spherical, and very densely stained nuclei containing from 6 to 8 conspicuous chromatin granules. (2) **Primary spermatocyte nuclei** are about twice the diameter of those in spermatogonia. They are located just internal to the spermatogonial layer and usually form about one or two layers of nuclei which are large, vesicular, and contain many intensely stained chromatic granules and threads. These are the most conspicuous and largest nuclei in the wall of the tubule. (3) **Secondary spermatocyte nuclei** are small, clear nuclei about the same size as the small spermatogonial nuclei, but they contain no distinct chromatin granules. Instead, they possess a single conspicuous, centrally located nucleolus. These are the most abundant nuclei in the wall of the tubule and comprise from 3 to 6 layers which make up most of the middle half of the tubule wall. (4) **Spermatid** and **spermatozoon nuclei** are extremely variable in appearance owing to the fact that these cells are in different phases of spermiogenesis in different tubules. They can be identified by the fact that they are the layer of nuclei adjoining the lumen of the tubule and are usually intensely stained, bar-shaped structures. (5) **Sustentacular cells of Sertoli nuclei** (i.e., nurse cells) are somewhat difficult to identify. The cell boundary, if stained, would reveal a flask-shaped cell attached at the periphery of the tubule and extending through the wall to the lumen. The nucleus of these cells is located near the attached basal end of the cell. They are the only oval nuclei in the wall of the tubule. Sertoli nuclei are quite large and are generally situated at about the level occupied by primary spermatocyte nuclei. The Sertoli nuclei, in addition to being oval, contain no conspicuous chromatin granules, but possess a large central nucleolus. These general characteristics of meiotic stage in spermatogenesis are diagrammed in Figure 2-7 D.

2. Guinea pig spermatozoa. Examine the smear preparation of guinea pig sperm under high power. The cells are very small and, for best results, sub-stage illumination should not be too intense. Note the very round globular **sperm head** which is easily recognized. It has two primary subdivisions: the **nucleus** will be rather darkly stained, and will have a crescent-shaped cap of lightly stained cytoplasm on the side opposite the tail; this cap, the **acrosome,** is involved in the mechanics of sperm penetration into the egg. The **neck** is a small bead-like expansion at the base of the **sperm tail** where the latter unites with the sperm head. The neck contains the sperm **centriole,** which cannot be distinguished as a discrete object without very high levels of magnification. Note that the sperm tail, or flagellum, is very long. In life it provides the mechanism for locomotion.

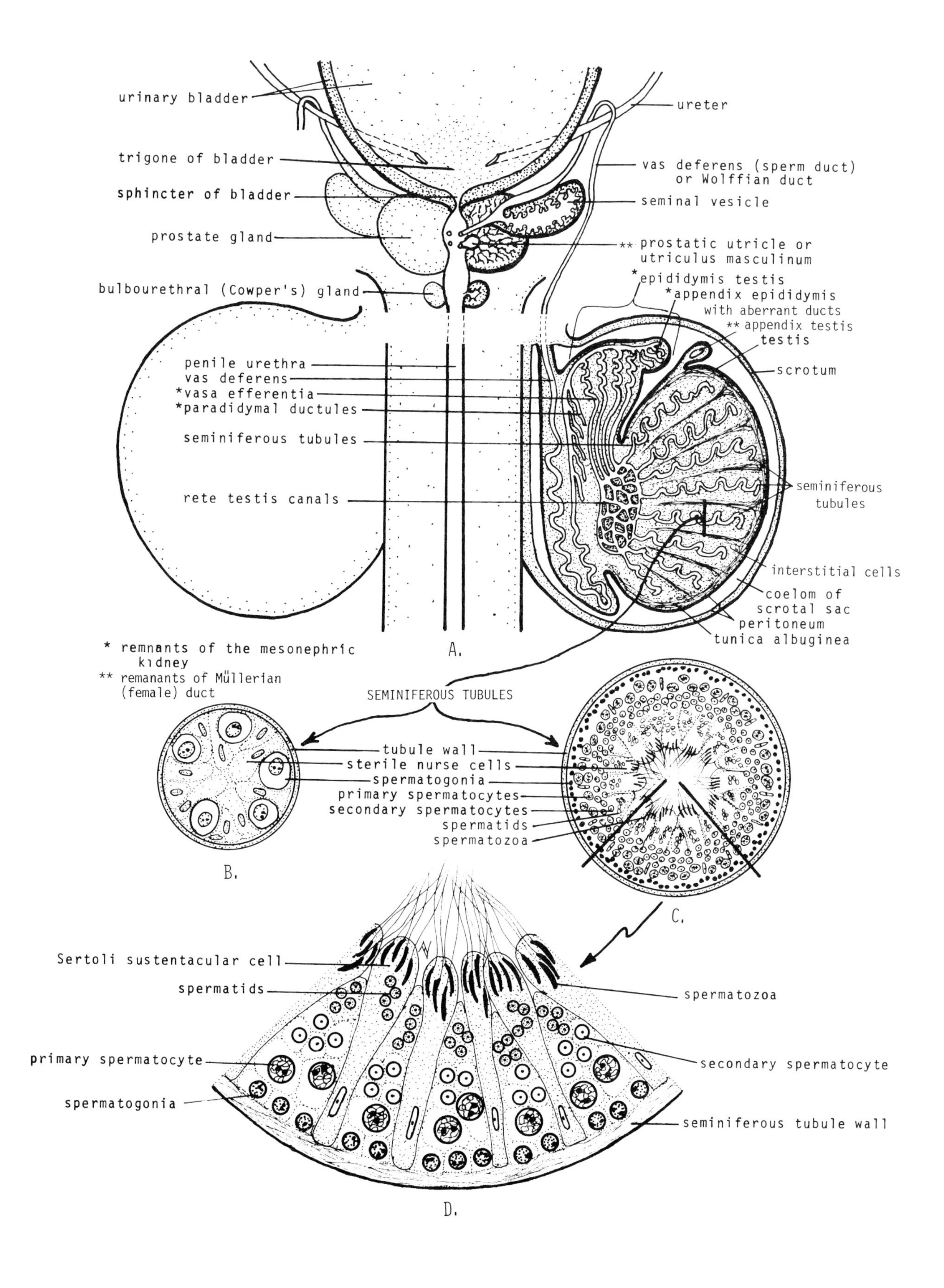

Figure 2-7. Diagrams of the structure of the mammalian testis. 2-7 A. General anatomy of the male reproductive organs. 2-7 B. Cross section of an immature seminiferous tubule. 2-7 C and D. Cross sections of mature seminiferous tubules at low and high magnifications, respectively.

If living testicular material is available from frogs, earthworms, aquarium snails, or sea urchins, make a living sperm suspension by smashing a small piece of testis in pond or seawater; use whichever is appropriate for the animal in question. Add a pinch of sodium bicarbonate to alkalinize the solutions slightly so as to bring the sperm to maximum levels of sperm motility. Observe the living sperm suspension. Add a drop of 0.1% Toluidine Blue to the edge of the cover and allow some of it to seep under and stain the sperm. Examine the borderline between stain and sperm and choose a favorably stained region for special study. Note the general parts of the spermatozoa and the differences in relative proportions and structures they reveal.

3. Maturation and nuclear union in fertilized eggs of Ascaris (Fig. 2-8). Ascaris is a parasitic roundworm from the intestine of the horse. It is a classic species for the study of meiosis and, indeed, is the species used by Boveri in his original description of the meiotic process. A highly schematic presentation of meiosis in Ascaris is provided in Figure 2-8. Ascaris has only two pairs of chromosomes (i.e., 2 N = 4) and the steps in meiosis are almost diagrammatically visible. It provides an ideally simplified example of chromosome behavior during the reduction process.

In Ascaris the sexes are separate and fertilization is internal. The female reproductive system (Fig. 2-8 A) is simple and consists of a pair of thread-like ovaries which have an enormous capacity for egg production; each liberates thousands of eggs per day. The ovaries open into a pair of large tubular uteri packed with eggs. Eggs are at various maturation and embryonic stages of development. The paired uteri are united at the distal ends. They open to the exterior through a short copulatory duct or vagina. After copulation, the sperm cells, which are non-flagellated amoeboid cells, migrate up the uterine tubes and fertilize the eggs as eggs are released from the ovaries. Fertilization occurs while the eggs are in the primary oocyte stage. After sperm entrance, a shell or chorion is formed around each egg. The sperm awaits within the egg cytoplasm for the egg nucleus to complete its maturation divisions before pronuclear union and cleavage can occur.

The slides to be studied have several longitudinal sections through an Ascaris uterus mounted in each preparation. The top section on the slide contains the part of the uterus nearest the ovary. This section will contain the eggs at the earliest maturation stages. Progressively older stages will be found in subsequent sections. The following directions for study refer to the eggs within these sections of the uterus. **Begin with the top section** and follow the maturation stages through the sections to the bottom of the slide. Some slides will not have all stages described because the uterine sections may not cover enough of the uterine length to complete them. Exchange slides with other members of the class to complete the record of your own observations. Identify each of the following stages.

Spermatozoa in the uterine cavity (Fig. 2-8 B). At the upper end of the uterus, sperm cells are numerous in the spaces between eggs. Each spermatozoon possesses an intensely stained spherical centriole near the broad base of the cone-shaped nucleus. Sperm cytoplasm usually stains poorly, but the amoeboid, irregular outline of the cell membrane can be distinguished in some instances.

Unfertilized primary oocytes (Fig. 2-8 C). These cells will not be found if the first section on your slide was not made immediately adjacent to the end of the ovary. Unfertilized eggs can be identified by the absence of a shell, or chorion, around the cell, and by the presence of an egg nucleus with a well formed nuclear membrane near the center of the cell. The chromosomes in the nucleus may, or may not, be distinct. If they are well formed, they will be represented by two sets of tetrads (i.e., 2 sets of 4 chromatids each). These, it will be recalled, represent two synapsed homologous chromosomes that themselves are diplotene.

Fertilized primary oocyte (Fig. 2-8 D). This differs from the preceding stage by the presence of a thick-walled, but nearly transparent, **chorion** surrounding the egg. The sperm nucleus, with its spherical centriole, occupies the center of the egg cell. The egg chromosomes will usually already be on the first maturation spindle in anticipation of the first meiotic division. When this is the case, there will be two sets of tetrads on the meiotic spindle located just under the cell membrane.

Secondary oocyte and first polar body (Fig. 2-8 E). The chorion and sperm nucleus are the same as described for the preceding stage. Secondary oocytes can be identified by the presence of two sets of dyads (i.e., 2 paired chromosomes, or a total of 4 chromatids on the second maturation spindle), and by the presence of the first polar body. It is located just outside the cell membrane and is quite small. The first polar body will contain two sets of dyad chromosomes (or a total of 4 chromatids).

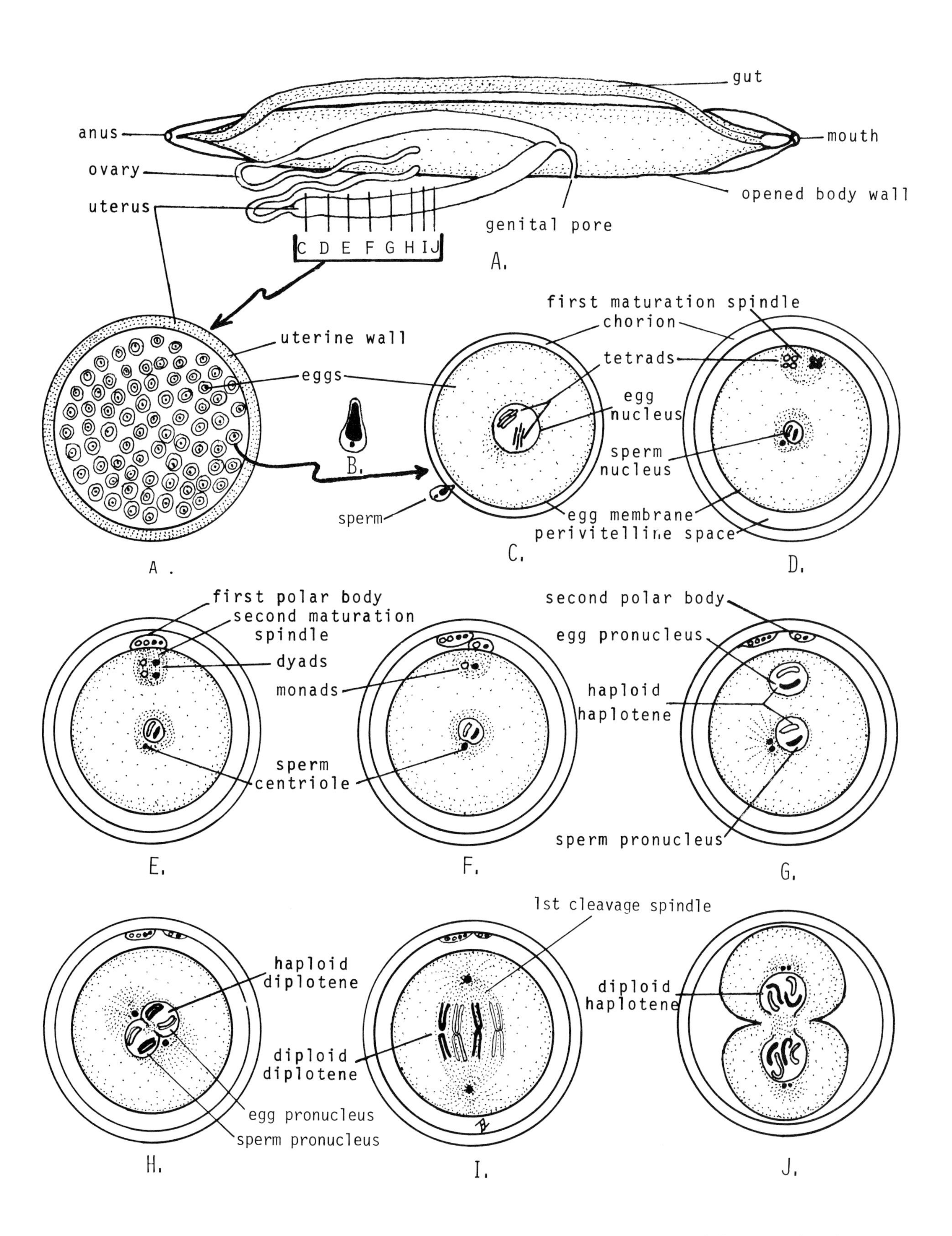

Figure 2-8. Meiosis and karyogamy in the roundworm, *Ascaris megalocephala.* 2-8 A. Diagram of the female reproductive system with guide lines to show levels of the uterus at which meiotic stages C through J are found. 2-8 A'. Diagrammatic cross section through the uterus. 2-8 B. Sperm cell. 2-8 C. Unfertilized egg at primary oocyte stage. 2-8 D. Metaphase of first maturation division. 2-8 E. Metaphase of second maturation division. 2-8 F. Reduced ootid. 2-8 G. Zygote with egg and sperm pronuclei. 2-8 H. Karyogamy of zygote. 2-8 I. Metaphase of first cleavage. 2-8 J. Telophase of first cleavage.

Ootid or mature ovum with first and second polar bodies (Fig. 2-8 F). Not many eggs will be found in which the plane of section is optimally oriented to show the sperm nucleus, the egg nucleus, and both polar bodies. However, a few eggs should be present on each slide which show this combination of features. The ootid nucleus at the end of the second maturation division will contain only two chromosomes (i.e., monads), and at this stage the haploid chromosome number, N = 2, is achieved. The second polar body, which is the sister cell of the ootid, also contains only two chromosomes. It will be recalled that the first polar body contains two sets of dyads (4 chromatids).

Zygote with egg and sperm pro-nuclei (Fig. 2-8 G). After the maturation divisions are complete, the egg and sperm nuclei form distinct nuclear membranes and become large and vesicular. The two chromosomes, in the egg and sperm nuclei, lose definition, and the egg cell appears to be binucleate. These are the haploid egg and sperm **pronuclei.** At first they are widely separate. However, they soon migrate toward the center of the cell and lie in contact with one another. It is no longer possible to distinguish between male and female pronuclei. The conspicuous sperm centriole divides and provides the two division centers that organize the aster–spindle system for the first cleavage mitosis.

Pronuclear union and first cleavage (Fig. 2-8 H, I, J.). Pronuclear union, called **karyogamy,** is the preeminently essential feature in all forms of sexual reproduction. By this event the diploid chromosome number is restored. At their points of contact, the egg and sperm pronuclear membranes breakdown, and the pronuclear chromosomes begin to condense again into chromosomes. There is no zygote, or fusion nucleus per se, since the pronuclear chromosomes go directly onto the spindle of the first cleavage mitosis. At this time the chromosomes are duplicated in anticipation of first cleavage. The nuclear membranes entirely disappear, and the two duplicated egg and two duplicated sperm chromosomes become oriented radially at the equator of the spindle. At this stage the zygote is at **metaphase** of the first cleavage division. During **anaphase,** the paired chromosomes separate, and one of each of the four pairs migrates toward the opposite ends of the first cleavage spindle. Finally the cytoplasm separates during **telophase,** and a nuclear membrane forms around the chromosomes in each diploid (2 N = 4) daughter cell. This division is mitotic; at the completion of the division the egg is at the 2-cell stage.

The development of eggs differs from the general scheme of gametogenesis in three primary respects. (1) The primary specialization of the egg occurs precociously in the **primary oocyte** stage when vast quantities of nutrients (yolk or deutoplasm) are stored for use by the developing embryo. (2) Extremely unequal cell divisions take place during the maturation stage, with the result that virtually all the sorted yolk remains in only one of the 4 daughter cells. Accordingly, the first maturation division produces one **secondary oocyte** and its sister cell is the minute **first polar body;** at the second maturation division the secondary oocyte divides to produce the **ootid** (ovum) and the **second polar body** (see Figures 2-6 and 2-8). Thus, one primary oocyte produces only one functional egg. (3) Last, functional specialization, involving physiological ripeness of the egg to receive and be activated by a sperm, may be achieved in eggs of different species at various stages in oogenesis. For example, almost all vertebrate eggs are ripe as secondary oocytes or at metaphase of the second meiotic division; *Ascaris* eggs are ripe as primary oocytes (see above), and sea urchin eggs become ripe only after meiosis is complete.

C. PHYSIOLOGICAL ACTIVATION: THE ASPECTS OF FERTILIZATION THAT INITIATE CAPACITIES FOR TOTAL DEVELOPMENT IN EGGS OF STARFISH AND SEA URCHINS

The essential feature that distinguishes sexual from asexual mechanisms of reproduction is the nuclear union of two physiologically ripe gametes, the egg and sperm, with the formation of a diploid **zygote,** or **fertilized egg,** which has total capacities for development appropriate to its species.

Gametic union accomplishes three primary goals, namely: (1) **Karyogamy** is the fusion of egg and sperm pronuclei; this restores the diploid chromosome number in the zygote and all cells of the embryo that are derived from the zygote by mitoses. (2) **Ooplasmic segregation** is the localization of special cytoplasmic components of the eggs which have morphodynamic function in initiating regional differences in cell specialization in the embryo. (3) **Physiological activation** of the egg is a release mechanism that transforms the dormant egg into an active synthetic system with the potential for complete embryonic development, juvenile growth, and sexual maturity.

Karyogamy has been demonstrated in the study of meiosis and fertilization in Ascaris. Ooplasmic segregation will be one of the major aspects to be covered in Chapter 3. For the present, emphasis will be given to the evidence for physiological activation of eggs in sea urchins and starfish.

1. Fertilization, activation, and membrane formation in the sea urchin (Fig. 2-9 and 1-20). The eggs of echinoderms are particularly favorable for the illustration of basic aspects of fertilization. Figure 2-9 diagrams the steps involved in sperm contact and sperm entry, and Figure 2-10 shows the origin of the fertilization membrane, hyaline plasma membrane, and plasma membrane at a point on the egg surface other than that at which the sperm enters. The fine details of structure in cortical granules, plasma membrane, vitelline membrane, and sperm head will be invisible at light levels of microscopy. The drawings are based on electron microscope studies of various species. The schematic presentation shown in Figures 2-9 and 2-10 is therefore to be considered general, rather than specific, for a particular egg or animal group.

a. The major aspects of sperm entry (Fig. 2-9) proceed as follows. First, external secretions of the egg activate the sperm acrosome and cause the **acrosome filament to form.** The acrosome filament consists of a central core, or axial filament, which projects out and carries with it an extension of the sperm plasma membrane. For effective fertilization, an acrosome filament must come in contact with the plasma membrane of a ripe egg. Note that the egg plasma membrane projects, as small microvilli, through the vitelline membrane surrounding the egg. Presumably the initial point of gamete fusion is between the cell membranes of the acrosome filament and the microvilli of the egg plasma membrane. These membranes fuse and, within a few moments, the hyaline cytoplasm of the egg cortex flows toward the sperm entry point and into the acrosome filament, causing it to enlarge greatly. The cone of cytoplasm that is formed is called the **fertilization cone.** In some means not fully understood, the egg cytoplasm draws the passive sperm nucleus into the fertilization cone, possibly by traction on the axial filament which is attached to the sperm nuclear membrane. Once inside the fertilization cone cytoplasm, the sperm nucleus rotates 180 degrees. It begins to enlarge and its centrioles begin to organize astral fibres that are probably involved in the migration of the sperm nucleus into the egg cytoplasm in search, so to speak, of the egg pronucleus.

b. The essential steps in fertilization membrane formation are shown in Figure 2-10; they are as follows. The fusion of egg plasma membrane and acrosome filament membranes, at the point of sperm contact, apparently initiates a **wave of activation** that passes over the egg surface in about 30 to 100 seconds. The passage of the activation wave initiates changes in the underlying cortical cytoplasm of the egg. The first noticeable one involves the movement of **cortical activation granules** to the surface and the fusion of their membranes with the plasma membrane of the egg. The activation granules then rupture and release their central substance called the "dark body" against the vitelline membrane. Dark bodies undergo partial dissolution and release a mucoprotein that has strong water absorbing properties. With the swelling of the dark body gel, the plasma membrane is stripped away from the vitelline membrane. The latter is then lifted off from the egg surface by the continued swelling of gel. The space created between the egg surface and the inner face of the vitelline membrane is the **perivitelline space.** By enzymatic action, the chemical nature of the vitelline membrane is changed, or "tanned," and becomes impervious to many ions and resistant to enzymatic digestion by trypsin. The tanned vitelline membrane is the **fertilization membrane.** During the time the fertilization membrane is forming, the alveoli of the cortical activation granules also undergo swelling but do not dissolve. They form a layer of mucoprotein closely adherent to the egg membrane called the **hyaline plasma membrane.** Both the fertilization and hyaline plasma membranes are non-living secretions; they both play significant roles in blocking polyspermy, and in stabilizing the early cleavage cells in a geometric pattern through the blastula stage. The **zygote membrane** is the original egg plasma membrane with additional contributions from the membranes of cortical activation granules.

During the period when the egg membranes are forming and the sperm nucleus is entering the egg (ca. 3 to 5 minutes), physiological measurements indicate that profound changes are taking place. These include changes in membrane permeability to inorganic ions, in the electrical charge on the egg surface, in the utilization of oxygen and production of carbon dioxide, and in the binding of amino acids into polypeptides and proteins. Within

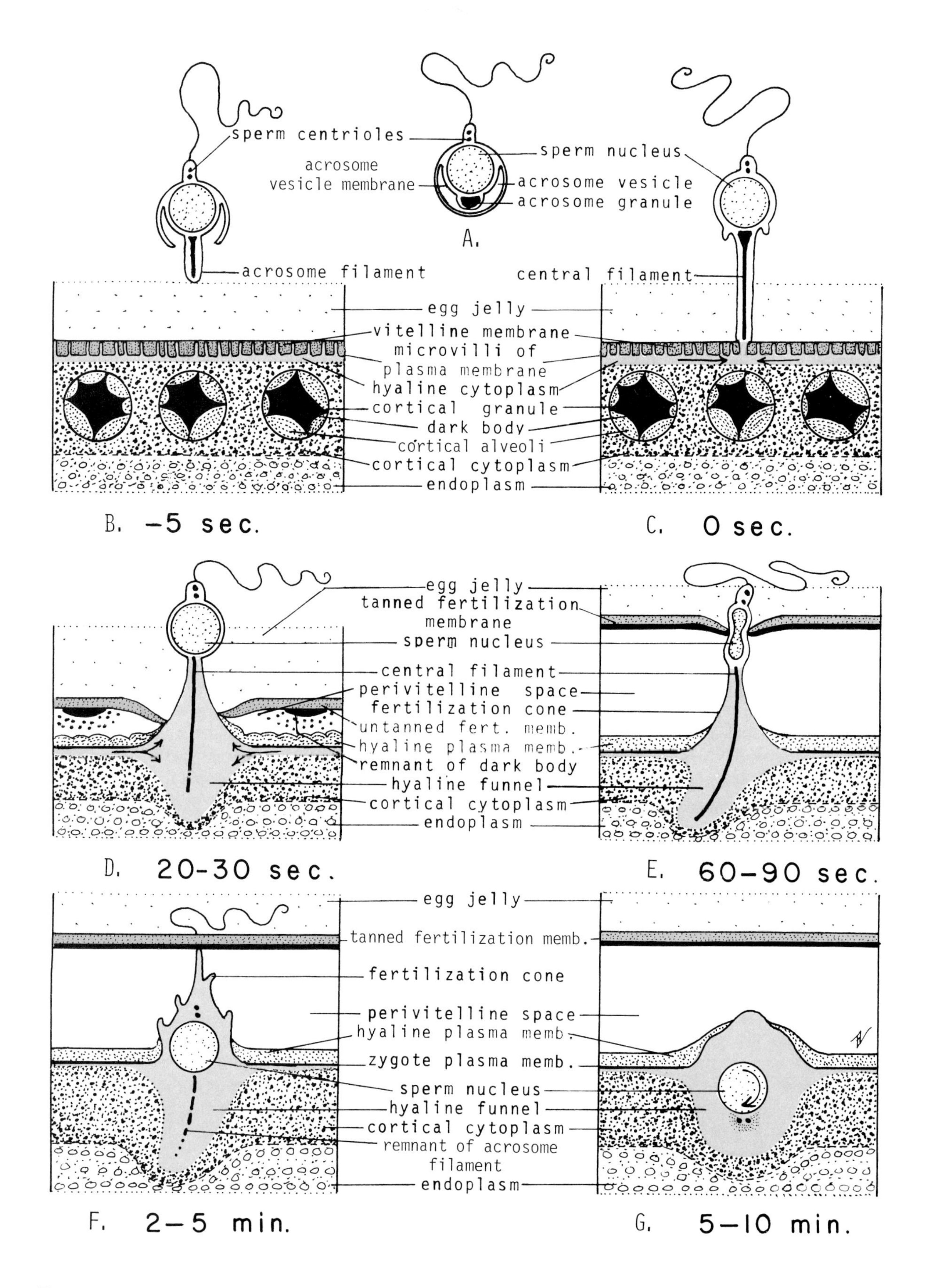

Figure 2-9. Generalized diagram of sperm entry in sea urchin eggs. 2-9 A. Mature sperm. 2-9 B. Acrosome activation. 2-9 C. Annealing of egg and sperm plasma membranes and primary activation of the egg cortical reactions. 2-9 D. Hyaline cytoplasm being recruited to the sperm contact point. 2-9 E. Fertilization cone and fertilization membrane elevation. 2-9 F. Entry of sperm nucleus into the fertilization cone. 2-9 G. Entry of sperm nucleus complete. Color code: blue — hyaline cytoplasm, red — vitelline membrane, yellow — cortical alveali and hyaline plasma membrane.

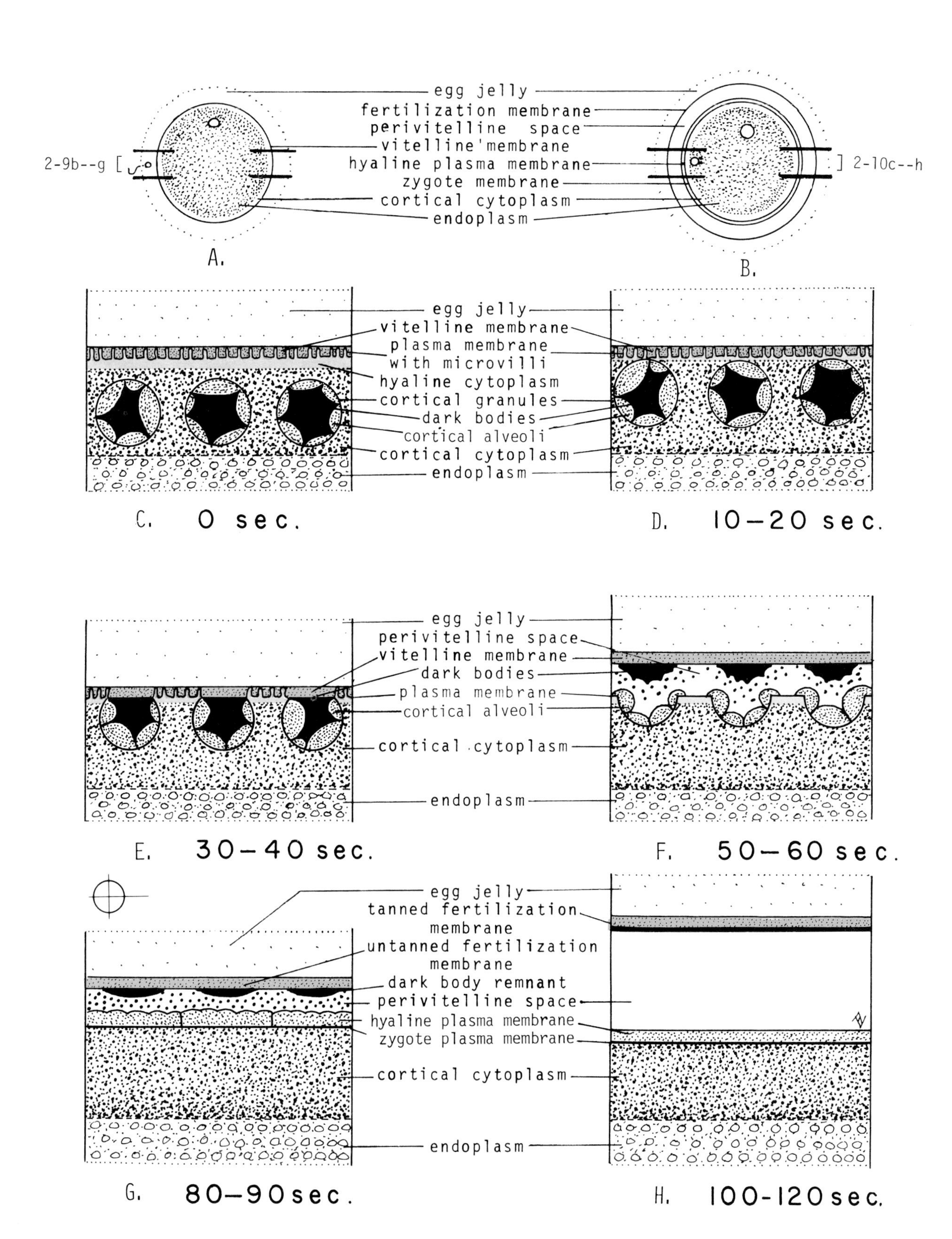

Figure 2-10. Generalized diagram of membrane formation in sea urchin eggs. 2-10 A and B. Unfertilized and fertilized eggs with guide lines to show regions magnified in 2-9 B–G and 2-10 C–H. 2-10 C. Cortex before activation. 2-10 D. Cortical granules move to egg surface. 2-10 E. Cortical granules fuse with plasma and vitelline membranes. 2-10 F. Cortical granules open and discharge contents. 2-10 G. Transient stage in membrane formation. 2-10 H. Fertilization, hyaline plasma, and zygote membranes fully formed. Color code: blue — hyaline cytoplasm, red — vitelline membrane, yellow — cortical alveoli and hyaline plasma membrane.

10 minutes following sperm contact, the egg is a fully energized and activated cell with all systems for synthesis, energy production, and cell division fully activated.

Time-lapse motion pictures of the fertilization reaction in echinoderms show the general nature of sperm entry and the elevation of the fertilization membrane most clearly. Films of these processes in eggs of sea urchins, starfish and sand dollars will be used to supplement our laboratory study.

Time-lapse motion pictures of the fertilization reaction in amphibians. The sequence of activation and sperm entry in the frog and salamander will also be shown. The precise details described above for sperm entry and for the cortical reaction at the egg surface are broadly applicable to these eggs as well as to those of echinoderms. A **fertilization membrane,** or chorion, is elevated from the surface of the amphibian egg. The asymmetrical distribution of heavy yolk at the vegetal pole permits fertilized eggs to orient freely with gravity so that dark (animal pole) hemispheres will be up and light (vegetal pole) hemispheres of the eggs will be down.

2. Fertilization of living sea urchin or starfish eggs. Students will work in pairs or in groups of four. Steps a through d in the procedure will have been done in advance of the class so that a maximum amount of time will be available for observing the phenomena of sperm entry and activation.

Procedures for obtaining and fertilizing echinoderm eggs.

Species for choice:

Sea urchins:

Lytechinus (available from May through October).
Arbacia (available at all seasons).
Strongylocentrotus (available at all seasons).

Starfish:

Asterias (available February through May or somewhat later).
Pateria (available February through June or somewhat later).
Astropecten (available January through May).

Differences between starfish and sea urchin fertilization involve the time at which the egg becomes physiologically ripe for activation. In sea urchins the egg is not ripe until both meiotic divisions are complete and the egg is in the ootid, or fully reduced, condition. Starfish eggs are almost unique in being physiologically ripe as soon as they are ovulated under the stimulation of a radial nerve factor. That is, they can be fertilized as primary oocytes, secondary oocytes, or as ootids; however, development is generally better in cultures that are inseminated after meiotic divisions are complete.

Starfish eggs are particularly desirable for laboratory study since ovulation can be induced by radial nerve hormone. The meiotic divisions will begin almost immediately and will be complete in about 25 minutes at room temperature. No other material so easily provides material for the observation of polar body formation in fertilized eggs. Eggs fertilized as primary oocytes will form the fertilization membrane. This then "traps" the polar bodies that subsequently are formed in the perivitelline space where they remain throughout cleavage and blastulation and provide a definite marker for the animal pole of the egg. For this reason, starfish fertilization is to be preferred for laboratory study if a choice of echinoderm eggs is available.

Materials for each working group: 1 liter filtered seawater, 4 deep-well depression slides, 8 Syracuse culture dishes, 4 Stender or covered culture dishes, 4 standard medicine droppers, 1 standard medicine dropper coded with a red band to be used only for unfertilized eggs, standard microscope slides and coverslips, small pillbox with fine washed quartz beach sand, 8 1-inch screw cap vials, 1 fingerbowl or waste jar.

a. Wash adult animals in fresh water and rinse in seawater. Place individuals in separate dishes with oral side down.

b1. For starfish: Break off an arm and remove the pair of highly divided gonads to the dish of seawater; rinse twice by pouring off seawater and replenishing it. Leave seawater about half an inch deep in the bowl. Induce shedding and meiosis by crushing the radial nerve of the starfish arm with a blunt dissecting needle or forceps; the radial nerve is located in the groove between the tube feet. Run the needle back and forth several times in the ambulacral groove with sufficient force to crush the tissues. Pipette seawater over the crushed groove and let the water drain into the dish with ovaries. Indeed, the arm can, to advantage, be left in the dish with the ovaries. Shedding will begin in ripe ovaries in about 10 to 20 minutes.

b2. For sea urchins: Induce spawning by injecting about 2.0 ml 4% KCl into the coelomic cavity. (If KCl is not available cut around the edge of the shell and remove the oral half. The gonads will be in the aboral half and will open to the exterior by gonopores arranged radially around the anus. Rinse out the coelomic cavity with seawater.) Place the animal aboral side down in a small beaker of seawater so that the animal is suspended by the edge of the glass and the aboral surface with gonoducts is submerged in seawater.

As soon as gametes begin to be shed, sex the individuals by the identification of eggs or sperm produced.

Remove males immediately to a separate covered culture dish lacking seawater and place them in a refrigerator until needed. Undiluted sperm remain useful for many hours and up to a few days if specimens are in optimal condition.

Allow females to shed out in their dishes (ca. up to half an hour).

c. Wash eggs by transferring them from the Stender dish to a 50 ml beaker filled with seawater. Let the eggs settle to the bottom and decant off the supernatant water. Fill the beaker with seawater and let the eggs settle again. These eggs are now ready to be fertilized. At room temperature they remain useful for about 4 to 6 hours. Their useful life can be prolonged by storing them in covered dishes at 6 to 10° C.

d. Prepare a sperm suspension just before it is needed by placing one drop of undiluted ("dry") sperm in a Stender dish containing 10 cc of seawater. Mix with a pipette until a uniform milky suspension is obtained.

e. Fertilize eggs by adding two or three drops of washed eggs to a clean Stender dish half full of seawater. Add three drops of sperm suspension to the eggs. Use a clean pipette to mix the eggs and sperm gently. Wait for exactly two minutes and then add two or three more drops of standard dilute sperm. The first drops of sperm usually will not activate all eggs; the 2 minute wait allows eggs that were activated to elevate their membranes before second insemination. By this double insemination method, virtually 100% activation can be obtained with virtually no polyspermic eggs. **Record the time of the first insemination** as the time of fertilization.

f. Check success of fertilization by **immediately** taking a sample of fertilized eggs and placing them in a deep-well depression slide. Fill the depression slide to brimming and carefully add a cover glass from the side so that no air bubble is trapped under it. Blot off the excess seawater from the slide before placing the slide on the microscope stage. Examine the eggs under medium power. Depression slides are too thick for high power. Motile sperm will be seen as small active dots in the solution and the individual eggs will be surrounded by a halo of sperm attached to the jelly coat surrounding the egg. If fertilization is successful, **fertilization membranes** surrounding the eggs should be visible within 2 to 3 minutes after the addition of sperm.

g. Record the percentage of fertilization 10 minutes after the addition of sperm by making a random count of 100 eggs and recording the number having fertilization membranes as the percentage fertilized.

h. Wash sperm out of the culture dish by pipetting off most of the seawater from the Stender dish and replacing it with fresh seawater. This should be repeated again after the eggs have settled.

i. Observe cleavage in these eggs during the remainder of this period and record your observations concerning times of cleavage.

j. Keep cultures away from heat and intense light; keep cultures covered when not observing them; use separate pipettes for each culture to avoid contamination.

k. Make fresh deep-well preparations frequently from stock cultures. Development will be repressed by lack of oxygen and/or overheating on the microscope stage. Therefore, cultures should be replaced after 15 to 20 minutes throughout the period of observation.

3. Observation of sperm entrance, fertilization cone formation, and fertilization membrane elevation (Figs. 2-9, 2-10).

For a more detailed observation of the fertilization reaction in sea urchin eggs, use the following procedure to **observe sperm entrance, fertilization cone formation,** and **membrane elevation** under high power. Place a few grains of sand on a microscope slide to act as a support for a coverslip. Place a drop of washed, unfertilized eggs on the slide and cover them. Place the slide on the microscope stage and focus on them with the high power lens. Without moving the slide or lens, add a small drop of sperm to one edge of the cover glass. Observe eggs near the added sperm. Watch the attachment of spermatozoa to the jelly coats surrounding eggs. Look for an egg with a fertilization membrane just forming. Note that the membrane does not elevate smoothly but arises as small blister-like elevations. For a time during the initial phases of membrane elevation, the egg surface is quite irregular in contour. The irregularities are associated with the invisible process of granule fusion with the plasma membrane and with the plasma membrane's separation from the vitelline membrane at the surface. As all contact points between the egg and vitelline membranes are broken, the egg will again assume a perfect spherical contour inside the fertilization membrane. Some eggs will show the activating sperm in profile attached to the egg surface. Watch the process of sperm entrance and observe the formation of the fertilization cone; it is an extension of the egg cytoplasm that remains attached to the fertilization membrane at the point of sperm contact. Observe the movement of the sperm head into the egg. This is accomplished by the egg cytoplasm; the sperm itself is a passive agent in sperm entry. The fertilization cone will often throw out pseudopodial extensions called **flame protuberances** before it finally separates from the fertilization membrane and subsides into the egg cytoplasm. If time is available (i.e., about 30 to 45 minutes), the fusion of the egg and sperm pronuclei can be observed with great ease in the transparent eggs to starfish or the sea urchin, Lytechinus. **Sketch** the events you see at two minute intervals to provide a time-lapse record.

At the end of the laboratory period transfer a sample from your best culture to a covered Stender dish half full of seawater. The number of eggs should not exceed about one-fourth the bottom area of the dish. Mark the dish with your seat number and initials and place the dish in the appropriately marked place in the refrigerator-incubator, or on the running water table provided to keep the cultures cool.

Fertilization of eggs of the grass frog, *Rana pipiens,* will be described at the end of Chapter 3 in association with the description of experimental work that can be done on living frog eggs.

D. DEMONSTRATION OF SELECTIVE GENE ACTIVITY IN THE GIANT POLYTENE CHROMOSOMES OF CHIRONOMOUS

Selective gene activity during early embryonic development is responsible for distinctive cell specialization in different tissues of the growing body. Gene activity is a characteristic of all stages in development. It is therefore of interest to demonstrate structural differences in chromosomes which are related to local regions of gene activity. This can be shown very easily in the chromosomes of salivary gland nuclei, in the mosquito-like midge, Chironomous. These chromosomes are among the largest in the animal kingdom. Their large size is due to repeated replications of DNA strands without separation of chromatids by mitotic divisions. Multistrand chromosomes are said to be **polytene,** as opposed to the haplotene (single strand), or the diplotene (double strand) chromosomes which are the general rule in ordinary cell cycles of most tissues. The giant salivary chromosomes of Chironomous larvae are amplified up to 10,000 times the haplotene DNA value. In polytene chromosomes all DNA strands are enclosed within the same protein envelope. Homologous genetic regions are aligned precisely so as to form horizontal plates at right angles to the length of the polytene chromosomes. The DNA of genetic loci that are not actively transcribing RNA remain tightly coiled. In this condition they can easily be stained with dyes having strong affinities for nucleic acid. Conversely, active DNA requires that sites be uncoiled; in this condition the staining of DNA is much more diffuse.

By using appropriate procedures, chromosome preparations can be made which clearly show a banding pattern of alternately darkly stained (inactive) and lightly stained (active) regions called **chromomeres.** These are essentially equivalent to functional gene loci on chromosomes. In addition, some chromosomes may undergo hyperactivity. In such cases the chromosome diameter becomes enlarged into a faintly stained **puff,** or in extremely active locations they may form enormously expanded **Balbiani rings,** as shown in Figure 2-11 D.

Materials needed. Compound and dissecting microscopes, fine dissecting needles, blotting paper, plastic Petri dish covers, fingernail polish with applicator brush, and a dropping bottle of the following solutions.

Acid Carmine: Dissolve 2 grams dry Carmine in 100 ml 45% acetic acid. It is best done by heating in a reflux condenser for two hours. Cool and filter each day before use.

OR:

Lacto-Aceto-Orcein: To 50 ml distilled water add 25 ml glacial acetic acid, 25 ml 85% lactic acid, and 2 g synthetic Orcein. Reflux 2 hours, cool, and filter before use.

50% acetic acid: To 50 ml glacial acetic acid add 50 ml distilled water.

Insect physiological saline: 9 g NaCl, 0.2 g KCL, 0.27 g $CaCL_2 \cdot H_2O$, 0.2 g $NaHCO_3$, 4 g glucose dissolved in distilled water to equal 1 liter. (Modified from J.W.S. Pringle, *Journal of Experimental Biology,* **15:**144, 1938).

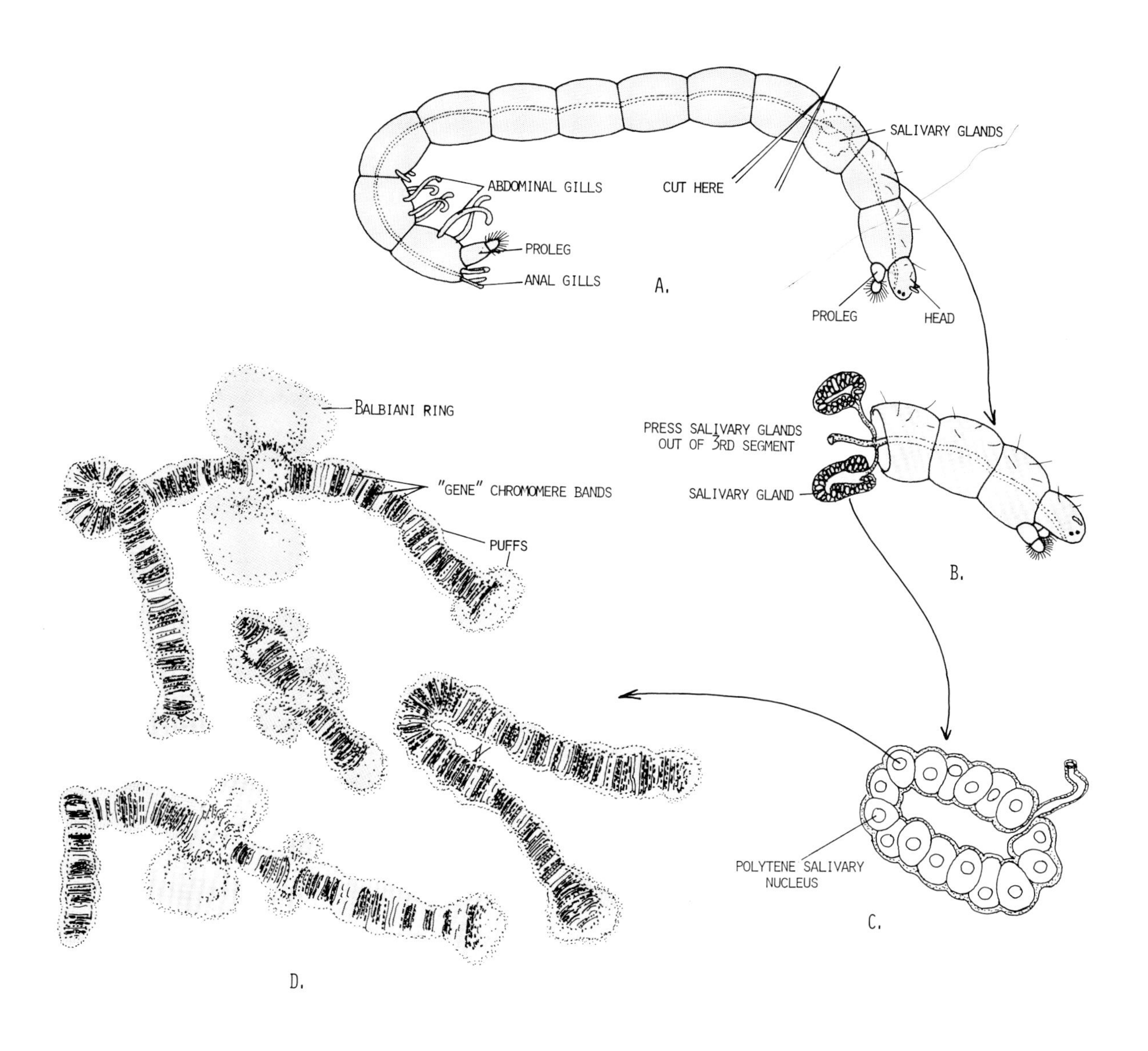

Figure 2-11. Chironomous polytene chromosomes. See explanation in text for methods of preparation. 2-11 A. An entire Chironomous larva with the location of the salivary glands in the third post-cephalic segment is shown, indicating where transection of the body should be made. 2-11 B. Anterior end of larva with salivary glands pressed out. 2-11 C. A salivary gland magnified. 2-11 D. Schematic drawing of stained, polytene chromosomes at high power.

a. Dissection of Chironomous salivary gland chromosomes. Chironomous larvae are about one centimeter in length and are bright red in color. The color is due to the presence of hemoglobin, an adaptive respiratory pigment which permits them to survive in ponds or streams that are relatively poor in oxygen. Obtain three or four living larvae for dissection. Removal of salivary glands is very simple and is partially diagrammed in Figure 2-11 A and B.

Place larva on a microscope slide. Blot off the excess water and examine the worm at low power with a dissecting microscope. Identify the body regions shown in Figure 2-11 A. The salivary glands are located in the third post-cephalic segment and are attached to the right and left sides of the gut. In the manner diagrammed, use two sharp, fine dissecting needles. Place one needle flat across the worm between segments three and four. Press the needle firmly to the glass to hold the worm securely in place. Then cut the body into two parts by drawing the second needle across the body behind the first needle. Discard the posterior part. Remove the salivary glands along with the other viscera from the anterior piece. This is done most easily by holding the piece down by placing a needle across the body between the head and first segment. Then, use the other needle to press out the viscera as you might squeeze a toothpaste tube. Most tissues pressed out will be colored red, brown, or amber. However, the salivary glands are very nearly transparent and can be easily identified.

b. Preparation and staining of salivary chromosome squashes. Proceed directly to free the salivary glands from their attachments to the gut. Isolate them on the slide from all other parts which should be wiped off the slide. If there is any delay in staining, add a drop of insect physiological saline to the salivary glands to prevent drying. Before staining, blot off all excess fluid. Then add two or three drops of Acid Carmine or Lacto-Aceto-Orcein directly to the tissue. It is suggested that at least three salivary gland slides be prepared and that one be stained for 30 minutes, one for 20 minutes, and one for 10 minutes. Place them on a piece of paper towel to act as a blotter in case the dye runs off. Cover the slides with a plastic Petri dish to prevent drying; additional stain can be added as needed during the staining period.

After the staining period, add a coverslip to the preparation. Cover the slide with a piece of paper towel. Then roll the ball of the index finger firmly over the preparation without sliding the coverslip under the blotter paper. This will smash the cells, spread the chromosomes, and squeeze out excess stain. Remove the blotting paper and add a drop of 50% acetic acid to the edge of the coverslip to chase out any air bubbles under the cover glass. Wipe the ends of the slide dry but, in so doing, do not move the cover glass. Let the slide air-dry. Then ring the cover glass with fingernail polish to prevent drying under the cover. These slides will keep in good conditions for several weeks if they are completely sealed.

As soon as the fingernail polish is dry, the slide is ready to examine with the compound microscope with high-dry magnification. Hunt for the salivary gland and for examples of spread chromosome figures. They should resemble Figure 2-11 D. At first the stain will color all tissues uniformly. Gradually, however, the stain will bleach out of regions poor in DNA. After one or two hours only the chromomeres of the polytene chromosomes will retain the stain and linear chains of genetic loci will become very clear. Puffs and Balbiani rings will be much less heavily stained than other parts of the chromosomes.

In the event that Chironomous larvae are unavailable for use, other tissues can be used to demonstrate chromosome activity. These include Drosophila salivary chromosomes, Tradescantia pollen mother cells from developing flower buds, tail tips of amphibian tadpoles, or onion root tips. However, these will all provide chromosomes that are much smaller than those of Chironomous, and will require oil immersion magnifications to reveal any regional differences in chromosome structure.

In concluding this chapter we will indicate that the preliminary events that provide the prologue for embryonic development have been reviewed briefly. Some of the most favorable material in the entire animal kingdom have been used to illustrate the essential points. Chapter 3 will be concerned with the origin of initial differences in the egg cytoplasm that will lead to the formation of axiate patterns in the developing embryo. These patterns will first be expressed in preliminary events that lead to specificity in primary embryonic germ layer formation.

CHAPTER III

ORIGINS OF POLARITY, BILATERAL SYMMETRY, AND PRIMARY GERM LAYERS DURING CLEAVAGE AND GASTRULATION

Illustrated in Molluscs, Echinoderms, and Chordates

TABLE OF CONTENTS

TABLES AND ILLUSTRATIONS

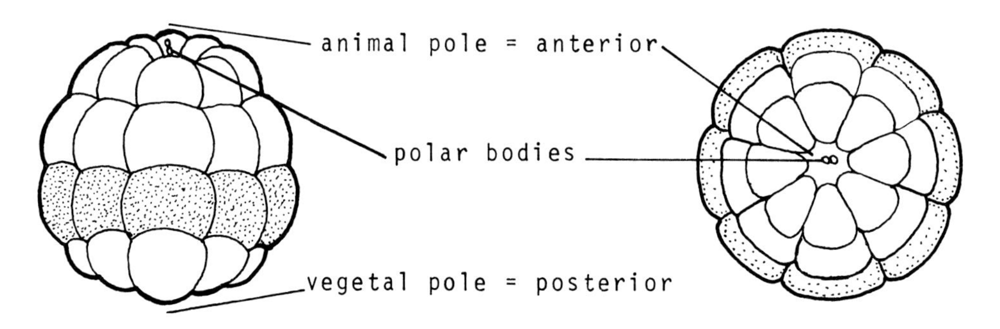

A. Radial cleavage from lateral and anterior polar view (based on echinoderms).

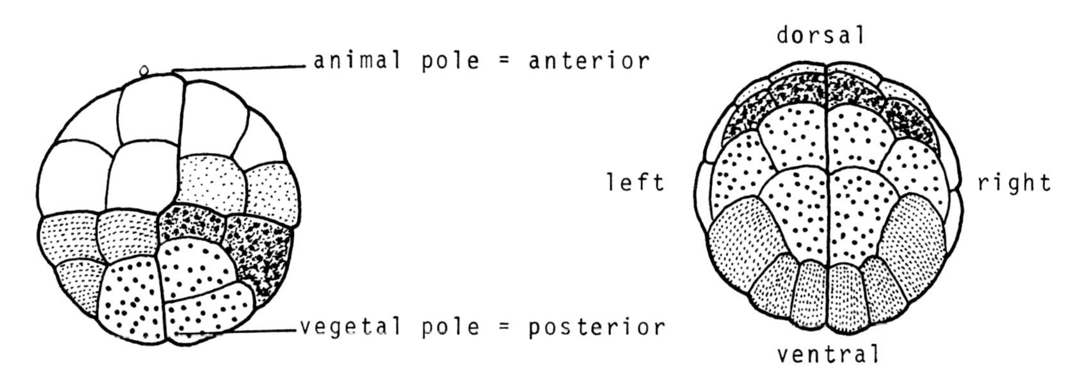

B. Bilateral cleavage from left lateral and posterior view (based on tunicates).

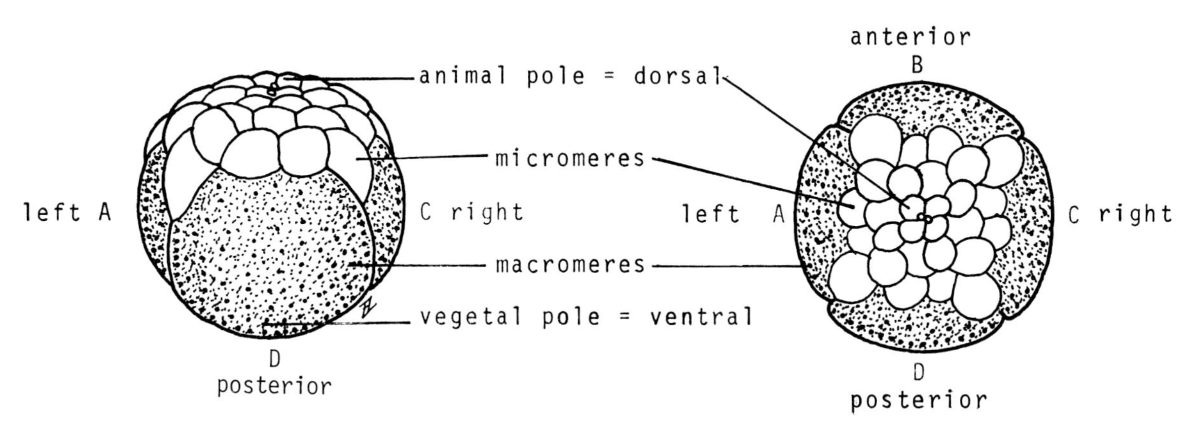

C. Spiral or oblique cleavage from posterior and dorsal view (based on gastropod molluscs).

Figure 3-1. Comparison of primary cleavage types in animal development.

CHAPTER III

ORIGINS OF POLARITY, BILATERAL SYMMETRY, AND PRIMARY GERM LAYERS DURING CLEAVAGE AND GASTRULATION

Illustrated in Molluscs, Echinoderms, and Chordates

Mitotic mechanisms assure that all cell nuclei in embryonic development will possess the same chromosomal and gene composition as those that formed in the zygote by karyogamy of egg and sperm pronuclei during fertilization. However, no such equality applies to the allocation of egg cytoplasm to the cell progeny of the fertilized egg.

Almost from the beginning of primary oocyte development, there are regional differences in the composition of egg cytoplasms with regard to either yolk, mitochondria, golgi apparatus, cortical granules, visible pigments, or macromolecules including very important categories of cytoplasmic RNAs and proteins. Because of regional differences in egg cytoplasms, cleavage cells will be unequal in the kinds of cytoplasm they receive. These differences play a significant role in establishing the fundamental polarity of the egg, and the planes of symmetry in the embryo that develops from it.

The primary polarity in eggs is associated with quantitative differences along the so-called animal-vegetal axis of the egg. The animal pole of the egg is the point on the egg surface at which the polar bodies are given off during meiosis. This position is usually determined by the attachment of the egg to the ovarian wall, or to the other extrinsic influences on the developing early primary oocyte. The vegetal pole is the point opposite the animal pole on the egg surface. The **animal — vegetal axis** is an imaginary line that passes through the center of the egg and connects the animal and vegetal poles. In most eggs there is a fixed relationship between the primary polarity of the egg and one of the major planes or axes of symmetry in the embryo and future adult body. For example, the animal pole will coincide rather closely with the anterior end of echinoderm, tunicate, Amphioxus, cyclostome, and amphibian eggs, but it will mark the dorsal pole of snail, reptile, and bird eggs.

A review of the general anatomical concepts of poles, axes, and planes of symmetry is given in Figure Int.-2. It is clear that establishing these geometric relations between parts of the embryo is one of the most important functions of early development. The mechanisms for determination of germ layers, and for their separation into spatially exact locations, must be highly precise if these steps in embryo formation are to result in a normal distribution of parts. It is almost axiomatic that if cleavage and gastrulation are not normal, nothing else that follows in development can be normal.

Mechanisms of cleavage and gastrulation vary with different phyla. The two most widely encountered patterns are the so-called protostomian and deuterostomian plans of development. The protostomes include the annelids, arthropods, molluscs, flatworms, and many lesser phyla; they are said to have a spiral plan of cleavage and gastrulation. The cleavage pattern in the common marine mud snail Ilyanassa will be used as a representative protostome. The **deuterostomes** include the echinoderms, chordates, and several lesser phyla; they are said to have radial and bilateral plans of cleavage and gastrulation. Examples will be given of the starfish, sea urchin, tunicate, Amphioxus, and amphibian as variations on the deuterostome plan of development. Collectively all actions that lead to the determination of adult planes of symmetry during embryonic development (i.e., patterns of cleavage, gastrulation, and germ layer separation) are called **symmetrization.** For a review of deuterostome and chordate classification, see Tables Int.-1 and Int.-2. Examples of radial, bilateral, and spiral symmetry in cleavage are diagrammed in Figure 3-1.

A. POLAR BODY FORMATION, OOPLASMIC SEGREGATION AND SPIRAL CLEAVAGE IN THE MARINE MUD SNAIL, Ilyanassa obsoleta (*Recently renamed* Nassarius obsoletus)

1. Procuring and handling living material. Living mud snails are the most common intertidal gastropod mollusc on the Atlantic coast. They can be collected in the lower intertidal zone in almost any quiet estuary or bay with a sandy-mud bottom. They will congregate in hundreds around dead or decaying vegetation or animal matter. They can be obtained from the Marine Biological Laboratory, Woods Hole, Massachusetts, for shipment inland.

Living snails can be maintained almost indefinitely in the laboratory in aquaria of seawater. They can be brought to spawning condition by weekly feeding on frozen clam meat, which should be shredded and placed in an enclosed bowl in the aquarium. Any uneaten food should be removed after six hours to avoid fouling the aquarium. To initiate spawning, place no more then 50 snails in a 5-7 gallon aquarium and keep them under constant illumination. Line the sides of the aquarium with plates of glass so that shed egg capsules can readily be removed for harvesting.

Egg capsules contain from 50 to 500 eggs and are about 1 to 2 mm in diameter. The capsules are best opened with minimum damage to eggs by snipping off the end of the capsule with finely sharpened scissors. The central cavity of the capsule is filled with an exceedingly sticky jelly that makes the eggs difficult to remove unless you wait for about 15 minutes for the jelly to dissolve in seawater. After that the eggs can be dumped out with little difficulty.

For quick optical staging of capsules, it is useful to know that young capsules have eggs that are snow white. As development progresses, embryos acquire progressively darker yellow and brown pigments. Table III-1 gives a time schedule of development for eggs developing at room temperature (i.e., 24-26°C). Development in eggs can be slowed almost to a halt by placing the capsules in seawater at 4-6°C in an ordinary refrigerator.

TABLE III-1
TIME SCHEDULE FOR NORMAL DEVELOPMENT OF ILYANASSA EGGS AT 24-26° C

Figure		Development Stages	Time After Laying
—	1.	First polar body given off & first pre-polar lobe formed	1 hours
3-2b	2.	Second polar body given off & second pre-polar lobe formed	1 1/2 hours
3-2c	3.	First polar lobe forms as a vegetal evagination	2 1/2 hours
3-2d	4.	First cleavage begins & first polar lobe is attached by a thin neck; this is the "trefoil or clover leaf" stage	3 1/2 hours
3-2e	5.	First cleavage complete & first polar lobe begins to flow back into the CD cell	4 hours
3-2f	6.	AB and CD cells round up into unequal spheres	4 1/4 hours
—	7.	Second polar lobe begins to form at vegetal end of the CD cell	4 3/4 hours
3-2g	8.	Second cleavage begins to divide AB and CD cells & second polar lobe is attached to the D cell	5 hours
3-2h	9.	Second cleavage complete with four macromeres: A, B, C & D and second polar lobe included with the D macromere	5 1/4 hours
3-2i	10.	Third cleavage complete with formation of first generation micromeres (1a, 1b, 1c, 1d) and macromeres (1A, 1B, 1C, 1D)	6 hours
3-2j	11.	Fourth cleavage of macromeres forming a 12 cell stage with a second generation of micromeres (2a, 2b, 2c, 2d) and second generation macromeres (2A, 2B, 2C, 2D)	6 1/2 hours
3-2k	12.	Late cleavage and stereoblastula stages	8-15 hours
3-2l	13.	Epibolic gastrula stages	20-30 hours
3-2m	14.	Veliger larval stages and hatching	3-5 days

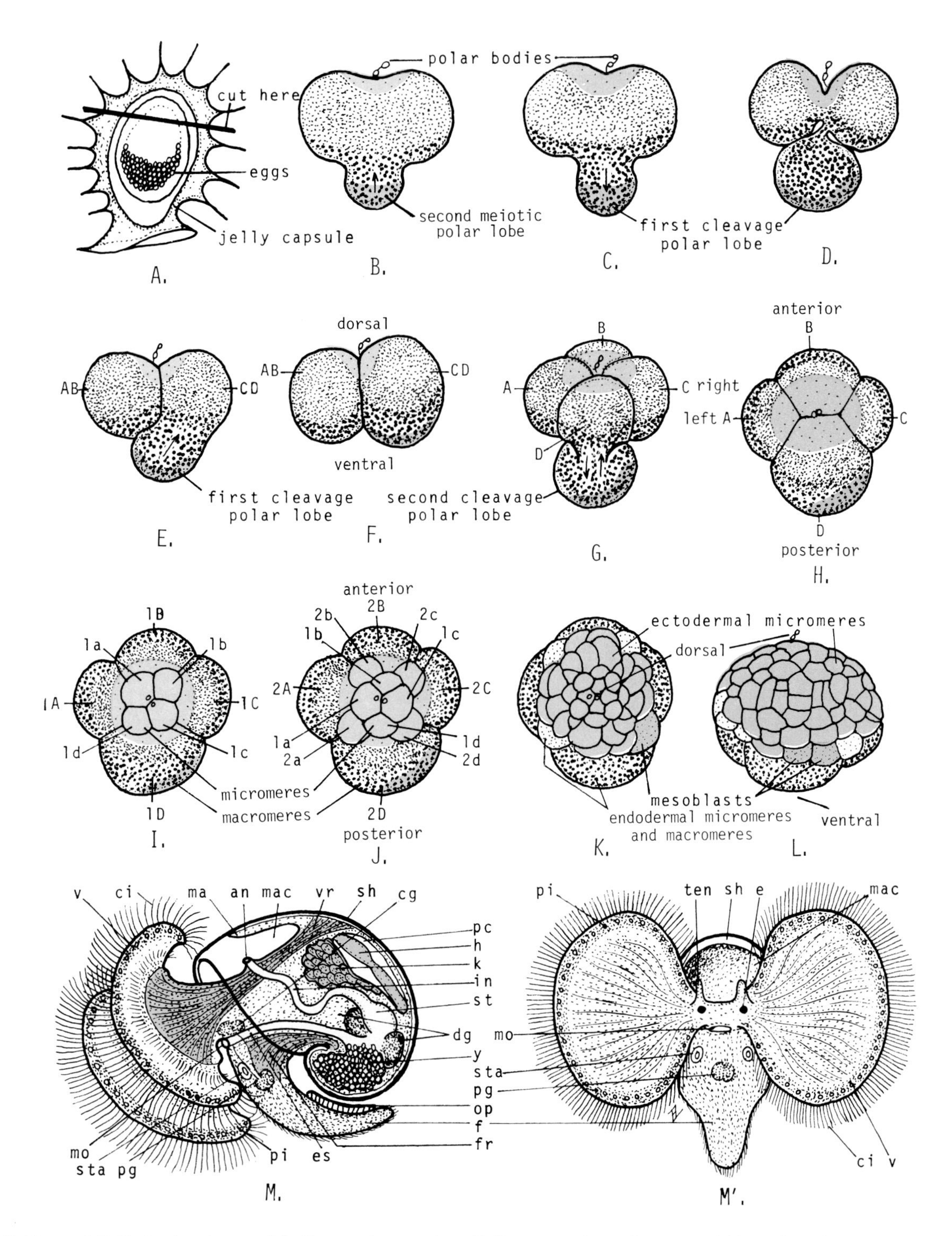

Figure 3-2. Early development in the marine mud snail, Ilyanassa (recently renamed *Nassarius obsoletus*). Figures are described in the time schedule of development in Table II-1. Color code: blue — ectoplasm and ectoderm, red — mesoplasm and mesoderm, yellow — endoplasm and endoderm.

an	anus	h	heart	pi	pigment cells
cg	cerebral ganglion	in	intestine	sh	shell
ci	ciliated bands	k	kidney	st	stomach
dg	digestive glands	ma	mantle	sta	statocyst
e	eye spots	mac	mantle cavity	ten	tentacles
es	esophagus	mo	mouth	v	velum
f	foot	op	operculum	vr	velum retractor muscle
fr	foot retractor muscle	pc	pericardial cavity	y	yolk reserve from egg
		pg	pedal ganglion		

They can be kept without harm for 24 to 48 hours in this way for future use.

Materials needed per person: 1 small culture dish, 1 deep-well depression slide, fine sand, 1 finely drawn out medicine dropper, 1 vial of seawater, standard slides and coverslips, 1 dropping bottle of 0.1% Toluidine Blue. Mixed samples of living eggs will be provided to each student in small culture dishes. From this sample, deep-well and standard microscope slide preparations can be made for study under medium microscopic magnification.

2. Observation of living Ilyanassa spiral cleavage (Fig 3-2). Each person will be provided with a small watch dish of seawater containing a mixture of embryonic stages. Make deep-well depression slide preparations and standard microscope slide preparations for observation of the eggs in the living condition under medium power. Use sand grains for cover glass supports on standard microscope slides so that the eggs will not be smashed under the covers. For deep-well preparations, make a brimming drop in the depression and add a cover from the side so that no air bubble is trapped under the cover glass.

Look for all stages shown in Figure 3-2 and described in Table III-1. Especially search for early cleavage stages and ones that show **polar bodies** attached at the animal pole. They can best be seen in profile when the animal and vegetal axis of the egg lies in the horizontal plane. Polar bodies will appear as minute glass-clear spheres attached to the animal pole by slender cytoplasmic threads.

3. Demonstration of ooplasmic segregation by vital staining with Toluidine Blue. Following maturation and fertilization, a considerable period of time elapses in most eggs before the first cleavage. During this time various components of the egg cytoplasm segregate into different localized regions. Many experimental studies of cleavage in different species have shown that these specific cytoplasmic components have "prospective organ significance." That is, the types of cytoplasm of the egg are allocated to different blastomeres, and they in turn possess unique capacities for special development. This re-assortment of the constituents of egg cytoplasm is termed **ooplasmic segregation.** When the early cleavage furrows consistently result in the separation of different organ-significant cytoplasm into specific blastomeres, cleavage is described as being **mosaic.** That is, the embryo is made up of separate cells, each of which has the capacity to form a special type of structure, or structures, in later development. Ooplasmic segregation and mosaic cleavage can be seen with particular clarity in the eggs of molluscs and annelids. The eggs of Ilyanassa provide one of the most extreme examples of visible ooplasmic segregation. For this reason, they are selected for special study.

Toluidine Blue is a polychrome stain that stains DNA green, RNA blue, and mucoproteins pink. If Ilyanassa eggs are lightly stained with Toluidine Blue, different parts of the egg that correspond roughly with the ectodermal, endodermal, and mesodermal components exhibit different staining properties, allowing recognition of these respective zones. Make a deep-well slide preparation for Toluidine Blue staining in the following way. First add a small drop of Toluidine Blue to the depression. Then shake out all the stain that you can with three hard shakes. Enough stain will be left in the depression to serve for vital staining. Then fill the depression to brimming with normal seawater (it should be tinged a light blue) and add a small sample of young cleavage stages. Cover the preparation, taking care to avoid trapping air bubbles. A brief description of the major aspects of germ layer determination by ooplasmic segregation in Ilyanassa is given in the following paragraph.

Following fertilization, the "mesoderm-forming" cytoplasm of the Ilyanassa egg, along with a large part of the heavily granular yolk, segregates toward the vegetal pole of the egg. Just before each of the two maturation divisions, and also at the first two cleavage mitoses, this mesoderm-forming cytoplasm and yolk constricts partially away from the rest of the egg. It forms a hernia-like evagination that equals almost one-third of the entire egg volume (Fig. 3-2). Evaginations are called **polar lobes.** At the end of first cleavage, all yolky polar lobe material is incorporated in only 1 of the first 2 cells. Accordingly, the one receiving the polar lobe material will be the larger of the first two cells. The smaller cell is named the **AB-cell,** and the larger cell containing **polar lobe** mesoplasm is the **CD-cell** (Fig. 3-2 D–F). At second cleavage, the polar lobe again goes entirely into only one of the blastomeres at the 4-cell stage; this cell is larger than the others and is the **D-cell.** The remaining cells are named in clockwise manner with the **A-cell** to the left, **B-cell** opposite, and **C-cell** to the right of the **D-cell.** The third cleavage is unequal in all four blastomeres. This division results in 4 small cells, the **micromeres** (1a, 1b, 1c, and 1d), and 4 large yolky cells, the **macromeres** (1A, 1B, 1C, and 1D; Fig. 3-2 I). During the 4th, 5th, and 6th cleavages, three additional generations of micromeres are also given off. The first three generations of micromeres all form various kinds of ectoderm. The 4th generation micromeres from the 4A, 4B, and 4C macromeres will form endoderm, and the 4d micromere from the 4D macromere will form all body mesoderm.

After completing the study of living Ilyanassa embryos, discard all embryos. They can be reared outside the egg capsule only if special precautions and sterile techniques are employed. If you so desire, older swimming veligers can be kept for observation at a later period, even though they will not be given any further special attention.

B. CLEAVAGE, GASTRULATION, AND GERM LAYER SEPARATION IN ECHINODERMS REPRESENTED BY THE STARFISH AND SEA URCHIN

The eggs of echinoderms are particularly favorable for illustration of early steps in development. They represent one of the more primitive types of cleavage and ooplasmic segregation found in the animal kingdom. Cleavage is **radial;** this means that cleavage furrows divide the egg either in the plane of the animal vegetal axis or at right angles to that axis. Moreover, the distribution of egg contents is also a simple gradient of quantitative differences along the animal-vegetal axis. Animal pole plasm will form ectoderm in normal development; vegetal pole plasm will form mesoderm, and vegetal equatorial plasm will form endoderm. However, there is considerable ability for different regions to regulate and form more than their normal fate might suggest. Hence these eggs are said to be labile, or **regulatory in their determination.**

Fertilization is the fusion of "physiologically ripe" gametes. It initiates sexual reproduction. Gametic union accomplishes three ends, namely: (1) **activation** of the egg into a metabolic system capable of total development characteristic of the species; (2) **nuclear fusion** of the egg and sperm pronuclei, which provides the basis for bi-parental inheritance; and (3) **ooplasmic segregation** of special materials in the egg which determine body axes and/or localize organ-significant materials in definite areas of the egg cytoplasm. In the broad sense, fertilization begins at the moment of contact between an egg and sperm and ends with prophase of the first cleavage mitosis. Fertilization was the subject of Chapter 2.

Cleavage is a period of rapid cell division following fertilization. It accomplishes the sub-division of the egg cytoplasm into discrete cellular units. With the increase in cell number, the early embryo first assumes the form of a solid ball of 50 to 100 cells. This later hollows out with all of the cells arranged peripherally around a central cavity. These two conditions are respectively called the **morula** and **blastula** stages of cleavage. During the cleavage period, very little morphodynamic rearrangement of the cell contents occurs. However, cleavage does permanently isolate the "organ significant" materials which were segregated earlier. They can then be moved in cells without "contamination" during the gastrulation period which follows.

Gastrulation is primarily associated with the formation of the **archenteron** (primitive gut), and the separation of prospective ectoderm, mesoderm, and endoderm into separate germ layers. This is accomplished primarily by mass movements of cells, rather than by mitotic increase in cell number. During this period there are widespread changes in the distribution of various parts of the egg. These are called **morphogenetic movements** since they precede and lead to the formation of structures which will develop at a later time.

Organogenesis, as the name implies, refers to the early formation of the primordia of various organs. This activity involves local differences in the development of the three germ layers that are established during gastrulation. During this period, definitive organs of the adult are mapped out in rough outline. Each organ rudiment is referred to as a **primordium** or **anlage** as soon as it is recognizable as a distinct regional structure, and before it carries out its definitive function in the adult.

1. Cleavage, gastrulation, and bipinnaria formation in the starfish. (Fig. 3-3). Slides containing whole mounts of embryos at all developmental stages from the unfertilized egg to the free-living bipinnaria larval stages are provided for study. The embryos are usually mixed at random on the slide and you should look for the best representatives of each stage in the preparations. Refer to the plastic models and Figure 3-3 for aid in identifying the stages and special structures described below. Identify the following representative stages in the development of the starfish. Reserve the use of colors to indicate germ layer derivation of special structures in gastrula and bipinnaria larval stages. Color **ectoderm** — blue, **mesoderm** — red, and **endoderm** — orange or yellow.

Unfertilized eggs (Fig. 3-3 A) become physiologically ripe during the primary oocyte stage but they can be fertilized at any time during the maturation stages of oogenesis. They are characterized by the absence of a fertilization membrane and, usually, by the presence of a very large nucleus, or **germinal vesicle.**

Fertilized eggs (Fig. 3-3 B). In preparing for pronuclear fusion and cleavage, the germinal vesicle disappears shortly after sperm entry and **no nucleus** is usually seen in these cells. These cells are characterized also by the presence of a fertilization membrane around the egg. This membrane in prepared whole mounts usually is shrunken and wrinkled and does not give the smooth spherical appearance of the membrane in living eggs.

2-cell, 4-cell, and 8-cell stages (Fig. 3-3 C–E). Identify these by cell count and compare the arrangement of the cells in later cleavages with that shown in plastic models.

Morula (16-cell, 32-cell, 64-cell stages; Fig. 3-3 F). These stages represent the transition from the solid mass of cells into a hollow ball that characterizes the next stage. The term "morula" is now little used and is of little developmental significance. The term "late cleavage" is more generally used.

Blastula stages (Fig 3-3 G–H). These are characterized by the arrangement of blastomeres into a hollow ball. The cells form one cell layer. In late blastulae, the wall is thinner at the animal pole (prospective ectodermal region) than at the **vegetal pole,** where invagination of mesoderm and endoderm will take place during gastrulation. The cavity of the blastula is the **blastocoele.** At the end of this stage, the embryo develops cilia; it escapes from the fertilization membrane and becomes a free-swimming embryo.

Gastrula stages (Fig 3-3 I-K). During the formation of the primitive gut and the separation of the 3 primary germ layers, the vegetal half of the embryo carries out the morphogenetic movement of **invagination.** These cells will give rise to mesoderm and endoderm. They stream into the interior and form a tube, the **archenteron.** The **ectoderm** carries out the movement of **epiboly.** This is a spreading action which results in the animal pole cells stretching and surrounding the invaginated cells. The opening of the archenteron is the **blastopore,** which will later form the anus. The cavity of the archenteron is the **gastrocoele.** The archenteron will give rise to the lining of the gut and the coelom. **Mesoderm** arises from the innermost end of the archenteric wall. In this region the gastrocoele expands greatly and the wall becomes very thin. At this stage the terminal expansion is called the **mesodermal** or **coelomic vesicle.** At the end of gastrulation, this vesicle is completely separated from the remainder of the archenteron. It will form a right and left **coelomic,** or **mesodermal pouch.** In this manner the mesoderm is separated from the endoderm. The posterior (vegetal) two-thirds of the archenteron wall will give rise to the endodermal organs of the body, namely, the lining of the esophagus, stomach, and intestine. The endoderm remains attached at the blastopore to the surface ectoderm.

Bipinnaria larval stages (Fig. 3-3 L–M). Following gastrulation, the mesodermal pouches expand greatly and, by a proliferation of cells, give rise to the skeletal, muscular, and peritoneal structures of the body. Shortly after gastrulation, the endodermal archenteron begins to bend toward one side and meets an in-pocketing of ectoderm called the **stomodaeum.** At the point of contact between ectoderm and endoderm, they fuse and perforate, thus forming the mouth and completing the basic organization of the digestive tube. From this time on, the embryo is called a **bipinnaria larva.** It is bilaterally symmetrical at first with regard to all its structures. The archenteron soon undergoes constrictions and enlargements and you can identify the following divisions of the gut: **stomodaeum** (mouth cavity lined by ectoderm); **esophagus** (endoderm), consisting of a thick-walled muscular tube connecting the mouth to the stomach; **stomach** (endoderm), a spherical enlargement of the archenteron; and **intestine** (endoderm), a small tube which runs obliquely forward from the posterior end of the stomach and opens to the outside by means of the **anus.** The latter is derived from the blastopore of the gastrula stage. Irregular bands of ciliated ectoderm pass over the surface of the embryo. The ectoderm in these regions expands so that the external body surface is thrown into bilaterally symmetrical folds and depressions. These **ciliated bands** are the locomotor organelles of the larva. Within 36 to 48 hours after fertilization, bipinnaria larvae are able to ingest food and become independent organisms.

In concluding the study of the early starfish development, it should be emphasized that the starfish illustrates in almost diagrammatic style one of the simplest and probably the most primitive living form of development among deuterostomes. These are phyla including the echinoderms and chordates in which the blastopore remains as the anus, or marks the location at which the anus will form; they are phyla in which the mesoderm forms from the wall of the archenteron and in

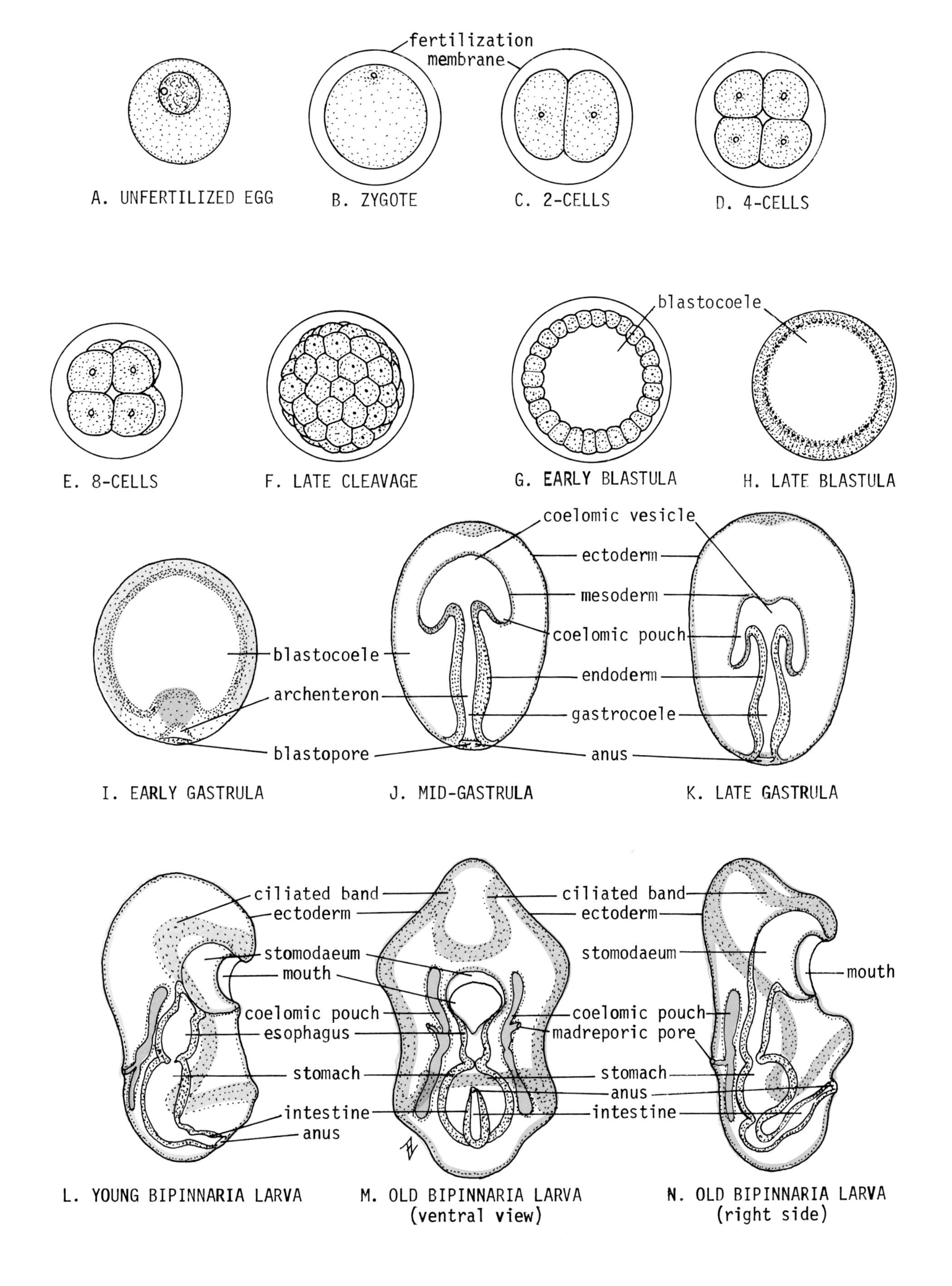

Figure 3-3. Normal development of the starfish. Color code: blue — ectoderm, red — mesoderm, yellow — endoderm.

which the coelom is primitively derived from the archenteric cavity as simple pouches. For a review of the deuterostome and chordate characters, see Tables Int.-1 and Int.-2.

2. Cleavage, gastrulation, and pluteus larval development in sea urchins (Fig. 3-4). In nearly all respects, the development of the sea urchin compares favorably with that outlined in detail for the starfish. The slide preparations available for study are of the same sort as those used for the study of starfish development. The sea urchin larva, however, is called a pluteus.

Examine the following developmental stages and note in particular the differences in the two echinoderm classes. Where differences occur, the developmental pattern in the starfish is considered to be evolutionarily the more simple.

Unfertilized eggs (Fig. 3-4 A). These differ from those of the starfish in that they do not become ripe until the ootid stage; hence there is no conspicuous germinal vesicle to distinguish them from fertilized eggs.

Fertilized eggs (Fig 3-4 B). As in the starfish, the major distinguishing feature that separates these clearly from unfertilized eggs is the presence of a fertilization membrane. If appropriate fixation has been used, a hyaline plasma membrane can also be seen adhering closely to the zygote membrane.

Cleavage stages through the 64 cell stage (Fig. 3-4 C–F). The first three divisions are similar to those in the starfish. However, when the 8-cell stage undergoes the 4th cleavage, the vegetal cells divide unequally and give rise to 4 micromeres at the vegetal pole. These **micromeres** represent ooplasmic segregation of that component of the mesoderm that will give rise to the skeleton. Micromeres are also called **primary mesoderm,** or **scleroblast mesoderm.** At the same division, the 4 cells at the animal pole divide equally and form a ring of 8 cells, all of which will form embryonic ectoderm in the normal course of development; these cells are therefore called **ectoblasts.** The four large cells near the vegetal equator will give rise to both endoderm and coelomic mesoderm; these cells, for the time being, are merely called **macromeres.** Later cleavages differ in no fundamental way from those described for the starfish.

Blastula stages (Fig. 3-4 G–H). A hollow coeloblastula is first formed by a hollowing out of late cleavage cells. The cavity is called the **blastocoele.** The mitotic products of micromere division

Legend for Figure 3-4. Normal development of the sea urchin.

A	Unfertilized egg	I	Early gastrula
B	Fertilized egg	J	Late gastrula with triradiate spicules
C,D,E	2, 4, 8 cell stages	K	Prism stage
F	16 cell stage with micromeres	L	Pluteus, ventral view
G	Prehatching blastula	L'	Pluteus, left side
H	Mesenchyme blastula		

Legend and checklist of structures to be identified

AN	anus	M1	primary mesoderm
AP	animal pole (anterior)	M2	secondary mesoderm
AR	archenteron	MAC	macromeres (mesentoblasts)
BC	blastocoele	MES	mesomeres (ectoblasts)
BP	blastopore	MIC	micromeres (primary mesoblasts)
CL	coelom	OE	esophagus
E	ectoderm	SK	skeletal rods
EN	endoderm	ST	stomach
FM	fertilization membrane	STD	stomodaeum
INT	intestine	VP	vegetal pole (posterior)
LA	larval arms		

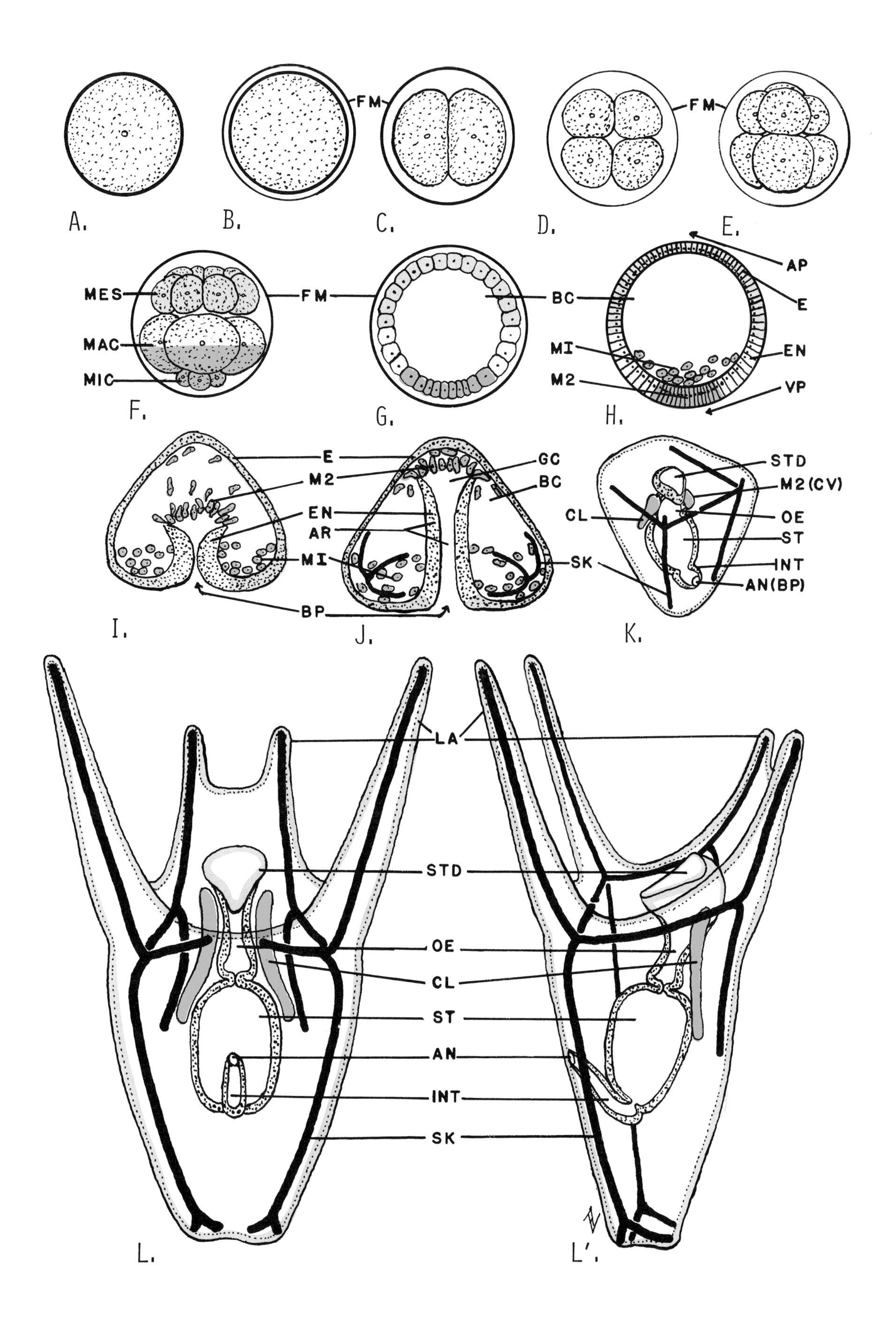

Figure 3-4. Normal development of the sea urchin. See legend. Color code: blue — ectoderm, red — mesoderm, yellow — endoderm.

result in a small pool of cells at the vegetal pole which will migrate individually into the blastocoele cavity after hatching by the blastula takes place. The late blastula with primary mesenchyme cells in the vegetal blastocoele cavity is called a **mesenchyme blastula.** It represents a precocious separation of part of the mesoderm.

Gastrulation (Fig. 3-4 I–J). The major difference between sea urchin and starfish embryos is found in the separation of coelomic mesoderm from the tip of the archenteron. In most sea urchins there is no distinct coelomic vesicle formed directly as an expansion of the gastrocoele cavity, as in starfish embryos previously described. Instead, the mesoderm cells at the end of the archenteron detach from the archenteron. Some of them remain dispersed and form the larval muscle and mesenchyme; others, near the apical inner end of the archenteron, will reassemble into a pair of coelomic pouches. The final effect is essentially the same, but the manner of coelomic pouch formation in sea urchins is more simple and direct than that exhibited by the starfish. Late in gastrulation, primary mesenchyme cells begin to organize the primary skeletal spicules. The skeletal primordia first form triradiate stars in which the three rays are united at a central point and each arm Is at right angles to the other two.

Prism and Pluteus larval stages (Fig. 3-4 K–L). The internal organizations of mesodermal pouches and gut are essentially the same as those of the bipinnaria larva of the starfish. The distinctive feature that characterizes the prism (pre-feeding larva) and pluteus larva is associated with the enormous growth of spicules which force the body epidermis into long arms supported by slender skeletal rods. Since the skeletal rods are calcium carbonate and many staining methods involve acids, many slide preparations will lack visible skeletal elements.

3. Observation of echinoderm cleavage and gastrulation in time-lapse motion pictures and in living cultures. Films of living sea urchin, starfish, and/or sand dollar development will be shown to illustrate with remarkable clarity the basic events of cleavage, blastulation, gastrulation, mesoderm separation, and larval structure.

Methods for artificial insemination have been described at the end of Chapter 2. If living, reproductively ripe adult echinoderms are available, cultures inseminated at 1, 2, 3, 6, 12, and 24 hour intervals before laboratory will be prepared in order that the major cleavage, blastula, gastrula and larval stages will be available for observation during the laboratory period. In particular, devote your time to the early cleavage stages and observe the processes of cell division as they occur in living eggs.

C. FUNDAMENTAL CHORDATE PATTERNS OF OOPLASMIC SEGREGATION, CLEAVAGE, GASTRULATION, AND GERM LAYER SEPARATION IN TUNICATES

In Chapter I the larval stages of tunicates and Amphioxus were described as part of the evidence for establishing the concept of the developmental common plan for the vertebrate body. We return to them again now to examine the processes underlying symmetrization as clues to the fundamental nature of these processes in vertebrates.

1. Patterns of symmetrization in tunicates. (Fig. 3-5). Slides may not be available for study of early tunicate development. However, tunicates occupy such an important position at the base of the chordate phylogenetic tree that their early developmental characteristics will be briefly reviewed and illustrated diagrammatically for general reference in the following studies.

Unfertilized eggs become ripe at metaphase of the second meiotic division. They are radially symmetrical, with a pool of prospective **ectoplasm** surrounding the meiotic spindle, with a cortical layer of mesoplasm surrounding the egg, and with a central core of yolky **endoplasm** (Fig. 3-5 A).

At fertilization sperm generally penetrate the egg from the vegetal hemisphere. The sperm entry point activates the egg to undergo an elaborate program of cytoplasmic movements which transforms the radial egg into a bilaterally symmetrical arrangement of parts. This ooplasmic localization of germ layers into specific regional potencies is shown in Figure 3-5 B. The egg is shown from the left side with the sperm entry point at the lower left side of the figure.

The first three cleavage planes (Fig. 3-5 C–E) coincide with the major planes of bilateral symmetry of the future embryo and adult. **First cleavage**

divides the egg along the median sagittal plane and results in the formation of a R (right) and a L (left) blastomere (Fig. 3-5 C). **Second cleavage** separates the first two cells in the frontal plane such that at the four cell stage there is an RV (right ventral), LV (left ventral), RD (right dorsal), and LD (left dorsal) cell. This is a determinate division in that the sister cells produced from the original R and D cells no longer have equal developmental competence; that is, the ventral cells each possess general muscle mesoderm, but both lack neural and chordamesodermal competence. The reciprocal is true of the two dorsal cells (Fig. 3-5 D). **Third cleavage** passes at right angles to the first two divisions and is in the equatorial plane of the egg. It corresponds with the transverse plane of the embryo. At the 8 cell stage, following third cleavage, there are the following sets of cells: RVA & LVA (right and left ventral anterior cells), RDA & LDA (right and left dorsal anterior cells), RVP & LVP (right and left ventral posterior cells), and RDP & LDP (right and left dorsal posterior cells). The third division is also a determinate one. Note in Figure 3-5 E that the RVA and LVA cells possess only general epidermal competence; the RDA & LDA cells have both epidermal and neural competence; the RDP & LDP cells have chordamesoderm and endodermal competence, and the RVP & LVP cells have muscle mesoderm and endodermal competence. Experiments in which blastomeres have been removed or isolated in culture reveal that their normal fates very closely coincide with their total capacities for development. That is to say, the cells for the most part are determined and unable to regulate toward the replacement of missing parts.

Late cleavage and blastulation are shown in Figure 3-5 F–G. Additional divisions result in finer refinements in the assignment of cell fates. By the time the blastula stage is reached, the program of development is sufficiently specific in all its parts to make the use of "mosaic" development a meaningful description of the processes that have taken place. By mosaic development we mean that each part of the embryo contributes one small bit of cell specificity to a total pattern, similar to the parts of a stained glass window or a mosaic tile floor. With the exception of the neural area, which acquires its specificity by inductive influence from the chordamesoderm cells between the 8 and 32 cell stage, all other parts shown in the diagrams of fertilization, cleavage, and blastulation (Fig. 3-5 B–G) represent **ooplasmic segregation** of preformed cytoplasms with specific cell competence in the uncleaved egg.

Gastrulation and neurulation (Fig 3-5 H–I) are simple and direct. They start with a collapse of the vegetal hemisphere of the blastula into the space formerly occupied by the blastocoele. In the **late gastrula stage** closure of the blastopore is accomplished primarily by a separation of the epidermal ectoderm from its junctions with muscle mesoderm and neural ectoderm. After this, the epidermis from both sides slips over the muscle mesoderm, chordamesoderm, and neural ectoderm in a zipper-like manner that progresses from the ventral lip anteriorly to the end of the neural plate. Thus, gastrulation moves directly into neurulation with no perceptible break. In the closure of the blastopore by epidermal over-growth, a neurenteric canal, continuous with the neurocoele, is present for a time. The neurenteric canal will disappear when growth of a tailbud begins.

The tailbud and larval stages (Fig. 3-5 J–L) entail elongation of the tail by the posterior growth of muscle mesoderm, chordamesoderm, and neural plate. In the head, differentiation of pharynx, gut tube, cerebral vesicle, and sensory organs occur in the manner already described in Chapter I, Part A which treats the tunicate tadpole and metamorphosis.

In concluding this brief account of early tunicate development, it should be emphasized that these eggs illustrate as great a precision in the pattern of ooplasmic segregation and cleavage symmetry as can be found in the entire animal kingdom. Tunicate development is a classic example of determinate bilateral cleavage; this is in contrast with the general statement that Deuterostomes (Echinoderms and Chordates) have indeterminate and regulatory eggs.

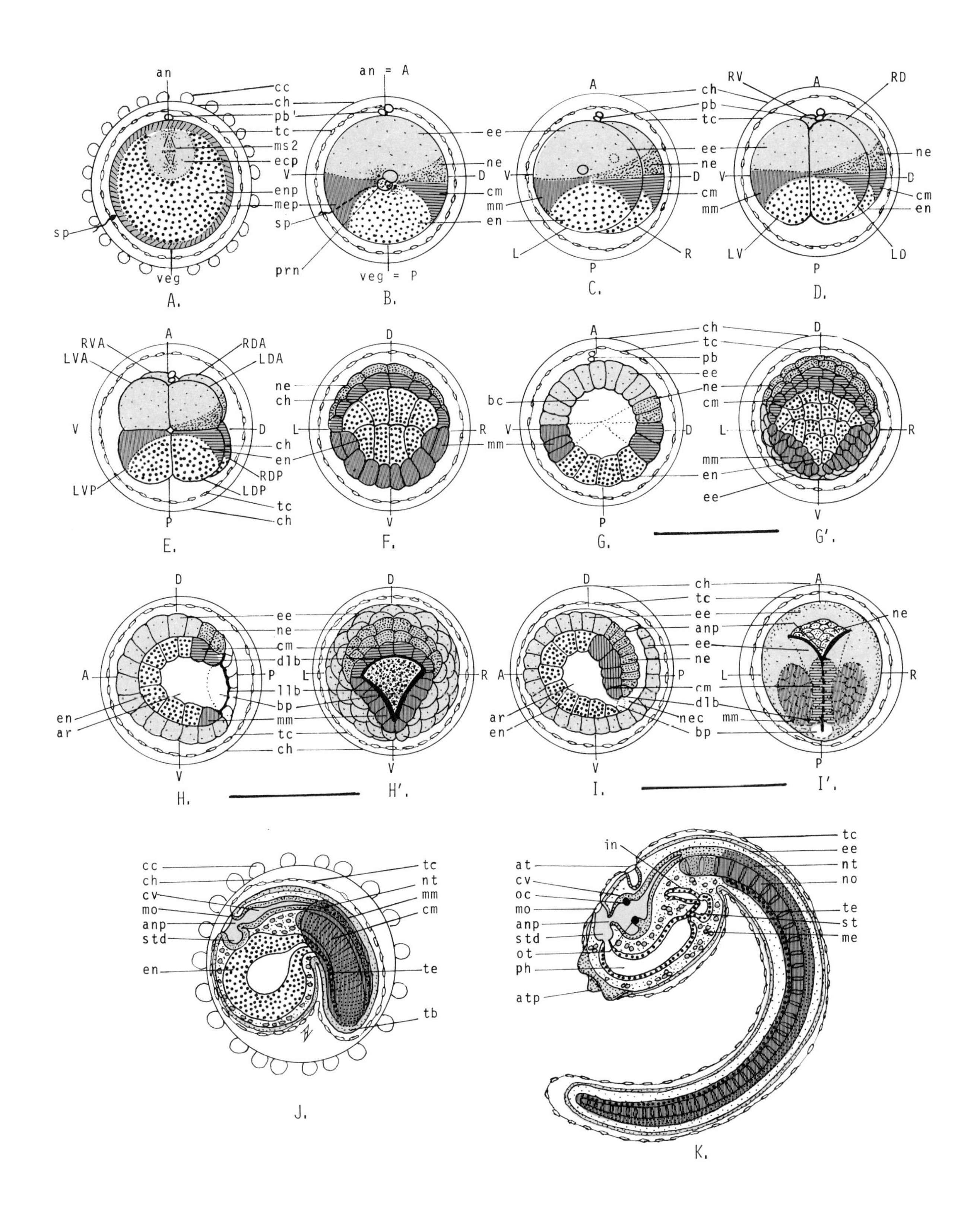

Figure 3-5. Fertilization, cleavage, gastrulation, and larval development in tunicates as represented by ***Ascidia nigra.*** Prospective germ layer regions are based on Conklin's observations on Styela (Cynthia). See legend. Color code: blue — ectoderm, red — mesoderm, yellow — endoderm.

Legend for Figure 3-5.

A	Unfertilized egg	H	Gastrula, left side
B	Fertilized egg, left side	H'	Gastrula, posterior view
C	2 cell stage, left side	I	Neurula, left side
D	4 cell stage, left side	I'	Neurula, dorsal view
E	8 cell stage, left side	J	Tailbud embryo
F	Late cleavage, vegetal view	K	Hatching larva
G	Blastula, left side		
G'	Blastula vegetal view		

See Chapter 1, Figure 1-3, for tunicate larvae and metamorphosis.

Legend and checklist of structures to be identified in Figure 3-5

A	anterior pole	LD	left dorsal cell	pb	polar bodies
an	animal pole	LDA	left dorsal anterior cell	ph	pharynx
anp	anterior neuropore	LDP	left dorsal posterior cell	prn	male & female pronuclei
ar	archenteron	llb	lateral lip of blastopore	R	right cell or right side
at	atrium, excurrent siphon	LV	left ventral cell	RD	right dorsal cell
atp	attachment papillae	LVA	left ventral anterior cell	RDA	right dorsal anterior cell
bc	blastocoele	LVP	left ventral posterior cell	RDP	right dorsal posterior cell
bp	blastopore	me	mesenchyme	RV	right ventral cell
cc	chorion cells	mep	mesoplasm of egg	RVA	right ventral anterior cell
ch	chorionic membrane	mm	muscle-mesenchyme mesoderm	RVP	right ventral posterior cell
cm	chordamesoderm	mo	mouth, incurrent siphon	sp	sperm, or sperm entry point
cv	cerebral vesicle	ms2	second meiotic spindle	st	stomach
D	dorsal pole	ne	neural ectoderm	std	stomodaeum
dlb	dorsal lip of blastopore	nec	neurenteric canal	tb	tailbud
ecp	ectoplasm of egg	no	notochord	tc	test cells, or larval tunic
ee	epidermal ectoderm	nt	neural tube	te	tailgut endoderm
en	endoderm or endoplasm	oc	ocellus, eye spot	V	ventral pole
in	intestine	ot	otocyst, balance organ	veg	vegetal pole
L	left cell or left side	P	posterior pole		

D. PATTERNS OF SYMMETRIZATION IN AMPHIOXUS

1. The study of normal development in Amphioxus (Fig. 3-6). This study will use charts, models, and demonstration slides of representative whole mount stages in fertilization, cleavage, blastulation, gastrulation, and neurulation. Emphasis will be placed on the similarities and differences that exist between tunicates and Amphioxus in their early developmental patterns.

Unfertilized eggs (Fig. 3-6 A) of Amphioxus become ripe at metaphase of the second maturation division. The distribution of animal pole **ectoplasm,** of cortical **mesoplasm,** and of internal **endoplasm** are as described for tunicates.

Fertilized eggs and early cleavage stages (Fig. 3-6 B–E) also very closely correspond with those in tunicates with regard to ooplasmic segregation of prospective ectodermal, mesodermal, and endodermal regions, as well as with respect to the precise bilaterality of the first three cleavage planes. The first furrow is in the median sagittal plane; the second is in the frontal plane, and the third cleavage furrow is in the transverse plane of the egg and future embryo. It is not surprising that the letter coding of blastomeres at the 2 cell, 4 cell, and 8 cell stages in tunicates and amphioxids is identical (Compare Fig. 3-5 C–E and Fig. 3-6 C–E). Tunicates and amphioxids may, however, differ in degrees of regulation and determination in early cleavage. In the species of Amphioxus used by Boveri at Naples, the first four blastomeres can each give rise to a complete embryo and the eggs therefore appear to be completely regulatory up to the four cell stage. Conklin, working with the same material, reported that Boveri's observations were highly exceptional and that, in his own experience, Amphioxus at Naples was essentially like tunicates in having bilateral cleavage from the beginning. Both agreed that by the third cleavage, the eggs of Amphioxus acquired bilateral symmetry and determinative cleavage of the type shown in the diagram Fig. 3-6 E. By the blastula stage the pattern of mosaic regions in Amphioxus and tunicates is essentially the same.

Late cleavage, blastulation, and early gastrula stages (Fig. 3-6 F–H) differ in no fundamental ways in tunicates and Amphioxus. In early gastrulation of Amphioxus eggs there is a general collapse of the vegetal pole cells in the interior, as in tunicates. Later there is also a rolling action of chordamesoderm cells to the interior in an action called **involution.** Involution can most easily be described as the movement of cells as though they were a belt passing over an invisible pulley. Involution involves, primarily, the cells of chordamesoderm which form the dorsal lip of the blastopore.

Late gastrulation and neurulation (Fig. 3-6 I–J) entail epidermal overgrowth of the blastopore from posterior to anterior. As in tunicates, this action is zipper-like and will result in the formation of a neurenteric canal. The epidermal overgrowth encloses first the blastopore and then, progressively, the posterior, middle, and anterior parts of the neural plate. However, an anterior neuropore will remain open until the tailbud and larval stages (see Fig. 1-4 and 1-5).

Tailbud stages (Fig. 3-6K) show the first real departures by Amphioxus embryos from the developmental pattern in tunicates. The differences are primarily associated with the formation of general **segmental mesoderm** or **somites.** Somites begin to form at the anterior roof of the archenteron, lateral to the middorsal chordamesoderm. Each somite first forms as a hollow evagination of the archenteron. The wall of these evaginations will form all types of mesoderm except chordamesoderm. The cavity of these evaginations is the **myocoele.** It is the primordium of the embryonic **coelom.** In actual fact only the first six to eight somite pairs form the myocoele from the archenteron as evaginations of the gastrocoele. The remaining sixty or more somites will develop their myocoeles by splitting solid blocks of somite mesoderm in the characteristic vertebrate way.

For larval development refer to Chapter I which gives detailed attention to larval characteristics.

Legend for Figure 3-6. Fertilization, cleavage, gastrulation, neurulation, and tailbud stages in Amphioxus diagrammed to show prospective germ layer regions. Guide lines and arrows on Figs. I, J, K, L show planes of section for Figs. I', J', K', L'.

Figures:

A	Unfertilized egg	I	Mid gastrula, sagittal section
B	Fertilized egg, left side	I'	Mid gastrula, transverse section
C	2 cell stage, left side	J	Neurula, sagittal section
D	4 cell stage, left side	J'	Neurula, transverse section
E	8 cell stage, left side	K	Early tailbud, longitudinal section
F	Late cleavage, left side	K'	Early tailbud, transverse section
G	Blastula, left side	L	Late tailbud, longitudinal section
H	Early gastrula, left side	L'	Late tailbud, transverse section

Legend and checklist of structures to be identified

A	anterior pole	llb	lateral lip of blastopore		
an	animal pole	lm	lateral plate mesoderm	pb''	second polar body
ar	archenteron	LV	left ventral cell	ph	pharynx, foregut
anp	anterior neuropore	LVA	left ventral anterior cell	prn	male & female pronuclei
bc	blastocoele	LVP	left ventral posterior cell	ps	perivitelline space
bp	blastopore	mep	mesoplasm of egg	R	right cell or side
ch	chorionic membrane	mg	midgut lumen	RD	right dorsal cell
cl	coelom	mo	mouth	RDA	right dorsal anterior cell
cm	chordamesoderm	ms2	second meiotic spindle	RDP	right dorsal posterior cell
D	dorsal pole	my	myotome	RV	right ventral cell
dlb	dorsal lip of blastopore	myc	myocoele	RVA	right ventral anterior cell
ecp	ectoplasm of egg	myc'	first myocoele, head coelom	RVP	right ventral posterior cell
ee	epidermal ectoderm	nc	neural canal	sm	somite mesoderm
en	endoderm or endoplasm	ne	neural ectoderm	sp	sperm or sperm entry point
enp	endoplasm of egg	nec	neurenteric canal	std	stomodaeum
fm	fertilization membrane	no	notochord	tb	tailbud
L	left cell or side	np	open neural plate	V	ventral pole
LD	left dorsal cell	nt	neural tube	veg	vegetal pole
LDA	left dorsal anterior cell	P	posterior pole	vit	vitelline membrane
LDP	left dorsal posterior cell	pb'	first polar body	vlb	ventral lip of blastopore

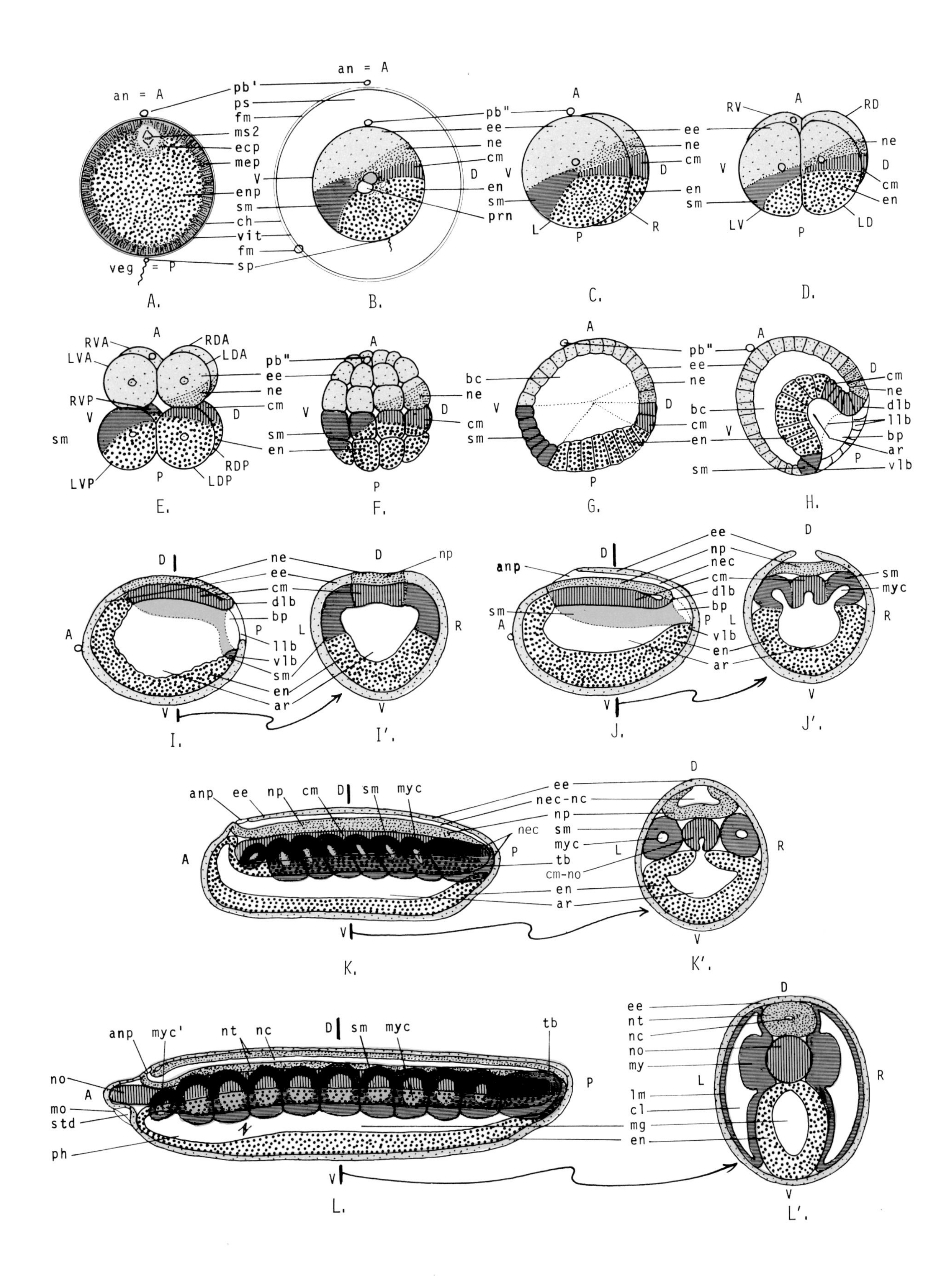

Figure 3-6. Fertilization, cleavage, gastrulation, neurulation, and tailbud stages of Amphioxus diagrammed to show prospective germ layer regions. Adapted primarily from descriptions by Hatschek and Conklin. For more advanced stages, see Figures 1-4 and 1-5. See legend. Color code: blue — ectoderm, red — mesoderm, yellow — endoderm.

2. Analysis of morphodynamic movements of gastrulation in Amphioxus (Fig. 3-7). There can be little doubt that the ancestral chordate mechanisms for creating bilaterality and axial organization during embryonic development must have been closely similar to those aspects of early development that are shared in common by tunicates and Amphioxus in their acquisition of symmetry. In addition, enterocoelous formation of the coelom and somites in the anterior segments of Amphioxus are also exceedingly ancient or primitive mechanisms that possibly antedate the chordates and go back to some common echinoderm-chordate invertebrate ancestor.

From the perspectives of contemporary embryology, the study of homologous structures is a somewhat sterile exercise unless it is correlated with the dynamic causes responsible for them. Accordingly, it is appropriate to reexamine the events of gastrulation in Amphioxus in the light of independent cell movements by respective germ layer regions that underlie the homologies in structure and symmetrization described above.

Figure 3-7 is a diagrammatic presentation of gastrulation movements in Amphioxus. The individual movements described below for specific germ layer regions are shown in the diagrams by correspondingly numbered arrows. They indicate the directions of cell movements at progressive stages in gastrulation and germ layer separation.

Endodermal morphodynamic movements:

1. Emboly (Fig. 3-7 D, E F) is the active movement of endodermal cells to the interior at the vegetal hemisphere of the blastula wall so as to replace the blastocoele cavity. Emboly has previously been described as a collapse of the blastula wall. It is one of the many forms by which invagination may occur.

2. Abscission of endoderm (Fig. 3-7 J, L) is the active separation of endoderm cells from somite mesoderm along their anterior-posterior borders in the dorsolateral wall of the archenteron. It is by this action that endoderm disengages from prospective mesoderm in the archenteron roof.

3. Tubulation of the gut (Fig. 3-7 M, N) is accomplished by a movement of the dorsolateral edges of endoderm toward each other until they meet in the dorsal midline. By this action the archenteron, which originally had only a ventral floor of endoderm, acquires a complete endodermal lining, and the gastrocoele cavity becomes a closed endodermal tube.

Mesodermal morphodynamic movements:

4. Involution (Fig. 3-7 E, F, I) involves movement of prospective chordamesoderm and general somite mesoderm to the interior. After endodermal emboly, the two mesodermal regions form the blastopore lips. The dorsal lip is formed by chordamesoderm, and the two lateral lips of the

Figure 3-7. Morphodynamic movements in Amphioxus gastrulation and neurulation. The arrows numbered from 1-10 refer to the following specific cell movements (see explanation in text).

1. **Emboly** of endoderm
2. **Abscission** of endoderm from mesoderm
3. **Tubulation** of endoderm
4. **Involution** of mesoderm around blastopore lips
5. **Axial elongation** of notochord
6. **Constriction and vesiculation** of somite mesoderm
7. **Internal spreading and expansion** of lateral plate mesoderm
8. **Abscission** of epidermal ectoderm from neural plate and somite mesoderm
9. **Epiboly** of general ectoderm
10. **Tubulation** of neural plate ectoderm

Legend and checklist of structures to be identified

A	anterior pole	ee	epidermal ectoderm	np	neural plate
an	animal pole	en	endoderm	nt	neural tube
anp	anterior neuropore	L	left side	P	posterior pole
ar	archenteron	ll	lateral blastopore lip	pb	polar body
bc	blastocoele	lm	lateral mesoderm	R	right side
bp	blastopore	my	myotome	sm	somite mesoderm
cl	coelom	myc	myocoele	V	ventral pole
cm	chordamesoderm	ne	neural ectoderm	veg	vegetal pole
D	dorsal pole	nec	neurenteric canal		
dl	dorsal blastopore lip	no	notochord		

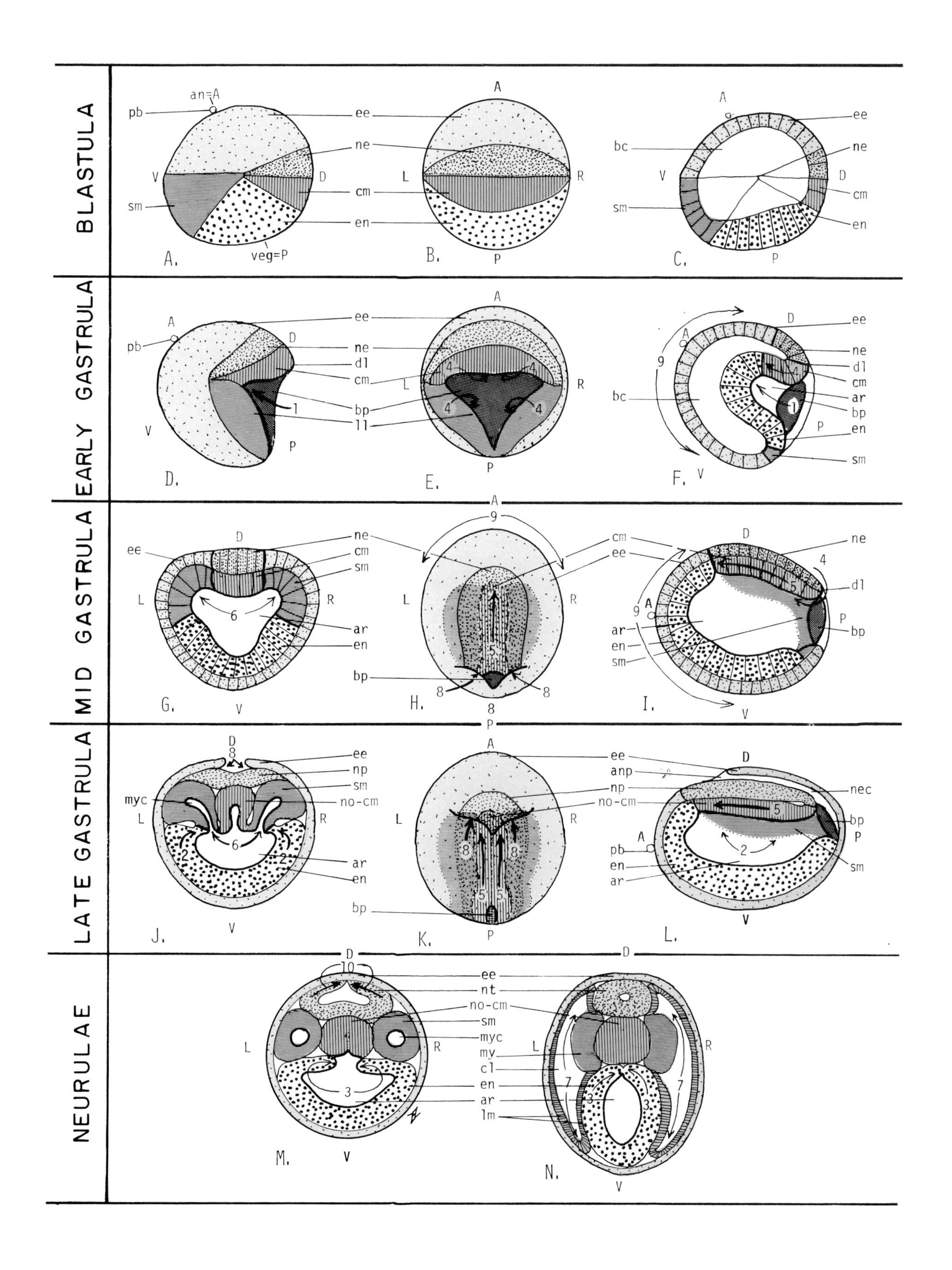

Figure 3-7. Diagrams of morphodynamic movements in Amphioxus gastrulation and neurulation. 3-7 A, B, C. Left, posterior, and sagittal views of blastulae. 3-7 D, E, F. Left, posterior, and sagittal views of early gastrulae. 3-7 G. K, I. Transverse, dorsal, and sagittal views of mid gastrulae. 3-7, J, K, L. Transverse, dorsal, and sagittal views of late gastrulae. 3-7 M. Transverse section of early neurula. 3-7 N. Transverse section of late neurula. See Table IV-3 for comparison of gastrulation movements in Amphioxus, amphibians and birds. Color code: blue — ectoderm, red — mesoderm, yellow — endoderm.

triangular blastopore are formed by somite mesoderm. There is no distinct ventral lip of the blastopore since it is merely the point where the right and left lateral lips meet in the posterior midline. The movement of involution is another form of invagination. It can be most easily described as a pulley-like in-rolling of cells around the blastopore lips so that the direction of cells is reversed by 180°.

5. Axial elongation (Fig. 3-7 H, I, K, L) involves primarily the chordamesoderm which, early in gastrulation, forms the dorsal blastopore lip and has a wide mesiolateral and short anteroposterior axis. During involution, these cells tend to crowd toward the dorsal midline with the result that, following involution, the long and short axes of chordamesoderm are reversed; that is, after gastrulation the AP axis is long and the ML axis is very short and reduced to the chorda cells in the dorsal midline. To a lesser degree, somite mesoderm (particularly the anterior somites) is also involved in axial elongation; however, note in Figure 3-7 D, E that the long axis of somite cells in the lateral lips, although oblique, is primarily anterior-posterior in orientation at the start of gastrulation.

6. Constriction and vesiculation (Fig. 3-7 G, J, M), involve the separation of somite mesoderm by constriction from the gut endoderm laterally and from the chordamesoderm middorsally. Constriction, therefore, is the mesodermal counterpart of abscission by endoderm. It accomplishes complete separation of mesoderm from endoderm. Once separate from chorda and endoderm, the somite mesoderm constricts itself into a series of hollow blocks; the walls of these vesicles are general somite mesoderm and their cavities are the myocoeles.

7. Internal spreading and expansion (Fig. 3-7 N) are characteristic movements of lateral plate mesoderm cells derived from the lateral walls of the mesodermal somites. These cells first spread laterally and ventrally by inserting themselves between the epidermis and endoderm. Concurrently the myocoele expands along with the spreading lateral plate mesoderm and thereby creates the general body coelom.

Ectodermal morphodynamic movements:

8. Abscission of ectoderm (Fig. 3-7 H, J, K) is the active separation of epidermal ectoderm from somite mesoderm while the latter is still located at the lateral lips of the blastopore. Abscission also takes place by the active separation of epidermis along its line of contact with the neural plate ectoderm.

9. Epiboly (Fig. 3-7 F, H, I) is the spreading of epidermal ectoderm over the mesoderm along the lateral lips of the blastopore; this closes the blastopore with a sheet of epidermal ectoderm. Epiboly continues over the neural plate, by the movement of epidermis from right and left sides toward the midline. The fusion of epidermis over the neural plate progresses from posterior to anterior; the last remaining opening is the anterior neuropore at the anterior end of the neural plate. It is by epiboly and the covering of somite mesoderm, blastopore, and neural plate, that the neurenteric canal is formed.

10. Tubulation of neural ectoderm (Fig. 3-7 M and contrast with J) involves the formation of a tubular dorsal nerve cord along the length of the neural plate. This is accomplished after the plate is covered by epibolic growth of epidermis. Tubulation takes place by a sinking of the floor of the neural plate and by the dorsal folding of its lateral edges until they meet and fuse under the epidermis in the dorsal midline.

Post-gastrula caudalization: Not generally included among gastrulation movements *per se* is tailbud formation and caudal axial elongation. These activities involve cells in the vicinity of the closing blastopore: i.e., the neural plate, chordamesoderm, and general somite mesoderm which unite to form a common cell mass, the **tailbud blastema.** Out of this primordium will be formed most of the trunk and tail neural tube, somites, and chordamesoderm posterior to the level of somite 8-10. In the posterior elongation of the tailbud a small amount of hindgut endoderm is passively carried for a short distance into the tail as a strand of tailgut endoderm.

In conclusion to our consideration of primary patterns of symmetrization in chordates, as represented by tunicates and Amphioxus, it now follows that we should look for evidences of these common plans in a primitive vertebrate. Cyclostome ooplasmic segregation, cleavage, and gastrulation would be the vertebrate class of choice for such a study (see Fig. 1-7). However, cyclostome eggs are exceedingly difficult to obtain for general study. Fortunately, amphibian development is remarkably close to that of cyclostomes, and the following section deals with amphibian symmetrization as representative of the primitive vertebrate plan.

E. PATTERNS OF OOPLASMIC SEGREGATION, CLEAVAGE, GASTRULATION AND GERM LAYER SEPARATION IN AMPHIBIA

The basic characteristics of fertilization, cleavage, gastrulation, and germ layer separation which have been illustrated in echinoderms and protochordates are also recognizable with modifications in the eggs of vertebrates. The major deviations in the early development of vertebrate classes are associated largely with differences in the relative amounts of inert yolk stored in eggs. **Isolecithal eggs** such as those of the sea urchin, starfish, tunicate, and Amphioxus have relatively little yolk and it is distributed more or less uniformly throughout the egg cytoplasm. **Telolecithal eggs,** such as those of fish, amphibians, reptiles, birds, and monotreme mammals, contain large amounts of yolk concentrated near the vegetal pole of the egg. In these eggs, the yolk acts as a physical impediment to cleavage and gastrulation. These activities can be expected to be altered and, therefore, to modify to some degree the primary chordate common plan of gastrulation.

Normal developmental series are given for the salamander in Figure 3-8, and for the frog in Figure 4-1. They summarize two variations in early amphibian development. Of the two forms, salamander development is evolutionarily more primitive than that in the frog. However, the mechanisms for ooplasmic segregation, fertilization, cleavage, and gastrulation are sufficiently similar that common directions for study can be given for both.

1. Patterns of symmetrization in amphibians (Fig. 3-8, 3-9, and 4-1). Amphibian eggs are moderately telolecithal and occupy an intermediate position, in this respect, between the isolecithal eggs of protochordates and the very strongly telolecithal eggs of sharks, bony fish, reptiles, and birds. The materials for study will include charts, models, and sections of eggs at various stages between fertilization and the completion of neurulation at the, so-called, neural tube stage of prehatching embryonic development. For later stages in the development of amphibians, refer to Chapter 4 for larval development in the frog, and to Chapter 5 for larval development in the salamander.

a. Unfertilized eggs (Fig. 3-9A) of the frog when fully ripe are shed as primary oocytes but shortly after entering the oviducts they will spontaneously undergo the first maturation (meiotic) division and will be arrested at metaphase of the second maturation division. Their distribution of cytoplasmic components is strictly radially symmetrical with no trace of bilaterality at the time of ovulation. There are marked quantitative differences in the amounts of different materials along the animal-vegetal axis. Cortical pigment, cytoplasmic RNA, general protoplasm, and the cell nucleus are eccentrically distributed toward the animal pole. Yolk protein, in the form of yolk spheres and glycogen bodies, are more abundant at the vegetal pole.

b. Fertilization and the origin of bilaterality (Fig. 3-9 B), as in the protochordates, are initiated by sperm entry. A redistribution of cytoplasmic constituents by ooplasmic segregation includes a shift, by about 15°, of the pigmented cortical cytoplasm of the animal hemisphere and the coarse yolky cytoplasm at the vegetal pole toward the sperm entry point. Concurrently, the subcortical RNA-rich **mantle cytoplasm** near the animal pole streams in the opposite direction and creates the visible, so-called, **gray crescent** near the equator of the egg, opposite to the sperm entry point. The midpoint in the gray crescent marks the middorsal pole in the dorsoventral axis of the future embryo. The animal-vegetal axis will coincide within 15° of the anteroposterior axis. With the AP and DV axes thus delineated, the axiate pattern of the egg becomes clearly bilateral. Fertilization also reactivates maturation and the second polar body of meiosis will be given off about one hour after sperm entry.

c. Cleavage (Fig. 3-9 C–D) is radial and unequal. That is, the cleavage furrows are located at random relative to the DV axis but run at parallel or right angles to the AP axis of the egg. The animal pole plasm is relatively rich in cytoplasm and poor in yolk. Mitotic divisions occur more rapidly in the animal half of the egg than in the more heavily yolky vegetal half. Accordingly animal pole cells are, in general, smaller than their vegetal counterparts.

d. Blastulation (Fig. 3-9 D–E) is anticipated by the early appearance of a segmentation cavity among cells at the late cleavage stages. A midblastula has a large, well defined, **blastocoele**

cavity. It is eccentrically placed with its vegetal wall three or four times as thick as the animal roof. In all regions the blastula wall is multicellular in thickness, averaging from about three cells deep at the animal pole to about a dozen cells thick at the vegetal pole. The surface of the blastula has been thoroughly mapped with regard to the normal germ layer fates of respective regions. A prospective fate map for the midblastula stage is shown Figure 3-10 A, B. Note the relative positions of prospective epidermal and neural ectoderm, chordamesoderm and other general mesoderm, as well as the location of endoderm. Note the marked similarity in positions that they bear to the blastula fate map of the Amphioxus egg in Figure 3-7 A, B, and the tunicate egg in Figure 3-5 B.

e. Gastrulation (Fig. 3-9 F–H). Gastrulation begins with the direct migration of many vegetal cells to the interior without blastopore formation. Slightly later the blastopore begins to form in the middorsal line near the boundary between prospective mesoderm and foregut endoderm. This is the dorsal lip region. Gradually invagination of mesoderm cells, which form the **blastopore lips,** progresses along the boundary between mesoderm and endoderm. When the lateral blastopore lips are formed, they give the appearance of a horseshoe bordering about two-thirds of the endodermal cells still on the surface. Finally, invagination of mesoderm will reach the posterior end of the egg, and invagination will take place around the entire endodermal mass which will fill the blastopore as the **yolk plug.** In median sagittal sections of middle and late gastrulae (Fig. 3-9 F, G), note that the **dorsal lip of the blastopore** is composed of cells that are continuous internally with the cells that form the roof of the archenteron. These cells are prospective **chordamesoderm cells.** Gradually the archenteron will replace the **blastocoele** cavity, the remnant of which can be found as a small space under the epidermis in the anteroventral midline. In addition, identify the **archenteron lumen, endodermal floor of the archenteron, yolk plug,** and **ventral lip of the blastopore.** Note that the invagination of ventral lip mesoderm cells does not entail the formation of an archenteron cavity as is the case with the dorsal lip.

Occasionally errors are made in orienting yolk plug stage late gastrulae for sectioning. It is not unusual to find sections among laboratory slide sets that have been cut in the frontal, rather than in the median sagittal plane. Frontal sections will pass through two lateral lips of the blastopore and not through the dorsal and ventral lips as described above for ideal sagittal sections. In frontal sections, both lateral lips will appear like the ventral lip in that no archenteron cavity is associated with them. Figure 3-9 I shows how frontal sections will appear. Compare this with Figure 3-9 H of a median sagittal section through the yolk plug of a late gastrula.

Legend for Figure 3-8. Normal stages in the development of the salamander, *Taricha torosa.* The numbers in the series correspond with standard developmental stages in salamanders.

Stage

1. fertilized eggs
2. 2 cells
3. 4 cells
4. 8 cells
5. mid cleavage
6. late cleavage
7. early blastula
8. late blastula
9. beginning gastrulation
10. early gastrula (crescent blastopore)
11. mid gastrula (horseshoe blastopore)
12. late gastrula (early yolk plug blastopore)
13. late gastrula (late yolk plug blastopore)
14. early neurula, flat neural plate
15. open neural plate with folds
16. shallow neural groove
17. deep neural groove
18. early closing neural groove
19. middle closing neural groove
20. late closing neural groove
21. early neural tube
22. neural tube with forebrain swelling
23. neural tube with optic vesicles
24. early tailbud
25. tailbud with gill plate
26. tailbud with pronephric swelling
27. tailbud with nasal placode forming
28. tailbud with stomodaeum invaginating
29. late tailbud with otic placode
30. late tailbud showing first muscle twitches
31. late tailbud with 2 gill swellings
32. late tailbud with 3 gill swellings
33. early larva with balancer buds
34. early larva with lens placode forming
35. early larva with heart beating and first pigment cells
36. swimming larva with tail fin forming
37. swimming larva with forelimb buds and simple unbranched gills
38. larva with branched gills
39. larva with retinal pigment forming
40. larva with conical limb buds

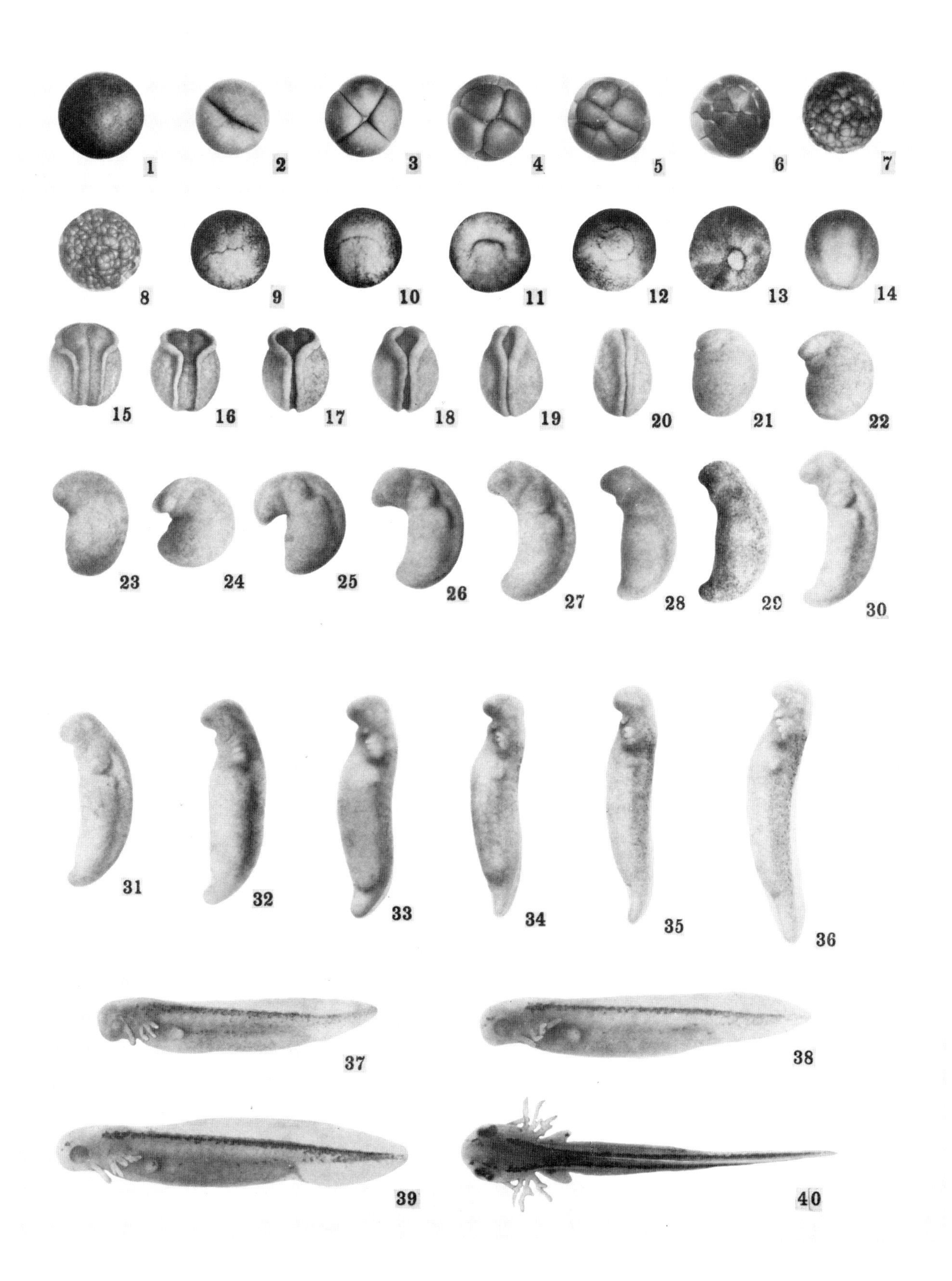

Figure 3-8. Normal stages in the development of the salamander, *Taricha torosa.* (Photographs by courtesy of V.C. Twitty, Stanford University.)

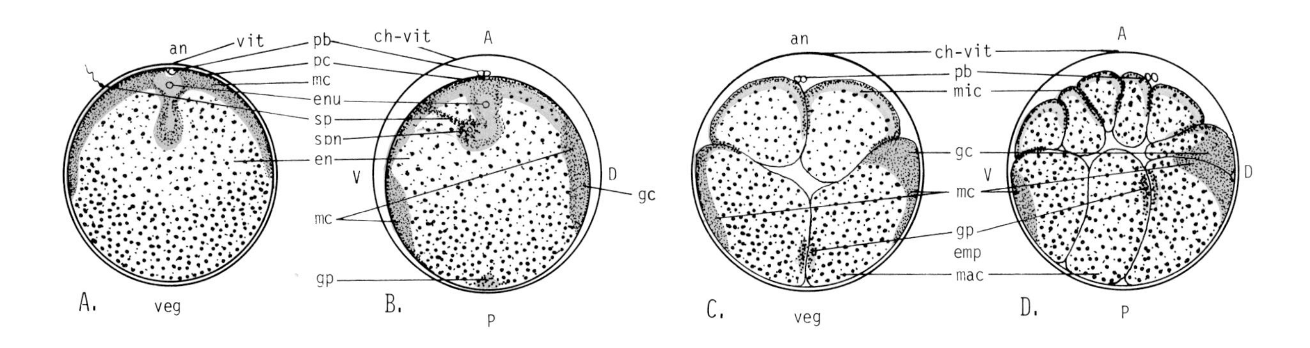

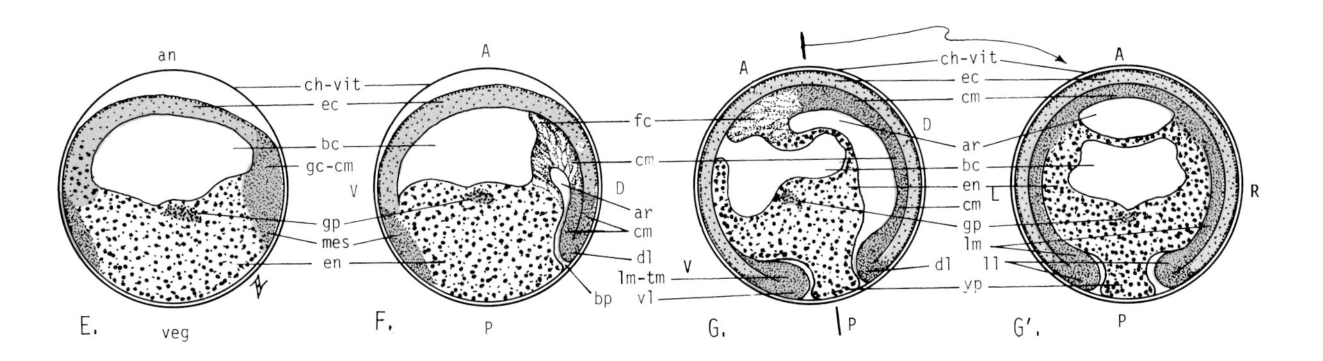

Figure 3-9. Diagrams of early development in sectioned frog eggs. 3-9 A–G' are above and 3-9 H–J' are on facing page. The legend for individual figures and checklist of structures labelled are in the legend below. Color code: blue — ectoderm, red — mesoderm, yellow — endoderm, green — primordial germ plasma or primordial germ cells.

A.	Stage 1, unfertilized ripe egg	G'.	Stage 12, late gastrula, frontal
B.	Stage 2, fertilized egg, sagittal	H.	Stage 14, early neurula, sagittal
C.	Stage 5, early cleavage, sagittal	H'.	Stage 14, early neurula, transverse
D.	Stage 7, late cleavage, sagittal	I.	Stage 15, late neurula, sagittal
E.	Stage 9, late blastula, sagittal	I'.	Stage 15, late neurula, transverse
F.	Stage 11, early gastrula, sagittal	J.	Stage 16. neural tube, sagittal
G.	Stage 12, late gastrula, sagittal	J'.	Stage 16, neural tube, transverse

See Table IV-1 for description of frog developmental stages.

Legend and checklist of abbreviations

A	anterior pole	gp	primordial germ plasm or primordal germ cells	nec	neurenteric canal
an	animal pole			nf	neural fold
anp	anterior neuropore	hb	hindbrain	no	notochord
ar	archenteron	hg	hindgut	np	neural plate
bc	blastocoele	hm	head mesoderm	nt	neural tube & canal
bp	blastopore	ht	heart primordium	P	posterior pole
ch	chorion	L	left side	pc	pigmented cytoplasm
cm	chordamesoderm	li	liver diverticulum	R	right side
D	dorsal pole	ll	lateral blastopore lip	sm	somite mesoderm
dl	dorsal blastopore lip	lm	lateral plate mesoderm	sp	sperm or sperm path
ec	ectoderm	mac	macromeres	spn	sperm nucleus
ee	epidermal ectoderm	mb	midbrain	tb	tailbud blastema
en	endoderm	mc	mantle cytoplasm	tm	tail mesoderm
enu	egg nucleus	mes	mesoderm	V	ventral pole
fb	forebrain	mic	micromeres	vit	vitelline membrane
fc	flask cells	mg	midgut	vl	ventral blastopore lip
gc	grey crescent	nc	neural crest	yp	yolk plug

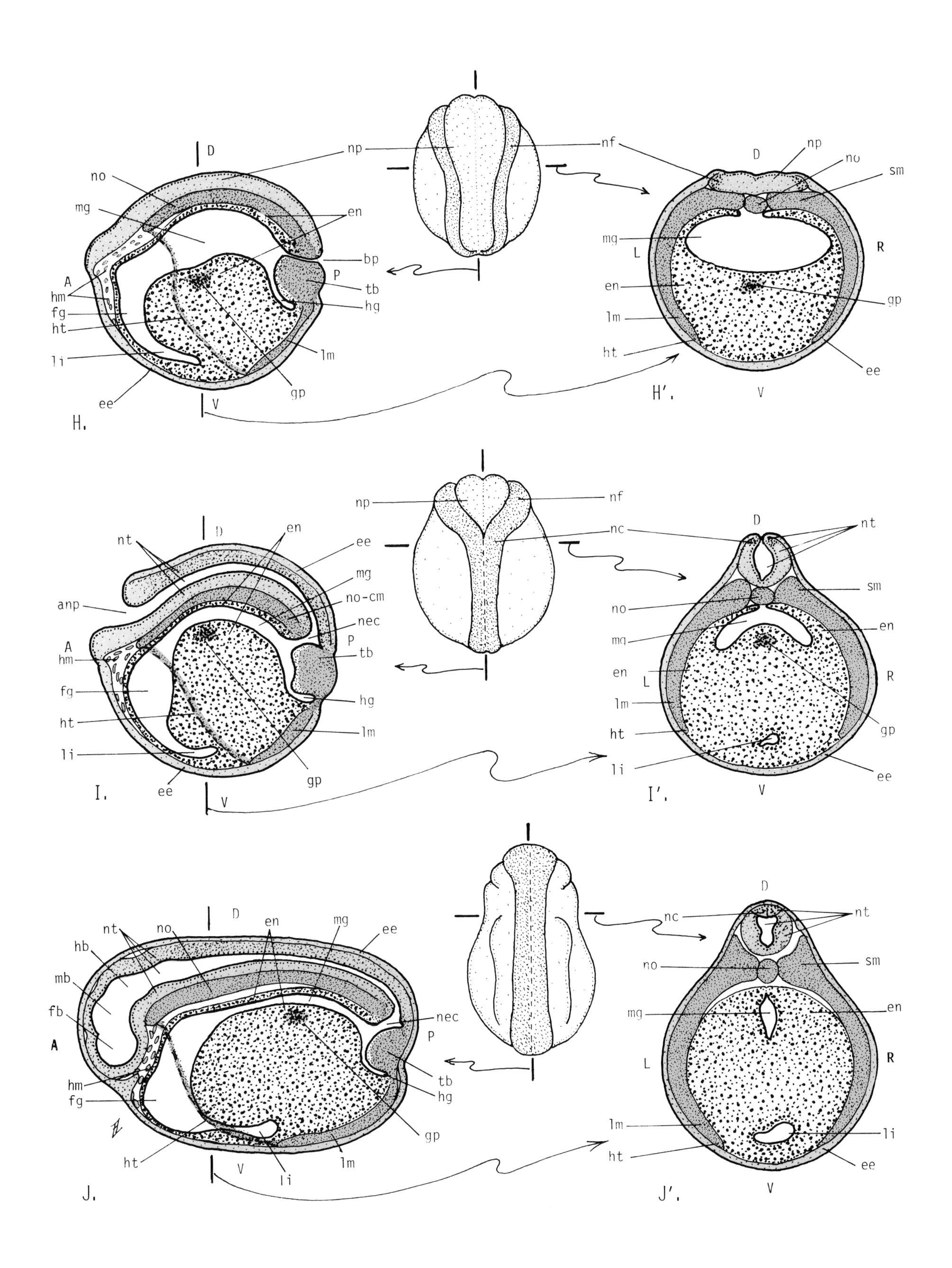

Figure 3-9 H–J'. Diagrams of early development in sectioned frog eggs, continued. Guide lines on central figures show levels at which sagittal sections at left and cross sections at right have been made. See legend on opposite page for description of stages and structures that have been labelled. Color code: blue — ectoderm, red — mesoderm, yellow — endoderm, green — primordial germ cells.

f. Neurulation (Fig. 3-9 H, I) is represented by early and late stages in the formation of the neural plate and neural tube as depicted in median sagittal and transverse sections. Identify the following germ layer regions in appropriately sectioned material: **epidermal ectoderm** is a thin surface layer surrounding the ventral and lateral parts of the embryo; **neural plate ectoderm** or **neural groove** is a thick, flat or curved, sheet of dorsal ectoderm; its edges are the **neural folds.** Internally the middorsal mesoderm in the roof of the archenteron has begun to condense into a recognizable **notochord.** On both sides of the notochord are poorly condensed masses of **somite mesoderm.** The **lateral plate mesoderm** extends ventrally between the epidermal ectoderm and the very thick endodermal lining of the gut. These cells will eventually insert themselves between the epidermis and endoderm all the way to the ventral midline. The leading cells at the ventral edge of the lateral plate will form primordia of the ventral mesentery and heart. Note that endoderm, which up to this stage has formed only the ventral and lateral floor of the archenteron cavity, is in the process of growing up along each side under the lateral plate and somite mesoderm and lead to the final completion of the endodermal lining of the archenteron roof.

g. The neural tube stage (Fig. 3-9 J) will be studied more thoroughly in Chapter 4. For the present rely on the figures provided to assist in identifying the major features of sagittal and transverse sections in later stages in the closure of the neural tube. Identify the following germ layer regions. In median sagittal sections, identify the **neural tube, neurenteric canal** and the fore-, mid- and hindgut regions of the gut. The latter two have very reduced cavities and thick ventral floors. The **foregut** lumen is greatly expanded and is thin walled. Note the **liver diverticulum,** which extends as a posterior projection of the foregut lumen into the ventral yolk mass. Between the ventral floor of the hindgut and the epidermis, identify the strand of **lateral plate mesoderm,** which is the primordium of the ventral mesentery and blood island.

The transverse plane of section will be particularly favorable for showing the cross section appearance of the **neural tube, neural crest, notochord** and, to a lesser degree, the differentiation of **somite** and **lateral mesoderms.** Note that there is no vestige of the primitive origin of a coelom directly from the archenteric cavity as there is in the anterior somites of Amphioxus (Fig. 3-6 J) and cyclostome (Fig. 1-6 I') embryos. Depending upon the level of section, it will be possible in some slides to identify the gut at levels of the fore-, mid-, or hindgut. Anterior sections will often show the liver diverticulum as a separate lumen as in Figure 3-9 J'. Sections through the ventral mesoderm will show little specialization in anticipation of the heart, blood island, or other structures that will be derived from it.

2. Analysis of morphodynamic movements of gastrulation in amphibians (Fig. 3-10).

A reflection on the nature of symmetrization during fertilization, cleavage, and gastrulation in Amphioxus and amphibians reveals striking similarities indicative of a developmental common plan based on genetic and evolutionary relationships. It is of interest, therefore, to pursue this comparison by an examination of the gastrulation movements characteristic of amphibians which are responsible for the structural homologies with protochordates.

Little change takes place in the arrangement of ooplasmic materials during cleavage and blastulation. The prospective germ layer regions are shown in Figures 3-10 A–C. Note that the animal third of the blastula will form ectoderm and, of that region, the dorsal third will form neural plate and the remainder will form epidermis. The equatorial belt of the blastula will form mesoderm. This belt is much wider at the dorsal side than it is ventrally. Much of the dorsal mesodermal region coincides with the gray crescent of the egg and will form the chordamesoderm and somite mesoderm of the embryo. The vegetal third of the egg is prospective endoderm. It is arranged from the dorsal to ventral side in sequence from the prospective foregut, to midgut, to hindgut regions. Although the egg is distinctly bilateral in its organization and symmetry, the distribution of parts can hardly be said to coincide with a recognizable vertebrate pattern of parts prior to gastrulation. It is perhaps the major function of the morphodynamic movements of gastrulation to rearrange the parts and bring them into a final pattern that will coincide with the embryonic developmental common plan. If the movements are not precisely coordinated, development following gastrulation cannot be normal. It would be difficult to over-emphasize the importance of the synchronized movements of gastrulation that are responsible for normal symmetrization. Figure 3-10 diagrammatically shows the gastrulation movements in amphibians, with the salamander chosen as the type-example for the class. In many ways eggs of salamanders exhibit what are believed to be evolutionarily more primitive characteristics in their gastrulation movements when compared with analogous activities in frog embryos. The gastrulation movements of respective germ layer regions are as follows.

Endodermal morphodynamic movements.

1. Polar ingression (Fig. 3-10 C) is a precocious movement of some endodermal cells from the vegetal pole to the interior before blastulation is complete. This does not involve blastopore formation and is analogous with polar ingression of primary mesoderm cells of the sea urchin which also enter the blastocoele in this manner.

2. Inversion (Fig. 3-10 F, I, L) involves all surface endoderm at the time the blastopore lips begin to form. These cells move in a linear direction toward the dorsal boundary between the foregut endoderm and dorsal mesoderm. At this point the endoderm cells dip to the interior and move dorsally and then anteriorly. As a result of this movement the surface endoderm cells move through an arc of 90° so that the original foregut–midgut–hindgut axis, which was dorsoventral, becomes anteroposterior. The foregut will then be at the anterior end and farthest from the blastopore. The last endoderm to enter the blastopore is the hindgut cells which are at the opposite end. During inversion, the center of gravity of the egg is changed by the internal redistribution of yolk. Before inversion the heavy posterior end of the egg will orient downward; after inversion, gravity will orient the ventral surface down, as is shown in Figures 3-10 J–O. Note that inversion results in the presence of endoderm only in the ventral and lateral walls of the archenteron. Inversion, although somewhat different, is probably the homologue of emboly in Amphioxus which also results in an archenteron having only an endodermal floor.

3. Tubulation (Fig. 3-10 I, L, O) involves the movement of the lateral edges of the endodermal floor up the sides of the archenteron roof to result ultimately in the completion of an endodermal lining in the lateral and dorsal portions of the gut tube. This movement excludes mesoderm from its original role as archenteron roof. This movement is entirely similar in action and function to tubulation of the gut in Amphioxus. These two movements may be considered entirely homologous in the two chordate groups.

Mesodermal morphodynamic movements:

4. Involution (Fig. 3-10 E–L) involves the movement of all mesoderm to the interior around the dorsal, lateral, or ventral lips of the blastopore. Involution begins at the dorsal lip and gradually progresses laterally and ventrally along the border between prospective endoderm and mesoderm; however, except for some foregut endoderm cells immediately at the dorsal lip, involution is initiated only by prospective mesoderm cells. Involution begins by the traction of special cells which begin to migrate to the interior while remaining in firm contact at their peripheral ends with other cells at the surface. By traction they become greatly elongated and "flask-shaped." These are the so-called **bottle cells** which initiate and lead involuting cells to the interior. The movement of cells by involution is most easily described as the movement of a belt over a pulley in which the blastopore lips remain stationary but the cells at the lips are constantly changing as those at the surface roll to the interior. In Amphioxus, flask cells have not been described; otherwise the movements are the same as those in amphibians.

5. Shearing (Fig. 3-10 H–I) is not an independent movement in its own right. Instead it is the consequence of the direction of cell movements created by endodermal inversion (Movement 2) and by involution of mesoderm (Movement 4). It will be noted that, along the lateral blastopore lips, mesoderm cells move to the interior at right angles to the direction of movement by endoderm cells which are directed toward the dorsal lip. As a consequence of this 90° difference at the lateral lips, and up to 180° difference in direction at the ventral lip, stresses are created which finally tear, or shear, the mesoderm from the endoderm along the lateral and ventral lips of the blastopore. Only at the dorsal lip do mesoderm and endoderm cells remain in contact after invaginating to the interior. This contact is maintained because the two cell regions move in the same direction in going to the interior. Note that through early and mid gastrulation stages, head mesoderm and chordamesoderm form an integral part of the roof of the archenteron and are continuous with the foregut endoderm in contributing to the archenteron wall (Fig. 3-10 F, I, L, O). Shearing bears a superficial resemblance to the movement of "abscission" by Amphioxus endoderm. However, shearing does not occur in amphibians without cell tensions resulting from divergent movement of mesoderm and endoderm; thus, shearing cannot be considered the homologue of abscission, even though the final consequence of these movements is the same and will result in the separation of mesoderm and endoderm.

6. Convergence Fig. 3-10 D, E, G, H) is a special behavior of chordamesoderm and somite mesoderm cells but does not involve lateral plate or tail mesoderm. If we consider first the direction taken by lateral plate cells on the surface toward the blastopore lip, it is fair to say that these cells move in as direct a manner as possible to the part of the blastopore lip nearest them. In the movement of convergence, however, we find that chordamesoderm and somite mesoderm crowd compactly toward the dorsal lip of the blastopore. Chordamesoderm, led by head mesoderm cells, involutes at the dorsal lip and somite mesoderm cells involute immediately lateral to the chorda cells in parallel streams as shown in Figure 3-10 N.

7. Axial elongation (Fig. 3-10 M-N) is primarily associated with chordamesoderm and somite mesoderm. These cells, after converging on the dorsal lip, undergo involution. However, instead of following a path internally that is a reciprocal of the one they used to reach the dorsal lip, the direction of cell movement is strictly in the anteroposterior axis. Thus, chordamesoderm cells form a long cell chain in the dorsal midline of the archenteron roof. Flanking rows of somite mesoderm bands lie parallel to the chorda cells along both sides of its entire length. By the combined movements of convergence and axial elongation, we can see that the long axis of chorda and somite cells, which before gastrulation was in the mesiolateral axis, has been altered by 90°. After gastrulation the long axes of somite and chorda mesoderms are in the AP axis. The latter state is one of the most characteristic features of the vertebrate body plan. Axial elongation of the chorda and somite mesoderm are sufficiently similar in amphibians and Amphioxus to be considered as essentially homologous movements.

8. Internal spreading (Fig. 3-10 M–O) involves the lateral plate mesoderm after it reaches the interior by involution around the lateral and ventral blastopore lips. As in Amphioxus, the lateral mesoderm inserts its way between the surface epidermis and endoderm of the gut. Internal spreading from right and left sides will finally bring cells to the ventral midline. These vanguard cells are primordia of the ventral mesentery and vascular system (i.e., blood island and heart). Internal spreading will not be complete until late neurula and early tailbud stages. This movement is fundamentally the same as that encountered in Amphioxus.

Ectodermal morphodynamic movements:

9. Epiboly (Fig. 3-10 D–L) Involves the entire ectoderm during the full course of gastrulation. Ectoderm undergoes radial spreading and, as the endoderm and mesoderm move to the interior, the epidermis passively replaces them on the egg surface. Epiboly is not achieved by any traction or tensions exerted on the ectoderm; instead, radial spreading is primarily achieved by deep cells in the prospective epidermal regions moving to the surface. After arriving at the surface, the cells become transformed into a thin epithelium with great covering capacity.

Legend for Figure 3-10. Morphodynamic movements in amphibian gastrulation. The arrows numbered from 1 to 9 refer to the following specific gastrulation movements (see explanation in text).

1. **Polar ingression** of endoderm
2. **Inversion** of endoderm
3. **Tubulation–internal spreading** of endoderm
4. **Involution** of mesoderm around lips of the blastopore
5. **Shearing** of lateral mesoderm from endoderm along lateral and ventral blastopore lips
6. **Convergence** of somite and chordamesoderm at the dorsal lip
7. **Axial elongation** of somite and chordamesoderm internally
8. **Internal spreading** of lateral plate mesoderm
9. **Epiboly,** radial external spreading of ectoderm to cover exterior

Abbreviations

A	anterior	F	foregut endoderm	P	posterior
AR	archenteron	H	hindgut endoderm	SM	somite mesoderm
BC	blastocoele	HM	head mesoderm	TM	tail mesoderm
BP	blastopore	LL	lateral lips of blastopore	V	ventral
D	dorsal	LM	lateral plate mesoderm	VL	ventral lip
DL	dorsal lip of blastopore	NE	neural ectoderm	YP	yolk plug (uninvaginated hindgut endoderm)
EE	epidermal ectoderm	NO	notochord or chordamesoderm		

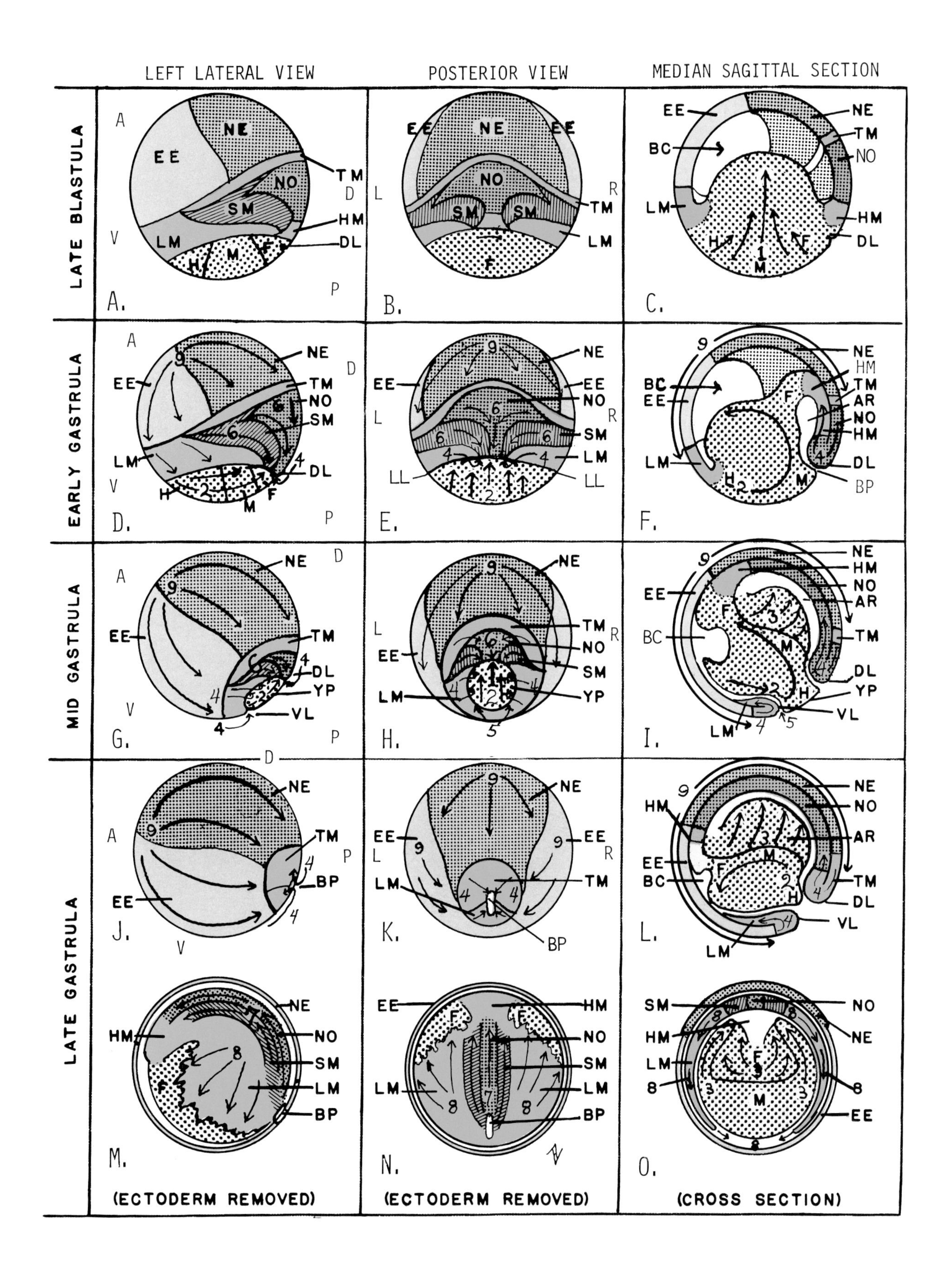

Figure 3-10. Morphodynamic movements in amphibian gastrulation. See legend on facing page. See Table IV-3 for comparison of gastrulation movements in Amphioxus, amphibians and birds. Color code: blue — ectoderm, red — mesoderm, yellow — endoderm.

Morphodynamic movements of neurulation. It is generally not customary to include the special movements associated with neural tube formation among the gastrulation movements of amphibians. However, since this was done for Amphioxus, in which gastrulation movements associated with germ layer separation occur concurrently with neural tube formation, we will also briefly consider amphibian neurula movements at this time.

10. Tubulation of neural ectoderm (Fig. 3-9 H-J') involves the sinking of the center of the neural plate and upward folding of its lateral edges. Gradually a neural tube is created directly by the final fusion of the right and left neural folds with one another.

11. Abscission of epidermal ectoderm (Fig. 3-9 I'-J') will separate the surface epidermis from the neural folds and then fusion of epidermis in a zipper-like action will completely enclose the neural tube inside the surface layer. As in amphioxus, this movement also closes the blastopore with the result that a neurenteric canal temporarily connects the hindgut lumen with the neural canal as is shown in Figure 3-9 I, J.

12. Caudalization (Fig. 3-9 J), as in Amphioxus, entails a post-gastrulation formation of the tailbud. This aspect of germ layer separation is not completed by amphibian embryos until the end of the neurula stage. Indeed, during the open neural plate stage the posterior one-sixth of the neural plate is actually composed of prospective tailbud chordamesoderm, somite mesoderm, and neural tube ectoderm. When this region finally invaginates through the blastopore, it will form the tailbud blastema. Only then will the closure of the blastopore be accomplished as the epidermis closes over the neural tube and blastopore, and temporarily produces a neurenteric canal which will be closed when the tailbud begins to elongate.

The only real differences in the neurulation movements of Amphioxus and amphibians lies in differences in the time of epidermal abscission and covering of the neural plate. In Amphioxus this occurs during gastrulation; in amphibians it occurs at the end of neurulation. Tubulation and caudalization are fundamentally the same in the two groups.

In concluding this account of morphodynamics in amphibian gastrulation movements, two aspects emerge as dominant generalizations. **First,** in spite of yolk accumulation, there is a very strong and conservative tendency for primitive gastrulation mechanisms to be preserved in the development of higher vertebrates. **Second,** certainly one of the major functions of gastrulation, in addition to archenteron formation and separation of germ layers, is the role gastrulation movements play in correcting the spatial errors in axial organization that are created by the basic plan of chordate ooplasmic segregation. Without the relocation of parts within the bilateral framework of the egg (a relocation achieved by such movements as inversion of endoderm, convergence and axial elongation of mesoderm, and ectodermal epiboly), the typical axiate common plan of the vertebrate body would never be achieved. Lastly, we must acknowledge the great debt of gratitude owed to the brilliant pioneer experiments by W. Vogt for the first detailed mapping of prospective germ layer regions in the amphibian blastula, and by J. Holtfreter for the primary analysis of autonomous gastrulation movements underlying all present understanding of processes in axiation and symmetrization in amphibians.

F. INDUCED OVULATION AND FERTILIZATION OF FROG EGGS

The following techniques are modified and abbreviated from procedures given in greater detail in the manuals of experimental methods by Rugh, Hamburger, and Wilt and Wessles (See Appendix I for bibliographic references).

a. Pituitary implantation and progesterone injection to induce ovulation and maturation in frog eggs. The grass frog, *Rana pipiens,* can usually be obtained from commercial vendors from March through June in good condition for induced ovulation and artificial insemination. During late summer, fall, and early winter, the material is less reliable. The standard technique for obtaining frog eggs for laboratory use is by implanting one or more living pituitary glands under the skin and by giving supplementary injections of progesterone as an adjuvant and inductor of maturation in ovulated eggs. Pituitaries from freshly killed animals can be removed in the following way. Insert the blunt point of a standard surgical scissors horizontally in the mouth of a frog. Then with a clean cut remove the top of the head at the level just back of the tympanic membranes. Turn the removed head so that the mouth cavity is up and the tip of the nose is pointing away from the operator. Insert the point of a fine scissors into the foramen magnum at the base of the skull and make two oblique cuts that form a V-shaped incision through the floor of the skull with the apex of the V at the foramen magnum. With a fine forceps, pick up the apex of the bone and lift it forward. This will expose the brain floor. The pituitary gland will either be attached to the brain floor just posterior to the optic chiasma, or the gland will remain attached to the reflected piece of bone. The pituitary is a small, bean-shaped pink body. It can be removed with a fine forceps and unless it is to be immediately implanted (see below), it should be placed in a small amount of physiological saline (i.e., Ringer's solution or 0.7% NaCl; see below) to prevent drying.

Fresh pituitary glands are implanted under the skin of the belly, or into the dorsal lymph spaces beneath the skin of the back. Cut a small hole in the skin with a sharp scissors; the slit should be about two times as long as the width of the pituitary gland. Lift the skin at the slit with a fine forceps and push a pituitary gland into the opening. Then, use the side of a dissecting needle to press the gland deep under the skin away from the opening. Usually one pituitary implant per day for two or three days is sufficient to bring gravid females to readiness for spawning within four or five days after the first implantation. Pituitaries from gravid females have twice the hormonal value of pituitaries from males or off-season females. The efficacy of pituitary implants or pituitary powder (from Connecticut Valley Supply Co., Southhampton, Mass.) can be greatly improved by the injection of 2.0 mg progesterone in 1.0 ml physiological saline (from Worthington, Sigma, or Nutritional biochemical companies). Injections of progesterone should be made one day after the first pituitary implantation.

Female frogs used in the following experiments have been treated in the above manner and their uteri should be filled with ovulated eggs ready for insemination. The first two steps in the following procedure will be carried out for you. **Beginning with step 6,** each pair of students will individually manipulate and observe their own egg sample.

Materials needed per pair: 1 fingerbowl and cover, 1 Stender dish and cover, 3 pieces of 15 cm filter paper, 1 small scissors, 2 pairs of forceps, 2 dissecting needles, 2 pipettes, hand lens or low power dissecting microscope, 500 ml spring water or boiled and cooled tap water, 1 wax-bottom Stender dish (i.e., covered with about one-quarter inch of paraffin), 1 Petri dish with cover, and 100 ml Ringer's solution (6.5 g NaCl, 0.14 g KCl, 0.12 g $CaCl_2 \cdot H_2O$, 0.2 g $NaHCO_3$,0.2 glucose; make up to 1L).

In the event that living material is not available, selected time-lapse motion pictures of amphibian fertilization, cleavage, and gastrulation will be shown.

b. Methods for artificial insemination of frog eggs entail the preparation of sperm suspensions 15 minutes in advance of fertilization by macerating 4 fresh testes in 10 ml spring water. Place sperm suspensions in fingerbowls to a depth not exceeding one-eighth inch.

1. Strip eggs from ovulating female by a firm, slow squeezing pressure applied to the upper abdomen. Eggs in the uteri will be forced out of the anus and should be allowed to fall directly into fingerbowls containing "ripened" sperm.
2. Rock the fertilization dish gently so that the sperm suspension washes over the eggs; the latter will adhere by their jelly coats firmly to the bottom of the dish. If there is a dense cluster of eggs, pipette sperm suspension over them.
3. After 10 minutes of insemination, flood the eggs with one-half inch of spring water and allow them to stand for 20 minutes.
4. Pour off the first spring water and sperm suspension and flood with spring water a second and third time. The eggs will remain in the third wash.
5. After 1 hour free the eggs in their swollen jelly masses from the bottom of the dish. This can be done by pushing a bent dissecting needle under them. **Do not** try to pull the eggs free with the forceps since this will distort and damage the eggs.
6. Check the number of eggs that are successfully fertilized by rolling the jelly mass containing the eggs upside down. Fertilized eggs will rotate so that the dark side is uppermost. Eggs that fail to rotate have not been fertilized.
7. After the jelly has been allowed to swell for 2 hours (after first flooding) cut the eggs apart so that they will have free access to oxygen on all sides.
8. Check the percentage of fertilization again by observing cleavage on each egg. Discard the unfertilized eggs. Cover the eggs and keep them for observation for **one week until they reach the tadpole stage.**

c. Artificial parthenogenesis induced in frog eggs by pricking. Frog eggs can be activated to develop in the complete absence of sperm by pricking them with a fine glass needle. This type of stimulation is called artificial parthenogenesis (i.e., "virgin development"). The embryos rarely advance to the tadpole stage; however, cleavage and blastulae can be obtained quite frequently when the following procedure is observed.

1. Place pipettes, forceps, and dissecting needles in boiling water for half a minute to kill all sperm that may be adhering to them from the preceding experiment. Wash hands thoroughly with hot soap and water.
2. Place a few fresh uterine eggs on a clean microscope slide and place it in a Petri dish lined with a wet piece of filter paper. Cover the eggs to keep them moist.
3. Open the body cavity of a freshly killed female frog and snip off the end of the ventricle. Pipette a small amount of blood over the eggs. Pricking will usually not activate eggs for total development unless they are contaminated with serum or blood.
4. Prick the center of each egg lightly with a fine glass needle. Flood the eggs by dropping the slide into a fingerbowl of spring water. Thereafter treat eggs as described for normal fertilization in steps 6, 7, and 8.

d. Observation of ovulation and coelomic transport of mature eggs in the frog. Mature eggs of the frog can frequently be seen emerging from their follicles onto the surface of the ovary in females at the height of ovulation. These eggs are shed directly into the coelomic cavity and are then transported by ciliary action of the peritoneal lining to the paired ostia of the oviducts, which lie at the anterior extremity of the body cavity near the base of the lungs and liver. The "coelomic eggs" are not fertilizable until progesterone has caused the first polar body to be given off. This does not normally occur until eggs have entered the oviducts and have acquired part of their jelly coats.

On the frog which was opened to get blood for parthenogenesis, quickly review the anatomy of the female reproductive system and identify the ovaries, oviducts, ostia, uteri (generally with eggs), and coelomic eggs free in the body cavity. Remove the ovaries by cutting the mesovaria which attach them to the dorsal body wall. Place the ovaries on clean pieces of moist paper. Remove and discard the digestive tube and wet the body cavity with a small amount of Ringer's solution.

Ovulation can be accelerated by cutting small pieces of ovary, consisting of from 10 to 20 eggs, and placing them in Petri dishes of Ringer's solution. The damage to ovarian tissue frequently results in follicle rupture and stimulates a contraction of smooth muscle in the ovarian walls. This causes the eggs to squeeze slowly out. Observe this activity under a low power dissecting lens. Use bright illumination.

Transport of coelomic eggs can be observed in the body cavity of the eviscerated frog. The movement of the coelomic eggs is slow, but if one egg is singled out and patiently watched, it can be seen to glide or rotate slowly in an anterior direction. If no coelomic eggs are in the body cavity, the ciliary motion can be demonstrated by placing a small fragment of paper or tin-foil ($1/2$ mm sq.), or some carbon dust from a pencil, on the parietal peritoneum of the body wall. Observe the direction and progress of these markers with the naked eye or with a low power dissecting lens.

e. Methods for culture and feeding of tadpoles under the laboratory conditions are exceedingly simple. Tadpoles should be kept in clean aquarium water. Standard tap water can be made biologically safe by letting it stand for 48 hours after drawing it from the tap. This permits all free chlorine to escape. If it is needed more rapidly, simply boil the water, cool it, and it will be safe to use.

Tadpoles can be fed on wilted lettuce or spinach leaves. Alternatively, a very nutritious pablum–agar mixture can be made and stored for use indefinitely. To make pablum–agar, proceed as follows. First make a 2% agar solution by boiling 2 g non-nutrient agar in 100 ml tap water until it is dissolved (i.e., about 5 minutes between simmer and rolling boil). Filter the agar through two layers of cheese cloth to remove foam and undissolved agar particles. When the agar is cool enough to touch but is still very warm, add 10% baby pablum by weight or volume. Mix agar and pablum until they make a smooth paste. Quickly pour or spread the pablum–agar out in a flat dish to a thickness of about one-quarter inch and allow it to cool and harden thoroughly. When solidified, cut the pablum–agar into one inch strips; place them on and cover them with Saran Wrap or other suitable plastic. Store the food in wide-mouth screw-cap collecting jars in the refrigerator. Pieces of food can be removed and placed in aquaria as needed for feeding tadpoles.

For additional methods involved in handling amphibian eggs, see the special sections at the ends of Chapters 4 and 5.

CHAPTER IV

DEVELOPMENTAL CONCEPTS OF EPIGENETICS AND INDUCTIVE DETERMINATION

Illustrated by Neurulation and Early Organogenesis in the Frog

TABLE OF CONTENTS

TABLES AND ILLUSTRATIONS

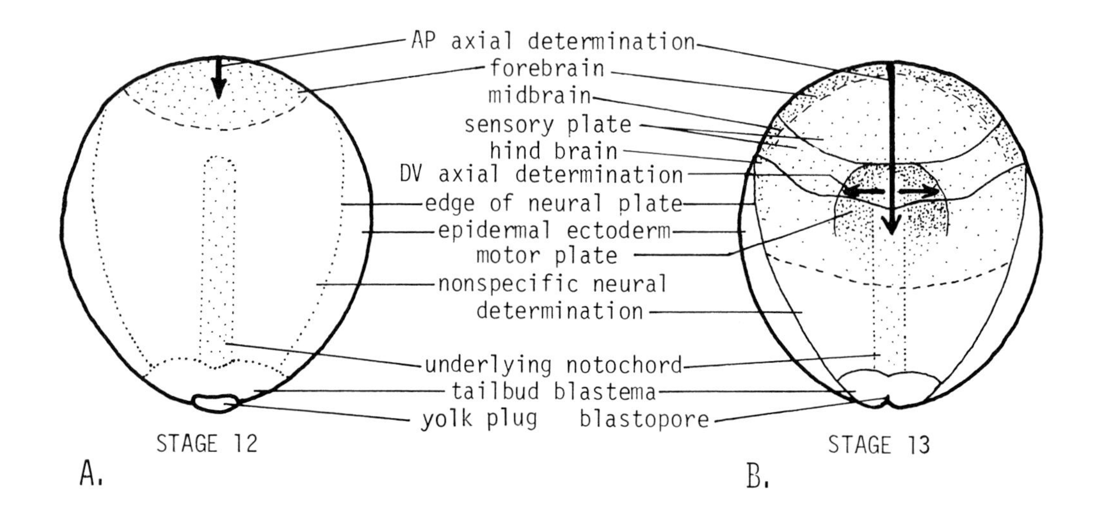

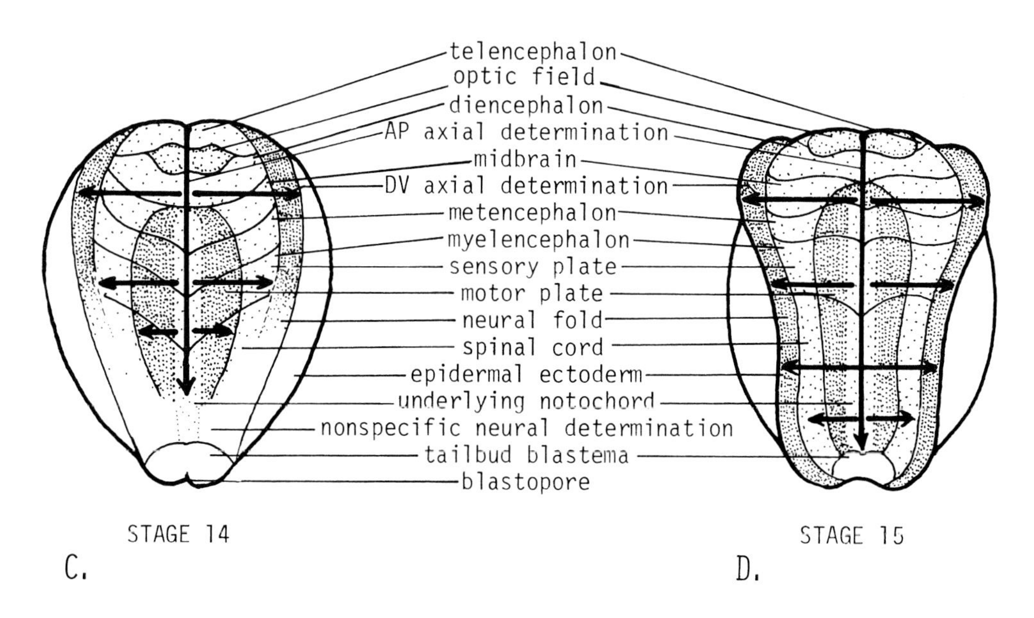

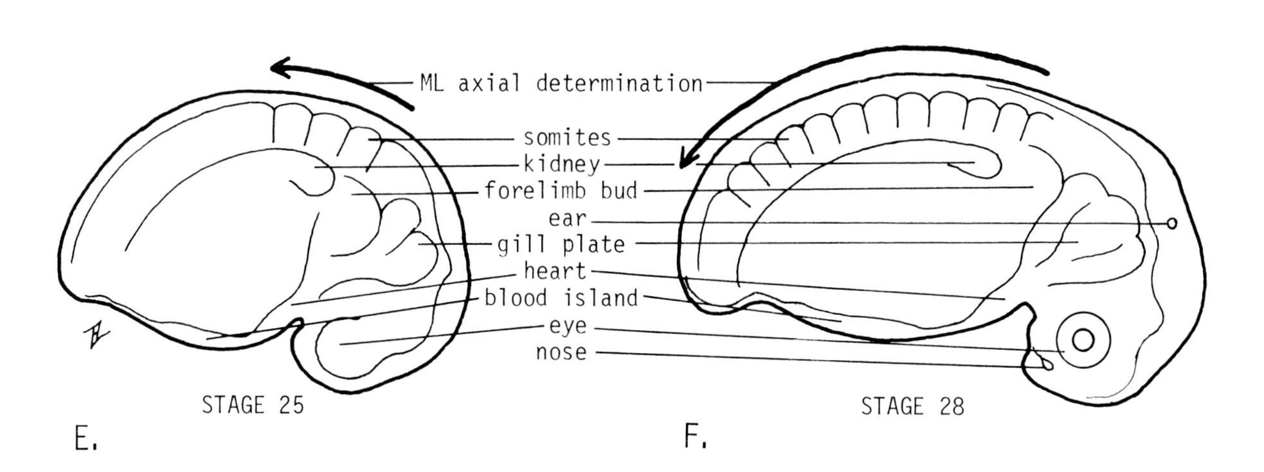

Figure 4-1. Diagrams of progressive determination of A–P, D–V, and M–L axes in the central nervous system of amphibians based primarily on ambystomid salamanders. See explanation in text.

CHAPTER IV

DEVELOPMENTAL CONCEPTS OF EPIGENETICS AND INDUCTIVE DETERMINATION

Illustrated by Neurulation and Early Organogenesis in the Frog

In Chapter 3 early development in amphibians, with special emphasis on salamanders, was given from fertilization through neurulation. In concluding the discussion of morphodynamic movements during gastrulation and neurulation, it was pointed out that one major function of these movements is in the role they play in "correcting the spatial errors . . . created by the basic plan of chordate ooplasmic segregation." These so-called errors are primarily associated with the fact that before gastrulation the long axes of chordamesoderm, neural ectoderm, and somite mesoderm are lateral instead of anteroposterior. In the present chapter attention will be given to another aspect of importance in gastrulation. It involves the altered spatial association of parts resulting from the repositioning of prospective germ layer regions during the process of axial symmetrization. This relates to the problem of embryonic determination by induction.

A. EMBRYONIC DETERMINATION BY INDUCTION

Embryonic determination can be described most simply as all events that lead ultimately to the assignment of a specific cell fate in structure and function to the various cells of a multicellular body. Since all cells of an individual contain the same basic genetic information in their nuclei, determination is all mechanisms that select a specific and limited part of the genome for expression in individual cells. The remainder of the genetic competence of the cell as a rule will never be expressed. Stated another way, although all genetic information needed to form all kinds of tissues of the species is present in all cells, an individual cell will express only one of the one hundred or more cell types that make up the metazoan body. Determination functions almost as an all-or-none commitment by a cell to form a single tissue type.

Determination results from external stimuli upon cell nuclei; this is the essence of the meaning of the term epigenetics. **Epigenetics** entails all causal factors operating from outside a cell which tend to regulate the phenotypic expression of its genes. Embryologists generally refer to phenotypic appearances of cells as cell differentiation. **Cell differentiation** is, therefore, the production of special products of gene activity including special proteins, pigments, secretions, cellular organelles, and whatever else is characteristic for special cell structure or function.

We have briefly defined determination and epigenetics, two key words in the title of this chapter; "inductive" remains as the final term to consider. **Embryonic induction,** without considering all its special variations, is most simply described as a process whereby one type of embryonic cell (the inductor) causes another cell type (the reactor) to embark on a path of cell differentiation that it previously was incapable of carrying out. **Inductor cells** are usually determined as to cell fate. They may induce reactor tissues to form the same cell type as the inductor, a case called **homogenic induction,** or the inductor may induce the reactor to form an entirely different tissue type by **heterogenic induction. Reactor cells** are always, to a degree, undetermined and incapable of self-differentiation. They are also said to be **regulatory;** that is, at a given period in development they are sensitive and responsive to stimuli from inductors. They may, and often do, permanently lose their capacity to respond to inductors after critical periods in development. Thus, inductive systems are highly stage-specific in development, and time itself plays a critical role in limiting the ability of cells to respond to inductors. The term, **competence,** is used to refer to all capacities a cell, tissue, or organ primordium has for specialization at a given time in development. In early stages competence is usually relatively great and there are still many options for specialization open to the cell and its descendants. During development with progressive determination taking place, competence becomes more and more restricted until in the end, usually only one capacity for specialization remains. At that point the cell, tissue, or primordium is fully determined but need not yet be fully differentiated.

Although determination is the immediate cause of differentiation in embryonic development, determination cannot itself be recognized without some evidence of differentiation taking place. Often differentiation is long delayed after the determining influences of epigenetic induction have taken place. Moreover, determination is often progressive, occurring in steps rather than as one cataclysmic event. This is because many inductive stimuli do not actually select a specific kind of determination. Instead, they often act by progressive elimination of alternative options until only one phenotypic expression is left for a cell to express. Cells that are some mid-point in their scale of determination are said to be in states of **labile determination.** That is to say, they do not have all of the options open that are available to the zygote before cleavage, nor have they been restricted to one and only one specific fate; more important, they still require external stimuli before they can self-differentiate in a particular way.

These general concepts can best be illustrated by a concrete example. A classic case of induction involves the well documented ability of prospective gastrula general ectoderm to be induced to form neural plate by the inductive action of the chordamesoderm during the period of gastrulation. The time period of ectodermal responsiveness to neural induction is highly critical. Blastula ectoderm prior to the crescent blastopore stage is too young and mid-neurula ectoderm is too old to respond to the chordamesoderm stimulus. After the mid-neurula stage, however, the chordamesoderm is no longer necessary for neural differentiation by the neural plate. Figure 4-1 summarizes the time sequence in the acquisition of determination by the neural plate in salamanders. Essentially the same conditions are believed to apply to the frog.

Stage 12 (Fig. 4-1 A) is the slit blastopore or late gastrula stage. By this time axial elongation of chorda- and somite mesoderms has placed them in the middorsal line. It is during their movement into this position that primary induction of the neural plate occurs. The chordamesoderm is the primary inductor, and the ectoderm is the labile neural reactor, which, until induction takes place, is incapable of any independent neural differentiation. At Stage 12 the entire prospective neural plate is non-specifically activated to form neural tube, but the exact kind of neural tissue is as yet not specified as forebrain, midbrain, hindbrain, or spinal cord. Thus, neural induction is at a labile stage of determination.

Stage 13 (Fig. 4-1 B) entails the beginning of anteroposterior axial determination of parts of the neural tube. AP axial determination begins at the anterior end with the forebrain first (prosencephalon) which soon becomes differentiated into two parts, namely, the telencephalon and diencephalon. Very soon the midbrain and hindbrain also are very generally determined. It is at Stage 13 that the outlines of the neural plate first become visible in viewing the living embryo.

Stages 14 and 15 (Fig. 4-1 C, D) are generally called the mid-neurula stages. During this period the neural folds are prominent ridges that mark off precisely the lateral edges of the neural plate. It is during this period that the various parts of the spinal cord are determined in the AP axis. It is also during this time that the motor (or basal) plate and the sensory (or alar) plate become regionally and functionally determined. Although the motor and sensory regions are in the ML axis of the open neural plate, they represent the dorsoventral axis of the neural tube after neurulation is complete. Determination of the definitive DV axis (sensory–motor axis) of the neural tube begins at the anterior end and moves progressively toward the posterior end. By the closing neural tube stage the neural tube is already a mosaic of determined and self-differentiating parts in both structure and function. By this it is meant that if one isolates any one of these regions in tissue culture, or transplants it to an atypical location in the body, it will differentiate exactly as though it were still a part of the normal part of the brain from which it originally came. It is no longer labile.

Tailbud stages 25 to 28 (Fig. 4-1 E, F) cover a period in early organ formation which, for the early neural tube, is also the time during which the definitive mediolateral organization of the neural tube is first determined. Mediolateral organization will entail the differentiation of three primary regions in the wall of the developing neural tube. The innermost region is the **ependymal layer** or **germinal epithelium.** This layer is responsible for the mitotic replication of neurones along the length of the neural tube. The middle element is the **mantle layer,** or **gray matter;** these are the neurones or nerve cell bodies containing cell nuclei. The outermost component is the **marginal layer,** or **white matter;** it consists primarily of axons and dendrites which grow out from neurones in the mantle layer. Determination of these elements in the ML axis occurs concurrently with the appearance of somites lateral to the neural tube. Thus ML determination will begin in the anterior trunk region in the vicinity of the hindbrain and will progress backwards to the tip of the tail as somites are added. Anterior to the first somite, ML determination in the midbrain and forebrain is not so simple. In the absence of somites beside the neural tube, the cerebellar region of the hindbrain and the entire midbrain, diencephalon, and telencephalon tend to have a reversed ML axial pattern. The mantle layer (gray matter) tends to take a position at the surface of the neural tube, that is to say a lateral position, whereas the marginal layer (made up primarily of nerve fibers) tends to crowd to the interior between the ependymal and mantle layers.

From the foregoing brief account of neural determination, we can see that it is a progressive story of continuing determination and specification of function. We can point out that a major function of gastrulation movements is to bring embryonic inductors in contact with appropriate embryonic reactors. Without the mesodermal morphodynamic movements of involution, convergence, and axial elongation occurring with a high order of spatial precision at an exactly correlated time, neural induction could never take place. To dramatize its importance, we need merely ask, "What kind of vertebrate would you have if there were no central nervous system?"

In concluding this general discussion of epigenetics, determination, and differentiation, it can be pointed out that one consequence of induction is that the reactor at one step in induction often acquires the capacities of being an inductor for a following stage. Inductive series built upon this principle can result in the precise positioning of parts in organs which are derived from different embryonic germ layers. Classic cases of induction series can be found in the development of the vertebrate sensory organs, as follows:

Representative Induction Series for Major Sense Organs

Induction series for the nose

1. Head mesenchyme induces neurula stage epidermal ectoderm to form nasal pit; then
2. Nasal pit ectoderm induces tailbud stage neural crest to form nasal capsule.

Induction series for the eye

1. Head mesenchyme or anterior chordamesoderm induces the optic field in the forebrain of the neural plate at the early neurula stage; then the
2. Optic field forms optic vesicles which induce lens placodes at early tailbud stages from the epidermis of the head; then the
3. Lens placodes form lens vesicles which induce head epidermis at the late tailbud stage to form cornea.

Induction series for the ear

1. Anterior chordamesoderm induces gastrula ectoderm to form hindbrain; then
2. Hindbrain–medulla induces neurula epidermal ectoderm to form otic vesicles (inner ear);
3. Otic vesicle induces head mesoderm to form an otic capsule (bony labyrinth).
4. Otic capsule induces the first pharyngeal pouch endoderm to form a middle ear cavity and induces the production of the eardrum from middle ear endoderm and head epidermis.
5. Later otic capsule (tympannic ring cartilage) induces the formation of the eardrum from late tailbud stage ectoderm and middle ear endoderm.

It is to be understood that the preceding synopsis of induction series in amphibian sensory organs is greatly abbreviated and does not necessarily apply with equal force to all vertebrates. Other induction series are also known for the circulatory, excretory, digestive, skeletal, muscular, and reproductive systems. These induction series will be mentioned at appropriate points in the discussion of embryonic development in this and remaining chapters. It is sufficiently relevant to the present discussion to indicate that one of the major embryonic devices for assuring that the various parts needed in the construction of complex organs is **serial induction.** By this means one part assures the addition of the next part by providing the inductive information for the step to follow.

B. EARLY ORGANOGENESIS IN THE FROG WITH EMPHASIS ON SENSE ORGAN FORMATION IN THE 4 mm TAILBUD EMBRYO, STAGE 17–18.

The present laboratory study will emphasize the multiple germ layer compositions of sense organs as an expression of epigenetic control in morphogenesis. Two embryonic stages have been selected for special study. These are the 4-5 mm, Stage 17–18 (2½-3 day tailbud stage) and the 6-7 mm, Stage 19–20 (3½-4 day) hatching tadpole. To show their relative positions in the total plan of normal embryonic development, a developmental series is given in Figure 4-2 along with a developmental time schedule in Table IV-1.

When we compare the ages of neurula and neural tube stage embryos examined at the end of Chapter 3, with the Stage 17–18 4-5 mm tailbud embryo (Fig. 4-3 A–G), we will find that no more than one day of development separates them. In these few hours some remarkable advances in organ building have taken place. Among the most readily recognized advances is the differentiation of axial myotomes to a level where muscle twitches can bend the body if the living embryo is touched gently. Another conspicuous progressive step is in the differentiation of the central nervous system into recognizable regions that will form the forebrain, midbrain, hindbrain, and spinal cord. The heart primordium and pericardial cavity have formed although the heart itself has not begun to beat. Of special interest for the present study is the evidence of induction in the development of sense organ primordia.

The materials for study will include models and preserved whole embryos which will be most useful in providing a three dimensional overview of the stage. In addition, each student will have two slides to examine. The first is a section in the **median sagittal plane;** it will show to advantage the relative position of median structures in the anteroposterior axis. Particularly favorable is the information sagittal sections give on the brain regions, heart, and gut relationships. The second slide will contain from four to six representative slices through the embryo in the **transverse plane;** these cross sections permit us to visualize structures that lie in the mediolateral and dorsoventral planes. Since the series in both sagittal and transverse planes are often incomplete there may be sizeable gaps and it will be an advantage to exchange slides with laboratory partners so as to fill in information that may be missing.

TABLE IV-1

TIME SCHEDULE FOR *Rana pipiens* EGGS. DEVELOPING AT 24° C

Shumway Stage	Description of Developmental Age	Average Age Hours
1	Fertilization, rotation of egg in chorion	0
2	Gray crescent formed and bilaterality established	1
3	First cleavage, 2 cell stage	2-2½
4	Second cleavage, 4 cell stage	3-4
5	Third cleavage, 8 cell stage	4-5
6	Fourth cleavage, 16 cell stage	5-6
7	Morula, 32 to 250 cell stages	6-10
8	Young blastula stage	10-12
9	Mid- to late blastula stages	12-14
10	Early gastrula, crescent blastopore stage	16-18
11	Mid-gastrula, horseshoe blastopore stage	18-22
12	Late gastrula, yolk plug blastopore stage	24-30
13	Post-gastrula, slit blastopore — pre-neurula stage	32-36
13+	First visible outline of neural plate	36-40
14	Open neural plate stage with elevated neural folds	40-45
15	Closing neural folds	45-50
16	Neural tube, pre-tailbud stage	50-60
17–18	Tailbud stages, first muscle twitches, 4 to 5 mm. stage	60-75
19–20	Late tailbud, first heart beat, external gills, hatching	85-100
21–22	Swimming stage, stomodaeum opens	4-5 days
23–24	Operculum forming and covering right gill	7-8 days
25	Operculum complete, feeding begins	9-10 days

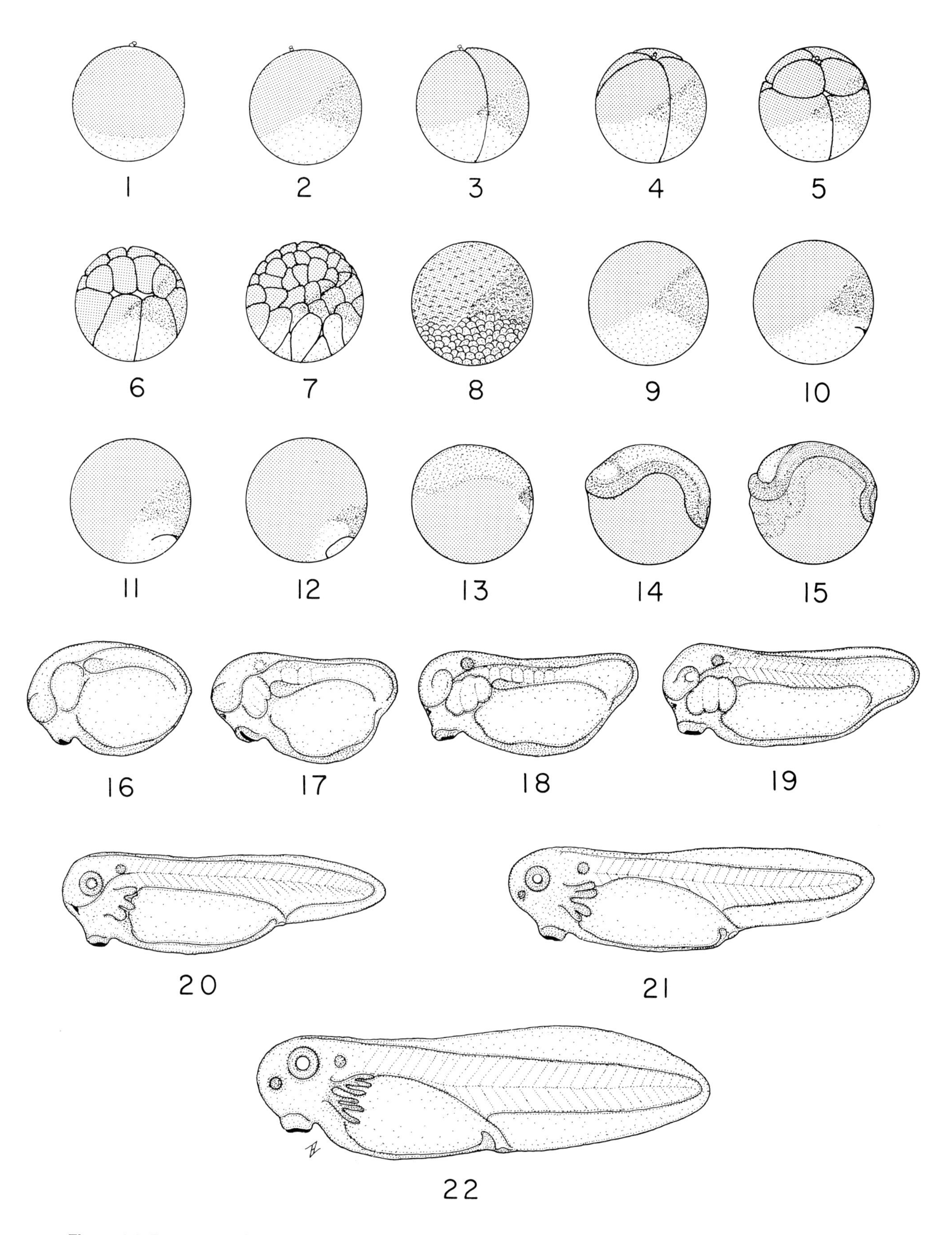

Figure 4-2. Representative stages in normal frog development. Numbers of figures correspond with Schumway stages given in Table IV-1. Drawings are from living stages of *Rana pipiens*.

1. Representative median sagittal sections (Fig. 4-3 A.). With the aid of models and illustrations provided, identify the following general derivatives of the three primary germ layers. **Epidermal ectoderm** covers the entire body surface and over much of this area the layer is two cells thick; the outer layer or **superficial epidermis** will form the embryonic epidermis of the skin. Most special epidermal derivatives come from the **deep epidermal layer.** The most conspicuous of the latter in median section is the primordium of the **hypophysis** or anterior pituitary primordium, which projects as a slender tongue of epidermal cells under the floor of the forebrain. Just ventral to the hypophysis the epidermis forms a very thick, solid cord of cells which fuses with the anterior wall of the foregut. This is the **stomodaeal plate.** At the posterior end of the body, the epidermis is invaginated a short distance as the **proctodaeum** which unites with, and opens into, the **cloaca,** or **hindgut.** The neural tube is clearly differentiated into a very large **forebrain.** Its roof contains a very small evagination, the **epiphysis,** the primordium of the pineal gland. The posterior ventral expansion of the forebrain is the **infundibulum** or primordium for the posterior pituitary gland. The landmarks that distinguish the **mid-brain** at this stage are the posterior margin of the epiphyseal evagination from the roof of the forebrain and the exceedingly thin roof of the hindbrain. Accordingly the midbrain will be identified as that part of the brain which has the thickest roof; this local thickening is the primordium of the **optic tectum** or future visual center of the brain. The diagnostic character for the **hindbrain** in median section is its very thin roof, the **posterior choroid plexus.** There is no distinctive landmark to separate the posterior hindbrain from the spinal cord. As one proceeds back into the trunk region, it is common to find that the plane of section will not coincide precisely with the median axis of the embryo. Oblique sections are common in the trunk region. An ideal median sagittal section is shown in Figure 4-3 A. Note that the spinal cord terminates in the **tailbud blastema** along with the posterior notochordal material which lies immediately under the neural tube. Note that the **notochord** extends anteriorly only to the level of midbrain and ends just posterior to the infundibular evagination of the forebrain. The three divisions of the gut can be easily recognized in sagittal section. The **foregut,** or **pharynx,** is enormously expanded. Its anterior wall fuses with stomodaeal ectoderm and forms the endodermal component of the oral membrane. At approximately the boundary between the anterior and middle thirds of the body, note that the expanded foregut lumen terminates in narrow dorsal and ventral cavities. The ventral one is the **liver diverticulum,** which ends blindly in the midventral part of the endodermal yolk mass. The cells that make up the ventral wall of the liver diverticulum are the primordium of the liver itself. The lumen of the diverticulum is the primordium of the common bile duct and gall bladder. The dorsal posterior extension of the foregut lumen is the **midgut.** It extends back to the posterior end of the body and terminates in the **hindgut** which, at this stage, is no more than a slight ventral deflection in the course of the gut lumen. At this stage, a dorsal extension of the hindgut may still be present as a remnant of the **post-anal gut** and **neurenteric canal.** The connection of the latter to the neural canal of the spinal cord was broken when the **tailbud blastema** started a vigorous phase in posterior elongation of axial structures in the trunk and tail. From the tail blastema are organized neural tube, notochord, and tailgut endoderm. Although not visible in median sagittal sections, somite mesoderm is also produced from the lateral tailbud blastema. In median section, the major mesodermal structure of interest is the notochord which lies between the floor of the neural tube and the roof of the gut. In the ventral midline are the ventral mesentery and its derivatives, the blood island, heart, and pericardial coelom. Unspecialized **ventral mesentery** hypomere mesoderm will be seen between the epidermis and floor of the posterior midgut. Note that this strand of thin mesoderm becomes very thick at the region of the midgut, where it forms the **blood island.** The single blood island is the primordium of all future blood cells and also contributes extensively to the formation of capillary walls throughout the body. At the level of the liver diverticulum, endodermal liver cells are migrating out into the midventral mesoderm. Anterior to the liver diverticulum, and under the floor of the gut, will be found the **heart primordium** and **pericardial cavity.** The heart is at a very early stage in development, and its regions cannot yet be identified by any distinguishing landmarks. The heart tube is forming as a dorsal invagination of the coelomic lining. This aspect of morphogenesis will be best seen in transverse sections.

After becoming thoroughly familiar with the overall structure of the embryo through the study of median sagittal sections and models, proceed to the study of representative transverse sections.

Figure 4-3 A is provided with numbered dorsal and ventral guide lines to indicate the levels at which each of the accompanying transverse section diagrams were made. By placing a straight edge across the diagram and connecting similarly numbered guide lines you can determine the plane of section at which Figures 4-3 B–G were cut. Emphasis will be given to sense organ development, but other general aspects of embryogenesis will be pointed out in passing.

2. Transverse sections through the eye cup, nasal placodes, epiphysis, and hypophysis (Fig. 4-3 B.). The first cross section of the representative series will pass through forebrain levels of the embryo. An ideal section would include the various structures shown in Figure 4-3 B. The **forebrain** should show three conspicuous evaginations, a middorsal **epiphysis** and paired lateral **optic vesicles.** Note that the latter press tightly against the lateral epidermis. It is at this period that secondary induction of the lens by the optic vesicle is taking place. In the ventral part of the section, identify a single median invagination and two lateral invaginations of epidermis. The median invagination, from the deep layer of the epidermis, is the **stomodaeal plate** which, at a later time, will split and hollow out to form the mouth cavity. At the innermost end of the stomodaeal plate can be seen a small knob of cells; it forms the hypophysis and is derived from the dorsal-most part of the stomodaeal plate. The inductive relationships begin with the foregut which induces the epidermis to form the stomodaeal plate from its deep layer. Secondarily the floor of the forebrain induces the stomodaeal plate to form the hypophysis. Just lateral to the stomodaeal plate you will find two thick crescents of epidermis; these are the **nasal placodes.** They include both deep and surface layers of epidermis in their formation. The nasal placodes are induced by the dense aggregates of head mesenchyme that immediately underlie them in the space between the brain floor and surface epidermis.

3. Transverse sections through the hindbrain, pharynx, heart, cement gland, and inner ear primordia (Fig. 4-3 C.). The second of the representative sections will usually pass through the posterior midbrain or hindbrain. The distinction between the two will rest primarily on the relative thickness of the brain roof. The **midbrain** roof is relatively thick, and the cross section of the **neural canal** in this region will show a brain wall of nearly uniform thickness. The roof is the **optic tectum.** At **hindbrain** levels the roof of the neural canal is paper-thin; this is the **posterior choroid plexus.** The latter level is shown in Figure 4-3 C. Ventral to the neural tube, note the **notochord** which in cross section will appear as a solid circle. Identify the paired **otic placodes** on both sides of the hindbrain. They are derived from the deep layer of epidermal cells, and their central cavity is closed to the exterior by the superficial epidermal layer that covers them. In the mesenchyme spaces between the dorsal wall of the gut and the neural tube, attempt to identify the early capillaries that are beginning to form. The appropriate locations for the anterior cardinal veins and internal carotid arteries are shown in Figure 4-3 C. The center of the section is dominated by the large cavity of the **foregut.** The dense masses of mesenchyme that fill the space between the lateral wall of the foregut endoderm and surface epidermis is **visceral arch mesenchyme** from which aortic arches, visceral cartilages, and muscles of the gills will later form. These cells are of mixed head mesoderm and neural crest ectoderm origin. Ventral to the foregut will be seen cross sections through the heart tube and **pericardial cavity.** These are just beginning to form from ventral mesentery mesenchyme. Usually two very thin-walled **endorcardial tubes** can be seen. The dorsal peritoneal lining of the pericardial cavity is reflected downwards to make room for them; this layer of peritoneal cells is the **myocardium.** Endocardial tubes are induced from lateral plate mesoderm by foregut endoderm. The endocardial tubes are the primordium of the heart lining. They also serve as the inductors for the transformation of pericardial peritoneum into myocardium, the primordium of heart muscle. The last structures to be observed are the **cement glands,** a pair of which will be seen in cross section on the ventrolateral surface of sections. They are derived from the surface layer of epidermis. They show considerable ability to self-differentiate from neurula stage epidermis and their determination must, therefore, be very early. It has been suggested that cement glands are induced directly by foregut endoderm or head mesenchyme. More recent evidence suggests a possible role of the anterior neural plate or neural crest cells in cement gland induction.

4. Transverse sections through the midgut, liver diverticulum, blood island, somites, and spinal cord (Fig. 4-3 D–E). One or more sections in the representative series will be through the

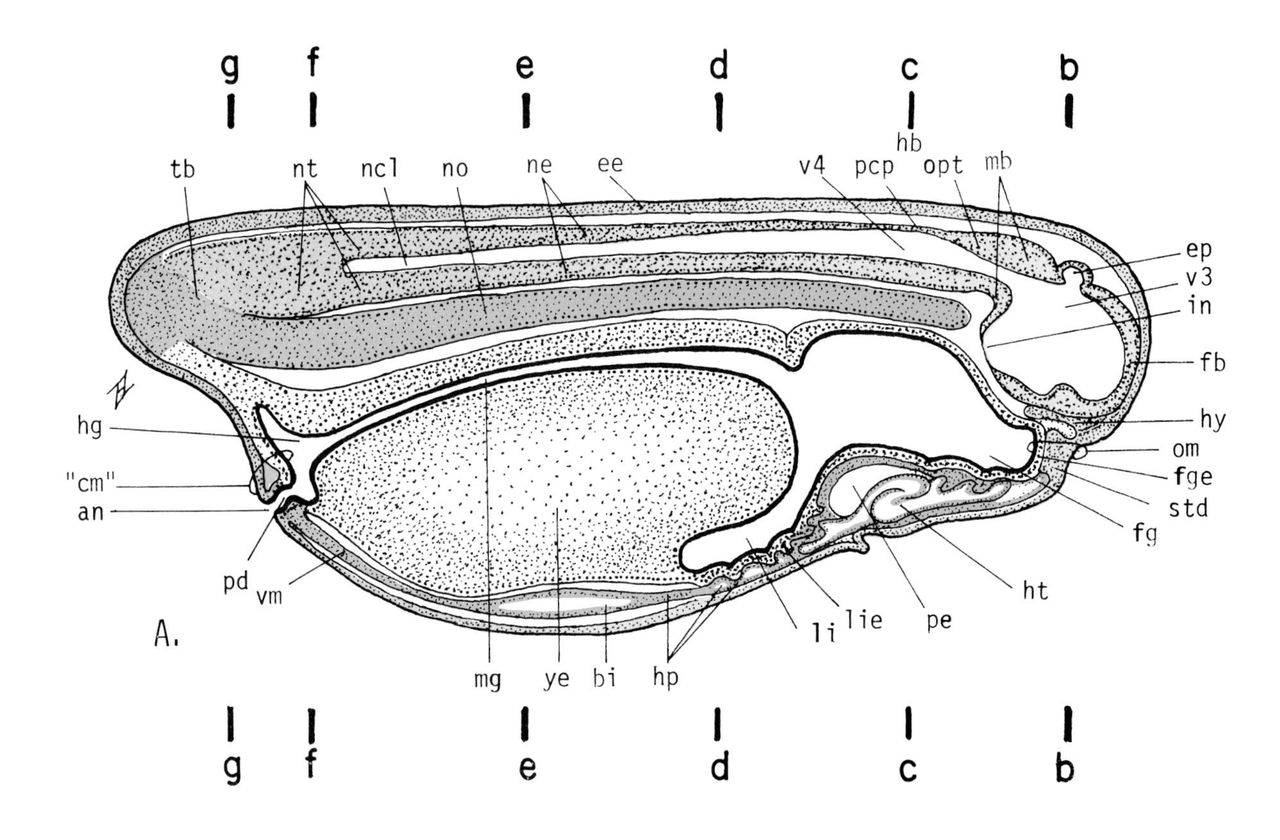

Figure 4-3 A. Median sagittal section with numbered guide lines to indicate the plane of transverse sections 4-3 B–G.

Figure 4-3. Structure of the 4-5 mm frog tailbud embryo. Stage 17-18.

Legend and checklist of structures to be identified

Color code: blue — ectoderm, red — mesoderm, yellow — endoderm.

ac	anterior cardinal vein
an	anus
bi	blood island
cg	cement gland
cl	coelom
"cm"	"cloacal membrane"
da	dorsal aorta
ee	epidermal ectoderm
eed	deep (sensory) epidermal layer
ees	superficial epidermal layer
end	endocardial tubes
ep	epiphysis, pineal primordium
fb	forebrain
fg	foregut, pharynx lumen
fge	foregut endoderm
hb	hindbrain
hem	hematoblasts of blood
hg	hindgut, cloaca
hm	head mesenchyme
hp	hepatic portal vein
ht	heart primordium
hy	hypophyseal ectoderm
ic	internal carotid artery
in	infundibulum
li	liver diverticulum
lie	liver endoderm
lm	lateral plate mesoderm
mb	midbrain
mc	myocardium
mg	midgut
mm	mesomere mesoderm, nephrotome
myc	myocoele
nc	neural crest
ncl	neural canal, neurocoele
ne	neural ectoderm
no	notochord
np	nasal placode
nt	neural tube, spinal cord
om	oral membrane
op	optic vesicle
opt	optic tectum, midbrain roof
ot	otic placode
pe	pericardial cavity
pcp	posterior choroid plexus
pd	proctodaeum
sm	somite mesoderm, epimere
std	stomodaeal plate
tb	tailbud blastema
v3	third ventricle of diencephalon
v4	fourth ventricle of hindbrain
vam	visceral arch mesenchyme
vm	ventral mesentery
ye	yolky endoderm

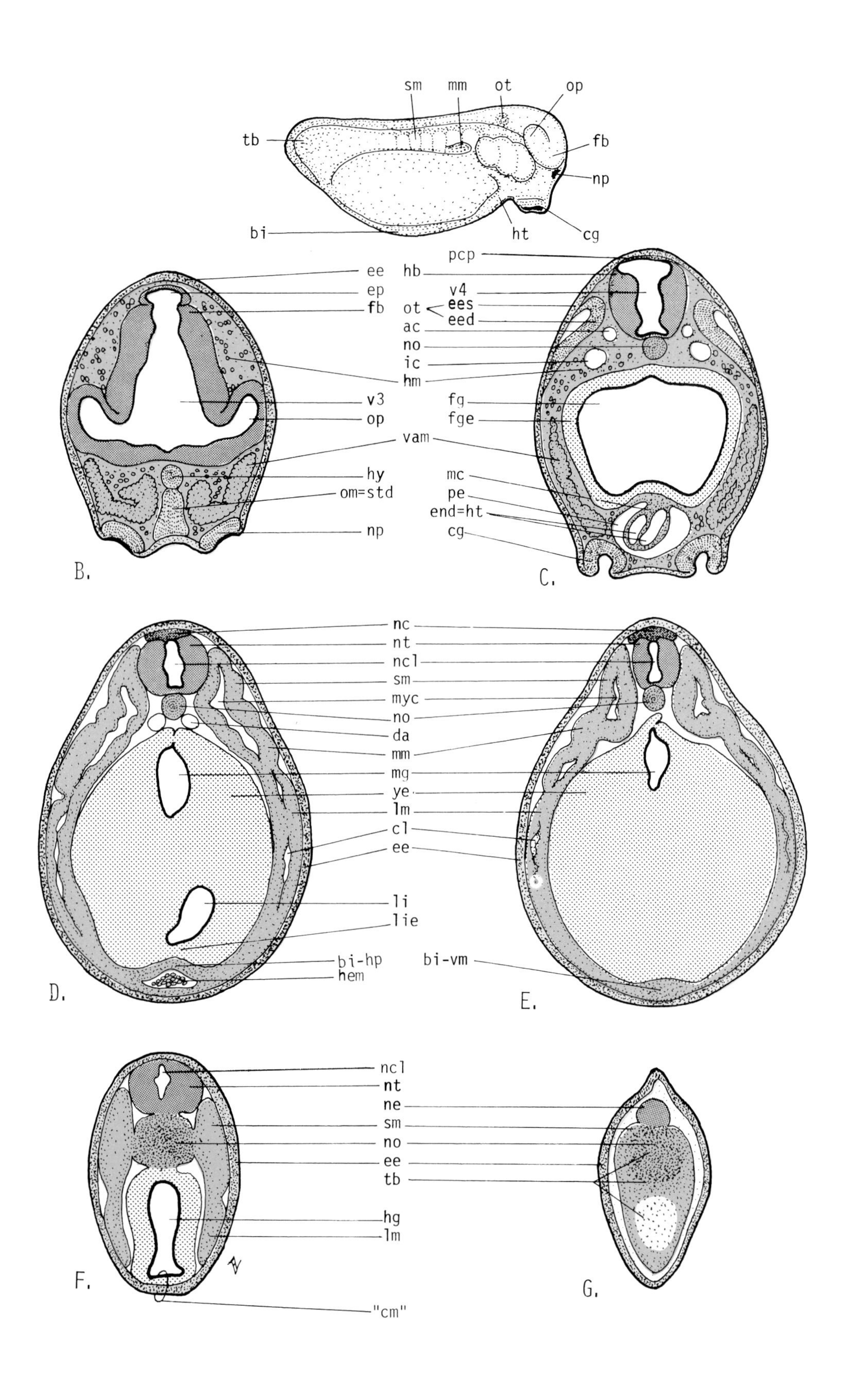

Figure 4-3 B–G. Representative transverse sections at levels indicated by guide lines in 4-3 A.

very extensive midgut region. They will show typical cross sections through the neural tube at the level of the spinal cord. At most of these levels, **neural crest** ectoderm (derived from the neural folds of the open neural plate) will be clearly distinguished as a cap of cells between the neural tube and the mid-dorsal epidermis. Mesodermal **somites** will usually be present on one or on both sides of the neural tube. However, since somites are periodic structures, some sections may entirely lack these structures by passing at levels that are intersegmental. Median sections through somites will show a **myocoele** cavity in the process of forming by a splitting of the mesodermal cord. As a rule, the **coelom** will still be very imperfectly formed in the lateral plate regions of mesoderm. However, by this time, the lateral plate has spread entirely around the embryo, and the primordium of the ventral mesentery will be visible in the ventral midline as a distinctly thicker region of mesoderm between the floor of the endodermal yolk mass and the surface layer of epidermis. At midtrunk levels, the ventral mesoderm becomes particularly thick at the region of the **blood island** (Fig. 4-3 D). An extensive vascular lumen may have formed, as the primordium of the **hepatic portal vein.** It will be filled with many spherical **hematoblasts,** the primordia of blood cells and capillary walls of the circulatory system. The majority of the middle region in these sections will be filled by **endodermal yolk cells.** The **midgut lumen** is strongly displaced toward the dorsal side of the yolk mass. Anterior sections through the midgut region will often include a section through the **liver diverticulum,** which will appear as an opening in the midventral yolk mass as is shown in Figure 4-3 D.

5. Transverse sections through the hindgut and tailbud (Fig. 4-3 F—G). A section from the posterior quarter of the embryo is almost always included in the representative sections although it is of relatively little intrinsic interest. Sections through the hindgut (Fig. 4-3 F) will often reveal a **cloacal,** or **hindgut,** lumen that approaches the epidermis but does not open to the exterior. In general this contact point is called a **cloacal membrane;** however, this is probably incorrect since the membrane by the hatching stage has ruptured and there is already an anal opening. Refer to the median sagittal section and note that a transverse section is not favorable for showing the relations of the anus, proctodaeum, and hindgut. More posterior sections through the tailbud region reveal an increasing fusion of neural tube, notochord, somite, and hindgut endoderm cells into the amorphous mass of **tailbud blastema** cells.

C. EARLY ORGANOGENESIS OF THE VERTEBRATE COMMON PLAN ILLUSTRATED BY THE 6 TO 7 mm FROG HATCHING LARVA, STAGE 19-20

Before beginning the special study of the 6 to 7 mm frog tadpole, again refer to the general discussion of the vertebrate common plan and the basic organization of the nervous, sensory, digestive, excretory, and circulatory systems that have been described in primitive chordate larval forms in Chapter 1. The following study will presuppose that the common plan of the early vertebrate body is clearly in mind. The overall plan will not be reviewed here in detail. Emphasis will be given to special features of frog development that are advances over the condition described for the 4 mm stage outlined in Section B.

The 6-7 mm tadpole corresponds with the Stage 19-20 hatchling in the time schedule for normal development that appears in Table IV-1. It is at the stage when the heart is beginning to beat and when muscles are sufficiently developed to permit a few flips of the tail as weak swimming movements. When we compare the level of organ development in the 4 mm Stage 17-18 embryo described in Section B with the 6-7 mm Stage 19-20 hatching larva, we cannot fail to be impressed by the progress in differentiation that has taken place between the third and fourth day of development.

The materials for study will include models and preserved whole embryos of material to aid in the visualization of three-dimensional relationships of organ parts within the early tadpole. In addition, each student will have two slides, one of a representative median sagittal section, and another of representative cross sections. Some demonstration slides will also be available for individuals who wish to review entire series that have been sectioned in each of the primary planes of symmetry — namely, sagittal, frontal, and transverse serial sections.

The illustrations provided in Figure 4-4 A–U provide representative sections in each major plane of body symmetry. Figure 4-4 A is a diagram of a whole mount embryo showing the major organ systems as they might appear if the embryo were transparent. Figure 4-4 B is an ideal median sagittal section corresponding more or less closely with your own slide which may not be exactly median at all body levels. Figure 4-4 C–F are frontal sections made at the levels indicated by similarly numbered guide marks at the anterior and posterior ends of Figure 4-4 A, B. The remainder of the series shown in Figure 4-4 G-U are transverse sections at levels corresponding to the appropriate numbers at the dorsal and ventral sides of Figure 4-4 A, B. The transverse series are made at all significant levels to show the major events in the development of the tadpole at this stage.

The representative series available for individual study will show only 4 to 6 representative sections, and these will vary on different slides. Match your sections with appropriate cross section diagrams and exchange slides with laboratory partners in order to obtain a broader understanding of structural development.

The following description will deal with organ system levels of development rather than with individual sections. You will need to go through the representative median sagittal and cross sections several times in the completion of the following study.

1. Ectodermal derivatives

a. Epidermal ectoderm surrounds the entire exterior of all sections. Over much of the body it is separated into a **superficial epidermal layer** and a **deep epidermal layer.** The latter is often called the sensory epidermal layer since it is associated, in most instances, with the formation of special sensory epidermal structures. Of these, the **nasal placodes** (Fig. 4-4 G), **lens placode** (Fig. 4-4 I), **lateral line placodes** (Fig. 4-4 J–S), **hypophysis** (Fig. 4-4 A, E, H–J), and **stomodaeal plate** (Fig. 4-4 H–J) show their relationship to the deep epidermal layer particularly clearly. A similar relationship was pointed out for the otic placode at the 4 mm stage; however, in the present series the ear primordium has moved away from the surface and has formed the **otic vesicles** (Fig. 4-4 L), pear-shaped hollow structures located ventrolateral to the myelencephalon. Two other well formed epidermal derivatives involve specialization of the superficial layer. These are the paired **cement glands** (often incorrectly called suckers) which form conspicuous cup-shaped invaginations from the ventral epidermis of the head (Fig. 4-4 J–L). Lastly, the **proctodaeal invagination,** which contributes the terminal segment of the gut, involves the surface epidermal layer which invaginates and fuses with the hindgut endoderm and forms the **anus** (Fig. 4-4 B, F, T). If tail sections are available for study, note that the tail epidermis is elevated into **dorsal** and **ventral fins** (Fig. 4-4 U). These are of obvious use to a swimming state, and it is of developmental interest that the fins are produced under the inductive influence of trunk neural crest cells acting upon late tailbud epidermis.

b. Neural ectoderm has differentiated into two primary components, namely, those that form various components of the **neural tube,** and those that are derived from **neural crest** and will contribute to the formation of the segmental nerves and visceral skeleton of the gills. The various regions of the neural tube are most clearly seen in the median sagittal section, but special details also are revealed to advantage in frontal and transverse sections. From anterior to posterior the brain divisions are as follows. The **telencephalon** (Fig. 4-4 A, B) is not differentiated in any special morphological way from the diencephalon immediately behind it. We can simply refer to the anterior brain wall as the primordium of the telencephalon. The **diencephalon** (Fig. 4-4 A, B, G–I) is clearly recognizable by its four conspicuous evaginations. Of these the **epiphysis** (Fig. 4-4 A, B, C, G) evaginates dorsally. The **infundibulum** (Fig. 4-4 B, J) evaginates ventrally from the diencephalon floor. It will contribute the neural or posterior lobe of the pituitary gland. The paired **optic cups** (Fig. 4-4 A, B, D, H, I) evaginate from the lateral wall of the diencephalon. The optic cups are derived from the optic vesicles. The lateral walls have invaginated into the lumen to form a two-layered cup. The inner layer of the optic cup is the **neural retina;** it will form the sensory layer of the eye. The outer wall of the cup is much thinner; it will form the pigmented layer of the optic cup. The connection of the optic cup to the brain wall is the **optic stalk;** it is a temporary structure that will degenerate into little more than the nerve sheath that guides the optic nerve from the retina back to the brain. Note the location of the **lens placode,** which is being delaminated from the deep layer of the epidermis into the eye cup. Note the close association of the infundibulum floor with the **hypophyseal epidermis** (Fig. 4-4 B, J). The **midbrain** or **mesencephalon** (Fig. 404 B, I–K) show little advance over that already described in the 4 mm embryo. It is most easily recognized by the more or less uniform thickness of its brain wall, the roof of which is the **optic tectum.**

The **hindbrain** is not clearly distinguishable into metencephalon and myelencephalon. However, the presence of the **posterior choroid plexus** as a broad thin roof of the myelencephalon does provide a reliable landmark for this brain region (Fig. 4-4 B, K–M). The neural tube posterior to the hindbrain has thick lateral walls and a thin floor and roof in the regions that correspond to the spinal cord (Fig. 4-4 N–U). Over most of this distance the **neural canal** tends to have the appearance of a key-hole with the lumen being somewhat wider in the dorsal half of the tube. The **neural crest** (Fig. 4-4 T, U) can be seen as a discrete aggregate of cells dorsal to the spinal cord only in the posterior quarter of the body. Elsewhere the neural crest has already migrated away from its primary middorsal position. Its derivatives are primarily associated with the formation of spinal ganglia (Fig. 4-4 D–Q) located to the dorsal half of the spinal cord at segmental levels of the body. Cranial neural crest gives rise to ganglionic components of the 5th, 7th, 9th, and 10th cranial nerves, the primordia of which have already made their appearance (Fig. 4-4 A, C, J—N). The **5th cranial nerve** will be associated with the first visceral arch into which it can be traced and in which the primordium of the visceral skeleton will be found as a dense cord of cells (Fig. 4-4 I, J). The 7th cranial nerve and primordia of visceral cartilages II can be identified in visceral arch 2 (Fig. 4-4 K, L). A similar association can be identified between cranial nerve 9, visceral cartilage 3, and visceral arch 3 (Fig. 4-4 M). The ganglia of the 10th cranial nerve have formed (Fig. 4-4 N), but posterior visceral arches have not formed, and therefore the 1.0th nerve cannot be traced to its final destination.

Legend for Figure 4.4. Structure of the 6–7 mm frog hatching larva, Stage 19–20.

Legend and checklist of structures to be identified

Color code: blue — ectoderm, red — mesoderm, yellow — endoderm.

V	fifth cranial nerve, trigeminus
VII	seventh cranial nerve, facialis
VIII	eighth cranial nerve, acousticus
IX	ninth cranial nerve, glossopharyngeus
X	tenth cranial nerve, vagus
aa 1-5	aortic arches 1 to 5
ac	anterior cardinal vein
an	anus
at	atrium
ca	conus arteriosus
cc	common cardinal vein
cg	cement gland
cl	coelom, general body cavity
da	dorsal aorta
di	diencephalon
dmc	dorsal mesocardium
ec	external carotid artery
ee	epidermal ectoderm
end	endocardial heart lining
ep	epiphysis, pineal primordium
fb	forebrain
fg	foregut lumen, pharynx
fge	foregut endoderm
fin	tail fins
gl	glomerulus
gp 1-4	endodermal gill plates 1 to 4
hb	hindbrain
hg	hindgut
hm	head mesenchyme
hp	hepatic portal (vitelline) vein
hy	hypophysis
ic	internal carotid artery
in	infundibulum
la	lateral abdominal veins
lb	lung bud
le	lens placode
li	liver diverticulum, gall bladder
lie	liver cord endoderm
ll	lateral line placodes
lm	lateral plate mesoderm, hypomere
lmc	lateral mesocardium
mb	midbrain, mesencephalon
mc	myocardium
mg	midgut lumen
my	myotome, axial segmental muscle
nc	neural crest
ncl	neurocoele, neural canal
no	notochord
np	nasal pit & nasal placode
nt	neural tube, spinal cord
om	oral membrane
op	optic cup
ops	optic stalk
opt	optic tectum, midbrain roof
ot	otic vesicle
pc	peritoneal (abdominal) cavity
pcp	posterior choroid plexus
pcv	posterior cardinal vein
pd	proctodaeum
pe	pericardial cavity
pnd	pronephric duct
pnn	pronephric nephrostome
pnt	pronephric tubules
pr	pigmented retina of optic cup
rt	sensory retina of optic cup
sc	scleratome, vertebral primordia
sg	spinal ganglia
sm	somite mesoderm, epimere
std	stomodaeal ectodermal plate
sv	sinus venosus
tb	tailbud blastema
te	telencephalon
tg	tailgut endoderm
th	thryoid diverticulum
v	ventricle of heart
v3	third ventricle of forebrain
v4	fourth ventricle of hindbrain
va	ventral aorta
vam 1-5	visceral arch mesenchyme
var 1-5	visceral skeleton primordium
vm	ventral mesentry of gut
ye	yolky endoderm

2. Endodermal derivatives

a. Foregut or pharyngeal derivatives (Fig. 4-4 A, B, D–F, J–P) are still at a very early and primitive level of differentiation. The pharyngeal **endoderm** lining and the large foregut cavity are among the most conspicuous aspects of the anterior third of the embryo. Anterior pharyngeal endoderm is in contact with the stomodaeal plate where it contributes to the transient oral membrane. The only other differentiated parts of the pharyngeal endoderm are in the lateral walls of the pharynx. At three locations on each side a solid sheet of endoderm grows out from the pharyngeal lining and fuses with the surface epidermis; these endodermal **gill plates** are the primordia of gill pouches. At a later time the sheets will split and form extensions of the pharyngeal lumen to the exterior. An endodermal gill pouch plate is shown in Figure 4-4 J. However, they are shown to best advantage in the frontal plane such as that shown in Figure 4-4 D–F where all three pairs of gill plates can be seen along with their relationship to the pharynx, epidermis, and the visceral arches that separate one gill plate from the next. The **thyroid diverticulum** (Fig. 4-4 A, B, F, P, Q) is just beginning to form as a ventral evagination near the anterior end of the foregut floor. A somewhat similar ventral outgrowth of endoderm is forming from the posterior foregut floor; this is the **lungbud** (Fig. 4-4 A, B).

b. Midgut derivatives (Fig. 4-4 A, B, D, E, Q–S) are little advanced over those already described for the 4 mm embryo. Note the dorsal position of the **midgut lumen,** and the ventral **liver diverticulum** which ends blindly in the general endodermal yolk mass. The lumen is the primordium of the gall bladder.

c. Hindgut derivatives (Fig. 4-4 A, B, F, O–Q) include the **cloaca** which unites with the **proctodaeum** and opens to the exterior by the **anus.** Of some interest is the fact that a solid strand of **tail endoderm** is present in sections of the tail; these cells are generated out of the tailbud blastema along with the posterior components of the neural tube, notochord, and somites. Tailgut endoderm has no function and will degenerate entirely soon after forming.

3. Mesodermal derivatives

a. Somite, or epimere mesoderm (Fig. 4-4 C, N–U) will form most axial muscles, connective tissue, and skeleton. At this stage the segmental **muscles** or **myotomes** are well differentiated along most of the trunk and tail region and occupy a distinctive position lateral to the spinal cord. It is of interest to note that myotomes provide the inductive influence that stimulates trunk neural crest cells to form segmental ganglia at appropriate

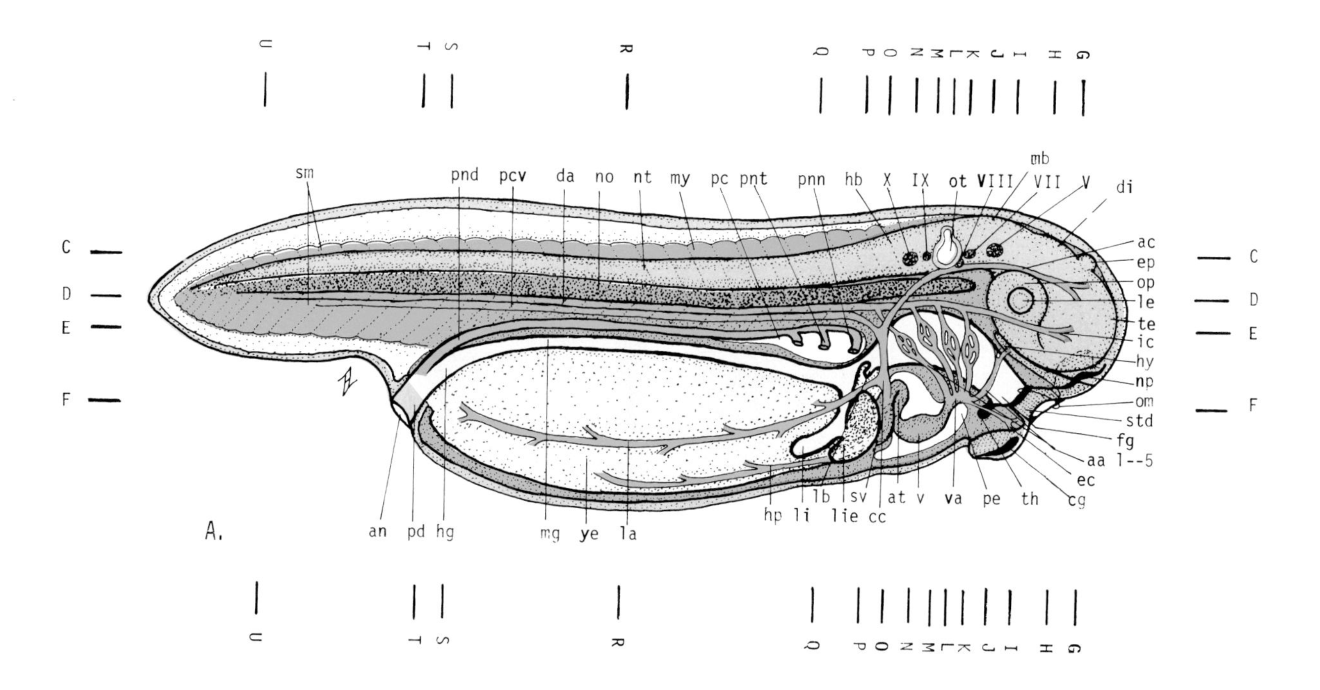

Figure 4-4 A. Whole mount reconstruction of a 6–7 mm frog hatching larva, Stage 19–20. Guide lines around the figure indicate the levels at which correspondingly lettered drawings in the remainder of this series have been cut. See the legend for abbreviations.

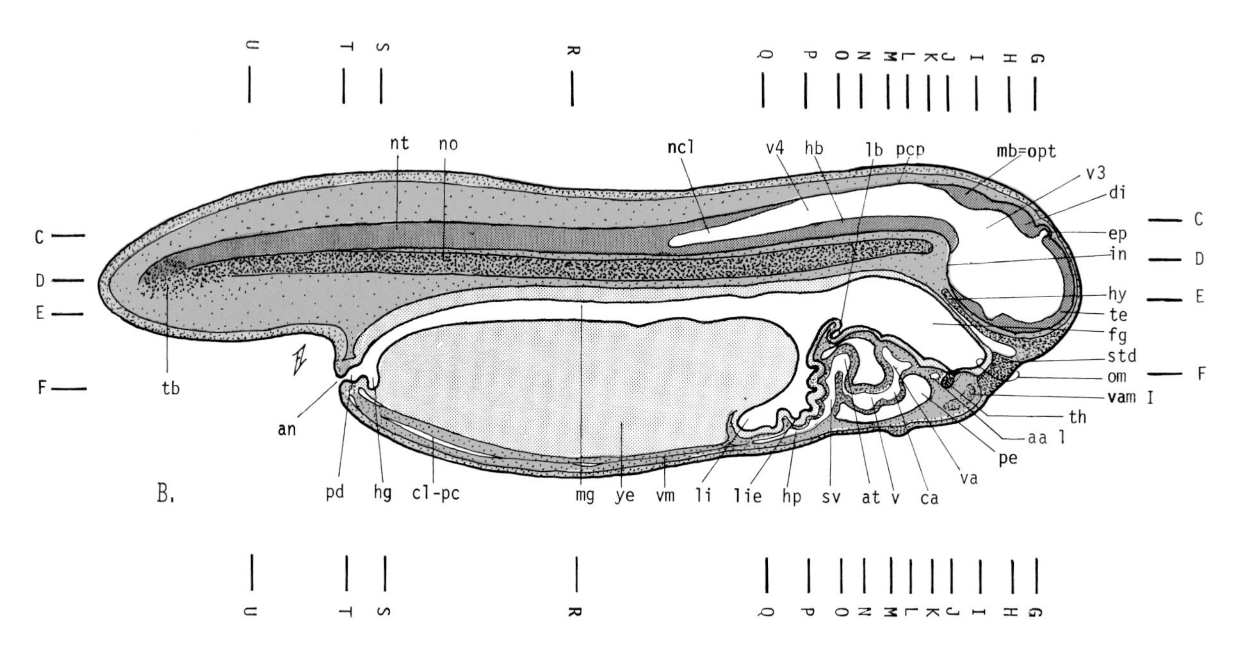

Figure 4-4 B. Median sagittal section of a 6–7 mm hatching larva, Stage 19–20. Guide lines around the figure indicate the level at which correspondingly lettered drawings in the remainder of this series have been cut. See legend accompanying Figure 4-4 A.

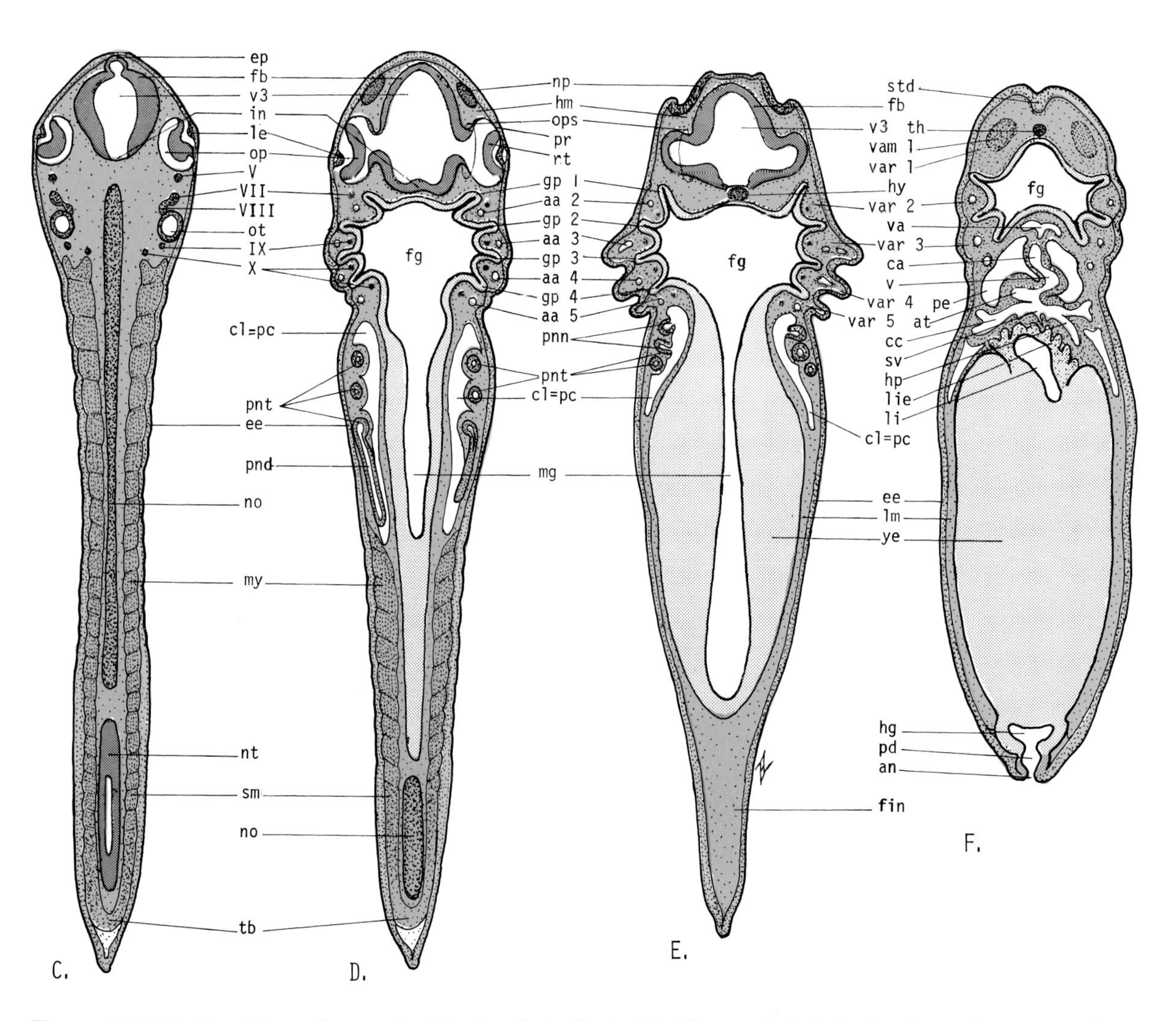

Figure 4-4 C–F. Frontal sections cut at the levels indicated in Figures 4-4 A, B. See legend accompanying Figure 4-4 A.

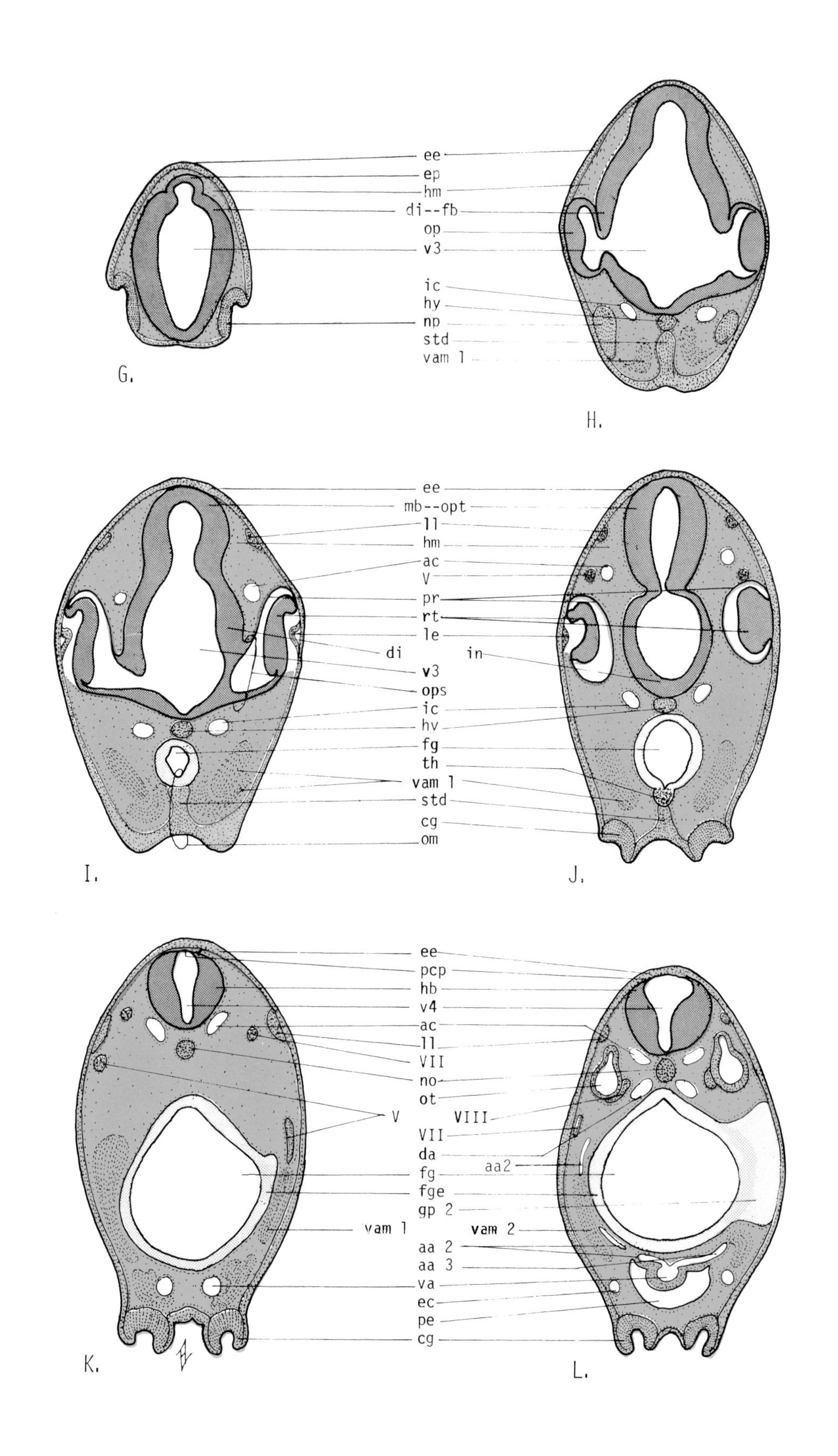

Figure 4-4 G–L. Representative cross sections of a 6–7 mm frog hatching embryo, Stage 19–20 at levels indicated by guide lines in Figures 4-4 A. and B. See legend accompanying figure 4-4 A.

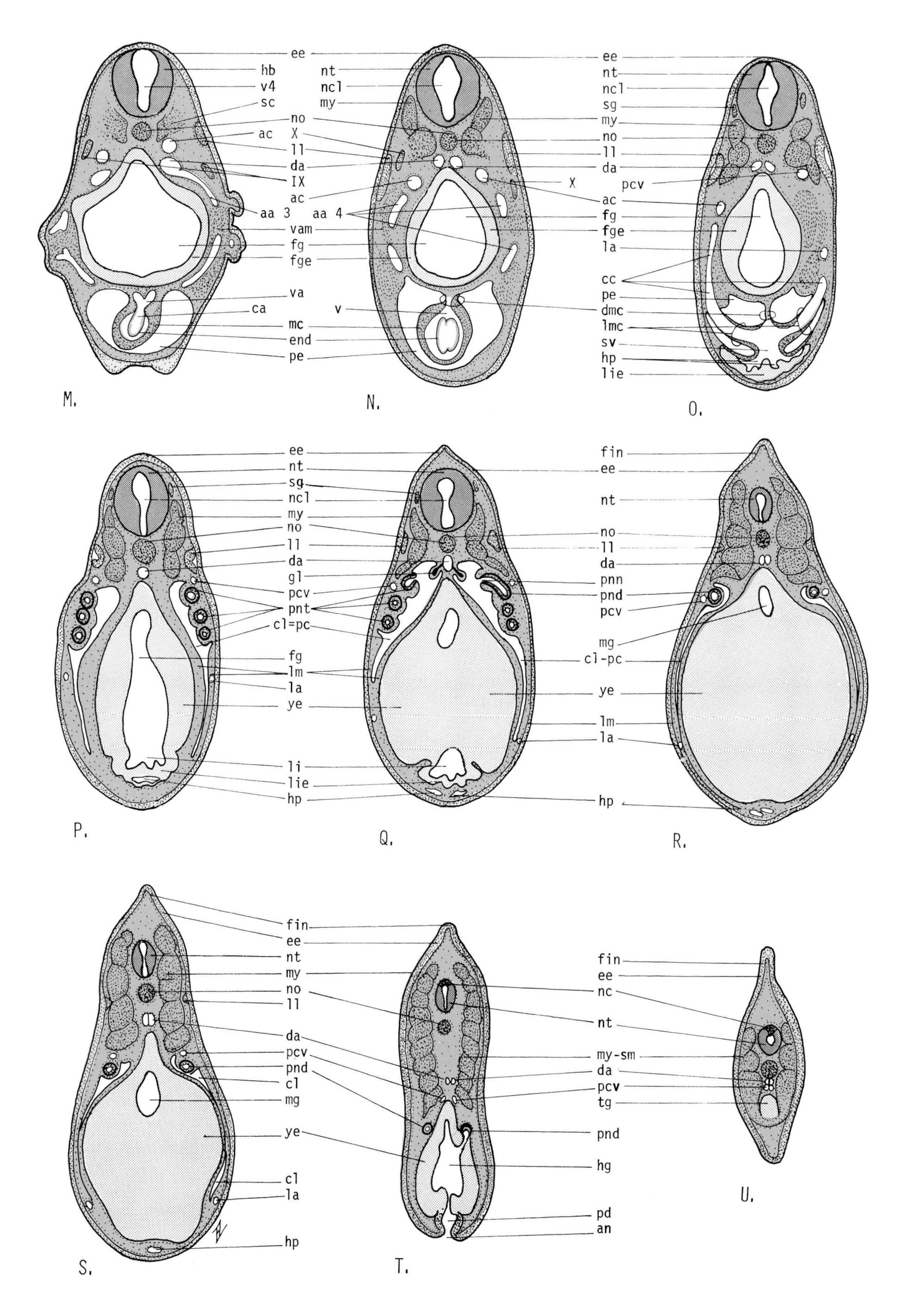

Figure 4-4 M–U. Representative cross sections of a 6–7 mm frog hatching embryo, Stage 19–20 at levels indicated by guide lines in Figures 4-4 A. and B. See legend accompanying Figure 4-4 A.

myotome levels. Much loose mesenchyme of the dorsal fin and dorsal axial regions is derived from **dermatome** and **scleratome** elements of the epimere; however, there is little evidence of their specialization at this time.

b. Nephrotome, or mesomere (Fig. 4-4 A, D, E, P–T) is in preliminary stages in the formation of three pairs of pronephric tubules from the intermediate cord of mesoderm that lies between the somites and lateral plate. Pronephric tubules are located at the anterior midgut level of the body; they will first form solid cords of cells that later hollow out and open into the anterodorsal peritoneal cavity. The openings are the **pronephric nephrostomes,** one of which is shown in Figure 4-4 A, E, Q. Posteriorly the three tubules will unite and form a common duct which probably possesses more synonyms than almost any other vertebrate structure; i.e., the **pronephric duct,** or **archinephric duct,** or **Wolffian duct,** to name only a few. The right and left pronephric ducts (Fig. 4-4 Q, T) then proceed to grow back along the ventral margin of the myotomes until they reach the level of the hindgut. At that point they will fuse with the endodermal wall of the hindgut destined to form the cloaca. At this stage under study the pronephric duct is in the middle of its posterior migration and may not have fused with the hindgut endodermal wall. It is of special interest that the pronephric duct, at this time, is in the process of inducing posterior mesomere to form mesonephric kidney tubules. The latter will not develop without the added inductive stimulus of the pronephric duct.

c. Lateral plate or hypomere (Fig. 4-4 P–S) is primarily of interest in association with its participation in **coelom** and **heart** formation. Aside from the small but well formed peritoneal cavities in the immediate vicinity of the pronephric nephrostomes, the peritoneal cavity is as yet very poorly defined. The lateral plate mesoderm in most posterior and ventral regions is not yet clearly split to form a coelom lined by a **somatic layer of hypomere** applied to the epidermis and a **splanchnic layer of hypomere** associated with the gut wall. Ventral to the foregut, and anterior to the liver diverticulum, will be found the **heart primordium** (Fig. 4-4 A, B, F, L–O). It is at an early stage of functional differentiation in which the first beats are being initiated. The blood will soon flow through a closed system of capillaries. The basic structure of the heart is most favorably shown in frontal sections such as that in Figure 4-4 F. It can be seen that the heart arises from the hepatic portal capillary bed of the yolk mass. The four primary regions of the heart, listed from posterior to anterior in the order in which the blood will flow, are as follows: **sinus venosus, atrium, ventricle,** and **conus arteriosus.** They are all suspended by a **dorsal mesocardium** in the pericardial cavity (Fig. 4-4 L–O). In cross sections, note again that the thin endocardial lining of the heart lumen is surrounded by the **muscular layer,** or **epimyocardium,** which is a direct continuation of the pericardial lining. The endocardial linings are the inductors of the muscular epimyocardial wall of the heart. All parts of the heart and its cavity are to be considered special derivations of the ventral mesentery of the lateral plate mesoderm. The remainder of the circulatory system is present as an exceedingly primitive capillary network. The primary elements that should be looked for include the following: The **ventral aorta** (Fig. 4-4 J–K) gives off **five pairs of aortic arches** (Fig. 4-4 A, D–F, I–O). Dorsal to the pharynx, the arches unite to form the **dorsal aorta** (Fig. 4-4 L–O) which is paired anteriorly but by fusion becomes a single vessel through most of the trunk and tail. Anterior to the first aortic arch, the dorsal aortae continue into the head as the paired **internal carotid arteries** (Fig. 4-4 A, H—J). This blood enters the capillary net of the head and is returned to the heart by the **anterior cardinal veins** (Fig. 4-4 A, I–O) which lie ventrolateral to the brain floor along most of their length. At the posterior level of the heart, the anterior cardinal veins enter the **common cardinal veins,** or **ducts of Cuvier,** which pass around the lateral wall of the pharynx and enter the lateral walls of the sinus venosus (Fig. 4-4 N). Blood from the posterior levels of the body returns to the sinus venosus by way of the **hepatic veins,** or the **lateral abdominal veins,** or the **posterior cardinal veins,** only parts of which are complete at this time (Fig. 4-4 O–U).

D. BASIC EXPERIMENTS RELATED TO THE INDUCTIVE CAPACITY OF CHORDAMESODERM TO DETERMINE THE EMBRYONIC AXIS

If living embryos at blastula and gastrula stages are available from natural sources, or form artificial insemination of frog eggs as described at the end of Chapter 3, the following experiments can be carried out. In the event that living materials are not available, selected motion pictures of experimental studies on amphibian gastrulation and induction will be shown.

Materials needed for each student. Dissecting microscope, 2 dissecting needles, fine forceps, 2 standard 15 cm Petri dishes, 2 covered Stender dishes or low wide-mouth collecting jars with screw caps, 1 flask of 250 ml boiled cool tap water, vaseline (preferable in a syringe gun prepared by filling a 5 ml plastic syringe with melted vaseline), standard micro-cover slips, and 3 small vials with 5% formalin for the preservation of embryos at the end of the experiment.

1. Inversion experiments on gastrulating eggs. Ordinarily gravity will cause eggs to orient with the yolk plug and blastopore lips directed downwards. If during gastrulation the egg is forced to remain in an inverted position with the vegetal pole up, this is often sufficient to prevent the normal internal movement of cells from taking place. When convergence and axial elongation of chordamesoderm is obstructed, various kinds of abnormality in neural induction and axial organization of the egg will occur. These abnormalities are usually revealed in tailbud and tadpole stages by doubling in the head or tail, by Siamese twinning, or by microcephaly (small head), and many other distortions of the normal body axis. Inversion experiments can be carried out in the following way.

a. With a vaseline gun place a dot of vaseline on each of the four corners of a standard micro-coverslip. Prepare one such cover for each egg to be inverted, i.e., 3 or 4.

b. Remove the external jelly from gastrula stage embryos. Use dissecting needles and fine forceps to remove the jelly. This can be done most simply with frog eggs by placing them on a piece of wet filter paper or paper towel and cutting off the jelly with needles. Sharp forceps will usually be needed to remove the jelly capsules from salamander eggs. Alternatively, jelly capsules can be removed enzymatically and chemically. The method recommended by D.D. Brown (in Wilt and Wessels, see Appendix I) uses a solution of 2% cysteine neutralized to pH 7.8 with NaOH and containing 0.2 percent papain (from Worthington, Sigma, or Nutritional Biochemical companies). Commercial "meat tenderizer" containing active papain can be used. Denuding eggs of jelly by this means will take from 10 to 30 minutes and can be assisted by mechanical means.

c. Transfer de-jellied eggs to a clean Petri dish with cool, boiled tap water. Also transfer one of the coverslips with vaseline drops upwards to the Petri dish and submerge it at the center of the optical field in the dissecting microscope. With a dissecting needle move an egg to the center of the cover glass and anchor it in place with a little vaseline transferred by needle to the egg.

d. Carefully cover the egg with a second plain cover glass. It should be carefully lowered into place in such a manner that its corners will rest on the vaseline drops at the corners of the lower cover glass. The end result will be a cover glass "sandwich" with the egg mounted between the two covers.

e. Now flatten the egg slightly. This can be done most effectively by pressing on the top cover with a glass needle until the egg is about 1 1/2 times its original diameter.

f. Invert the double cover glass sandwich. This can be done most easily by sliding the sandwich to the edge of the dish, inserting a dissecting needle under one corner, and then lifting the covers up and over. You should see the vegetal half with endodermal field and the blastopore lips clearly outlined.

g. Watch the preparation for a few minutes under the microscope to see whether or not the egg will still be able to rotate by gravity. If the egg does rotate, more pressure can be applied so that the egg will remain permanently inverted when the preparation is turned upside down again.

h. The preparation should remain inverted through all of gastrulation, i.e., for about 24 hours. After gastrulation, the cover glass should be loosened and the egg removed with care to a covered Stender dish of cooled boiled tap water.

i. Evidence of inductive malfunction will become visible in tailbud stages. Embryos inverted from crescent blastopore to mid-neurula stages will often show a doubling of the head and complete Siamese twinning. Eggs inverted from horseshoe or yolk plug blastopore to neurula stages will usually show normal heads but a doubling of the trunk or tail. Compare your results with those of others in the class.

2. Experiments involving pressure on the dorsal lip of the blastopore. When pressure is applied directly to the dorsal lip of the blastopore, convergence and axial elongation of chordamesoderm cells in the middorsal axis of the embryo will usually be deflected. As the cells continue to move to the interior, the chorda cells will form right and left streams. Each of these chordamesoderm bands can act separately as neural inductors within the same embryo. As a result there will be two neural plates and two sets of axial structures in the future embryo. If pressure is applied to the dorsal lip at the crescent blastopore or yolk plug stage, doubling of the head will result. If pressure is applied at the mid-ventral blastopore lip at yolk plug stages or mid-neurula, then doubling of the tail is most likely to result. In the following experiments cover slip sandwiches of the type described above will be prepared with one additional variation described as follows.

a. First prepare vaseline cover glasses in the manner described above. Mount a bottle-brush bristle across the middle of the cover by tacking it down with a dot of vaseline at each end.

b. Prepare de-jellied eggs in the manner described above.

c. Transfer a bristle-cover glass and de-jellied egg to a standard Petri dish. The dish should be about half full of cool boiled tap water.

d. Mount an egg in such a way as to straddle the bristle; vaseline will be needed to secure the position of the egg.

e. Add the second cover glass in the manner described before in making sandwiches; be exceedingly careful, however, to avoid displacing the egg from its position over the bristle.

f. Apply pressure very carefully to the top cover so that the egg will form a compressed saddle over the bristle. Patience and a number of tries will probably be required because the egg will tend to slip off center.

g. Invert the slide and check the orientation of the blastopore lip to the long axis of the bristle. At random, you can expect some eggs to be oriented with the AP-axis obliquely, transversely, or parallel with the pressure point.

h. Return the egg to its normal orientation by turning the sandwich back to its original position. Allow the egg to continue developing under pressure until the mid-neurula stage.

i. By the mid-neurula stage the egg should be freed and transferred to a covered culture dish.

Tailbud embryos and tadpoles should be observed daily and evidence of axial anomalies noted and sketched. At the end of the experiment, which should be terminated shortly after feeding begins, embryos showing axial abnormalities can be preserved for future reference in 5% formalin (i.e., 5 ml commercial formalin in 95 ml distilled water). Other preserving fluids for special use are given in Table V-1.

In conclusion, this chapter has given special attention to evidence for the primary mechanisms responsible for creating the axiate pattern in vertebrate embryos. We have seen that the chordamesoderm plays a primary role in starting a series of chain reactions that will ultimately lead to the characteristic placement of all embryonic germ layer derivatives in the correct three dimensional framework of the vertebrate body plan.

We are now in a position to examine a very old theory of development in the context of modern experimental ideas on induction and induction series. Chapter 5 will explore the recapitulation theory of embryonic development in amphibians and will provide additional opportunities to work with induction systems and embryonic patterns of differentiation in living amphibian embryos and larvae.

CHAPTER V

THE RECAPITULATION THEORY OF DEVELOPMENT

Illustrated by a Comparison of the Salamander Tadpole with the Ammocetes Larva of Cyclostomes

TABLE OF CONTENTS

ILLUSTRATIONS

NORMAL SALAMANDER LARVAE

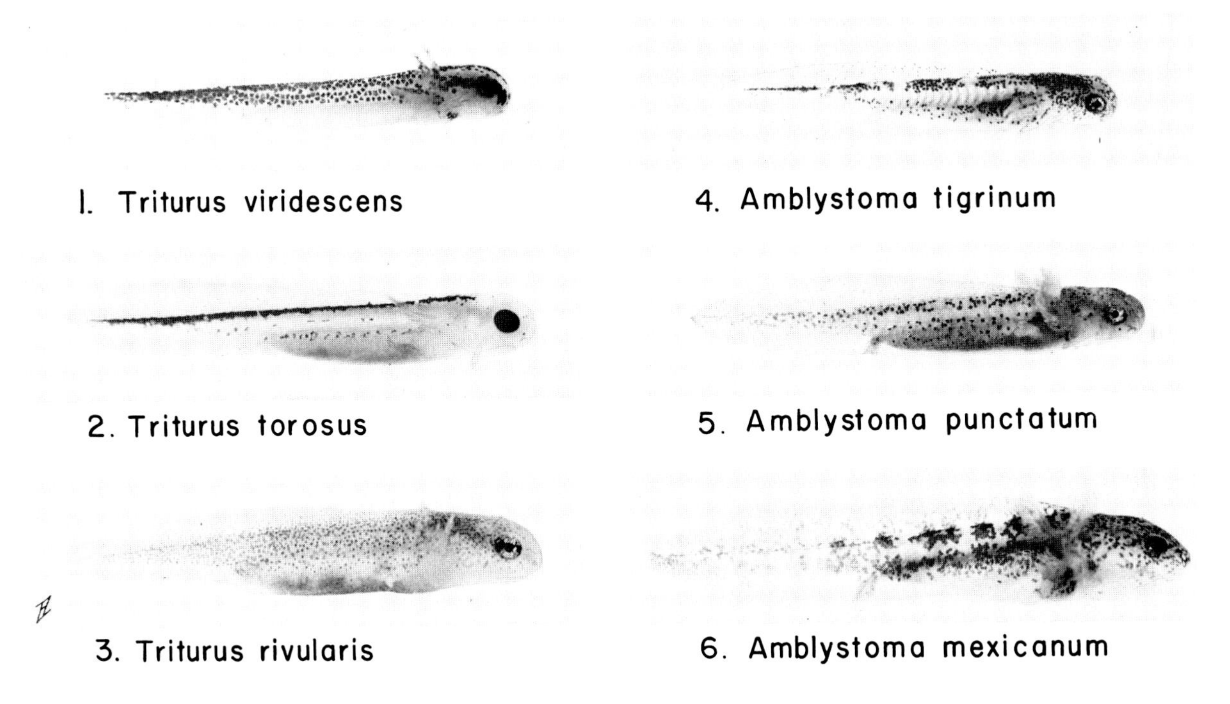

Figure 5-1 A. Normal salamander larvae. Pre-feeding larval stages. The names given are the ones by which they are generally known in the experimental embryological literature. The more correct names respectively are: 1. *Notopthalmus viridescens;* 2. *Taricha torosa;* 3. *Taricha rivularis;* 4. *Ambystoma tigrinum;* 5. *Ambystoma maculatum,* and 6. *Siredon mexicanum.*

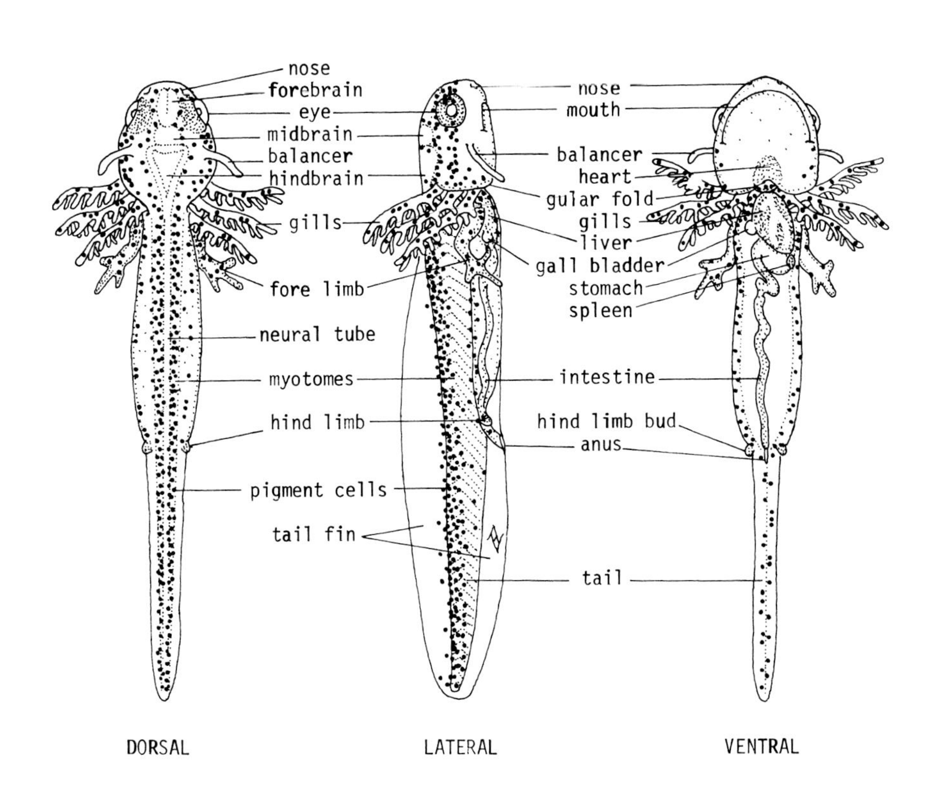

Figure 5-1 B. Normal salamander larvae. Topographic structural characters of salamander tadpoles as seen from dorsal, lateral, and ventral surfaces.

CHAPTER V

THE RECAPITULATION THEORY OF DEVELOPMENT

Illustrated by a Comparison of the Salamander Tadpole with the Ammocetes Larva of Cyclostomes

Shortly after the publication of Darwin's great book which presented the *Theory of Natural Selection* to the world in 1859, attention was directed to the importance of embryology as a source of information about evolutionary relationships among species. The most dramatic interpretation of embryological similarities was given in Ernst Haeckel's *Recapitulation Theory* in 1867. The central theme of this theory is captured in the phrase, *ontogeny recapitulates phylogeny.* That is to say, life histories (and in particular embryonic stages) summarize evolutionary histories of races. This theory, when it was first proposed, attracted wide attention. It played a major part in stimulating research that, by the end of the past century, led to knowledge of normal development in all major animal taxa.

Not all information on embryology, by any means, supports the underlying idea in the recapitulation theory. However, some examples can be found in which it does appear that the embryo repeats ancestral adult stages during the completion of its life cycle. One classic case cited in support of the theory is the presence of a fish-like tadpole stage in the life history of amphibians and, in particular, in the salamander tadpole. These larvae possess many very primitive characters that strongly call to mind the basic structure of the cyclostome Ammocetes larva which was described in some detail in Chapter I.

A. THE CYCLOSTOME LARVA AS REPRESENTATIVE OF THE VERTEBRATE ARCHAETYPE

As a prerequisite to the following study of the salamander larva, the discussion of structure in the cyclostome Ammocetes larva, and the concept of archaetypes and developmental common plans in Chapter I, should be reviewed. With information on primary vertebrate structure again clearly in mind, proceed to the study of the salamander tadpole. Give special attention to the many ways in which the two developmental stages resemble one another in the general plan of parts and the basic structure of organ systems.

B. THE STRUCTURE OF THE YOUNG SALAMANDER LARVA AS REPRESENTATIVE OF THE VERTEBRATE ARCHAETYPE

The basic organization of embryos of all terrestrial vertebrates is essentially fish-like at some period following gastrulation. The validity of this statement will be made abundantly clear in the study of early development of the chick in Chapters 7 and 8. However, except for amphibians, the so-called fish-stage is passed while the embryo is contained in its extraembryonic envelopes. Only in the amphibia is the fish-stage represented by a free-living larva that is adapted for survival in the primitive aquatic habitat for which a gilled respiratory system is admirably suited.

The structure of the salamander tadpole is much closer to the vertebrate common plan than is the greatly compressed and foreshortened head and trunk that is so well known in the tadpoles of frogs and toads. For these reasons the salamander larva is the form of choice for the following study.

1. The living or preserved whole salamander larva (Fig. 5-1, 5-2, and 5-7 A). Obtain a living or preserved specimen of an early salamander tadpole at an age just after the yolk has been resorbed and feeding has begun. The preserved specimen is in a preservative, clearing solution composed of 9 parts 5% commercial formalin in water and 1 part glycerin. The latter acts as a clearing compound so that the tissues will tend to remain soft and more or less translucent as they are in life. If living material is available for study, it will be necessary to narcotize it. Two agents are admirable for this purpose: A. 10% saturated aqueous **chloretone** in fresh water will put animals to sleep in about 20 minutes; they can be kept in it without harm for 24 hours. They can be awakened by a return to standing tap or aquarium water. B. 0.005% **Finquel** (Tricain methanosulfonate, Ayerst Laboratory, N.Y.C.) is a special amphibian narcotizing agent with similar or even superior qualities to chloretone. Use a dissecting needle to turn the specimen which should be **kept entirely submerged** in a culture dish during observation at low magnification. Use reflected light on the surface of the animal since it is opaque and will not transmit light from beneath.

a. General topography. Use the information in Figure 5-1 B to identify the following parts. It is immediately apparent that the salamander larva has a distinctly fish-like overall appearance. The body is clearly divided into a **head, trunk,** and **postanal tail.** The presence of three pairs of feathery **external gills** mark the dividing point between the head and trunk. The gut runs almost in a straight line to the posterior end of the trunk where the **intestine** bends sharply toward the ventral side and opens to the exterior by the **anus.** The **ventral fin** begins just back of the anus and continues to the tip of the tail. There it meets the dorsal fin, which runs forward throughout the entire length of the tail and trunk to the base of the head. Note the generous sprinkling of black, orange, and silver **chromatophores,** or pigment cells, that are scattered over the head, trunk, and tail. These cells are of neural crest origin. Their patterns of distribution are usually diagnostic for the species to which the salamander belongs. Several common species of salamander tadpoles are shown in Figure 5-1 A. They give some idea of the range in diversity of larval pigment patterns. Now proceed to the identification of special structures in the major body regions.

b. On the head, note the paired **external nares** or nostrils. They open at the anteroventral extremity in front of the mouth and connect by a short canal with the mouth cavity. Note the **eye** and its transparent **cornea** and **lens.** The **pigmented retinal coat** is an intense black; the lateral part of this coat, which surrounds the lens, is the iris. Highly branched silver pigment cells will be seen applied to the surface of the iris.

View the larva from the dorsal surface and note the position of the paired **inner ears** placed dorsolateral to the hindbrain. In living larvae the inner ears can best be identified by the white calcium carbonate crystals (otic sand) contained within the **sacculus-utriculus.** In materials preserved in acidic fixatives the otic sand will have been dissolved away. Try to identify the major divisions of the brain. The **myelencephalon** (medulla) is most conspicuous. It can readily be identified by the "V" shaped **4th ventricle** which is covered by a transparent posterior choroid plexus which enables you to see into the brain floor. The inner ears lie just lateral to this brain region. The **metencephalon** (cerebellum) is the narrow band of tissue that forms the anterior boundary of the 4th ventricle. Anteriorly the mesencephalon and diencephalon can be recognized by prominent constrictions in the brain stem. At the anterior end of the brain, the paired lobes of the **telencephalon** (cerebrum) can be seen from the anterodorsal aspect.

View the tadpole from the ventral surface and notice the wide mouth and lower jaw. Posterolateral to the angle of the jaws will be found a pair of small, thread-like **balancers** which project obliquely posteriad from the side of the head. These are temporary structures which drop off just before feeding begins, and after the fore-limbs are sufficiently developed to support the body. On the ventral surface of the head, note the gill cover, or **gular fold.** It extends ventrolaterally up the sides and hides the openings from the gill clefts into the pharynx. The gills arise from within this cover. Their attachments to the pharyngeal wall cannot be observed directly.

c. View the trunk and tail from the ventral surface. If living specimens are available, the heart can be identified readily by its contraction. The **heart** lies in the "V" created by the right and left gular folds as they pass toward the ventral midline. In preserved specimens, the heart will be most readily detected by the very high concentration of pigment cells that accumulate in the wall of the pericardial cavity and often obscure any structural details of the heart. Identify the **liver** which lies just back of the heart. At its posterior end will be found the large **gall bladder** on the right side. At about the same level at the left, identify the somewhat smaller **spleen** that will be lying close to the wall of the **stomach** which is sharply bent to the left. In living specimens the gall bladder will be a greenish yellow and the spleen a brilliant crimson. The anterior end of the intestine, where it unites with the stomach, is bent in a sharp S-shaped coil; this is the **duodenal intestine.** The **pancreas** lies in this loop of the gut and, in living specimens, will be a pale pink within the otherwise yellowish coil of the gut wall. Posterior to the duodenum, the **intestine** passes straight back through the **peritoneal cavity** to the **hindgut,** or **cloaca,** and **anus.**

When the trunk and tail are viewed from the

sides, the **fore-limbs** can be seen. Dorsal to the attachment of the limb to the pectoral girdle, a small swelling marks the position of the **pronephros** which lies internal to the girdle and dorsal to the coelom. Note the clear segmental arrangement of the **myotomes** which extend the full length of the trunk and tail. Of special interest are the **lateral line organs** which will appear as a row of small papillae running along the middle of the somites from the level of the gills to the tail. These are special sensory organs found in primitive aquatic vertebrates which are extremely sensitive to stimuli of pressure, vibration, and possibly chemicals as well. The lateral line organs are used as much for feeding as are the eyes and nose. Indeed, larvae with both eyes removed, or animals kept in complete darkness, can feed as well on actively moving food organisms as can animals with full vision.

d. Attempt to identify the major elements of the circulatory system. It is composed of capillaries at this time. Even under the best of circumstances, the details of arterial and venous pathways can be deciphered only with difficulty. In preserved specimens it is impossible, but in living individuals, in which the red blood with visible corpuscles is moving, it can be highly aesthetic and instructive to observe. Use the diagram provided in Figure 5-2 as an aid in identifying as many of the major vessels as possible. Try to trace the blood from the anterior end of the heart into the **ventral aorta** and from thence, into the gills by the **afferent branchial arteries** which carry blood into the **gill capillaries** where it is oxygenated by pharyngeal ventilation. **Efferent branchial arteries** carry oxygenated blood from the gills to the paired **dorsal aortae.** The circulation through these 3 gills represents functional 3rd, 4th, and 5th embryonic aortic arches. Other vessels that can readily be seen include the **lateral abdominal vessels** that parallel the path of the lateral line system; **intersegmental arteries** and veins that can be seen between the myotomes; and, lastly, the **subintestinal — hepatic portal vein** that goes into the liver.

2. The study of serial sections of the young salamander larva (Fig. 5-7 A–U). The illustrations for this study were prepared from salamander tadpoles at the beginning of feeding. The species used was *Taricha torosa,* a common newt of the Pacific coast range. However, the diagrams are also applicable to those of most other species. Some variations can be expected to occur, in accord with differences in stored yolk, which will allow species to advance to different developmental ages before demands for food require the initiation of the larval stage.

The remaining study will be devoted to sectioned material. Emphasis will be placed on those structures which cannot be demonstrated in strictly embryonic stages of avian and mammalian development.

a. The nervous system. The basic components of the neural tube and cranial nerves are well formed and histologically differentiated as neurons and fibers. They illustrate the distinction between the inner nucleated **mantle layer** which will form the gray matter, and the outer fibrous **marginal layer,** or white matter, area. On the whole mount and median sagittal diagrams, identify the major brain regions and their cavities. Locate them in cross and sagittal sections. Begin at the anterior end and proceed posteriorly.

The **telencephalon** (Fig. 5-7 C–F) is distinctly lobed into right and left **cerebral hemispheres.** The lateral walls of the hemispheres are much thicker than the mesial walls. Note the slit-like lateral ventricles. Trace them posteriorly until they unite to form the 3rd ventricle of the **diencephalon.** This communication point between each lateral ventricle and the 3rd ventricle is the **foramen of Monro.**

The **diencephalon** (Fig. 5-7 E–H) shows several specializations. It is distinguished from the remaining brain regions by its dorsoventrally elongated appearance in cross sections. It extends for a short distance anteriorly between the two cerebral hemispheres (Fig. 5-7 E, F). In this region the roof plate of the diencephalon forms a glandular epithelium which invaginates into the 3rd ventricle and, in section, appears as spongy shreds of tissue. This invaginated roof plate is the **anterior choroid plexus.** It functions in the secretion of the cerebrospinal fluid which fills the neural canal. There are two median dorsal evaginations from the diencephalon roof. The anterior one is the paraphysis and the posterior one is the **epiphysis.** These are exceedingly ancient vertebrate organs which originated in association with the median parietal eye of lobe finned fishes. Their exact functions in vertebrates remain in question. In many lower vertebrates they retain a slight residual photosensitive role that seems to be involved in the photoperiodic seasonal controls of reproductive cycles wherever they are found in vertebrates. In the floor of the diencephalon, near the posterior level of the eyes, will be found the attachment of the **optic nerves** to the brain floor (Fig. 5-7 H). This region, where the right and left optic nerves enter the brain wall, is the **optic chiasma.** Posterior to the optic chiasma, the floor of the diencephalon projects back under the mesencephalon. This can best be seen in median sagittal sections (Fig. 5-7 B). This posteroventral extension is the **infundibulum** (Fig. 5-7 I–J). At the posterior end of the infundibulum is a small dense mass of glandular tissue. This is the **pituitary gland** (Fig. 5-7, A, B, J). The bulk of the pituitary arises from stomodaeal ectoderm, and a relatively small part is derived from the infundibulum itself.

The **mesencephalon** (Fig. 5-7 I–J) is located dorsal to the infundibulum and, in contrast to the diencephalon, is circular in cross section. Its cavity, the **aqueduct of Sylvius,** connects the 3rd and 4th ventricles of the fore- and hindbrain respectively. Optic muscles have not developed, to any great extent, in the early salamander larvae. Consequently the 3rd and 4th cranial nerves, which innervate most optic muscles, have not yet made an appearance; nor have the 6th cranial nerves appeared which arise from the medulla and also go to certain muscles of the eyes.

The **metencephalon** or cerebellum (Fig. 5-7 B, J) is a slightly thickened plate at the anterior end of the **4th ventricle.** Its widest extent is in the lateral plane. This brain region is seen to best advantage in sagittal sections. In cross sections the metencephalon appears in only a few sections at the posterior end of the mesencephalon where it projects laterad beyond the widest part of the midbrain (Fig. 5-7 J).

The **myelencephalon** or medulla (Fig. 5-7 K–M) is primarily distinguished from the other brain regions by the thin **posterior choroid plexus** which covers the greatly expanded **4th ventricle** of the neural canal. The choroid plexus is, in essence, a greatly enlarged roof plate of the neural tube. The posterior end of the medulla merges imperceptibly with the spinal cord. This junction can be only approximately identified by the gradual reduction in the lateral extent of the choroid plexus and the reduction of the neural canal to a narrow dorsoventrally oriented slit.

From the mesencephalon, posteriorly through the **spinal cord,** the sensory and motor areas of the central nervous system can be identified by the presence of a lateral groove in the wall of the neural canal. This is the **sulcus limitans** (see Fig. 5-7 I) which marks the boundary between the **alar plate** (dorsal sensor column) and the **basal plate** (ventral motor column). Identify the thin **roof** and **floor plates** which make up the dorsal and ventral walls of the neural tube and note their different degrees of differentiation at various levels of the brain and cord. In posterior regions of the body at the level of the spinal cord, identify the paired **spinal ganglia** which appear in cross sections as oval masses compressed between the neural tube and myotomes.

For special references to the 5th, 7th, 9th, and 10th cranial nerves, see the discussion of the visceral arches.

b. The sense organs and epidermal derivatives. The unicell layered epidermis with gland cells is distinct in all sections and has no special features of note in it.

Nose. The nasal canals (Fig. 5-7 C–E) open externally by the **external nares.** They are located lateroventrally at the anterior end of the head. The **olfactory epithelium,** which lines the nasal canals, is of epidermal placode origin and is very greatly thickened mesiodorsally. Along much of their course the walls of the canals are in direct contact with the ventrolateral walls of the **telencephalon.** It is usually not possible to identify the **olfactory nerve** fibers which originate from the sensory epithelium of the nasal canals and enter the floor of the cerebral hemispheres. Internally the nasal canals open into the roof of the **oral cavity.** These paired openings are the **internal nares** (Fig. 5-7 E).

Eye. (Fig. 5-7 F–I). A median section through the eye at the level of the diencephalon will show most histological structures of the adult eye. The major omissions are the absence of eye muscles and of a well formed connective tissue capsule that forms the sclera, choroid, and inner parts of the cornea. The **cornea,** at this stage in development, consists simply of a very thin layer of superficial epidermis, the **conjunctiva.** The transformation of this epidermis from a thick glandular epidermis covering the rest of the body into a thin transparent membrane is brought about by the inductive action of the underlying **lens** and **eye cup.** Note the spherical lens and identify the thin epithelial **lens capsule** which surrounds the central core of **lens fibers.** Both of these components are derived from the epidermal lens vesicle that was seen at early stages of development in Chapter IV. The space separating the lens from the cornea is the **aqueous humor,** and the internal chamber between the lens and retina is the **vitreous humor.** The embryonic optic cup gives rise to the **pigmented retina, sensory retina,** and

iris; these structures are therefore of neural ectoderm origin. The **pigmented retina** is a single layer of epithelial cells containing dense accumulations of black melanin granules which completely obscure details of cell structure. The function of this layer is to prevent light from entering the inner cavity of the eye by any means other than through the pupil. The pigmented retina is derived from the outer wall of the embryonic optic cup. The **sensory retina** is very thick and is clearly differentiated into 4 distinct layers. Beginning with the deepest layer that is applied to the inner face of the pigmented retina, these layers are as follows. The **sensory epithelium** of the retina consists of a single layer of **rod and cone cells.** This is the light-sensitive layer of the sensory retina. These cells in life are closely applied to the pigmented retina; however, in the preparation of slides, there is usually a small amount of shrinkage and a small space frequently separates the sensory cells from the pigmented retina. The remaining 3 layers of the sensory retina are composed of nervous tissue. Next to the sensory epithelium is the deep ganglion layer containing many nuclei. This layer is separated from the **superficial ganglion layer** by the **median fibrous layer.** Toward the posterior margin of the eye, a connection between the fibrous layer and the **optic nerve** will be found (Fig. 5-7 H). Trace the pigmented and sensory retinal layers laterally around the lens. Note that the sensory retinal layer becomes very thin lateral to the lens. This part of the optic cup is the **iris,** a pigmented contractile membrane. It regulates the amount of light entering the inner chamber of the eye by controlling the size of the pupil.

Inner ear (Fig. 5-7 K–P). The entire inner ear is derived from the otic vesicle and consequently is of epidermal ectoderm origin. In the salamander larva, the inner ear shows many structural features of the adult ear. Three **semicircular canals** can be identified which arise from the central cavity, now called the **sacculus-utriculus.** Each semicircular canal is more or less at right angles to the others. They are the anterior-vertical canal, the median-horizontal canal, and the posterior-vertical canal. Each canal possesses a thickened patch of sensory epithelium near its ventral attachment to the sacculus-utriculus. These are the **sensory crista.** They are stimulated by body motion. The sacculus-utriculus is the sensory organ of static equilibrium or balance; in life the cavity contains calcium carbonate crystals termed **otoliths,** or otic sand, which rest on patches of sensory epithelial cells called **maculae** situated in the floor of the sacculus-utriculus. Changes in the body axis result in a shifting of the otic sand. This results in a change in the stimuli to the maculae. These changes are interpreted by the metencephalon in terms of orientation or position. Otic sand will not be seen in sections since an acid fixative and stain were used and they have dissolved the crystals away. The **8th cranial nerve** innervates the crista and maculae. Ganglion cells of the 8th cranial nerve are derived from the otic vesicle; they are therefore of epidermal placode origin; the only other epidermal nerves are the first cranial nerves, the olfactories. All other nerves are of either neural plate or neural crest origin. Prior to metamorphosis, the 8th nerve (auditory) functions exclusively as a sensory nerve of static and active equilibrium. The lagena (cochlea), which will develop as a sensory organ of hearing, has not begun to develop in the young larva.

The lateral line system of sensory pits and nerves from a branch of the 10th cranial nerve is not specifically shown in any figures. The system can be identified in individual sections as small pits in the surface epidermis with elevated edges somewhat as a cross section through a volcano might look. They are most abundant in the posterior parts of the head near the ear and the base of the gills, or along the median flank region close to the course of the lateral abdominal vein.

c. Endodermal and gut derivatives. The mouth, pharynx, and pharyngeal clefts of the foregut will be discussed in association with a general consideration of visceral arch structure and relationships. At this point the post-pharyngeal gut will be treated as a unit. Posterior to the last gill arch, the gut narrows very appreciably as the base of each gill finally fuses with the lateral body wall (Fig. 5-7 P). At this level note the midventral evagination from the floor of the gut; this is the **laryngotracheal groove.** It can be followed back under the **esophagus,** which is the posterior extension of the gut behind the pharynx. The laryngotracheal groove very soon will branch into right and left **lung buds** (Fig. 5-7 Q). They are just in the process of expanding into functional respiratory air sacs and there is often great individual variation in their size and posterior extension. If they were inflated with air at the time of fixation, they frequently are the most conspicuous visceral organs to be seen in the body cavity at levels represented by Fig. 5-7 Q–R. Follow the esophagus posteriorly into the **stomach,** a widely expanded portion of the gut. The **spleen** is just beginning to form and will appear as a very dense oval mass of cells in the left lateral wall of the stomach. Note that the stomach is suspended by a

dorsal mesentery into which the external glomeruli of the pronephroi will project (Fig. 5-7 H). Identify the **liver** and **pancreas** with the aid of Figure 5-7 S. They are evaginations of the gut tube that have special glandular functions associated with digestion. Note that the liver lies in the **ventral mesentery of the gut.** The pancreas is predominately located in the dorsal mesentery. Toward the posterior end of the liver, a large **gall bladder** will be found. Both the liver and pancreas show glandular differentiation. The cells of these two organs can be distinguished by the very strong tendency for pancreatic cells to accumulate the red cytoplasmic stain used in these preparations. The **intestine** is a simple tube suspended in the peritoneal cavity by a dorsal mesentery. There are no features of special interest associated with it. It can be traced directly into the **cloaca** and **anus** (Fig. 5-7 T–U).

d. Muscles. Three categories of non-cardiac muscles deserve attention. The first of these is the **visceral musculature** of the gill arches. These muscles are derived from splanchnic mesoderm of the pharynx and will be discussed later in connection with the gill arches. Their function is associated with the breathing movements of the gills during larval development. Second, the **hypobranchial muscles** are derived from occipital myotomes; they migrate into the floor of the mouth and attach to the basibranchial cartilages of the tongue (Fig. 5-7 G–L). After metamorphosis they will aid in carrying out swallowing and breathing movements. In birds and mammals the hypobranchial muscles are innervated by the 12th cranial (hypoglossal) nerve. Last, the **segmental muscles** are derived from embryonic myotomes from trunk and tail body levels. Each myotome sends a muscle bud into the lateral body wall and becomes divided into a dorsal and ventral component. The dorsal components form **epaxial muscles** situated lateral to the spinal cord and notochord; these are clearly segmental along the entire length of the trunk and tail. The ventral components of myotomes are located between the belly epidermis and parietal peritoneum (Fig. 5-7 Q–U). These are the **hypaxial muscles.** These relationships are best seen in sections through the middle of the trunk region. Both epaxial and hypaxial muscle masses contribute to the muscles of the fore and hind limbs.

e. Mesomere and kidney. The early kidney of the salamander larva at hatching consists of three pairs of functional **pronephric tubules** (Fig. 5-2, 5-7 A). No vestige of the mesonephros has made an appearance. However, in later larval development, it will appear and replace the pronephros as the functional kidney of the late larva and adult. The pronephric tubules are located in the dorsal wall of the peritoneal cavity at the level of the stomach (Fig. 5-7 R). They are greatly coiled tubules which open by means of **nephrostomes** directly into the anterodorsal wall of the **peritoneal cavity.** On each side, the 3 pronephric tubules unite to form the paired **pronephric ducts** which grow posteriorly close to the dorsal mesentery of the gut. They finally open into the cloaca (Fig. 5-7 S–U). Wastes from the bloodstream reach the pronephric tubules by an indirect route. Wastes are filtered out of the blood from capillary tufts which arise from the dorsal aorta and project into the dorsal mesentery of the gut (Fig. 5-7 R). These capillary tufts are called **external glomeruli.** Wastes pass from the glomeruli into the peritoneal cavity. Peritoneal fluid containing dissolved waste is swept into the pronephric tubules by ciliary action of the nephrostomes. The wastes are slightly concentrated in the pronephric tubules and are then transported by the pronephric duct to the cloaca where they are discharged through the anus. In later development the pronephric tubules disappear entirely. Only the pronephric duct remains as the functional duct of the mesonephros. As will be recalled, there is a truly remarkable similarity in pronephric and mesonephric development as seen in larvae of amphibians and cyclostomes. About the only real difference seems to be that amphibians have only three pairs, whereas the Ammocetes larva has five pairs of pronephric tubules.

f. Circulatory system. Details of the circulatory system cannot be seen very clearly. The vessels are very small and are difficult to trace in serial section; however, by careful observation, the entire circulatory plan shown in Figure 5-2 can be traced. This may be done if interest and time permit.

Refer to Figure 5-7 L–Q for representative cross sections through the heart. This organ is a simple S-shaped tube similar to that in adult fishes. The main divisions of the heart, the distribution of the **anterior** and **posterior cardinal veins,** and the entrance of the **common cardinal veins** into the heart, are essentially the same as will be described for the 48 and 72 hour chick embryos. The **hepatic vein** arises in the liver and enters the posterior end of the sinus venosus. The hepatic vein derives its blood from the gut by the **hepatic portal vein. Lateral abdominal veins** which enter the common cardinals are homologous to the allantoic veins of bird and mammal embryos.

g. Skeletal system. The common plan for the vertebrate skeletal system is given in Figure 5-3. In addition to the notochord which is the primary axial skeleton of chordates, vertebrates have three additional supporting cell types, namely, connective, cartilage, and bone tissue. Most adult skeletal parts arise in the embryo, first, by being roughed out by condensations of connective tissue. Then they are molded in cartilage, and, last, they are permanently cast in bone. In addition to the distinctions in tissue type, it is customary to divide the skeleton into four primary divisions based on fundamental differences in germ layer origin. These are:

Axial skeleton: Derived from scleratome mesoderm and consisting of vertebral column, ribs and sternum. The cranium or skull is usually also included as part of the axial skeleton; however, it is of complex germ layer composition.

Appendicular skeleton: Derived from lateral plate mesoderm and consisting of bones of the pectoral and pelvic girdles and appendages.

Visceral skeleton: Derived from neural crest ectoderm and forming the skeletal supports for the gill arches.

Dermal skeleton: Derived from dermatome mesoderm and consisting of scale plates of the skin and their various derivatives in diverse vertebrate groups.

Figure 5-3 shows the primary elements of the skeleton in their simplest form. The skeleton of the young salamander larva is of particular value from the comparative and evolutionary point of view because it illustrates, with almost diagrammatic simplicity, many aspects of the ancestral common plan for the origin of the cranium, vertebrae, appendicular skeleton, and visceral arches. The only skeletal division that is entirely absent is dermal skeleton, no parts of which have yet made their appearance in the early feeding larva.

(1). Cranium and axial skeleton. The primary components of the head skeleton of the early salamander larva are shown as diagrammatic reconstructions in Figure 5-4. Different styles of shading are used to indicate the germ layer origin of various cartilaginous components which underlie the brain. These elements are collectively called the **brain pan,** or **chondrocranium** (meaning cartilaginous skull). It will be noted that scleratome, head mesoderm, and neural crest ectoderm participate in cranium formation at this time; later on, dermatome will contribute most of the cranial roof, or **calvarium.** The chondrocranium begins its development as two pairs of parallel cartilaginous bars that underlie the brain. These are the **trabeculae** which underlie the forebrain, and the **parachordal cartilages** which underlie the midbrain and most of the hindbrain. The parachordals lie lateral to the anterior end of the notochord. The other primary components of the chondrocranium are the three pairs of **sensory capsules** which surround primordia of the nose **(nasal capsules),** eyes **(sclerotic capsules),** and ears **(otic capsules).** Extensive expansions and fusions between cartilaginous bars and sensory capsules will result in the formation of a deep cartilaginous brain pan with only minor openings to admit the passage of nerves and blood vessels.

Identify the following cranial components in the cross sections of the salamander larva. The **nasal capsules** (Fig. 5-7 C) are cartilages which partially enclose the nasal canals. The **ethmoid plate** is an anterior median fusion of the two nasal capsules. This plate of cartilage is only beginning to appear in preparations available for study. Posteriorly, the cartilage of the ethmoid plate and nasal capsules is continuous with the paired **trabecular cartilages** (Fig. 5-7 C–J). They run ventrolaterally under the telencephalon and diencephalon. The space between the two trabeculae is called the **hypophyseal foramen** or fenestrum. The infundibulum projects down into this space from the floor of the diencephalon. The trabeculae, at the level of the optic nerve, have an opening called the **optic foramen,** through which the optic nerve passes from the eye cup into the optic chiasma on the floor of the diencephalon. Trabeculae, anterior to the optic foramen, as well as the nasal capsules and ethmoid plate, are of mesectodermal (i.e., neural crest ectoderm) origin.

Posterior to the optic foramen, the trabecular cartilages are derived from head mesoderm. The paired trabecular cartilages unite with the paired quadrate cartilages laterally and the paired parachordal cartilages posteriorly, at the level of the midbrain and anterior hindbrain (Fig. 5-7 I–K). The paired parachordal cartilages lie just lateral to the anterior end of the notochord (Fig. 5-7 L–O). Laterally, at the level of the inner ear, the parachordal cartilages are fused with the **otic capsules.** These surround the inner ear and form the supporting cartilaginous labyrinth. Head mesenchyme gives rise to the trabeculae posterior to the optic foramina, to the parachordal cartilages, and to the otic capsules.

At the posterior end of the medulla, three of the four **occipital vertebrae** will be found (Fig. 5-7 P, Q). These cartilages are derived from sclerotomes of the first 4 somites. In later development, they will fuse with the parachordal cartilages and otic capsules to form the occipital region of the skull. In the young tadpoles studied here, the occipital cartilages consist only of the **neural arches** of the occipital sclerotomes. No centra have yet developed around the notochord. The **notochord** itself is the primary axial skeleton of the trunk and tail region at this stage in development. However, it will later be replaced as the vertebrae of the trunk and tail region begin to develop.

(2). Appendicular skeleton. The components of the appendicular skeleton are shown in the diagram of the common plan of the skeleton in Figure 5-3. As indicated earlier, the bones and cartilages of the pelvic and pectoral girdles and appendages are derived from the parietal layer of hypomere, or lateral plate, mesoderm. One exception to this is found in the pectoral girdle of tetrapod vertebrates; this is the clavicle which is derived from dermal bone; it often replaces the procoracoid in the anteroventral part of the pectoral girdle.

In the young salamander larva, the hindlimb primordium is represented by a very small limb bud on both sides of the cloaca. No regional differentiation can be distinguished in it. In contrast, the forelimb is well developed and shows the basic plan of the primitive tetrapod limb at a cartilaginous stage in its development. The forelimb is attached just posterior to the gill chamber. At this level, the primordium of the **pectoral girdle** is represented by a well developed **coraco-scapular cartilage** (Fig. 5-7 R). This is a slightly curved bar of cartilage located in the lateral and ventral body wall. The articular surface of the cartilage, which receives the head of the humerus, is the **glenoid fossa.** The ventral part of this cartilage bar will form the coracoid and procoracoid bones. The midventral fusion of the coracoid processes will also contribute to the formation of the sternum. The dorsal component of the coraco-scapular cartilage will form the scapula and supra-scapula of the adult. Cross sections through the forelimb will show the **humerus** in the upper arm, and the **radius** and **ulna** (Fig. 5-7 S) in cross sections of the lower arm. Additional cartilages of the wrist and digits can be made out in the terminal parts of the appendage.

(3). Visceral skeleton and visceral arches. The primary components of the visceral skeleton are shown for the vertebrate common plan in Figure 5-3, for the salamander larva in Figure 5-4, and for an adult mammal in Figure 5-5. In addition to consideration of skeletal elements that support the gills, it is appropriate to consider the basic structure of a gill arch in its entirety. It is of course preferred that they be demonstrated in a functional gill rather than in a pre-functional embryonic condition. The structural relationships of all elements of a gill are demonstrable with classic simplicity in the young salamander larva.

Typically, the pharynx of vertebrates is perforated laterally by a series of gill slits. Technically they are usually called **visceral** or **branchial clefts.** In the salamander tadpole there are six gill arches, but there are only four complete clefts. They are between the 2nd–3rd, 3rd–4th, 4th–5th, and 5th–6th arches. The first cleft does not perforate, but the pharyngeal component of the first slit does form and will persist as the cavity of the middle ear. The bars of tissue that separate visceral clefts are visceral arches. The relationships of visceral clefts and visceral arches are shown in Figure 5-6.

Each visceral arch typically contains a **visceral cartilage,** an **aortic arch,** and **branchial muscles,** and is innervated by a **cranial nerve.** The cartilage and nerve are of **neural crest origin,** and the aortic arch and muscles are of **head mesenchyme origin.** Each entire arch structure is induced by the visceral cartilage component of the arch. The remaining structures fail to develop if the cranial neural crest is removed during the neurula or tailbud stage. The **branchial muscles** are considered to be serially homologous with the smooth muscles of the gut and hence are also termed **visceral muscles.** They function in moving the gills or gill derivatives. The aortic arches in functional gills, such as those found in the salamander larvae, differ from those seen in the early embryonic stages. In the functional gill, the **aortic arch** has a capillary network interposed between the ventral and dorsal halves of the aortic arch. The ventral half of each arch arises from the ventral aorta and carries blood to the gill capillary bed. This part of an arch, the **afferent branchial artery,** is numbered in a manner appropriate to the gill arch it supplies. The dorsal half of the aortic arch returns blood from the gill capillaries to the dorsal aorta. It is called an **efferent branchial artery.**

In the following directions for study of the gill arches in the salamander larva, specific references are given to figures of sectioned material. You are encouraged to make use of general information in Figures 5-2, 5-4, and 5-6 to clarify special details.

First visceral arch (mandibular arch): This arch lacks an aortic arch and is specifically modified for the support of the lower jaw. The first visceral cartilage has two derivatives. These are **Meckel's cartilage** (Fig. 5-7 J) which is the horseshoe-shaped support of the lower jaw, and a pair of **quadrate cartilages** (Fig. 5-7 I, J). The quadrates are fused with the posterior end of the **trabecular cartilages** and with the otic capsules at their dorsal ends (Fig. 5-4 A, B). The ventral end of the quadrate provides the articulating surface for the posterior ends of Meckel's cartilage. The quadrate and Meckel's components of arch I correspond respectively with the epibranchial and hypobranchial components of a typical aortic arch (see Figs. 5-3 and 5-6 A). Of the four to six distinct kinds of jaw articulations in vertebrates, the type in the salamander larva is a, so-called, **autostylic jaw suspension.** This means that the jaw is part of the first arch and articulates within itself. This is the most primitive kind of vertebrate jaw suspension. It appeared with the evolution of the first jaws in extinct acanthodian placoderm fish. Various forms of autostylic jaw suspensions are retained in adult crossopterygian fish, amphibians, reptiles and birds. It will be recalled that in mammals the articular ends of the quadrate and Meckel's cartilages are diverted to auditory functions. The quadrate will form the **incus,** and the articular end of Meckel's cartilage will form the **malleus** in the chain of three middle ear bones.

Note the small pointed teeth embedded in the upper jaw, in the roof of the mouth, and along the lower jaw. Under high power it will be seen that they consist of an outer sheath of **enamel** secreted by **tooth germs** of stomodaeal ectoderm origin. The enamel cap has a dense **dentine** inner lining which surrounds the **pulp cavity** of the tooth. The dentine is formed from neural crest ectoderm, and the pulp cavity is primarily formed from wandering head mesenchyme cells. The induction system of the teeth involves the visceral cartilage (Meckel's, quadrate, or trabecular cartilages, all of neural crest origin) which is the inductor that causes stomodaeal ectoderm to form tooth germs. The tooth germs then secrete the enamel and induce the formation of their dentine and pulp cores from surrounding neural crest and head mesenchyme. Visceral muscles of arch I function primarily in moving the lower jaw; of these muscles, the **masseters** are the most conspicuous. They extend from the posterior part of the lower jaw to the anterodorsal edge of the quadrate cartilage (Fig. 5-7 J). These muscles are innervated by the **5th cranial nerve,** the trigeminus. The ganglion of the trigeminus will be found between the dorsal end of the quadrate cartilage and the lateral wall of the medulla (Fig. 5-7 J).

The remaining visceral arches, numbered 2 through 6, support functional gills; however, the second and sixth are rudimentary. The three gills which appear as feathery lateral head appendages, are extensions of the third, fourth, and fifth arches. As in bony fish, the gill chamber is covered by an additional fold of skin that arises from the head and opens behind the gill chamber. This covering is the **operculum,** or **gular fold.** It hides the gill slits from view. The cavity enclosed between the operculum and body wall is the **gill chamber** or **opercular cavity.** Note that visceral cartilages 2 through 6 all are attached in the ventral midline to the **basibranchial cartilage** (Figs. 5-3 and 5-4 B). This cartilage lies in the floor of the mouth and pharynx, and extends from the base of the tongue posteriorly to the anterior margin of the pericardial cavity. In later development, the basibranchial cartilage will contribute extensively to the body of the definitive hyoid cartilage and larynx. In the larva, it provides the ventral attachment for the **hypobranchial muscles** which are employed in swallowing. Trace the structure of the 2nd through 6th visceral arches through the serial cross sections.

Second visceral arch (hyoid). The second visceral arch will be seen at levels comparable to those shown in Figure 5-7 G–L. The second aortic arch has already disappeared. However, the paired **2nd visceral cartilages** are well formed and can be traced from their points of origin, from the basibranchial cartilage, around the lateral wall of the pharynx, to the dorsolateral wall of the pharynx (Fig. 5-7 T–L). The dorsal end of each **2nd visceral cartilage,** or **hyomandibular cartilage,** will form the single ear bone of the amphibian middle ear, namely, the **columella.** In mammalian development, their homologues will form the **stapes.** The ventral halves of the 2nd visceral cartilages will form the lesser horns of the definitive hyoid cartilage which supports the tongue. The muscles of the 2nd visceral arch are part of the tongue musculature. They are innervated by the **7th cranial nerve,** the facialis. The ganglion of this nerve is closely applied to the ventral wall of the otic capsule and is closely associated with the ganglion of the **8th cranial nerve,** the acousticus (Fig. 5-7 L).

Third, fourth, fifth, and sixth visceral arches. The 3rd, 4th, and 5th arches are the functional gills of the larva. The 6th arch is much reduced but has a vestigial respiratory function. The 3rd, 4th, and

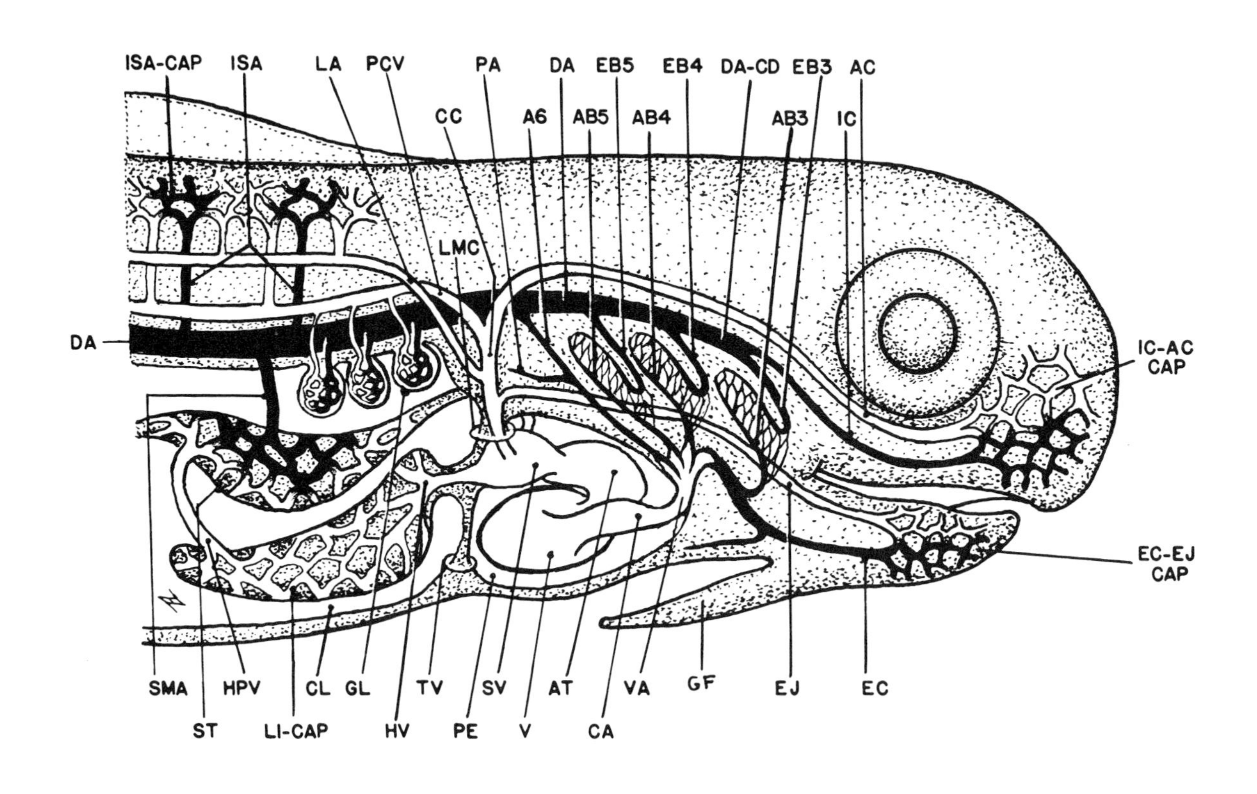

Figure 5-2. Diagram of the anterior circulatory plan of the young salamander tadpole. Legend and checklist of structures to be identified.

AB3 & EB3	afferent and efferent branches of aortic arch 3
AB4 & EB4	afferent and efferent branches of aortic arch 4
AB5 & EB5	afferent and efferent branches of aortic arch 5
A6	aortic arch 6, pulmonary arch
AC	anterior cardinal vein
AT	atrium of heart
CA	conus arteriosus
CAP	capillary bed
CC	common cardinal vein
CD	carotid duct of aorta
CL	peritoneal coelom
DA	dorsal aorta
EC	external carotid artery
EJ	external jugular vein
GF	gular fold
GL	external glomerulus
HV	hepatic vein
HPV	hepatic portal vein
IC	internal carotid artery
ISA	intersegmental artery
LA	lateral abdominal vein
LI	liver
LMC	lateral mesocardium
PA	pulmonary artery
PCV	posterior cardinal vein
PE	pericardial coelom
SMA	superior mesenteric artery
ST	stomach
SV	sinus venosus
TV	transverse septum
V	ventricle
VA	ventral aorta

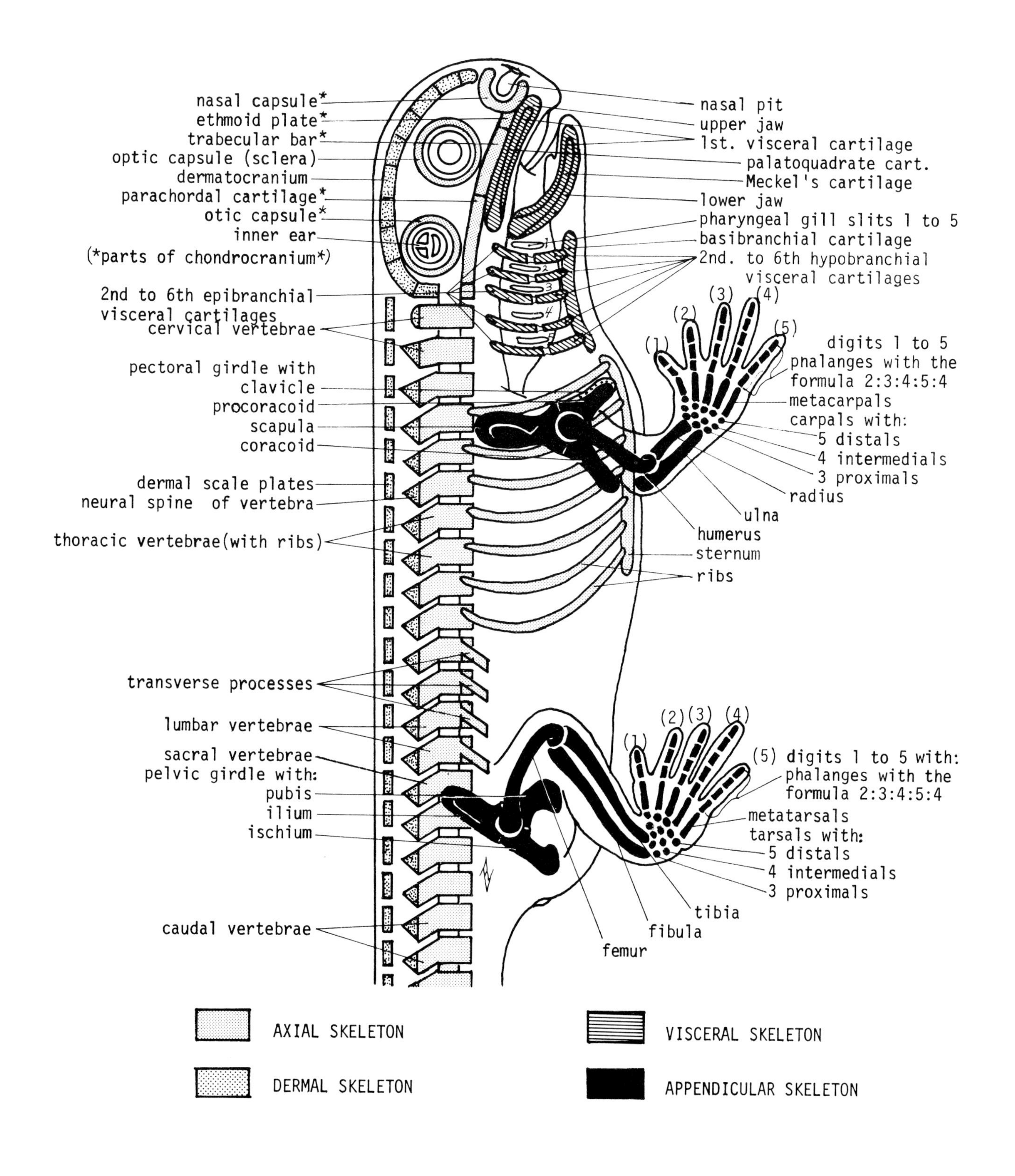

Figure 5-3. Diagram of the common plan of the vertebrate skeleton.

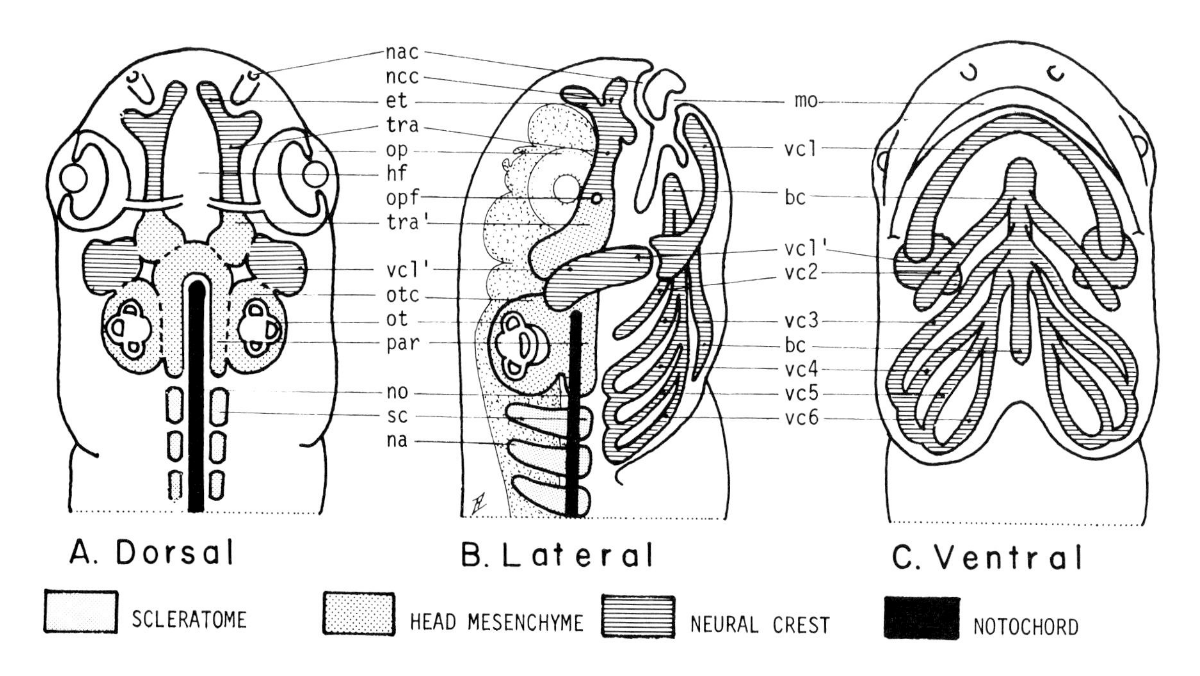

Figure 5-4. The chondrocranium and visceral skeleton of amphibian larvae. Modified from S. Horstadius, *The Neural Crest,* Oxford University Press, 1950. **Legend:**

bc	basibranchial cartilage
et	ethmoid plate
hf	hypophyseal fenestra
mo	mouth
na	vertebral neural arch
nac	nasal canal
ncc	nasal capsule
no	notochord
op	optic cup, eye
opf	optic foramen
ot	otic vesicle
otc	otic capsule
par	parachordal cartilage
sc	scleratome
tra	anterior trabecular cartilage
tra'	posterior trabecular cartilage
vcl	Meckel's cartilage of visceral arch I
vcl'	quadrate cartilage of visceral arch I
vc2	hyomandibular cartilage of visceral arch II
vc3, 4, 5, 6	visceral cartilages of arches III, IV, V, and VI.

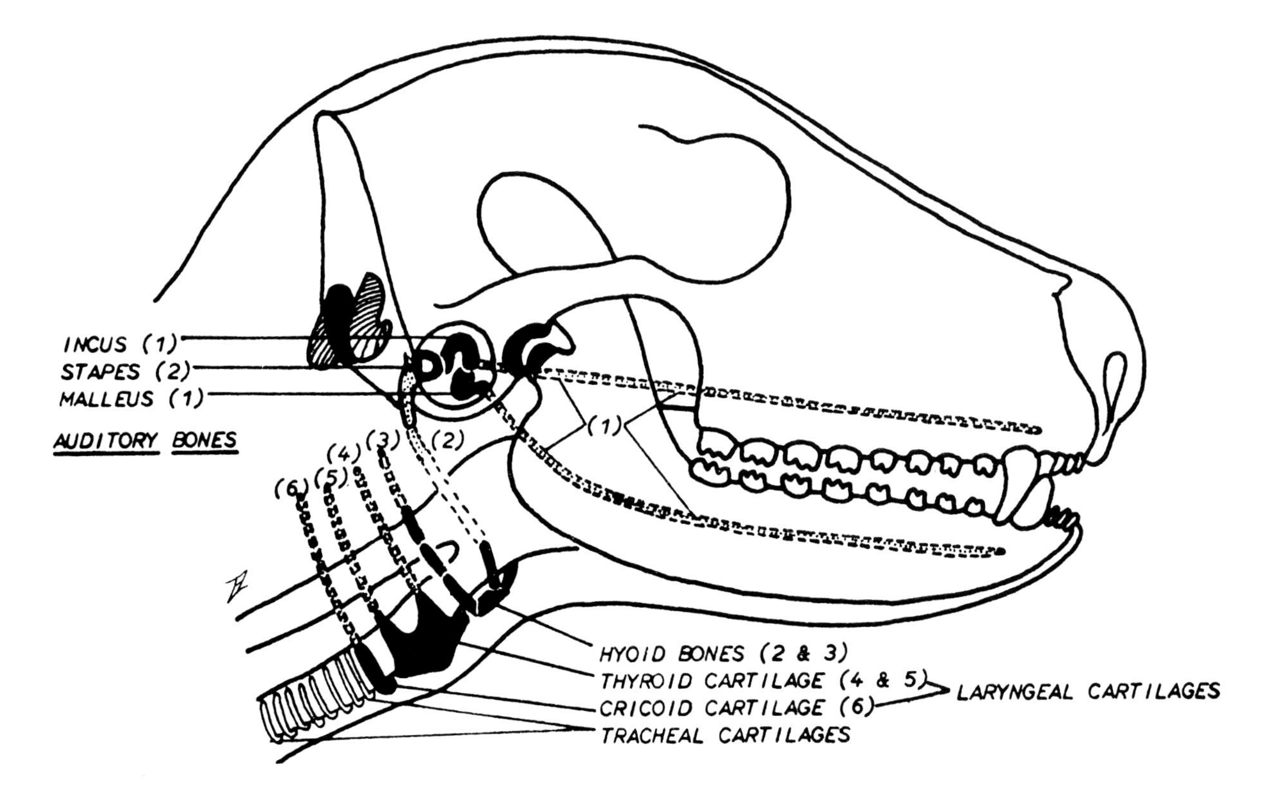

Figure 5-5. Diagram of the mammalian visceral skeleton as seen in a generalized mammal. Numbers in parenthesis refer to embryonic gill cartileges. Broken outlines indicate parts of gill cartilages which degenerate during embryonic development.

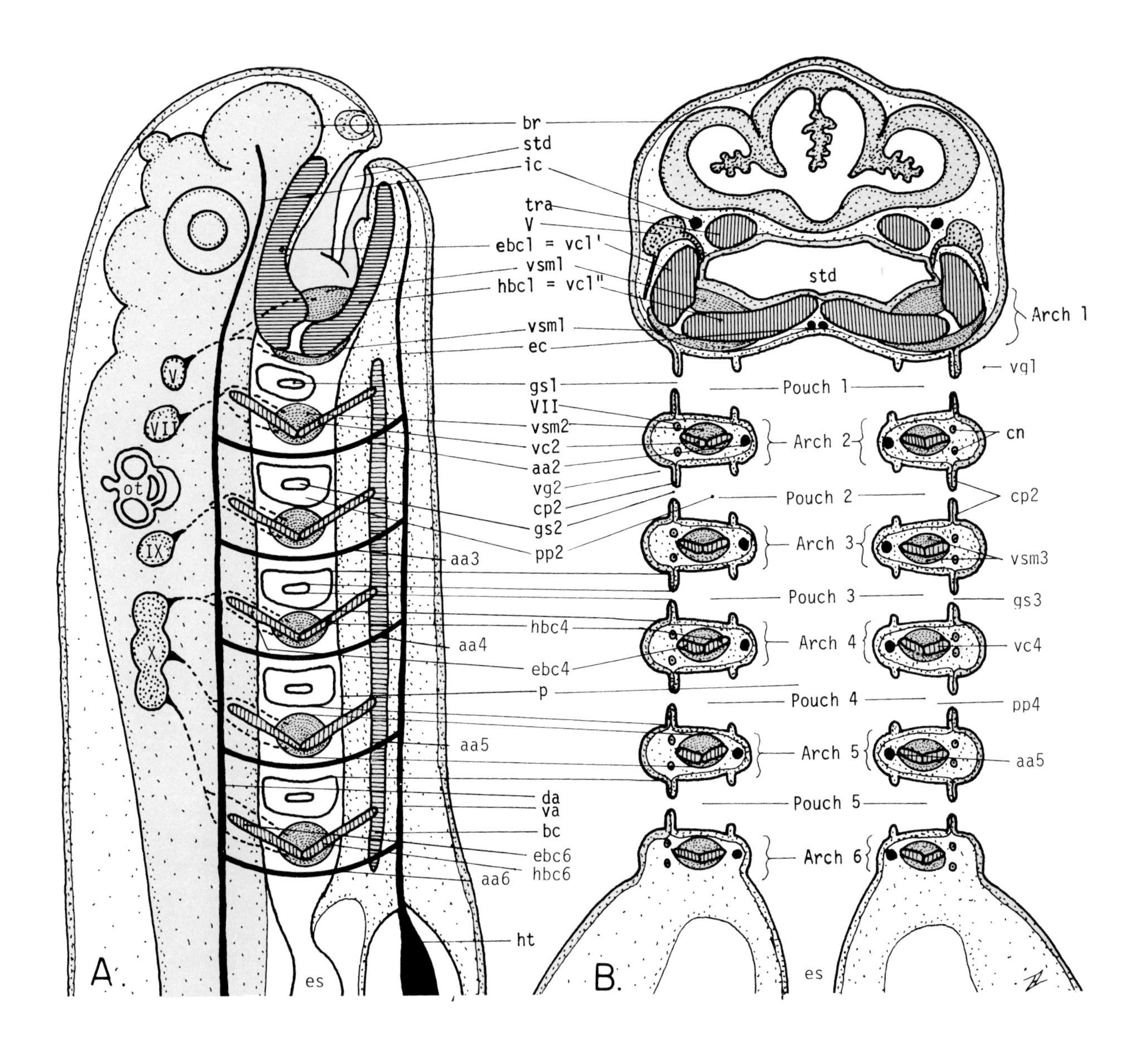

Figure 5-6. Structural relationships of components in visceral arches.

Figure 5-6 A. Lateral reconstruction of the pharyngeal region in a generalized vertebrate.

Figure 5-6 B. Frontal section through the pharynx and visceral arches of a generalized vertebrate. **Legend:** Postscript numbers 1 through 6 refer to corresponding visceral arches. Color code: blue — ectoderm, red — mesoderm, yellow — endoderm.

Legend: Postscript numbers 1 through 6 refer to corresponding visceral arches.

Color code: blue — ectoderm, red — mesoderm, yellow — endoderm.

V	5th cranial nerve
VII	7th cranial nerve
IX	9th cranial nerve
X	10th cranial nerve
aa 1-6	aortic arches
bc	basibranchial cartilage
br	brain
cn	cranial nerve
cp 1-5	gill pouch closing plate
da	dorsal aorta
ebc 1-6	epibranchial cartilage
ec	external corotid cartilage
es	esophagus
gs 1-5	gill slits
hbc 1-6	hypobranchial cartilage
ht	heart
ic	internal carotid artery
p	pharynx
pp 1-5	pharyngeal pouch
ot	otocyst, ear
std	stomodaeum, mouth
tra	trabecular cartilage
va	ventral aorta
vc 1-6	visceral cartilage
vcl'	quadrate cartilage
vcl''	Meckel's cartilage
vg 1-6	visceral groove
vsm 1-6	visceral arch muscle

Figure 5-7. Structure of the 12–15 mm salamander larva. Legend for figures on the following pages.

5-7A. Whole mount reconstruction of a 12–15 mm larva.
5-7B. Median sagittal section of a 12–15 mm larva.
5-7C–U. Representative cross sections through a 12–15 mm larva at levels indicated by correspondingly letter guide lines in 5-7 A and B.

Legend and checklist of structures to be identified

Color code: blue — ectoderm, red — mesoderm, yellow — endoderm.

Cranial nerves:

2 II, optic
5 V, trigeminus
5G semilunar ganglion
7 VII, facialis
7G geniculate ganglion
8 VIII, acousticus
9 IX, glossopharyngeus
9G superior-petrosal ganglion
10 X, vagus
10G jugular-nodes ganglion

General characters:

AA6 6th aortic arch
AB3, 4, & 5 afferent branchial arteries of 3rd, 4th, & 5th visceral arches
AC anterior cardinal vein
ACP anterior choroid plexus
AH aqueous humor of eye
AN anus
AP alar (sensory) plate
AQ aqueduct of Sylvius, or mesocoele
AT atrium of heart
BA balancer
BC basibranchial cartilage
BP basal (motor) plate
CA conus arteriosus
CC common cardinal vein or duct of Cuvier
CL coelom, peritoneal cavity
CLO cloaca
CO cornea, conjunctiva
CS coraco-scapular cartilage
CV caudal vein
DA dorsal aorta
DF dorsal fin
DI diencephalon
DM dorsal mesentery
EB3, 4, & 5 efferent branchial arteries of 3rd, 4th, & 5th visceral arches
EC external carotid artery
ED endolymphatic duct
EG external gills
EJ external jugular vein
EM epaxial myotome
EN external (naris) nostril
EP epiphysis, pineal gland
ET ethmoid plate cartilage
FAL falciform ligament of ventral mesentery
FL fore limb
FM foramen of Munro
FP floor plate
GB gall bladder
GF gular fold, operculum
GL external glomerulus
HM hypaxial myotome
HU humerus
HV hepatic vein
HY hypobranchial muscle
IC internal carotid artery
IN infundibulum
INA internal (naris) nostril
IR iris
IT intestine
LA lateral abdominal vein
LB lung buds
LE lens
LI liver
LJ lower jaw
MAN mantle (gray) layer
MAR marginal (white) layer
MAS masseter muscle of jaw
MB midbrain, mesencephalon
MT metencephalon, cerebellum
MY myelencephalon, medulla
NA neural arch of occipital vertebrae
NAC nasal canal
NAS nasal sensory epithelium
NCC nasal capsule cartilage
NO notochord
NT neural tube, spinal cord
OE esophagus
OP eye, optic cup
OPC opercular (gill) cavity
OPF optic foramen in the trabecular cartilage
OPR optic recess
OR oral cavity, mouth
OT inner ear, otic vesicle
OTC otic capsule cartilage
P pharynx, foregut
P 1, 2, 3, 4, & 5 pharyngeal pouches, or gill slits
PA pulmonary artery of 6th aortic arch
PAN pancreas
PAR parachordal cartilage
PCP posterior choroid plexus
PCV posterior cardinal vein
PE pericardial cavity
PN pronephric kidney
PND pronephric duct
PNN pronephric nephrostome
PNT pronephric tubule
PRF paraphysis
RF roofplate
RP anterior lobe of pituitary, Rathke's pouch
RTF median fibrous layer of retina
RTG' deep ganglion layer of retina
RTG'' superficial ganglion layer of retina
RTS sensory epithelium of retina
RU radius–ulna
SCC semicircular canal
SCR sensory crista of semicircular canal
SG spinal ganglion
SL sulcus limitans
SM sensory macula of sacculus–utriculus
SP spleen
ST stomach
SU sacculus–utriculus
SV sinus venosus
T teeth
TE telencephalon, cerebrum
TO tongue
TR trachea
TRA trabecular cartilage
V ventricle of heart
V1-2 lateral ventricle of telecephalon
V3 third ventricle of diencephalon
V4 fourth ventricle of myelencephalon
VC1 Meckel's cartilage of visceral arch 1
VC1' quadrate cartilage of visceral arch 1
VC2 hyomandibular cartilage of visceral arch 2
VC3, 4, 5, & 6 cartilages of 3rd, 4th, 5th, & 6th visceral arches
VF ventral fin
VM ventral mesentery
VSM 1,2,3,4,5, & 6 muscle of 1st, 2nd, 3rd, 4th, 5th, & 6th visceral arches
VTH vitreous humor of eye

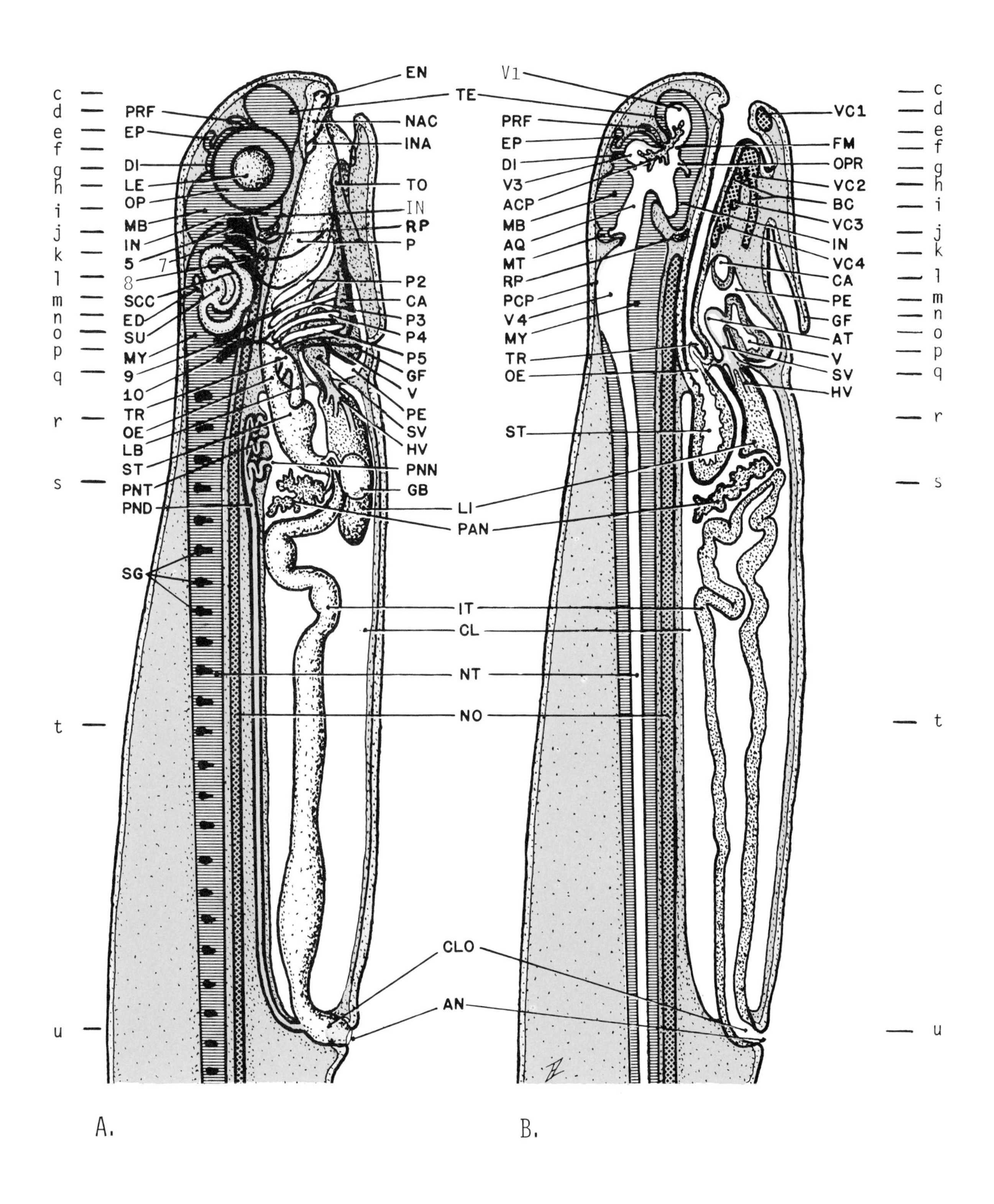

Figure 5-7 A. Whole mount reconstruction of a 12–15 mm salamander larva.
Figure 5-7 B. Median sagittal section of a 12 mm salamander larva. See legend.

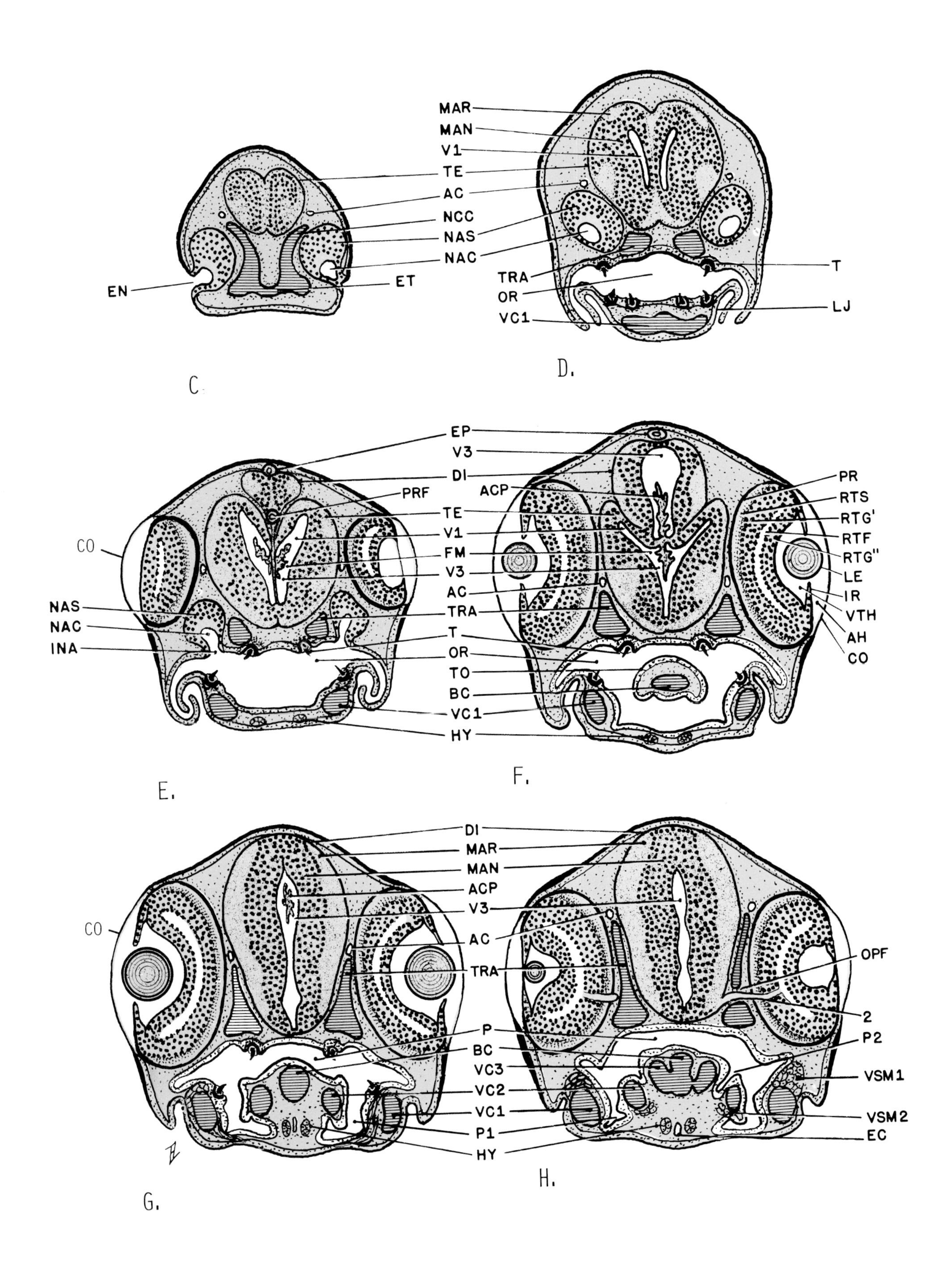

Figure 5-7 C–H. Representative cross sections through a 12–15 mm salamander larva, continued. See legend.

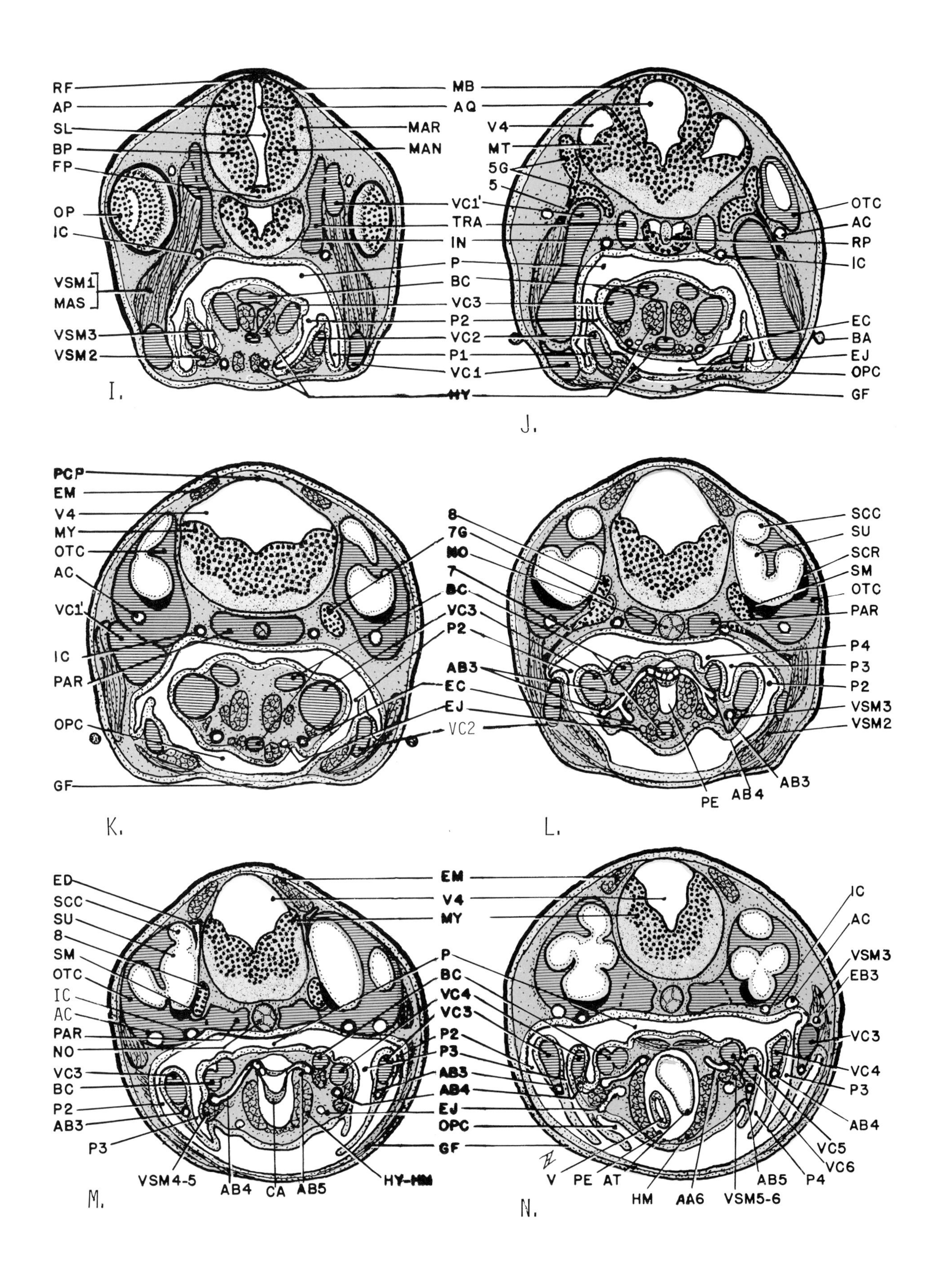

Figure 5-7 I–N. Representative cross sections through a 12–15 mm salamander larva, continued. See legend.

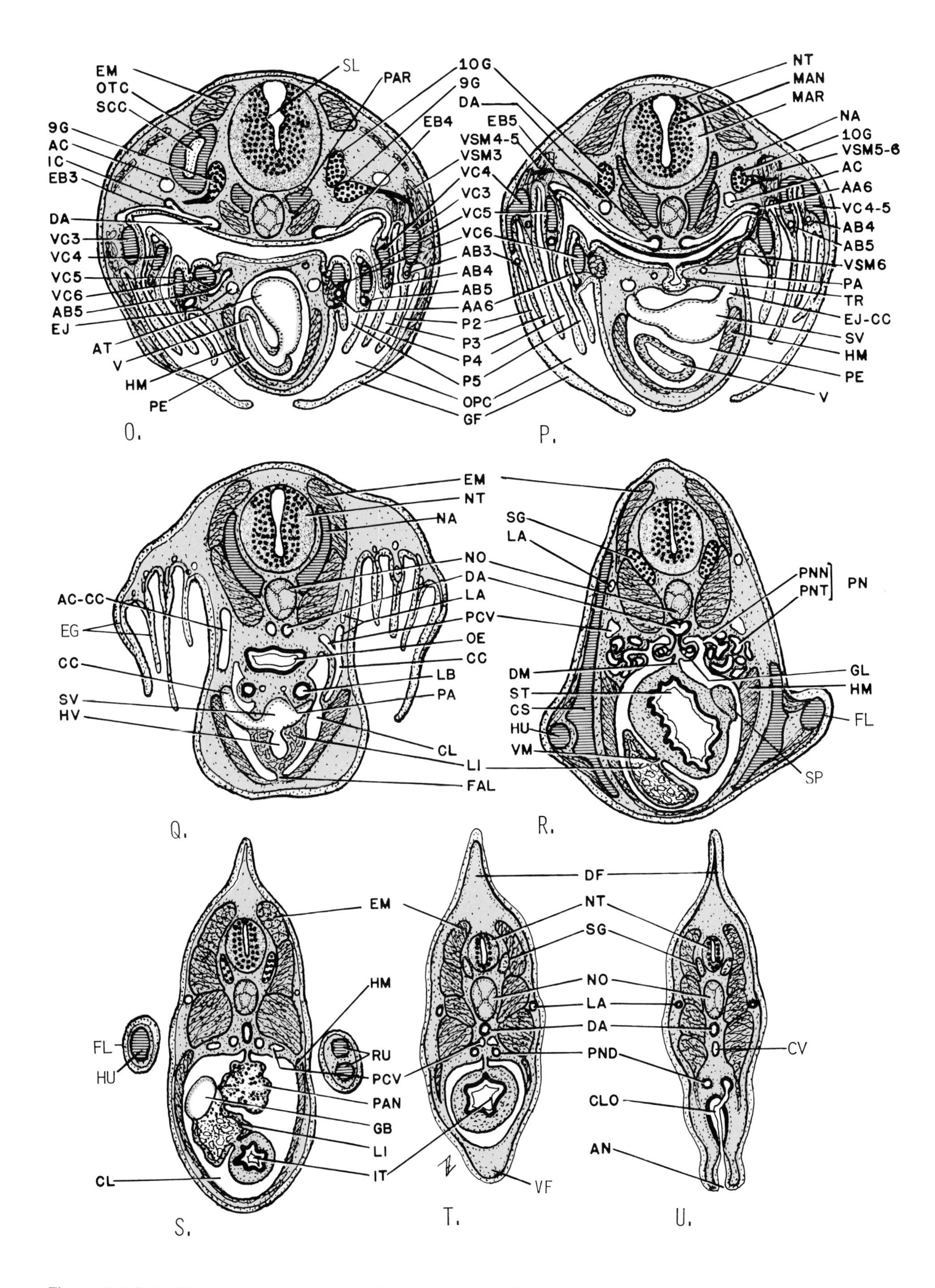

Figure 5-7 O–U. Representative cross sections through a 12–15 mm salamander larva, continued. See legend.

5th arches are separated from the body wall by complete 2nd, 3rd, 4th, and 5th **pharyngeal clefts** that open from the pharynx into the opercular cavity, which, in turn, opens posteriorly to the exterior. The 6th gill arch is continuous with the body wall posterior to the pharynx, and no cleft opens posterior to it. Trace the course of the last 4 gill arches through the serial sections from their ventral origins in the floor of the pharynx around the lateral wall of the pharynx.

The 3rd visceral cartilage arises independently from the basibranchial cartilage, whereas the 4th, 5th, and 6th cartilages have a common origin from the basibranchial rod (Fig. 5-4 B, C). The visceral arch musculature of the 3rd through 6th arches functions in the tadpole in the movement of the gills. This is analogous to the "breathing" movements of terrestrial air-breathing forms. Homologous muscles in the definitive mammal will supply the pharynx, larynx, and trachea. Note (Fig. 5-7 O) that the **9th cranial nerve** (glossopharyngeal) innervates the muscles of the 3rd arch. The **10th cranial nerve** (vagus) is not conspicuous, but nerve branches can be traced to the visceral muscles of the 4th, 5th, and 6th arches if care is taken in following them through the serial sections (Fig. 5-7 P).

The **aortic arches** and gill capillary networks are best seen in whole mount injected specimens (see above). However, with care these very small vessels can be traced through the gills. Reference to Figures 5-7 L–Q will show that the **afferent branchial arteries** of the 3rd, 4th, and 5th aortic arches lie ventral to the visceral cartilage. The **efferent branchial arteries** are situated dorsal to the gill cartilage. The **6th aortic arch** has a reduced gill circulation but does not have clearly defined afferent and efferent roots. It will later send a branch to the developing lung. This will be the functional **pulmonary artery.** The major extent of the 6th aortic arch, as seen at this stage of development, corresponds with the ductus arteriosus in foetal mammalian development.

In conclusion, the study of the 12 mm salamander larva has revealed a general and comprehensive structure that is more primitively fish-like than any adult living fish. It is not surprising that early embryologists saw evidence of recapitulation of ancestral phases in cases such as this. At the very least, the structure of the larva more clearly reproduces the ancestral developmental plan than does the adult stage. The same can be said for the Ammocetes larva and the adult cyclostome. **Indeed, it is in the larval stages from these two vertebrate classes that evidence for homology is strongest.** The comparison of adult cyclostomes and amphibians, or of larval amphibians with adult cyclostomes, provides appreciably less striking degrees of correspondence. We will return to further consideration of the questions raised by the recapitulation theory in Chapter 8 and at that time, will view these facts in an alternative light.

C. SOME BASIC TECHNIQUES FOR THE STUDY OF EMBRYONIC DETERMINATION AND DIFFERENTIATION IN LIVING AMPHIBIAN EGGS

The recapitulation theory plays a lesser role in embryological thought at the present time than it did at the peak of evolutionary dominance over biological theory. At the present time in embryology, more attention is directed toward identifying and explaining mechanisms that determine embryonic cells. Induction series, underlying the organogenesis of homologous organs, reveal many remarkable similarities in species from different vertebrate Orders or Classes. This must mean that there is a genetic basis for inductive stimulus and response.

We can view similarities in embryos of different vertebrates as being homologies based on similar inductive series. In this sense we can view the evidence for recapitulation as also being evidence for the activation of similar gene sequences during ontogenic cycles. From both the old evolutionary and the new genetic frame of reference, homologies in development seem to arise from legacies of common ancestral origins.

If living tailbud embryos of frogs or salamanders are available from natural or artificial insemination (see Chapter 3), the following simplified experiments can be carried out on embryonic primordia of different ages to test the time sequence in determination and differentiation of embryonic organ rudiments.

In the event that living materials are not available, selected motion pictures of experimental techniques and results will be shown.

Materials needed at each operating station. A dissecting microscope and focusable microscope lamp for reflected illumination are essential. In addition, on both sides of the microscope, for use with right and left hands, the following instruments and dishes are needed: 1 fine dissecting needle, 1 hair loop, 1 sharpened watchmaker forceps, 1 rounded glass probe, 1 wide-mouth curved embryo pipette, 1 wide-mouth standard medicine dropper, and, if available, a pair of iridectomy scissors, or 1 sharp glass needle. The above instruments should be arranged, in that order, on an instrument tray made with notches 1 1/2 inches apart to hold the instruments ready for use. In addition, you will need a flask of 100 ml sterile Niu-Twitty operating fluid (alternatively, half-strength Ringer's serves quite well), 2 wax or agar bottom operating dishes, 2 wide-mouth tumblers or beakers filled to a depth of 3 inches with 70% alcohol, 6 agar coated culture dishes for operated cases, 6 wide-mouth jars or fingerbowls to cover culture dishes, 1 vaseline syringe gun, 1 500 ml flask of cool boiled tap water. See Figure 5-9 for a suggested arrangement of these items on the table.

1. Preparation and use of instruments. Microsurgical instruments for the most part need to be manufactured individually or, at the very least, commercial products need to be sharpened. Figure 5-8 illustrates most of the following instruments.

a. Fine dissecting needles. These are made by mounting No. 000 stainless steel insect pins in bacteriological Pasteur pipettes into which melted paraffin has been drawn. Insert the pin head-first while the paraffin is still liquid, and let it harden. These needles are used for coarse cutting and transfer of embryonic parts. Alternatively, needles can be sealed in place with any of a variety of polymerizing plastic glues.

b. Fine glass needles. These can be made most easily by starting with a bacteriological Pasteur pipette and melting the tip of the pipette to form a solid bead of glass at the end. Then take two such closed pipettes and melt them in a fine Bunsen flame until they just begin to glow an orange yellow; touch the two melted beads together to fuse them. Then remove them from the flame and quickly pull. You should get a sharply tapering point that ends about 3/8 to 1/2 inch from the tip of the pipette. These needles are used for fine cutting and for isolating parts. Expect to pull many unsuitably long "whips" at first; they can be broken off, melted down, and pulled again until proper points have been obtained.

c. Hair loops. These are made by bending a piece of fine hair (preferably baby hair) so that both ends are pushed into the tip of a bacteriological Pasteur pipette which has been pulled down into a narrow tip that is only slightly larger in diameter than is necessary to accept a double thickness of hair. The hair loop is anchored in place by drawing melted paraffin into the capillary and letting it harden. Clean the excess paraffin from the hair by pulling the hair several times between the thumb and forefinger. Hair loops are used to move embryos around and to hold them in place while microsurgery is being performed. Alternatively, hair loops can be mounted with polymerizing plastic glue.

d. Watch maker forceps of stainless steel can usually be obtained with the assistance of local jewelers or from hobby shops. They must be sharpened under the microscope with a fine Arkansas stone so that both tips will end at the same level and in perfect points. These are used in moving tissues from one location to another and are generally useful wherever exact movement of parts is necessary.

e. Wide-mouth curved embryo pipettes. These are made from 8 mm soft glass tubing, or from Pasteur pipettes, by heating and drawing out the glass and bending it in a quarter circle arc before it hardens. The tip of the pipette should be broken off and fire polished so that the final pipette tip has a diameter of about 2 mm, or sufficient to allow a frog embryo to enter easily. These are used in transferring embryos or parts of embryos from one dish to another. In using the pipettes, suck up the embryo into the pipette and then turn the bent tip upwards so that the embryo will rest in the bow of the pipette. In this way the embryo can be kept from coming into contact with the air–water interface at the pipette mouth. It is particularly important to transfer operated cases in this way because the interfacial tension at the water surface will generally tear the embryo apart if injured epidermis passes through, or even touches, the water surface film.

f. Standard wide-mouth medicine dropper. These are prepared by breaking off the tip of a standard pipette after scoring it with a diamond chip pencil. The cut surface should then be flamed to give a burnished smooth edge. These are used in transferring embryos, still in their jelly coats, from dish to dish.

g. Round glass probes. These are prepared by drawing out glass rods to various diameters from 0.5 to 2.0 mm. Cut them off with a three-quarter inch taper; the cut ends should then be fire polished until they form smoothly rounded hemispherical ends. These will be needed in various sizes for making depressions in wax operating dishes used to hold embryos stationary during microsurgery.

h. Vaseline syringe gun. Vaseline is used for various purposes in sealing cultures and support-

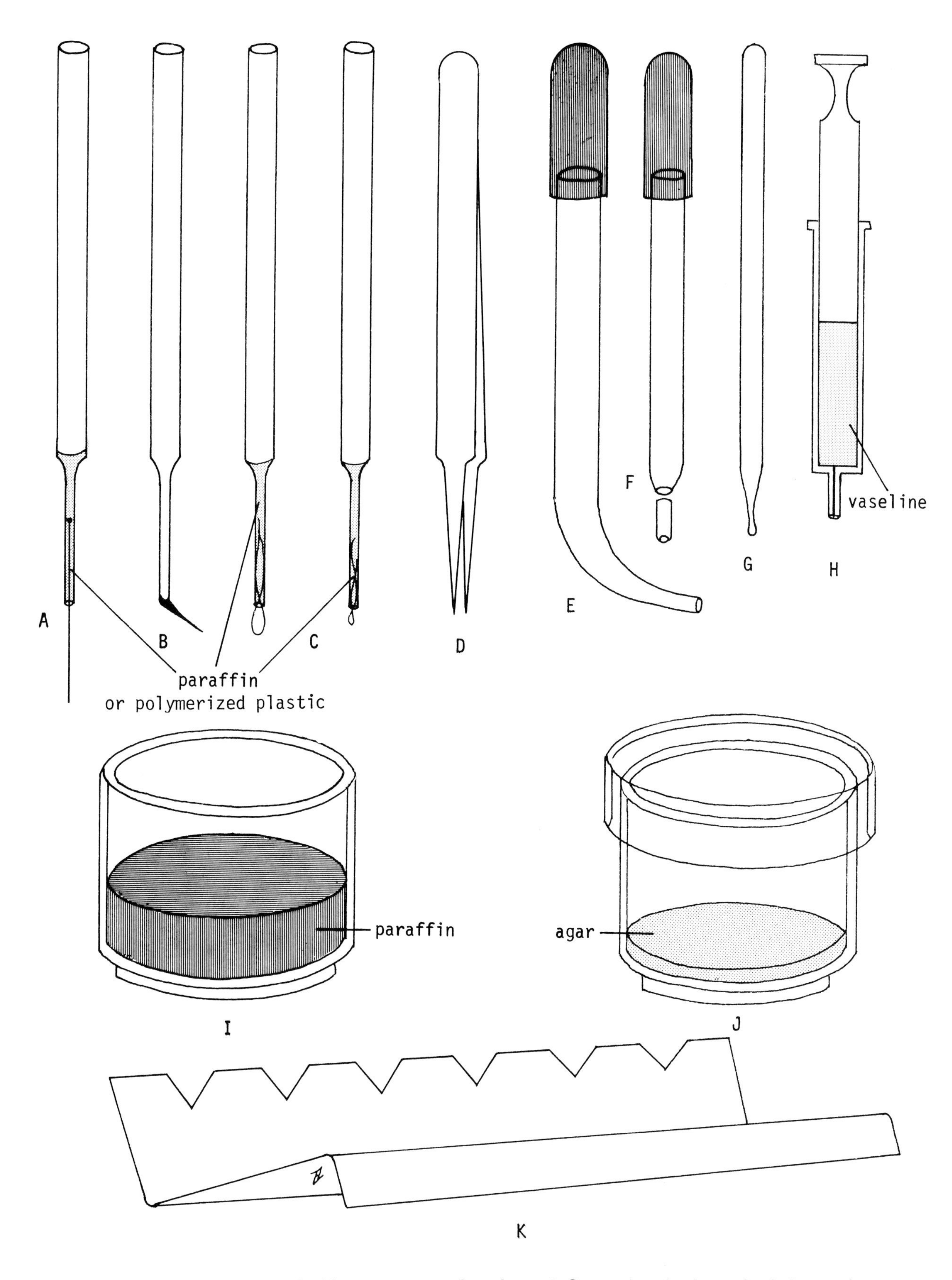

Figure 5-8. Diagrams of microsurgical instruments and equipment. See explanation in text for their manufacture and use.

ing coverslips or slides for special operations. Vaseline is most conveniently handled by filling a plastic disposable hypodermic syringe half full of melted vaseline and allowing it to solidify; 5 ml syringes without needles are a favorable size to use.

i. Paraffin bottomed operating dishes. Two inch Stender dishes or Carolina culture dishes about 1 1/2 inches deep are ideal. Ordinary paraffin, or paraffin blackened with lamp black or colored crayons, should be melted and poured into culture dishes to which a large iron washer or lead fish-weight has been added. The latter should be large enough to keep the paraffin from floating when the culture or operating medium is added. Paraffin should be about one-third to one-half the depth of the operating dish.

j. Agar bottomed wide-mouth jars. Non-nutrient agar should be made up, at a concentration of 2 to 3% by weight, in distilled water by boiling and filtering through cheese cloth to remove undissolved lumps. While still hot, pour enough agar into small Syracuse dishes or 2-inch wide-mouth collecting jars to a depth of about 3 mm. While the melted agar is still liquid, rotate the jar at an angle so it will form a thin film up the side of the dish. Then let the dishes stand until the agar is completely hardened. Dishes should be covered for storage to prevent drying and contamination. Jar lids, aluminum foil, or small plastic covers can be used. The dishes will be used for culturing experimental animals and small isolated organ primordia enclosed in epidermis.

k. Instrument trays. These can be made from tall disposable aluminum soft drink cans. Remove the top and bottom and cut the wall so that a flat sheet of metal will be obtained. It can be squared with scissors or paper cutter. Make right angle bends in the metal at opposite sides and in opposite directions. The width of the bend at one side should be about 3/4 inch, and about 1 1/2 inches wide at the other. In the latter make a series of V-shaped notches about 1 1/2 inches apart to hold instruments securely with their working tips supported in air. The notches should be widely enough spaced to permit you to pick up individual instruments with ease as needed.

2. Preparation of culture media and preserving fluids. Table V-1.

a. Operating and culture fluids. Among the most generally useful solutions for amphibian operations and tissue culture are Ringer's, Holtfreter's, and Niu-Twitty solutions. Formulae are given in Table V-1, with instructions for their preparation. Additional instructions for their handling and use are as follows:

For sterilization: All solutions can be briefly boiled or autoclaved in appropriately stoppered flasks. Niu-Twitty solutions A, B, & C should be heated separately and mixed when cool.

For bacteriostatic solutions for culturing embryos: Add 0.2 g sulfadiazine and/or 0.02 g streptomycin per liter of diluted medium, per instructions below.

Solutions for tissue culture: Use 1/2 strength Ringer's, or full strength Holtfreter's, or full strength combined Niu-Twitty solutions A, B, & C.

Solutions for microsurgery and transplantations: Use 1/4 strength Ringer's or 1/2 strength Holtfreter's, or 1/2 strength combined Niu-Twitty solutions A, B, & C.

Solutions for healing after operating: Use 1/8 strength Ringer's, or 1/4 strength Holtfreter's, or 1/4 strength combined Niu-Twitty solutions.

Solutions for culturing embryos or isolated epidermally covered parts: Use 1/20 strength Ringer's, or 1/10 Holtfreter's, or 1/10 combined Niu-Twitty solutions.

For dilutions of solutions: Use only distilled water, preferably water redistilled in Pyrex glass.

b. Preserving fluids for special use with amphibian eggs and embryos. Amphibian eggs are relatively yolky and pose special problems for preservation. **Smith's fluid** is recommended if the embryos are to be used in preparing microscopic sections. **Bouin's fluid** is a very good general fixative for morphology and is good for advanced stages after most yolk has been absorbed.

Glycerin–formalin is useful as a general preservative for whole mounts but is not recommended for special histological preparations. Formulae for these preserving fluids are given in Table V-1.

3. Methods for operating on amphibian eggs. Read through the following directions and be sure that all materials are ready and at hand before you begin to operate. It is essential that all materials, instruments, dishes, and solutions be kept as clean as possible and in a sterile condition. In the first 3 days of embryonic development, the embryos are particularly susceptible to bacterial infection. After the tailbud stage, and after healing is complete, embryos become quite hardy and bacteriostatic precautions are no longer so important. Follow the directions given below and **be particularly attentive to precautions against contamination.**

TABLE V-1
AMPHIBIAN OPERATING SOLUTIONS, CULTURE MEDIA, AND PRESERVING FLUIDS

A. Operating solutions and culture media

RINGER SOLUTION:[1]
6.5 g NaCl
0.14 g KCl
0.12 g $CaCl_2$
0.2 g $NaHCO_3$
H_2O to make 1 liter

HOLTFRETER SOLUTION:[1]
3.5 g NaCl
0.05 g KCl
0.1 g $CaCl_2$
0.8 g $NaHCO_3$
H_2O to make 1 liter

NIU–TWITTY SOLUTION:[2]

Solution A:
3.4 g NaCl
0.05 g KCl
0.08 g $Ca(NO_3)_2 \cdot 4\ H_2O$
0.1 g $MgSO_4$
H_2O to make 500 ml

Solution B:
0.11 g Na_2HPO_4
0.02 g KH_2PO_4
H_2O to make 250 ml

Solution C:
0.2 g $NaHCO_3$
H_2O to make 250 ml

Boil or autoclave Solutions A, B, & C separately and combine when cool.

Optional additives to Ringer, Holtfreter, and Niu–Twitty solutions:
0.2 g glucose & 10 drops 0.1% Phenol Red pH indicator dye

The pH should be 7.8: Visualization of pH can be provided by adding 10 drops of 0.1% Phenol Red indicator per liter solution; the correct pH will show a tangerine orange-pink color (too acidic will be yellow; too alkaline will be lavender); adjustments in pH can be made by drop-wise addition of 0.1 N HCl or 0.1 N NaOH as needed.

B. Preserving fluids

SMITH'S FLUID FOR HEAVILY YOLKY STAGES[3]
(Make up immediately before use.)
0.5 g Potassium bichromate
2.5 ml Glacial acetic acid
10.0 ml Commercial formalin
75.0 ml Distilled water

Fix for 24 hours, wash 12 hours in tap water and then preserve in 5% formal in or 70% alcohol.

BOUIN'S FLUID FOR OLDER STAGES[3]
(Will keep almost indefinitely.)
75 ml Saturated picric acid
25 ml Commercial formalin
5 ml Glacial acetic acid

Fix for 12 to 48 hours; wash 12 hours in tap water and then preserve in 5% formal in or 70% alcohol.

GLYCERIN–FORMALIN. The above fixing fluids are recommended if histological sections are planned; however, a mixture of 1 part glycerine to 9 parts 5% formalin is excellent for preserving whole mount materials indefinitely.

---

1. Formulae as given in V. Hamburger (1942) *A Manual of Experimental Embryology,* University of Chicago Press.
2. Formula as given in M.C. Niu and V.C. Twitty, *P.N.A.S.,* 39: 985–989.
3. Formulae as given in G.L. Humason (1962) *Animal Tissue Techniques,* W.H. Freeman Co.

a. Washing embryos. With steel needles and watch maker forceps remove the jelly from embryos that have been transferred to a Syracuse or Stender dish and flooded with cold, boiled tap water. With great care transfer the denuded embryos using a bent-tipped, wide-mouth pipette, to a sterile wide-mouth collecting jar that is two-thirds full of cold, boiled tap water. See Chapter IV, Section D for methods to use in removing egg jelly.

By careful decanting, pour off all but about 5 ml of the water containing the embryos. Carefully fill the jar with sterile tap water and decant again in the manner just described. Repeat the washes in this way through 6 changes of water. By this dilution technique the bacterial count is reduced to a level that will allow the antibiotics, sulfadiazine and streptomycin, added to the culture medium to keep the bacterial count low until the embryos reach a resistant age.

b. Preparation of the operating area and sterilization of instruments. Place the dissecting microscope in the middle of the working area about two inches from the edge of the desk. Place your hands on the microscope stage and your arms and elbows on the table. The arm space beside the microscope should at no time have any objects placed in it. This is an essential free area. In front of the arm area, on each side of the microscope, place a beaker of 70% alcohol. A one-inch-thick pad of cotton should be in the bottom of the beaker to act as a buffer and prevent the breaking of needle points, or dulling of forceps. Beside the alcohol bottles, place an instrument tray at the right and left sides. If you are right-handed, place the culture dish of washed eggs at the side of the left instrument tray, and place the sterile agar-covered culture dishes at the side of the instrument tray on the right. Immediately in front of the instrument tray, on the right side, place the sterile flask of Niu-Twitty or Holtfreter's solutions. Everything should be placed within easy reach and should always be kept in its own place for instant use (See Figure 5-9).

Proceed through the following checklist of steps:

(1). Sterilize the wax bottom operating dish by removing the cover and pouring in a small amount of 70% alcohol. Swirl the dish so that alcohol covers the sides and rim of the dish. Use a cotton swab and sterilize the inside of a fingerbowl with 70% alcohol. Pour out the excess alcohol from the operating dish and cover it with an inverted sterile fingerbowl swabbed with alcohol.

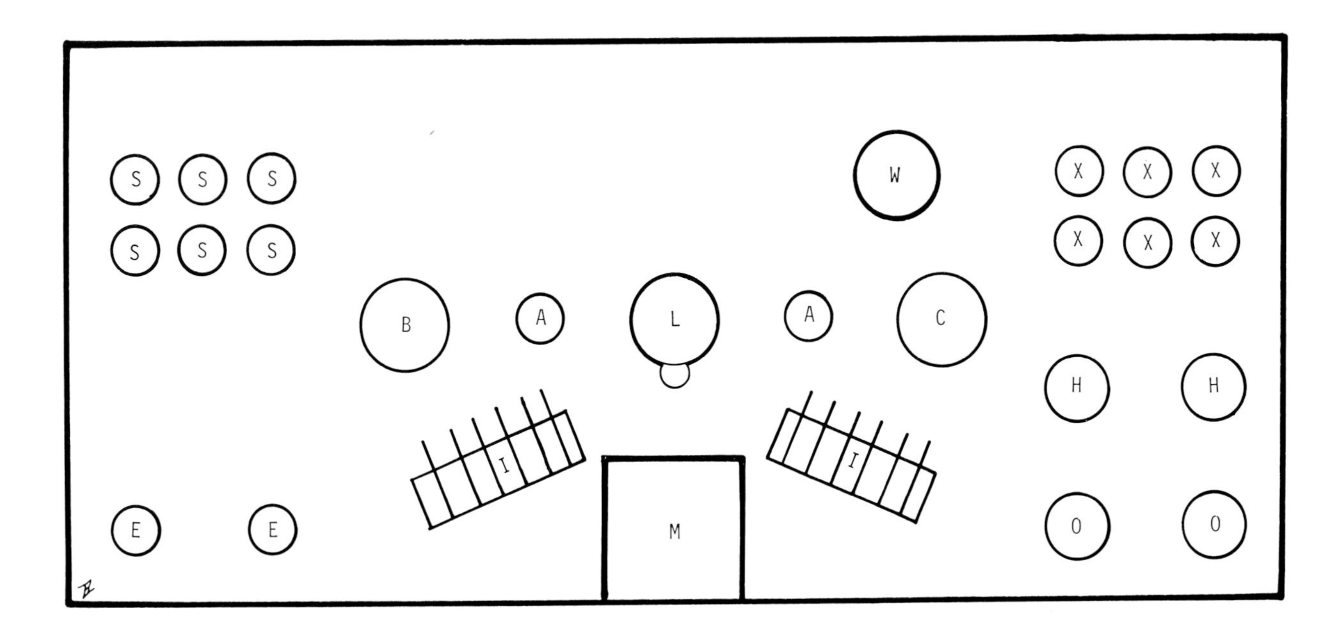

Figure 5-9. Plan for the arrangement of an operating area. Legend: A. 70% alcohol in beakers with cotton, B. Boiled cool tap water, C. Sterile culture medium, E. Washed donor and host embryos ready for operating, H. Healing embryos in operating dishes with quarter-strength operating solution, I. Instrument trays with instruments, L. Light source for microscope, M. Dissecting microscope, O. Sterile operating dishes ready for use, S. Sterile embryo dishes with agar floor ready for use, W. Waste jar or bowl, X. Operated cases in quarter-strength, or tenth-strength medium as needed for their stage in recovery, healing and/or development. See explanation in text.

(2). Mix Solutions A, B, and C of Niu-Twitty operating solution by pouring them together in flask No. A. Remove the aluminum foil covers of the flasks in such a manner that it will be possible to re-use the foil to cover the flasks after each use. This is "full strength operating solution." Check the pH and adjust it to 7.8, if necessary.

(3). Sterilize needles, hairloops, and forceps by dipping them in 70% alcohol. Then place them on the instrument tray which has also been swabbed down with 70% alcohol.

(4). Sterilize the wide-mouth, straight and curved pipettes by removing the rubber bulb and immersing the glass tubes completely in 70% alcohol. Shake off excess alcohol and replace the rubber bulbs.

Check the entire working area and see that all materials are at hand and ready for use before proceeding to the actual manipulation of embryos. Washed embryos, sterile agar culture dishes, and tissue culture tubes should all be ready and in place.

(5). Wash out alcohol from the wax bottom operating dish with three changes of cool, boiled tap water. Swirl water in the dish for 15 to 30 seconds at each change to remove all traces of alcohol. Discard all three changes of water.

(6). Fill the washed operating dish 3/4ths full of half strength operating fluid (i.e., 1:1 Niu-Twitty and boiled tap water). Place the operating dish on the microscope stage and center the microilluminator above the stage so that it will focus a bright spot of light on the middle of the optical field. Cover the dish at all times with an alcohol-swabbed fingerbowl when the operating dish is not in actual use.

(7). Make embryo depressions in the wax bottom dish with a round glass probe by pushing the probe back and forth across the bottom of the dish until a shallow groove is created. It should be about 1 1/2 times as long as the embryo and half as deep as its dorsoventral thickness. Be sure to burnish the sides of the groove so that there are no sharp wax edges to cut or tear the embryo.

(8). Transfer two or three embryos to the operating dish with a large-mouth, bent pipette; however, before doing so, rinse out the pipette with 2 or 3 squirts of sterile water in which the washed eggs are kept. Re-cover the dish of washed embryos.

(9). If embryos are old enough to show muscle movements, they will require an anaesthetic. The best one is Finquel (also known as MS-222, or tricain methanosulfonate, available from Ayerst Laboratories, N.Y.C.). For use, a stock solution of 0.5% Finquel-MS-222 can be made up in distilled water and sterilized (preferably by bacteriological filtration since it is not very heat stable). The addition of one full pipette squirt (ca. 1 ml) to the operating dish is usually sufficient to narcotize embryos in about 15 minutes. They can be kept in the solution for an hour or more without harm; however, as a rule healing should be done in solutions lacking narcotic. Alternatively a crystal of chloral hydrate, or chloretone, can be added directly to the operating dish of culture medium with the same general results.

c. Extirpation experiments on embryonic primordia. Extirpation is the removal of embryonic tissues from an embryo. The following suggested experiments can be done with glass or steel needles by micromanipulation while watching through the dissecting microscope. It is suggested that two or three examples of each experiment be carried out if sufficient material is available.

(1). Decapitation. Use steel needles to cut off the heads of embryos. Transfer both the heads and trunk parts to 1/4 strength healing solution (see below). Discard any trunk halves if the yolk mass is broken and pouring out; these rarely heal and almost always die.

(2). Endoderm-less embryos. Slit open the mid-belly region with steel or glass needles. With a hair loop sweep out all of the light yellow yolky endoderm cells that fill most of the middle of the body. Transfer the "axial embryo" to healing solution, and discard all of the yolk cells by pouring out the operating solution and replenishing it for the next experiment.

(3). Removal of an eye cup. With fine glass needles cut through the side of the head just in back of the eye cup. Hold the embryo in place with a hair loop. Carefully dissect the eye cup away from the side of the head until it is completely free from the head. Transfer the embryo, and the eye cup, to a dish of healing solution.

(4). Removal of the heart. With the aid of the diagrams of normal embryonic stages (Fig. 4-2), identify the heart swelling under the head and pharynx. With steel or glass needles cut this primordium out of the embryo with as little damage to the gill and pharyngeal region as possible. Transfer the embryo and heart primordium to a dish of healing solution.

(5). Removal of blood island. Slit open the skin of the mid-belly region with a fine glass needle. Carefully sweep out the light gray cells that lie just under the epidermis. Do not injure the yellow midgut endoderm that lies immediately underneath. The light gray cells are the primordium of ventral mesentery and vascular tissue. Transfer the embryo to a dish of healing solution.

(6). Remove and isolate any other primordia; however, it will be necessary that the embryos all contain an ample epidermal component so that they will heal and become completely enclosed in epidermis. Otherwise they will not survive when transferred to culture dishes of healing solution.

d. Care and healing of operated cases. Young embryos are very soft and must be handled as gently and with as little bruising as possible. Transfer operated embryos and pieces to a labelled agar bottom culture dish of 1/4 strength culture medium (i.e., 1 part Niu-Twitty + 3 parts boiled tap water). In this medium, the wound areas will be closed by epidermal spreading in from 30 minutes to 6 hours. Older stages take longer to heal than younger ones. Embryos can be left in 1/4 strength healing solution for 24 to 48 hours but, after healing is complete, they should be transferred to more dilute solutions. In general, this can be done by decanting half to three-quarters of the original solution and replacing it with boiled tap water down to a dilution of about one-tenth the strength of the operating solution. Streptomycin and/or sulfadiazine can be added as indicated in the formulary in Table V-1.

Daily attention to embryos is needed. All dead and dying tissues or embryos should be discarded, and embryos should be transferred to clean dishes with fresh dilute culture medium if there is the slightest evidence of bacterial growth as revealed by cloudy solutions. It is advisable to keep cultures in a cool place between 18 to 22°C; however, they can survive at ordinary room temperature, if not too warm (i.e., 20°C = 68°F).

e. A simplified method for parabiotic fusion of embryos. Parabiosis is the fusion of two embryos to simulate Siamese twins. First prepare the operating area as instructed in steps 1 through 9 in **Section b** of the preceding instructions.

(1). Wash the operating dish and fill it with half strength Niu-Twitty solution to about two-thirds full. Add Finquel as an anaesthetic if motile stages are being used.

(2). Transfer 2 embryos to the operating dish and remove lateral epidermis from the flank regions on the right side of one embryo, and on the left side of the other. Use glass needles and watch maker forceps. A clean cut can most easily be made with a glass needle by inserting it under the epidermis and then stroking the area with a hair loop until pressure cuts through the overlying epidermis. Make all four incisions on both embryos to outline the area of epidermis to be removed before trying to loosen the epidermis. After making all incisions, lift one corner of the epidermis with a glass needle and strip the epidermis away with the aid of a watch maker forceps. Take care to avoid damage to underlying mesoderm and endoderm.

(3). Transfer the embryos to an agar bottom culture dish containing 1/4 strength **Niu-Twitty solution.** Be particularly careful to avoid embryo contact with air–water interfaces, or they will surely be torn apart by the interfacial tensions upon contact with a wound area.

(4). Place the culture dish on the microscope stage and cut a square block out of the agar in the bottom of the dish. The sides of the square could be slightly longer than the length of the embryos. With a hair loop push the embryos into the depression in the agar. Adjust the embryos so that their wound areas are in contact with each other. Then cut out another small piece of agar and slide it into position like a bookend so that the two embryos are gently held together. Allow them to heal for at least 2 hours. After complete healing, the parabiotic embryos can be freed from the depression. Do not try to transfer them to another dish for at least 24 hours.

(5). After 24 hours transfer the parabiotic embryos to a clean dish with 1/10th strength culture medium for continued culture. See **Section d** for directions on care and healing of operated cases.

f. A simplified method for the transplantation of embryonic primordia. Transplantation entails the removal of a primordium from one embryo, called the **donor,** and the transfer of the primordium to another embryo, called the **host.** Two transplantation experiments will be described which will entail slightly different techniques, namely: 1. **implantation,** the insertion of a primordium under the epidermis or into the coelomic cavity, and 2. **grafting,** in which a primordium is precisely positioned in a predetermined specific location in the host embryo.

In both types of experiments, prepare the operating area as described in **Section b.**

(1). Implantation of the heart primordium under lateral epidermis. Transfer two tailbud stage embryos to the operating dish and select one as donor and one as host. First use Figure 4-2 to locate the heart primordium in the ventral head region of the **donor embryo.** Remove the epidermis from the heart. With needles dissect out the heart which will already appear as a small S-shaped structure; it may even have begun to beat. Remove the heart primordium and transfer it to a position close to the host embryo. Next, use a glass needle to make a longitudinal incision through the epidermis of the **host embryo** along the lateral boundary between the somites and bulging yolk mass. With the steel needles gently pry the epidermis up from the ventral yolk mass so that a pocket is created from the point of incision down the side under the epidermis about half the distance to the ventral midline. Last, pick up the isolated donor heart tissue with the glass needle or watch maker forceps and transfer it to the incision area on the host. With the glass needle, tuck the heart implant into the epidermal pocket and push it as deeply into the recess as possible. Transfer the embryo to an agar bottom culture dish half full of quarter strength Niu-Twitty solution. Cut a small square depression in the agar bottom suitable for the embryo to rest in. Under the microscope check to see that the heart primordium is still securely in the epidermal pocket. Then, with a hair loop, move the embryo into the slot in the agar. With a hair loop, adjust the embryo so that it is resting on its ventral surface and slide the agar block back until it gently presses against the embryo and holds the graft in place with the incision closed. Be most careful not to exert any more pressure on the embryo than is necessary merely to immobilize it. Any excess pressure will damage the embryo. See the directions for care and healing of embryos in **Section d** in the preceding directions. Allow healing to continue for at least two hours before transfer to dilute 1/10 culture medium.

(2). Grafting an eye into the mid-flank region. This experiment will entail the transplantation of a complete eye, including the optic cup, lens placode, and head epidermis, from an older tailbud donor embryo onto a young tailbud embryo. **First prepare the host area** for grafting by cutting a square in the host lateral belly ectoderm which is about 1 1/2 times as broad as the eye cup itself. This is most easily done by inserting a glass needle under the epidermis and stroking it with a hair loop until the epidermis over the needle is cut. Make all four cuts in this manner and then, with glass needle and watch maker forceps, strip away the epidermis. **On the donor** quickly cut out an entire eye vesicle by the use of glass needle, hair loop, and watch maker forceps. Transfer the donor eye cup and host to an agar bottom culture dish and cut out a depression in the agar to accept the host embryo. With a glass needle and watch maker forceps, transfer the eye to the region of the graft that has been prepared on the host and press it in place. As quickly as possible move the agar block back into position to hold the graft in place. See the directions for care and healing of embryos in **Section d.** Allow healing to continue for at least two hours before transferring the host embryo to dilute 1/10 culture medium.

NOTE: As an alternative to using agar blocks as a method for holding grafts in place during healing, small strips of No. 2 thickness cover glass can be cut with a diamond pencil and used as transparent bridges to hold tissues in place during the healing process.

g. A simplified method for tissue culture in capillary tubes. In addition to preparing the operating area, culture media, and glassware for sterile technique as described in **Sections a and b,** it will also be necessary to sterilize small glass Petri dishes to which have been added two to four pieces of blood coagulation capillary tubing cut to a length of 1 1/4 inches. The sterile dishes, culture media, and instruments must be kept covered at all times when not actually in use. **Sterile methods are essential to success.** It is also desirable to wash embryos eight times by the decanting–dilution method so as to assure a very low possibility of bacterial contamination being introduced by the embryos. Proceed then as follows.

(1). Place a donor embryo in the operating dish. Hold it in place with a sterile hair loop. With a sterile needle or watch maker forceps, pull the epidermis off the head and expose the side of the brain and the heart primordium.

(2). Isolate pieces of heart, neural tube, optic cup, and any other primordia that you choose to use. Cut them out in blocks of a size small enough to go into a capillary tube. Trim the tissue blocks to a desired size with steel or glass needles.

(3). Prepare a sterile Petri dish for culture in capillary tubes by placing a narrow line of vaseline across the middle of the culture dish. Then flood the dish half full with full strength, sterile Niu-Twitty solution.

(4). Load capillary tubes with culture fluid as follows. With a sterile forceps, pick up a piece of capillary tube and touch one open end of the tube to the dish of culture medium. Capillarity will immediately fill it with fluid. Then submerge the capillary tube in the culture medium.

(5). Load the capillary tubes with tissue fragments under the microscope. To do this, transfer the tissue isolates from the operating dish to the Petri dish with capillary tubes. Hold the capillary with a watch maker forceps in one hand, and push a tissue block into the open end of the capillary with a glass needle, or with a bent steel needle. Move the tissue as deep into the tube as possible.

(6). Seal the two ends of the capillary tube with vaseline in the following way. Use the watch maker forceps to lift the capillary from the culture medium. Then very carefully and very quickly touch the open edge of the capillary to the water surface while holding it in a vertical position. This will drain some of the fluid out of the capillary and leave an air space. The tissue will move with the fluid. No more than one-quarter of the capillary should be filled with air. The remainder should contain culture medium. Then, resubmerge the capillary with the end containing the air bubble inserted in the water first to prevent the capillary from filling again. Last, push both ends of the capillary into the vaseline to plug both ends with a vaseline seal. The capillaries can now be anchored in the bottom of the culture dish with vaseline. They can be kept and observed without fear of further contamination, so long as the vaseline seals are intact.

h. Care and observation of operated embryos and cultures. The general culture methods outlined in Chapter 3 for the maintenance of normal embryos and tadpoles applies to experimental embryos. In general, the major precautions in early stages entails the solicitous transfer of embryos to new, clean, culture media at the first hint of bacterial contamination. Also, it is very unwise to crowd experimental animals in culture dishes because, if one dies, they usually all die. One dish per embryo is best, but no more than 4 embryos per culture dish is good standard practice.

Daily notes should be made on the progress of experimental embryos. These notes should contain a statement of developmental age, a sketch of the appearance of the embryo, and particularly, the appearance of the extirpation and transplant areas.

When normal control embryos reach the feeding stage, it is wise to preserve experimental embryos because they usually do not feed well and will soon starve. Use any of the fixing fluids that are recommended in the formulary provided at the beginning of this section.

For those who wish to undertake more advanced levels of experimental work with amphibian eggs, refer to additional experimental technique and type-experiments given in the manuals by Rugh, Wilt and Wessels, and in, particular, by Hamburger. Complete bibliographic citations for these works are given in Appendix I.

CHAPTER VI

RECOGNITION OF PALEOMORPHIC AND CAENOMORPHIC ASPECTS OF EMBRYONIC DEVELOPMENT

Illustrated by Gastrulation and Neurulation in 18, 24, and 33 Hour Chick Embryos

TABLE OF CONTENTS

TABLES AND ILLUSTRATIONS

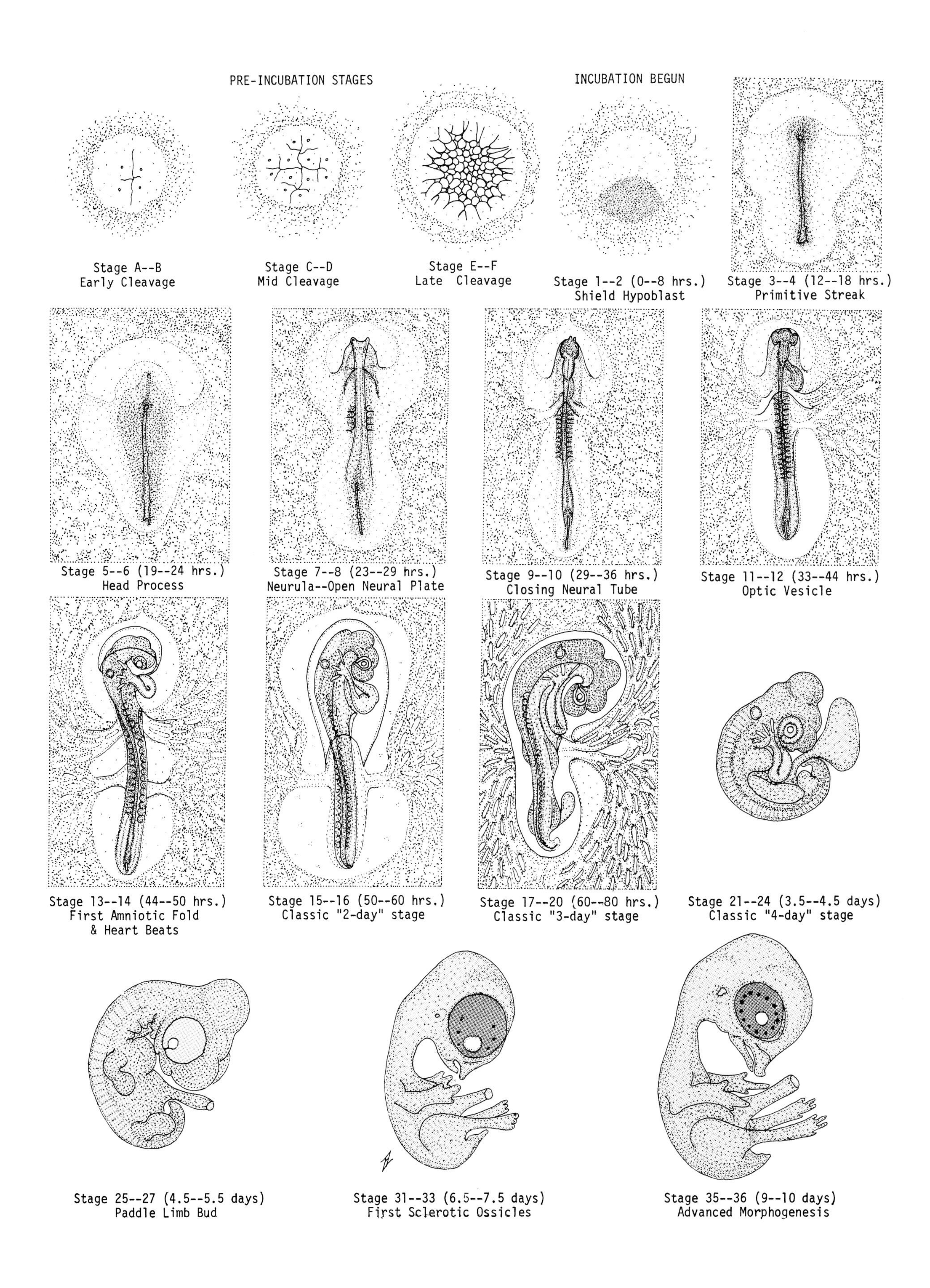

Figure 6-1. Representative stages in chick embryonic development. For legend, see Table VI-1.

CHAPTER VI

RECOGNITION OF PALEOMORPHIC AND CAENOMORPHIC ASPECTS OF EMBRYONIC DEVELOPMENT

Illustrated by Gastrulation and Neurulation in 18, 24, and 33 Hour Chick Embryos

Two terms used in the title were defined in the Introduction. Paleomorphic is a word derived from the Greek roots that mean "ancient form and shape." In embryonic development, **paleomorphic structures** are those in modern living species which repeat, with reasonable faithfulness, the ancient ancestral patterns of development that typify the most primitive members of the phylum, class, or other taxonomic level under consideration. Chapters 4 and 5 have emphasized paleomorphic aspects of Amphioxus, cyclostome, and amphibian development, as those which reveal the vertebrate ancestral common plan of basic structure and manner of embryonic development. These paleomorphic characters in development cause the common

TABLE VI-1
TIMETABLE OF EARLY STAGES IN CHICK EMBRYONIC DEVELOPMENT

Stage	Ages	Major Distinctive Structures
	Hours Post-fertilization	
A	2.5	First cleavage, 2 cell stage
B	3.5	Second cleavage, 4 cell stage
C	4.0	Third cleavage, 8 cell stage
D	4.5	16 to 32 cell stage
E	6-9	Late cleavage stage, about 100 cells
F	10-14	Mid-blastula stage, about 200 to 1000+ cells
	Hours of Incubation	
1	0.4	Early shield stage, hypoblast delamination
2	6-8	Early gastrula stage, first trace of primitive streak
3	12-15	Mid-gastrula stage, early streak with Hensen's node
4	17-18	Mid-gastrula, late primitive streak
5	19-22	Late gastrula with early head process forming
6	22-24	Late gastrula with head process and early head fold
7	23-25	Early neurula with first somite and neural folds
8	24-29	Mid-neurula with 4 somites, "classic 24 hour stage"
9	29-33	Late neurula with 7 somites and forebrain closed
10	31-36	Neural tube stage with 10+ somites and heart tube forming
11	33-38	Neural tube stage with 13 somites, "classic 33 hour stage"
12–13	38-46	Neural tube stage with 16 to 19 somites and first heart beats
14–16	46-56	Tailbud stage with 22 to 28 somites, "classic 48 hour stage"
	Days of Incubation	
17-20	2.5-3.5	Tailbud stage first allantoic evagination, "classic 72 hour stage"
21–24	3.5-4.5	Crescent limb buds, "classic 96 hour stage"
25–27	4.5-5.5	Paddle limb bud stage
28-30	5.5-6.5	Limb buds with first digit formation
31–33	6.5-7.5	Eyes with first sclerotic ossicles forming
34–46	8 through 21 days	One stage per each additional day up to hatching

Stages are based on H.L. Hamilton, *Lillie's Development of the Chick,* H. Holt, 1952.

plan to be formed. It is called the common plan for the phylum because it presumably applies to a way of development that is characteristic for all members of the phylum. It has evolutionary significance in pointing out the original or ancient ways in which form and structure arose. Paleomorphic form and pattern are, therefore, clearly the embryonic basis for determining homology and relationships among species.

When we superficially compare the normal developmental stages of the bird in Figure 6-1, with the comparable series for the amphibian (as represented by the salamander in Figure 3-7 or frog in Figure 4-2) instead of seeing clear similarities, we are faced with such great differences in early stages as to render a search for homologies in cleavage, blastulation, and gastrulation almost futile. The shelled egg of the chick, with its enormous accumulation of uncleaved yolk, and its voluminous extraembryonic areas which will all be discarded at the time of hatching, has no counterparts in the amphibian egg. The latter is modestly yolky; yet it undergoes total cleavage and all parts of the egg will be directly involved in forming the embryo and tadpole, with no lost parts.

Beginning with the neurula stage, however, some reasonably familiar landmarks become recognizable in the chick which bear a resemblance to neural plate formation as described in the amphibian. As development continues in the bird, the processes in the formation of the brain, heart, circulatory vessels, and axial structures of gut, muscle, skeletal, and excretory systems will be found to be increasingly more similar to those observed in more primitive vertebrate classes.

Briefly, then, what we observe is a wide departure from the so-called common plan or paleomorphic manner of development in the bird at its beginnings, and a somewhat delayed return to the common plan at tailbud stages. Later, of course, there will again be a divergence as the special characters that typify birds (as a class distinct from fish, Amphibia, etc.) will finally be added to the overall pattern of bird development.

We are now faced with a problem in interpreting the comparison of early cleavage and gastrulation in the amphibian and bird. Which of the two patterns of development is primitive (i.e., paleomorphic), and which is secondarily derived? For the latter type of developmental pattern, we will use the second special term in the title of this chapter, namely, caenomorphic, derived from Greek terms meaning "recent form and shape." **Caenomorphic structures** or patterns of development are those aspects of development that are secondary modifications, or additions, to the ancient common plan of development. In general, caenomorphic structures are associated with adaptations for embryo survival under special conditions. Examples are found in the shelled eggs of reptiles and birds, or the evolution of intrauterine development in mammals, neither of which could represent an ancient ancestral adult condition.

The solution to the problem of distinguishing between paleomorphic and caenomorphic aspects of development is, in part, provided by the solution to a similar problem that arises in comparative anatomy. Paleomorphic patterns in development bear much the same relationship to comparative embryology that **homology** bears to comparative anatomy. Both are concerned with the recognition of true evolutionary bases for structural similarities and differences among species. One (paleomorphosis) deals with similarities in the way homologous structures that result from evolutionary modification of the same ancestral part by different descendants from the same common ancestor. Caenomorphic aspects of development have nearly the same relationships in comparative embryology that **analogy** (similarity in function) bears to comparative anatomy. In both cases biological structures reveal convergent similarities, or divergent differences, based on function or adaptive usefulness.

The criteria for determining adult homologies (as opposed to analogies or adaptive convergent similarities for like functions), in the main, are also the criteria that apply to the recognition of embryological paleomorphosis (as opposed to caenomorpic adaptive modifications). The criteria for homology have been stated most clearly by Remane. A very brief summary of his criteria along with the special additional criteria that apply to developing systems is given in Table VI-2.

It is clear, from the above statement of criteria for determining homology and paleomorphosis, that the solution to our original problem will not be simple and cannot be expected to be absolute in all circumstances. However, if we examine the differences between cleavage, gastrulation, and neurulation for the chick and compare the special details with what has been reviewed for lower vertebrates, then, it should be possible to separate those aspects of chick development that are paleomorphic from those that are secondary, caenomorphic, novel improvisations of great adaptive value to the embryo, but of relatively little evolutionary value in revealing common plans in early vertebrate development. In general, adaptations

TABLE VI-2
CRITERIA FOR DETERMINING HOMOLOGY AND PALEOMORPHOSIS

The probability that two structures showing structural similarities in different species are homologous increases: —

Primary criteria of Remane:*

1. As similarity of position with regard to other body parts increases;
2. As possession of common special structures within the part increases;
3. As examples of transitional intermediates in other species increases;

Secondary criteria of Remane:*

4. As the presence of the structure with other known homologies increases;
5. As the percentage of related species with the same character increases;
6. As the absence of the structure from unrelated forms increases;

Ontogenic criteria for recognizing paleomorphic characters, Lehman:

**A. As similarity in the manner of embryogenesis and morphogenesis increases;
**B. As its similarity in germ layer derivation during development increases;
**C. As similarity in its epigenetic roles in determination and differentiation increases.

*A. Remane, *Die grundlagen des naturlichen Systems der vergleichenden Anatomie und Phylogenetik.* Ed. 2, Geest & Partig, Leipzig, Germany, 1956.
**A. Von Baer's third corollary or law (see Table VIII-1P.
**B. Von Baer's law of similarity in germ layer origin of homologous parts.
**C. F. Baltzer's postulate that structures are homologous if they have the same roles to play in inductor — reactor systems in development.

that provide more protection and nutrients for individual embryos are caenomorphic. They will tend to shortcut the ancestral paleomorphic characters and make the evolutionary sequence less clear. We should also add that what might be a caenomorphic improvisation in one taxonomic group (such as the shelled egg of early reptiles) will, of course, become a paleomorphic character in all the descendants of that group that retain the character (i.e., the amnion and yolk sac of human embryos are a paleomorphic evidence of reptilian ancestors; reptiles first evolved the structures caenomorphically in their early derivation from ancestral amphibians).

With the foregoing criteria for homology, and the special criteria for developing systems that permit us to recognize paleomorphic characters, we will now proceed to the examination of special conditions of cleavage and gastrulation in birds. The goal will be to distinguish ancient evolutionary patterns of development from the adaptive changes that are peculiar to the developing bird egg.

First we will examine the normal patterns of cleavage and gastrulation in birds by models, diagrams, charts, and slide preparations of embryos during the first 20 to 22 hours of incubation. With the descriptive facts of early avian development in mind, we will return again to the problems of judging paleomorphic and caenomorphic components of early avian embryology.

A. AVIAN GASTRULATION ILLUSTRATED BY THE PRIMITIVE STREAK AND HEAD PROCESS STAGES 16-22 OF INCUBATION

1. Early cleavage stages and timetable of development. Figures 6-1 and 6-2 give an overview of chick development including some early cleavage stages, most of which take place before the egg is laid. Eggs of birds are fertilized within the oviducts before albumen, egg membranes, and shell coats are added. For the completion of the latter processes, eggs remain in the oviducts and uterus for variable periods of time from 12 to 24 hours. During this time internal incubation takes place. There can be up to 12 hours difference in the developmental age of embryos at the time of laying. Accordingly, the clock-time of external incubation that takes place after laying permits a wide range in individual variation among eggs. This accounts for the different levels of structure that are observed in specimens that are all said to be of one specific incubation age. Table VI-1 gives the average incubation time and developmental stage for early chick development.

Individual drawings in Figures 6-1, 6-2, and 6-4 have been given stage numbers to correspond with those given in Table VI-I.

2. The study of gastrulation as seen in whole mount preparations of 16 to 22 hours chick blastoderms (Fig. 6-2). These preparations will all be marked 17 or 18 hour chick blastoderms, but in actual fact will represent a spread of development that ranges from Stages 3 through 6 on Table BI-1. Accordingly, first identify the developmental stage of the preparation, and then exchange slides with others in the laboratory that have either older or younger stages so that experience can be gained from the study of a series of pregressive times during gastrulation.

Entire blastoderms (i.e., the cleaved cellular portion of the egg, not including the uncleaved yolk mass) have been mounted on the slides. Identify the following regions: **area opaca** is the densely stained tissue that forms the outer, darker ring of cells in the blastoderm. The cells contain intracellular yolk. The **zone of junction** or **marginal periblast** is the outer edge of the area opaca adjacent to the uncleaved yolk in which many nuclei are in open contact with the yolk mass. The **area pellucida** is the central, lightly-stained area which is generally ovoid or pear-shaped. The wider (anterior) end is usually distinctly less heavily stained than the narrower (posterior) end. The **primitive streak** is homologous with the fused, lateral blastopore lips of the amphibian egg. The streak extends as a thickening in the anteroposterior median axis of the area pellucida. It is somewhat more darkly stained than surrounding regions. Identify the following special structures within the primitive streak: **Hensen's node** is a mass of invaginated notochordal mesoderm which appears as a club-shaped thickening at the anterior end of the streak (Hensen's node corresponds with the dorsal lip of the amphibian blastopore). The **primitive pit** and **primitive groove** correspond with the blastopore itself although they do not actually form an opening to the interior. The primitive pit is located just posterior to Hensen's node. It is the anterior end of the primitive groove which runs the length of the primitive streak, between the right and left **primitive folds.** The latter are comparable to the right and left lateral lips of the amphibian blastopore. In the chick, the primitive folds and Hensen's node represent areas where cells from the surface are involuting to the interior in a manner to be described in the following section. Identify the **anterior margin of invaginated mesoderm.** It will appear as a faint line extending laterally from just in front of Hensen's node across the area pellucida to the area opaca. If, as is often the case in these preparations, your specimen is at Stage 5 or older (19+ hours of incubation), the posterior regression of Hensen's node may have already begun (see the following description of axial elongation by chordamesoderm during gastrulation), in which case the embryo will be at the so-called **head process**

Legend for Figure 6-2. Blastula, primitive streak, and head process stages in chick embryonic development.

Stage 1.	Embryonic shield.
Stage 2.	Trace primitive streak.
Stage 3.	Early primitive streak.
Stage 4.	Late primitive streak.
Stage 5a.	Head process whole embryo.
Stage 5b.	Head process cross section.
Stage 5c.	Head process sagittal section

Legend and checklist of structures to be identified

Color code: blue — ectoderm, red — mesoderm, yellow — endoderm, green — uncleaved yolk.

am	anterior margin of mesoderm
ao	area opaca
ap	area pellucida
ar	archenteric cavity
bc	blastocoele
ee	epidermal ectoderm
en	endoderm
hn	Hensen's node, dorsal lip
hp	head process, chordamesoderm
m	mesoderm
no	notochord, chordamesoderm
pa	proamnion
pf	primitive fold, lateral lip
pg	primitive groove, blastopore
pp	primitive pit, anterior blastopore
ps	primitive streak
s	embryonic shield, hypodermis
y	uncleaved yolk
zj	zone of junction

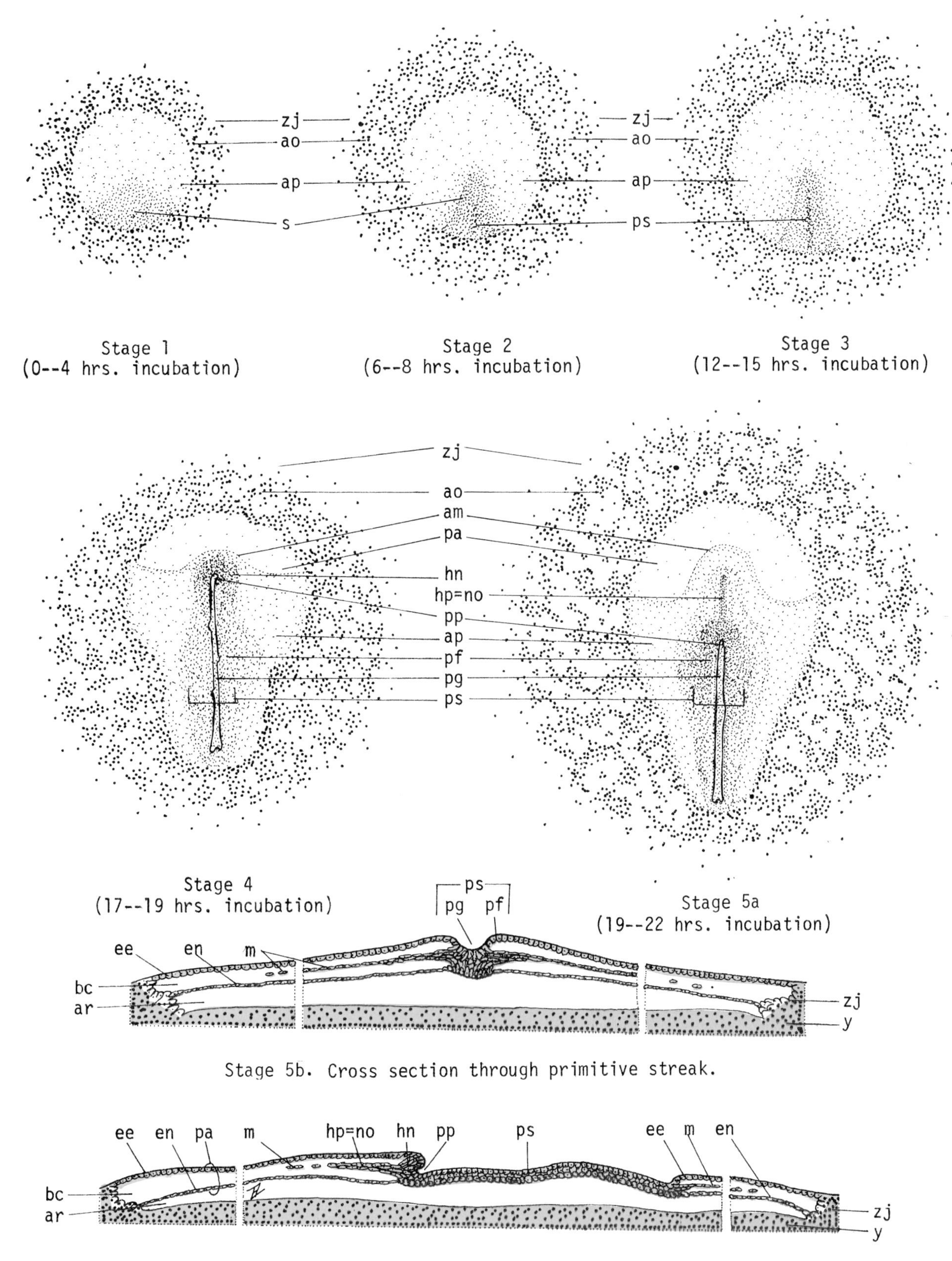

Figure 6-2. Blastula, primitive streak, and head process stages in chick embryonic development.

stage. In this event a thin strip of notochordal mesodern will be visible just anterior to Hensen's node as is shown and labelled in Figure 6-2, Stage 5a.

The last two diagrams in Figure 6-2 show a cross section and a median sagittal section through a head process stage embryo to show the relationships of primary germ layers in the flat blastoderm of the chick. A cross section through the primitive streak will be available for study in the 24 hour neurula stage. Details of its structure will be given at that time.

In preparation for the next part of this study, review the amphibian gastrulation movements in Chapter 3. This will serve as a reasonably good model of the paleomorphic, common plan for gastrulation in vertebrates.

3. Analysis of fate maps and gastrulation movements in the chick (Fig. 6-2, 6-3 and 6-4). The earliest stages in cleavage and gastrulation in birds have been reviewed in the preceding section. Early development in the highly telolecithal avian egg, at first sight, seems to bear little relation to cleavage and gastrulation in amphibian eggs. The latter undergo total cleavage, whereas cell divisions in the chick egg occur only in a small disc of cytoplasm situated at the animal pole of the egg. Cleavage furrows at first extend only a short distance into the yolk mass. The yolk will never become completely divided into cells. This is termed partial or **meroblastic cleavage.** By the time the egg is laid, mitoses have resulted in a flat disc of cells about 3 mm in diameter that floats as a multilayer of cells on the surface of the yolk mass. This is the **blastodisc,** and the cells at its outer perimeter are in direct cytoplasmic contact with the yolk. However, the more central cells have complete cell membranes and are separated from the underlying yolk (Stage F, see Fig. 6-2, 6-3, 6-4). The blastodisc is said to correspond with the blastula stage in lower vertebrates. The liquified pool of yolk immediately under the blastodisc is the homologue of the **blastocoele.** Figure 6-3 gives a series of diagrams that attempt to explain these differences and point out the basic similarities that permit this interpretation to be made.

The first assumption is that most modifications in early chick development result from the enormous accumulation of yolk that not only distorts the shape of the egg but renders normal cleavage and gastrulation mechanically impossible since the yolk itself never becomes cellular and cannot participate in autonomous gastrulation movements. The **second assumption** is that the genetic heritage of a vertebrate ancestry in birds will tend to make the patterns of development conform as closely as possible with ancestral or paleomorphic patterns as are permitted by the yolk problem.

The chick and amphibian can be homologized by the simple expedient of removing all the uncleaved yolk mass of the chick egg and keeping only the blastoderm and its yolk-rich cells of the zone of junction. After this, a "chick blastula" can be produced by drawing the cut edges of the blastoderm down and fusing them together to create a circular blastula structure. The sub-blastodermal fluid then becomes the blastula cavity. We would conclude that the 2-dimensional blastodisc is caenomorphic in shape, but that it is paleomorphic in having the area pellucida represent the yolk-poor animal pole cells, and the yolk-rich area opaca cells surrounding it representing the yolk-rich cells of a typical amphibian egg. This seems to dispose of the blastula problem reasonably satisfactorily.

The problem of paleomorphosis versus caenomorphosis in bird gastrulation is, however, considerably more complex. This is partly due to the fact that, in spite of the many studies of early bird development that have been carried out, there are still questions as to the actual timing and paths that cells take during this critical period of early development. The following discussion of gastrulation movements in birds must, therefore, be considered to be provisional and subject to revision pending new information. In the main, an attempt has been made to rationalize several views that may be in conflict in minor details and to give a generalized explanation that seems to be consistent with most observations.

We will first consider the separate gastrulation movements that are believed to take place in birds during the processes of gastrulation and germ layer separation. After this, comparisons with amphibian gastrulation will be made. Use the models, charts, and illustrations given in Figures 6-3 and 6-4 as aids in visualizing the dynamic movements that are described in the following discussion.

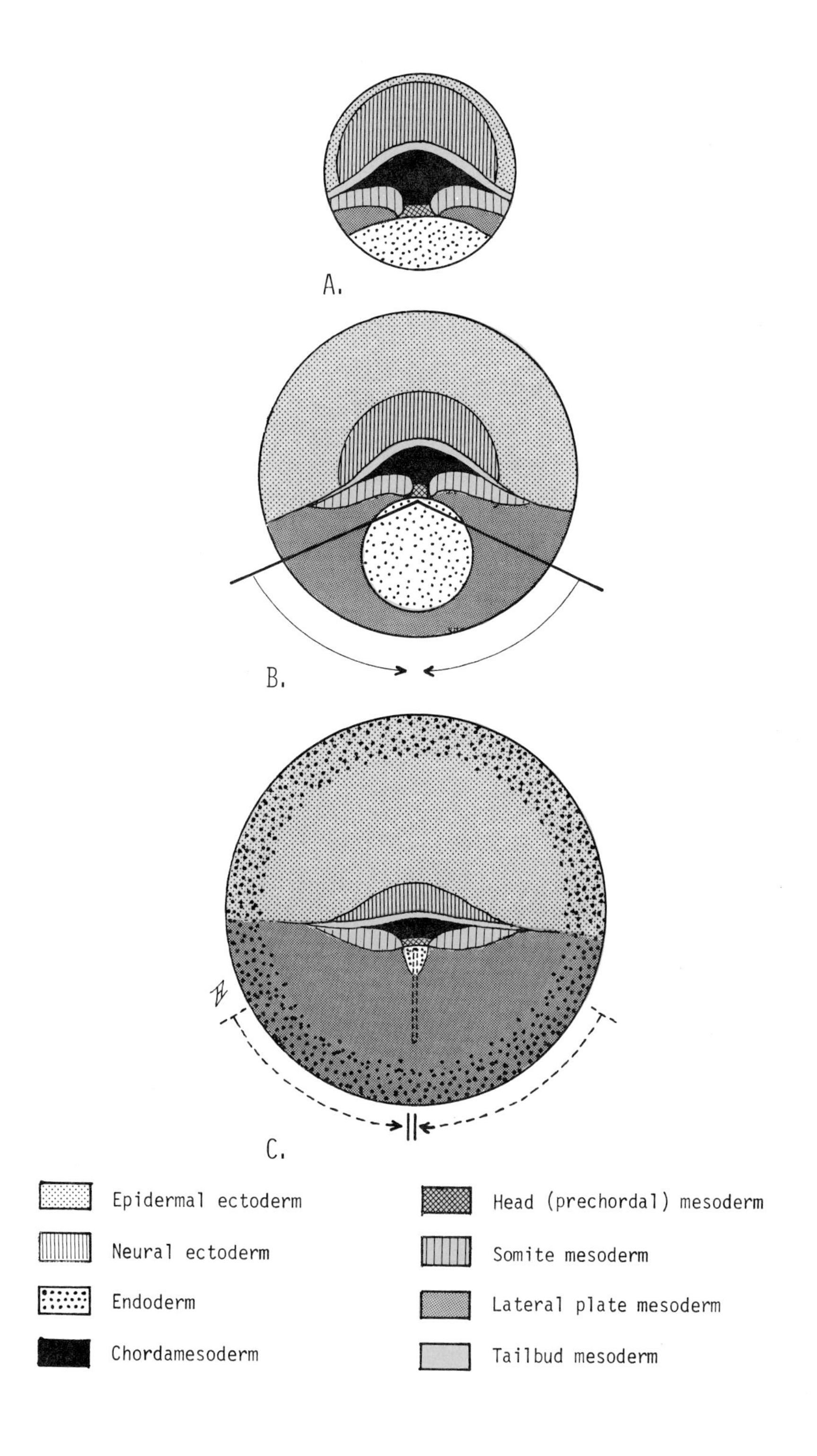

Figure 6-3. Homologies between amphibian and chick blastula stages showing the organization of prospective germ layer regions on their blastoderms. 6-3 A. Amphibian blastula fate map as viewed from the posterodorsal surface (compare it with Fig. 3-10 B). **6-3 B.** An hypothetical amphibian blastula fate map of germ layer regions drawn as a two-dimensional "world map" simultaneously showing the relationships of all prospective germ layer regions. **6-3 C.** Chick blastula fate map as viewed from the surface of a Stage 1, 0–4 hour, blastoderm. The heavy lines and arrows on Figure 6-3 B indicate the parts of the amphibian blastoderm that would need to be changed to convert an excessively yolky amphibian spherical blastula map into a typical, flat, chick blastoderm. This would entail the ingression, or delamination, of most surface endoderm, and drawing the lateral blastopore lips together in the midline to form a homologue of the chick primitive streak. The small amount of endoderm shown at the anterior end of the primitive streak in Figure 6-3 C will give rise to most embryonic endoderm of the chick.

Summary of Gastrulation Movements in Birds

Figure 6-4 diagrammatically shows early chick blastoderms at blastula and gastrula stages with the normal fate of regions indicated by special colors. The data presented were accumulated by marking experiments accompanied by time-lapse motion pictures. Other observations have permitted us to trace the paths of tissue movements during gastrulation. For each stage given in the figure, numbered arrows refer to the specific cell movements of gastrulation numbered and described below. In the pairs of figures for Stages 2, 3 −, 3 +, 4, and 5, the drawing at the left shows the distribution of materials on the surface, and the drawing at the right is of the same stage with the surface layers removed to show the deeper layers that have already moved to the interior.

Endodermal movements:

1. Delamination (polar ingression; Fig. 6-4, Stages F and 1). The early pregastrula or embryonic shield stage in chick development is one during which the prospective endoderm cells are at first localized in the posterior half of the area pellucida of the blastodisc. During the shield stage, most endodermal cells separate from the overlying layer and drop into the blastocoele cavity where they quickly assemble into an epithelial sheet at the posterior third of the blastoderm. The upper layer of the blastoderm is the **epiblast.** It contains cells that will form ectoderm, mesoderm, and embryonic endoderm. The deep layer which delaminated (split) from the original blastodisc is the **hypoblast.** It will form the cells of the **extraembryonic** (yolk sac) **endoderm.** The delamination of hypoblast endoderm is considered to be homologous with the polar ingression of vegetal endoderm in amphibians since it entails the migration of individual cells to the interior without any invagination of an epithelial sheet.

2.Internal spreading (Fig. 6-4, Stages 1-7). Almost immediately after delamination, the hypoblast endoderm cells undergo rapid cell multiplication and spread out as a sheet toward

Legend for Figure 6-4. Morphodynamic movements in gastrulation of chick embryos. Individual figures are numbered so as to coincide with corresponding developmental stages given in Table VI-1.

Stage F (unincubated). Diagrammatic sagittal section of a mid-blastula showing the presence of prospective endoderm cells in the posterior blastoderm.
Stage 1 (0-4 hr incubation). A. Surface view of whole blastoderm. B. Median sagittal section showing delamination of endoderm or hypoblast.
Figure A, Stages 2 through 5. Prospective fate maps of germ layer regions on the surface of blastoderms during gastrulation.
Figure B, Stages 2 through 5. Prospective fate maps of germ layer regions that have invaginated or delaminated to the interior during gastrulation.
Stage 2 (6-8 hr) Pre-gastrula or trace streak prospective fate map.
Stage 3- (8-12 hr) Early primitive streak fate maps.
Stage 3+ (12-15 hr) Mid-primitive streak fate maps.
Stage 4 (17-19 hr) Late primitive streak fate maps.
Stage 5 (19-22 hr) Head process or regression of Hensen's node fate maps.

Legend and checklist of structures to be identified

Color code: blue — ectoderm, red — mesoderm, yellow — endoderm, green — uncleaved yolk.

Arrows numbered 1 through 7 refer to gastrulation movements described in text:

1. **Delamination** of endoderm from epiblast
2. **Internal spreading** of endoderm under the epiblast
3. **Involution** of all mesoderm through the primitive streak
4. **Convergence** of chordamesoderm towards Hensen's node
5. **Internal divergence** of mesoderm between ectoderm and endoderm
6. **Axial elongation** posteriorly by regression of Hensen's node (notochord)
7. **Epiboly** by spreading of ectoderm so as to cover entire blastoderm
8. **Proliferation** by mitotic multiplication of cells in all regions but, in particular, at the primitive streak and around the zone of junction

a	anterior	en	general endoderm	no	chordamesoderm
ao	area opaca	epb	epiblast	p	posterior
apl	area pellucida	hm	head mesoderm	pa	proamnion
ar	archenteric cavity	hn	Hensen's node	ps	primitive streak
bc	blastocoele	hpb	hypoblast	so	somatic mesoderm
bd	blastoderm	lm	lateral plate mesoderm	tm	tail mesoderm
e	ectoderm	m	mesoderm cells	y	uncleaved yolk
ee	epidermal ectoderm	ne	neural ectoderm	zj	zone of junction
een	embryonic endoderm				

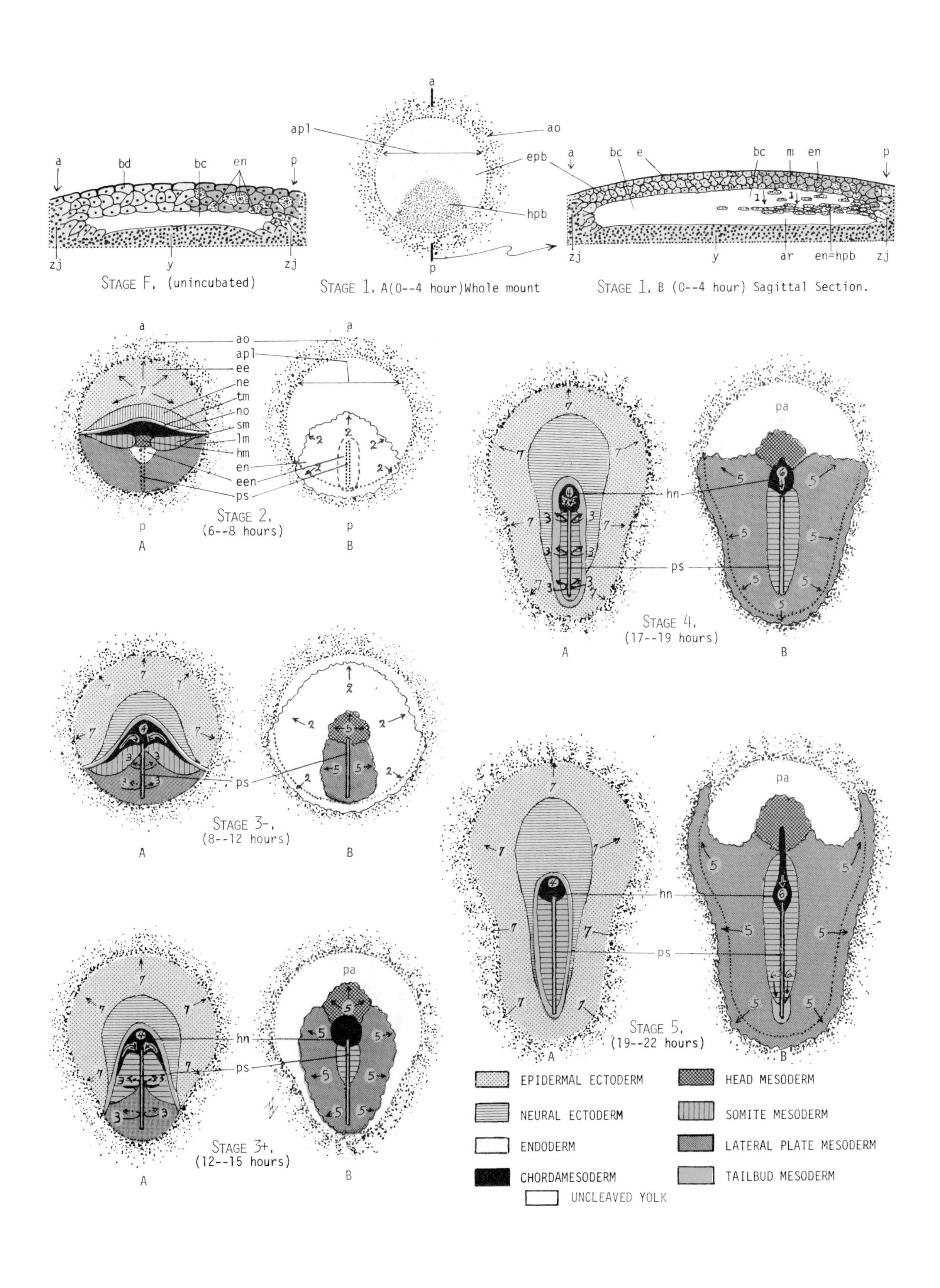

Figure 6-4. Morphodynamic movements in the gastrulation of the chick embryo. See legend. Compare also with gastrulation movements in Amphioxus (Fig. 3-7) and Amphibian (Fig. 3-10), and Table IV-3.

the anterior and lateral sides of the blastoderm. The cells at the advancing hypoblast perimeter will eventually reach the zone of junction in the area opaca of the blastoderm. In this region they unite with the ectodermal surface layer. Both layers will continue to spread over the yolk surface until the yolk sac is completed many days later. Internal spreading and fusion with the area opaca will result in the separation of the original blastocoele cavity into two parts. The space between the hypoblast and epiblast is still called the blastocoele, whereas the space beneath the hypoblast, filled with liquified yolk, now is called the **archenteron** or **gastrocoele.** Internal spreading in the chick is considered to be homologous with the same movement in amphibians. However, note that in the bird the endoderm **first** forms a roof for the archenteron, and only much later will it acquire an endodermal floor; whereas, in amphibians, the early endoderm **first** forms a floor for the archeneteric cavity and internal spreading proceeds dorsally until the archenteron roof is completed.

Mesodermal movements:

3. Involution (Fig. 6-4, Stages 2-7; Fig. 6-2). This is simply the movement of surface cells to the interior around the right and left folds of the primitive streak. The cell movement can be most clearly described as similar to the movement of a belt around a pulley. Involution is carried out by embryonic endoderm and all mesoderm as these cell regions move from the surface to the interior. At Stage 2, almost half of the area pellucida epiblast is endoderm and mesoderm. The beginning of involution becomes visible as the trace streak stage is reached. The general movement of involuting cells as they move toward the primitive streak can be most easily described by saying that they approach the streak much as the ribs of a fan will approach each other as the fully open fan is gradually closed. The final cells to reach the streak (i.e., the backs of the fan) lie along the horizontal dividing line between embryonic ectoderm and mesoderm which, at Stage 2, very nearly transects the area pellucida into anterior (ectoderm) and posterior (mesoderm-endoderm) halves. By Stage 5, the "fan is closed" and ectoderm now covers the entire surface of the blastoderm except for a small amount of tail mesoderm that flanks the edges of the primitive streak. Mesodermal involution in the chick and amphibian is essentially similar. The only difference between the two groups lies in the presence of embryonic endoderm in the immediate vicinity of the anterior primitive streak. This endoderm will involute immediately behind, or rather under, the chordamesoderm. This embryonic endoderm in the epiblast is homologous with the yolkplug endoderm of the amphibian egg; however, instead of forming only the mid- and hindgut regions as in amphibians, it will form the entire **embryonic endoderm.**

4. Convergence (Fig. 6-4, Stages 2 and 3). Convergence entails the angular movement of cells toward the dorsal lip, or, in birds, toward Hensen's node. The cells undergoing convergence crowd toward the dorsal lip as they approach it from the exterior. After reaching the dorsal lip, or Hensen's node, they will undergo involution just as any other streak material will do. In birds, convergence apparently entails only the movement of chordamesoderm cells. Somite cells (which in amphibians also converge on the dorsal lip region) do not undergo convergence but merely enter the interior by the primitive streak (blastopore lips) that happen to be closest to their linear path of movement. It is possible that convergence also is involved in embryonic endodermal movement. However, the mapping and extent of the embryonic endodermal area in the epiblast is still so poorly known that it seems best at present to say that these cells, which are from the beginning at the region of the anterior streak, simply undergo involution without any other special directed cell movements.

5. Internal divergence (Fig. 6-4, Stages 3-5). As in amphibians, this is a movement primarily of lateral plate mesoderm, although head mesoderm also participates in this activity to a modest degree. Lateral plate mesoderm is the first region to reach the primitive streak and after involuting, it rapidly migrates away from the midline. This is internal divergence. These cells displace the early blastocoele cavity that lay between the epiblast and hypoblast. Internal divergence will eventually carry lateral plate mesoderm to all parts of the area pellucida except the region immediately in front of the embryo. This region, the **proamnion,** will be permanently devoid of invaginated mesoderm and will retain the primitive two germ layer condition described for the early blastoderm when there were only epiblast and hypoblast layers. The anterior margin of diverging lateral plate mesoderm forms the border of the proamnion and, as the divergence continues, the area of the proamnion will become smaller. Lateral mesoderm also spreads out under the area opaca for an appreciable distance but does not quite reach the zone of junction where ectoderm and endoderm are in contact with the yolk mass.

6. Axial elongation (Fig. 6-4, Stage 5). This is the primary movement that will result in the axial organization characteristic of vertebrates. In it the chordamesoderm will lie in the midline and will be flanked on either side by somite mesoderm, followed by mesomere and lateral mesoderm in progressively more lateral positions. When we examine the distribution of chordamesoderm and somite mesoderm at Stage 4 (Fig. 6-4), we will immediately see that somite mesoderm is in the dorsal or median position at the sides of the primitive streak and that all chordamesoderm is at Hensen's node at the anterior end of the streak. Axial elongation is the movement that corrects this "error" of development. Axial elongation entails the posterior regression of Hensen's node, the deepest layer of which is embryonic endoderm. As the node moves posteriorly, it leaves a strand of embryonic endoderm and chordamesoderm behind it. By this means, chordamesoderm is somewhat belatedly interposed between the right and left masses of somite mesoderm.

In later stages, axial elongation also entails further regression of chorda, embryonic endoderm, and somite mesoderm in the posterior direction. This movement probably accounts for axial orientation back to the level of the 12th to 14th somite. Behind this point there is apparently the inductive organization of posterior blastodermal areas into posterior embryonic regions with appropriate axial structures including chordamesoderm, neural plate, somite mesoderm, and embryonic endoderm as is the case for the growing tailbud blastema of amphibian embryos.

Ectodermal movements:

7. Epiboly (Fig. 6-4, Stages 3-7). Epidermal and neural ectoderm (as in the case of amphibian gastrulae) undergo radial spreading in all directions. By this movement, prospective ectoderm will replace those regions on the surface formerly held by endoderm and mesoderm. At the end of gastrulation, all parts of the embryo will be covered by ectoderm except for a small area of tail mesoderm which continues to involute around the primitive streak for many hours after gastrulation is generally thought of as being complete.

8. Proliferation of endoderm, mesoderm, and ectoderm. A final difference between amphibian and avian gastrulation concerns the relative importance of mitotic divisions in these classes. Proliferation, or cell multiplication, contributes very little to the gastrulation movements in Amphibia, whereas, in birds, all germ layers are rapidly dividing and much of the internal spreading, divergence, and elongation or external epiboly is, in part, due to the actual increase in cell number. Indeed, the primitive streak itself should probably be considered as much as a center of mitotic activity as an area of involuting cells.

Comparative gastrulation in amphibians and birds. In spite of superficial differences shown in fate maps (Fig. 6-3), we discover that most mechanisms for germ layer separation employed by birds and amphibians are generally the same in the dynamics of cell movements. Some differences do exist, but most of them appear to be easily correlated with accommodating vast amounts of stored yolk reserves in the bird egg. It is clear that, wherever differences do appear, the bird is the modified or caenomorphic form in mechanisms for gastrulation.

Figure 6-3 A, B, C includes diagrams that are intended to make the homologies between birds and amphibians more evident. Although it is an obvious oversimplification, it is easiest to make these comparisons by imagining that the bird egg is a flat, two-dimensional projection of the amphibian egg, much in the same way that we are accustomed to thinking of the map of the spherical earth as a flat, two-dimensional, map of the world. The center for the projection of the gastrula map will be the dorsal lip region of the blastopore. There will be less distortion closest to the center and more distortion as we move away from the projection center. Figure 6-3 A shows the amphibian egg from the dorsal side as a hemispherical projection. The Stage 3 chick epiblast map is shown in Figure 6-3 C. A hypothetical, two-dimensional projection of an amphibian egg is given in Figure 6-3 B. It could also represent a hypothetical amphibian egg with a massive amount of yolk. It is given as an aid to visualizing the consequences of yolk volume on gastrulation movements in the bird. Despite some difficulties that still exist, we must be impressed by the fundamental similarities that have suddenly become apparent in these eggs that first seemed so different. It is reasonably safe to conclude that agreement between them will be evidence for the presence of paleomorphic characters. The presence of disagreement will be evidence of caenomorphic divergence in one or the other from the ancestral common plan. In this particular case, the heavily yolky bird egg, which is adaptive for development in a shelled egg, clearly represents a caenomorphic divergence from the more paleomorphic amphibian model of an archaetype for vertebrate gastrulation, as a naked egg developing in an aquatic environment.

B. EARLY NEURULATION ILLUSTRATED BY THE 24 HOUR CHICK EMBRYO STUDIED IN WHOLE MOUNTS AND SERIAL CROSS SECTIONS

During the 6 to 8 hours of incubation which separate 17–18 hour and 24 hour embryos, considerable differentiation of both embryonic and extraembryonic areas takes place. Before proceeding to the detailed study of structures, determine the approximate age of your own preparation. This can be done by adding 20 to the number of somite pairs observed in the embryo. The sum obtained is roughly equivalent to the number of hours of incubation.

The materials for study will include a whole mount of a 24 hour embryo and a complete set of serial sections of the 24 hour embryo on one or more slides. **Serial sections** are prepared by preserving, dehydrating, and embedding an embryo at a given stage in paraffin. The paraffin is then sectioned on a precision slicing machine called a microtome. Each slice is then mounted on a slide in numerical sequence and stained in some appropriate way. Hold a serial section slide up to the light, or place it on a white piece of paper. The individual slices are arranged in several rows along the length of the slide. "Read" the serial sections much as you would read the words on a line of the printed page. That is, begin with the section at the upper left end of the row on the slide, then proceed section by section across the top line toward the right. At the end of that line return to the left end of the slide and begin the second row of sections and on until all sections are studied. In "reading" serial sections, however, it is customary to proceed in a "jump forward — jump back" sequence so that we rarely proceed straight through a series without many referrals back and forth in the series.

1. Study of the 24 hour chick whole mount embryo (Fig. 6-5 A–B). Identify the following extraembryonic regions first, and note the advances over comparable parts that were seen in the primitive streak stage. Only part of the **area opaca** will be seen because the blastoderm has undergone very rapid growth and now occupies too great an area to be mounted conveniently on the slide without trimming off much of the membrane. With the aid of Figure 6-5 identify the following subdivisions of the area opaca. The **zone of junction** is the outermost edge of the blastoderm where cells are in actual contact with the yolk mass. The **area vitellina** is a broad area coarsely granular in appearance. It runs around the entire blastoderm and lacks mesoderm. The individual endoderm cells of this region contain many yolk granules and give the area vitellina its distinctive granular staining properties. The **area vasculosa** is the innermost part of the area opaca and is also the most darkly stained portion. It contains invaginated lateral mesoderm. The area vasculosa is much wider posteriorly and laterally than it is anteriorly; indeed, it is frequently incomplete anteriorly if the expansion of invaginated lateral plate mesoderm has not spread far enough to complete this zone. The area vasculosa has a coarse granular, or mottled, appearance due to the early differentiation of lateral mesoderm into primordial clusters of vascular cells. The mottled areas represent mesodermal masses of blood forming cells called **blood islands.** These isolated areas will later anastomose to form the capillary network of the yolk sac. They will eventually fuse with the vitelline veins and arteries that grow out from the embryo. The **area pellucida** is a relatively clear zone entirely surrounding the central **embryonal area.** The latter will form the embryo proper. The anterior extremity of the area pellucida is particularly transparent and represents the region of the blastoderm not yet invaded by the divergent spreading of lateral plate mesoderm. This anterior clear region of the area pellucida is the **proamnion.** It is composed only of extraembryonic ectoderm and endoderm (Fig. 5-6 A–D).

Identify the following structures and regions in the embryonal area; begin at the anterior end. The **head fold** is a cone-shaped elevation at the anterior end of the embryonal area. It is externally lined by **epidermal ectoderm** which can be seen in whole mount preparations as the **margin of the free head.**

Neural ectoderm at the 24 hour stage shows all steps in the formation of the neural tube. As in all organ systems, neural differentiation takes place more rapidly anteriorly than in more posterior regions. Begin at the anterior end of the head fold and identify the **anterior neuropore,** a persistent open neural groove at the extreme anterior end of the body. Just behind this, the **neural folds** are usually in contact with one another and may even be fused to form a complete **neural tube.** The cavity of the neural tube is the **neurocoele.** Posteriorly in the region of the open midgut the neural folds have not fused and an **open neural plate** is still present. Compare these stages in neural tube formation to those observed in your study of neurula and neural tube stages in Amphibia in Chapter 4 and note the unmistakable similarities between them (Compare Figs. 4-1 D and 6-5 A).

Within the head fold, identify the **margin of the foregut** lined by endoderm. Its relation to the open midgut is most clearly shown in Figure 6-5 B, a diagrammatic median sagittal section. Note that the foregut opens posteriorly into the open midgut, or general archenteron, underlying the blastoderm. The opening into the foregut from the midgut is the **anterior intestinal portal.** In the whole mount, it can be identified as a lateral crescent-shaped shadow at the posterior margin of the head fold. The anterior intestinal portal is more clearly seen when viewed from the ventral side as can be done by inverting the slide on the microscope stage. Between the head ectoderm and foregut endoderm, identify the tenuous strands of **head mesoderm.** These cells will later form most of the head skeleton and musculature. The mesoderm lateral to the anterior intestinal portal appears somewhat darker than elsewhere. This mesoderm will form the heart and is called **cardiac mesoderm.** In the open midgut region of the embryonal area, the archenteron lacks a cellular floor. In this region identify the 3 to 6 pairs of **mesodermal somites** appearing as ovoid blocks on each side of the neural tube. The **notochord** can be seen as a condensed median band of tissue lying between the neural folds. Identify also the thickened mass of tissue extending posterior to the somites which will gradually differentiate additional somites; this is **unsegmented somite mesoderm.**

At the posterior end of the embryo, identify the **primitive streak** in the posterior quarter of the embryonal area. Identify again **Hensen's node,** the **primitive folds, primitive groove,** and **primitive pit** on the basis of descriptions already provided for them in association with the study of the 18 hour chick embryo.

2. Study of serial cross sections of the 24 hour chick embryo (Fig. 6-5 C–I). The study of serial cross sections of the 24 hour chick embryo should be followed in reverse order from posterior to anterior. The anterior end develops appreciably more rapidly and, in a sense, represents an "older stage" morphogenetically than do the more posterior regions. By progressing through the series from posterior to anterior, it is possible to see progressive steps in the development of such structures as the neural tube, coelom, gut, somites, and notochord.

In the following directions, a regional approach has been taken in describing the structures to be identified. Make constant reference to whole mount and sagittal diagrams in your study of cross sections. Note that the levels of cross sections in Figure 6-5 C–I are shown by the guide lines at the right and left beside Figure 6-5 A and 6-5 B. By placing a straight edge across these drawings, and connecting corresponding letters at the right and left, you can approximate the level at which each of the cross section diagrams was made.

a. Cross section through the primitive streak or Hensen's node. (Fig. 6-5 I). Germ layer relationships seen at the posterior end of the 24 hour chick embryo correspond closely to those that obtain throughout the entire 17 hour embryo. The ectoderm, mesoderm, and endoderm are revealed in their simplest conditions at this level. In cross sections, identify the **primitive streak** composed of the primitive groove flanked on each side by a **primitive fold.** Note, in particular, that the **invaginating mesoderm** forms a dense mass under the primitive groove and that cells from this region extend laterally (by internal divergence) as a strip of loosely connected amoeboid cells. Note that the **endoderm** forms a complete membrane which extends from the midline to the outer edges of the section; these cells are greatly thickened in the **area vitellina** owing to large accumulations of yolk in the endoderm cells. Only in the midline, under the primitive streak, is the endoderm firmly attached to invaginated mesoderm. This is the **embryonic endoderm,** in contrast with the more laterally placed **extraembryonic endoderm** of hypoblast origin.

b. Section through the open midgut and open neural tube (Fig. 6-5 G–H). Identify the **open neural plate** composed of greatly thickened ectoderm, and the neural folds which make up the lateral edges of the neural plate. The **notochord** appears as a round mass of mesoderm just under the neural plate. On each side of the notochord is a dense mass of **somite mesoderm,** or epimere. Just lateral to each somite is a slender continuation of mesoderm which will later condense to form the kidney; this region is not differentiated clearly at this time but is, nevertheless, called the **intermediate (nephrotome) mesoderm.** Laterally, in the extraembryonic regions, the **lateral plate** is in the process of differentiating blood vessels and forming the coelom. The latter is accomplished by a splitting of the lateral plate into two layers. The space between them is the **coelom.**

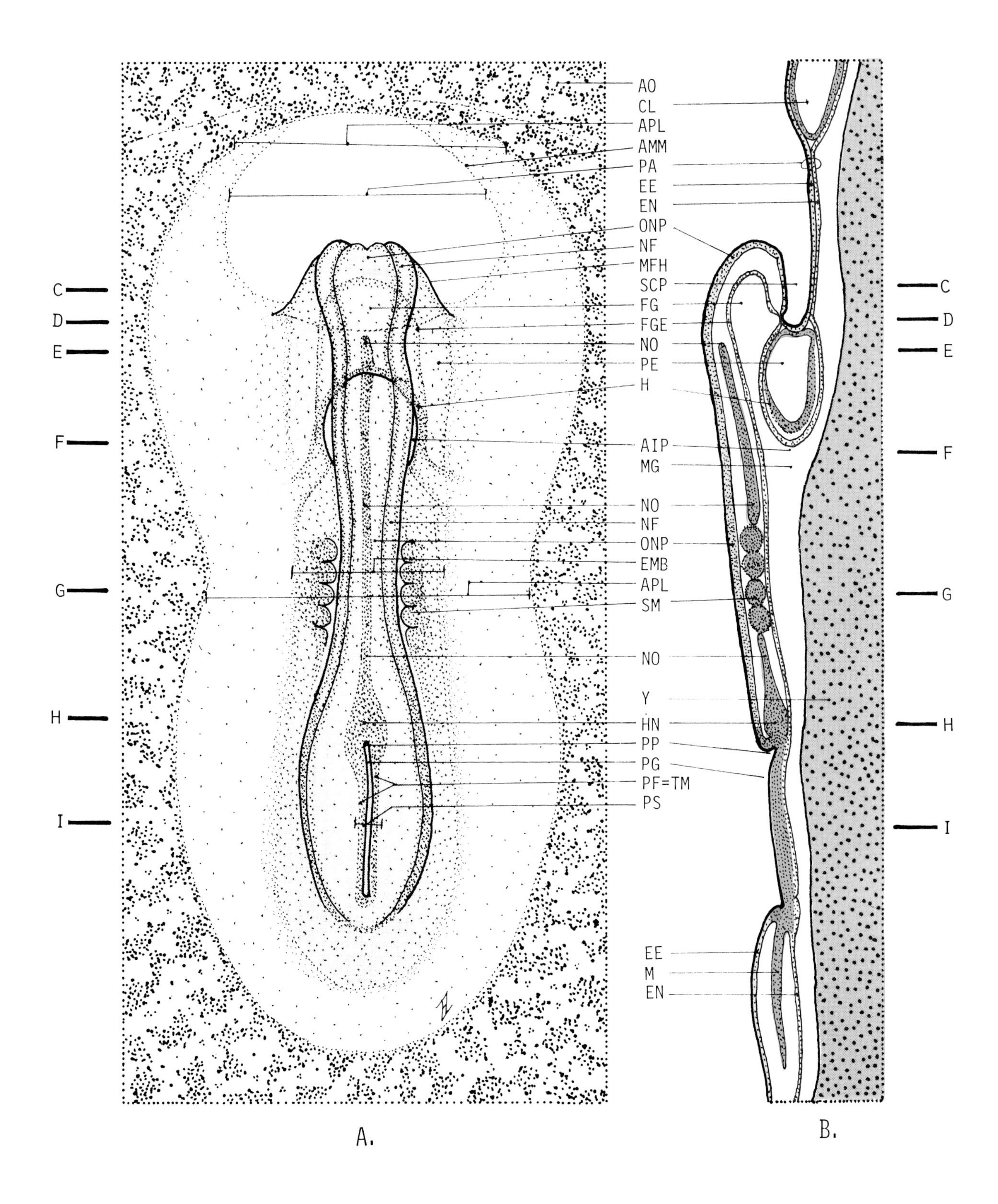

Figure 6-5. Structure of the 24 hour chick embryo. 6-5 A. Whole mount diagram. 6-5 B. Median sagittal section. See legend.

Legend for Figure 6-5. Structure of the 24 hour chick embryo.

Figure 6-5 A. Whole mount diagram of the 24 hour chick embryo.
Figure 6-5 B. Median sagittal section of the 24 hour chick embryo.
Figures 6-5 C–I. Representative cross sections at levels indicated by correspondingly lettered guide lines on Figures 6-5 A & B.

Legend and checklist of structures to be identified

Color code: blue — ectoderm, red — mesoderm, yellow — endoderm, green — uncleaved yolk.

AIP	anterior intestinal portal
AMM	anterior margin of invaginated lateral mesoderm
AO	area opaca
APL	area pellucida
CL	coelom
E	ectoderm
EE	epidermal ectoderm
EMB	embryonal area
EN	endoderm
FG	foregut lumen
FGE	foregut endoderm
H	heart primordium
HM	head mesoderm
HN	Hensen's node
LBF	lateral body fold
LM	lateral plate (hypomere) mesoderm
M	mesoderm
MFH	margin of free head
MG	midgut open to yolk
MM	mesomere, nephrotome, intermediate mesoderm
NCL	neurocoele, neural canal
NF	neural fold
NO	notochord
NT	neural tube
OM	oral membrane
ONP	open neural plate
PA	proamnion
PE	pericardial coelom
PF	primitive folds
PG	primitive groove
PP	primitive pit
PS	primitive streak
SCP	subcephalic pocket
SM	somite (epimere) mesoderm
SO	somatic lateral plate (hypomere) mesoderm
SOP	somatopleure (extra-embryonic ectoderm & somatic mesoderm)
SP	splanchnic lateral plate (hypomere mesoderm)
SPP	splanchnopleure (extraembryonic ectoderm & splanchnic mesoderm
TM	tailbud mesoderm
VA	ventral aorta capillary bed
Y	yolk

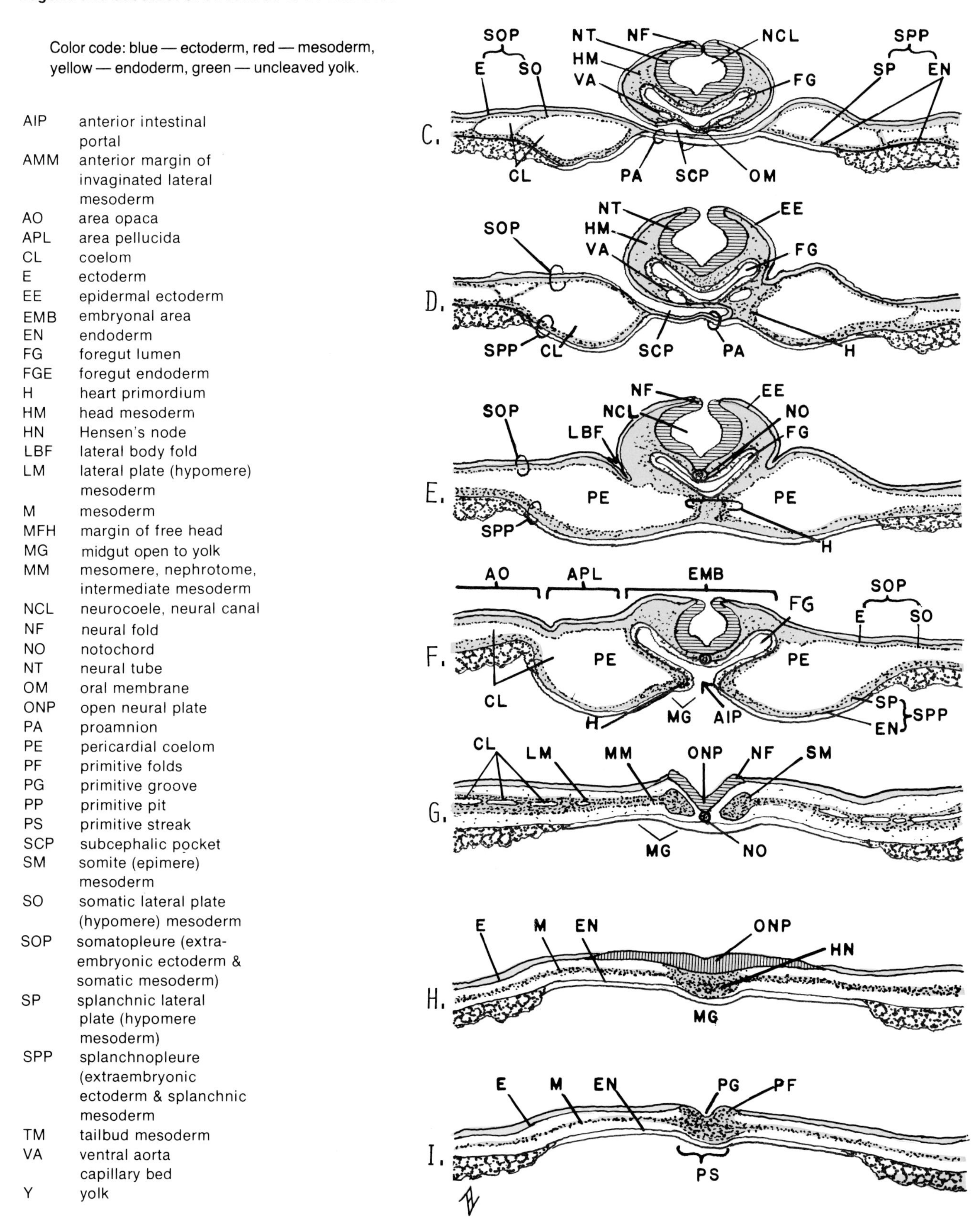

Figure 6-5. Structure of the 24 hour chick embryo, continued. 6-5 C–I. Representative cross sections at levels indicated by guide lines in 6-5 A. and B. See legend.

c. Section through the anterior intestinal portal and heart primordium (Fig. 6-5 E–F). At this level, note and identify the further differentiation of the **neural plate, neural folds** and **notochord.** In particular, note the very great expansion of the **coelomic cavity.** Trace the mesoderm laterally from the body and note that one layer of mesoderm will closely adhere to the embryonic and extraembryonic epidermal ectoderm. This mesoderm is the **somatic mesoderm;** the ectodermal and mesodermal layers together form a compound membrane called the **somatopleure.** The **extraembryonic coelom** is the cavity formed by a splitting of the lateral plate. The lower layer of lateral plate mesoderm is termed the **splanchnic mesoderm.** It becomes closely applied to the **extraembryonic endoderm.** This combination of mesoderm and endoderm is called **splanchnopleure.** Splanchnopleure will form the **yolk sac** and **allantois** in later development. Somatopleure will form the **amnion** and **serosa.** These will be discussed later in connection with the formation and function of extraembryonic membranes in Chapters 7 and 8. The coelom at this level of the 24 hour chick embryo will form the spacious cavity surrounding the heart; it is called the **pericardial cavity.** The most important feature to note is the manner in which the embryonic gut-cavity acquires an endodermal floor. It will be noted that the splanchnopleure layer bends mesioventrally in such a manner as to close off an endodermally lined pocket, the **foregut**. Follow the foregut anteriorly into the head fold. The splanchnopleure, at the level of the anterior intestinal portal, contains an exceptionally thick layer of splanchnic mesoderm; this mesoderm will form the heart; it is the **cardiac mesoderm.** The right and left cardiac primordia will meet in the ventral midline and fuse to form the tubular heart. The cardiac mesoderm is continuous laterally with the extraembryonic splanchnic mesoderm. At this level, note that small **blood vessels** and **blood islands** are forming in the lateral extremities of splanchnic mesoderm but that they are not present in the overlying somatic mesoderm layer. These capillaries are in the **area vasculosa.** The **area pellucida,** at this stage, can be distinguished by the absence of blood capillaries in its splanchnic mesoderm layer.

d. Section through the free head and foregut (Fig. 6-5 C–D). The head fold is entirely separate and elevated above the remainder of the blastoderm. Identify the **proamnion** underlying the head process. It is composed of extraembryonic ectoderm and endoderm. The head fold is surrounded by **epidermal ectoderm** and the space between the head epidermis and proamnion is the **subcephalic pocket** (Fig. 6-5 C–D). The **neural tube** may, or may not, be entirely closed by the fusion of the **neural folds.** The cavity of the neural tube is the **neurocoele.** Identify the **notochord** underlying the neural tube and the crescent-shaped **foregut lumen,** lined by endoderm, located ventral to the notochord. Note that in the ventral midline, the foregut endoderm comes into very close contact with the epidermis of the head, and that in this region both layers are thickened. This is the **oral membrane;** it marks the location at which the mouth will later form. If present, the **ventral aortae** will appear as a pair of very thin-walled tubes lying lateral to the oral membrane between the floor of the foregut and the ectoderm of the free head. Last, identify the scattered cells of **head mesoderm** which fill most spaces between the head epidermis and the internal structures just mentioned in the free head.

C. THE NEURAL TUBE STAGE IN CHICK DEVELOPMENT ILLUSTRATED BY THE 33 HOUR CHICK EMBRYO

Emphasis, in the study of the 33 hour chick embryo, will be given to those regions and organ primordia that are most conspicuously advanced over the level found in the 24 hour embryo. Materials for study will include prepared slides of whole mounts of blastoderms, and a complete series of serial sections which may require more than one slide to complete.

1. Study of the 33 hour chick whole mount preparation Fig. 6-6 A, B). Before studying the embryonal and extraembryonic areas, make a count of the number of pairs of somites and estimate the age of your preparation (7 somites = 29-31 hours; 9 somites = 32-33 hours; 11 somites = 34-35 hours). Identify and note the developmental advance that has taken place in the following structures since the 24 hour stage; **head fold, margin of the free head, margin of the foregut, anterior intestinal portal, somites, notochord, neural tube,** and **primitive streak.** Particular attention should be directed toward the nervous and circulatory systems which have undergone the most rapid differentiation since the 24 hour stage.

a. Extraembryonic blastoderm. Much of the outer blastoderm may have been trimmed away so as to permit the embryo to be mounted conveniently under a coverslip. Much of the **area vitel-**

lina and **zone of junction** may be absent in all except the anterior part of the preparation. Identify these regions and the **area vasculosa** which now contains many anastomosing capillaries forming the yolk sac circulatory network. **Yolk sac capillaries** are best formed near the area pellucida mid-way between anterior and posterior ends of the blastoderm. **Blood islands** are particularly well formed in the posterior and lateral regions of the area vasculosa. Capillaries fuse along the outer margin of the area vasculosa; here they will form a circular vein, the **sinus terminalis.** It later will collect blood from the yolk sac and return it to the heart by means of anterior and posterior vitelline veins which have not yet made their appearance. Most yolk sac capillaries, at this time, converge toward the middle of the **embryonal area** where they are beginning to make contact with the paired **vitelline veins** that enter the posterior end of the **heart.** The **area pellucida** is most clearly seen anterior and posterior to the embryonal area. In the midregion, the area pellucida is less distinct owing to the large number of yolk sac (vitelline) capillaries traversing the area in their course toward the heart. At approximately this developmental stage, we no longer refer to the area pellucida as a distinct region. The region of the **proamnion,** located just anterior to the head fold of the embryo, still lacks mesoderm and is the most transparent part of the blastoderm. Mesoderm has not expanded around the proamnion. The anterior **margin of invaginated mesoderm** makes very distinct lateral and anterior boundaries to the proamnion. **Amniotic fold** formation may have begun at the anterior end in front of the proamnion. This fold involves only the somatopleure and, in many preparations, this fold is merely a slight swelling of the surface layer that is not recognizable in whole mounts. The formation of the amnion and the other extraembryonic membranes will be considered briefly in Chapter 7 and more fully in Chapter 8.

b. Nervous system. Begin with the head fold and note the **anterior neuropore.** It is a vestige of the open neural plate condition that persists as a very small opening at the anterior extremity of the head. The **neural tube** is now closed along the body axis as far posteriorly as the last somite. Within the head fold, a number of swellings are present in the neural tube. These anticipate the formation of the three primary brain regions. From anterior to posterior, these are the **forebrain, midbrain,** and **hindbrain.** The **forebrain** or **prosencephalon** is expanded into right and left **optic vesicles** which are *anlagen* (primordia) of the retina. In the floor of the forebrain, a slight ventral evagination can be seen; it is the **infundibulum.** It contributes to the formation of the pituitary gland. The second major expansion of the neural tube is the **midbrain,** or **mesencephalon.** In it, on whole mount preparations a faint median line can be seen; it is the **notochord.** It underlies the neural tube and ends anteriorly at the boundary between the first and second brain regions. Behind the midbrain are a series of about 4 to 6 minor swellings in the brain wall; all are included in the **hindbrain,** or **rhombencephalon.** The small swellings are **neuromeres,** and the last one at this stage is at the level of the anterior intestinal portal. Posterior to the last neuromere, the closed neural tube will form the **spinal cord.** The neural tube is closed to the level of the last somite. Posterior to this level, and extending back to **Hensen's node,** the neural plate is open and is rather inappropriately named the "sinus rhomboidalis" which means "diamond-shaped cavity." In the region of the open neural plate, identify the **notochord** which extends under the neural ectoderm forward from Hensen's node to the level of the midbrain. Identify the **unsegmented** somite mesoderm posterior to the somites and lateral to the neural tube. It will continue to form additional somites as development proceeds. The structure of the 33 hour embryo, at the level of the open neural plate and primitive streak, is smaller but nevertheless is similar in all basic relationships to the structure of comparable regions in the 24 hour chick embryo.

c. Circulatory system. The heart and blood vessels are in a prefunctional stage of development. Blood has not yet begun to flow. Most structures are present in only partially completed outline. Figure 6-6 B shows a lateral reconstruction of an embryo and illustrates the vessels and heart chambers that are developing at this time. It will be advantageous to invert slides and examine the whole mount preparation from the ventral surface since the heart is superficial in ventral view. Begin your study with the paired **vitelline veins** which enter the heart just in front of the anterior intestinal portal. The median point of fusion of the vitelline veins is the **sinus venosus.** It is directly continuous anteriorly with the **atrium.** No structural landmark is present at this stage to permit a clear distinction between these two heart chambers. The atrium bends strongly toward the right and opens into the **ventricle,** which is expanded and somewhat crescent-shaped. Anteriorly, the ventricle bends back to the midline and again becomes sharply constricted. This narrow anteromedian part of the heart tube is the **conus arteriosus.** The heart at this stage consists of a simple, slightly bent tube lying in the **pericardial coelom.** Throughout its length, the heart is suspended from the floor of the foregut by a part of the ventral mesentery of the gut called the **dorsal mesocardium.** The latter structure cannot be identified in whole mount preparations but should be recalled in the examination of the serial cross sections. Anterior to the pericardial cavity, the conus arteriosus communicates with the paired **ventral aortae.** They run forward under the foregut. Toward the anterior end of the foregut, the ventral aortae break down into capillary beds which anticipate the formation of the major intraembryonic pathways of early development. Although not visible in most whole mount preparations, these vessels are the paired **internal carotid arteries** to the head, **paired dorsal aortae,** and paired **first aortic arches which connect the ventral** and **dorsal aortae.** In addition to these arteries, the capillary beds associated with the formation of the **anterior, posterior** and **common cardinal veins** are also taking form. For the relationships of these vessels, see Figure 6-6 B which shows a diagrammatic reconstruction of the circulatory system and then compare the early circulatory plan of the developing chick embryo with the vertebrate common plan shown in Figure 7-5.

2. Study of serial cross sections of the 33 hour chick embryo (Fig. 6-6 C–P). The study of the serial sections should not be started before you are thoroughly familiar with the general topography and structure of the whole mount. The structures that have already been pointed out should be identified in the sections with the aid of Figure 6-6 C–P. All labelled parts should be identified and correlated with their counterparts shown in the whole mount, Figure 6-6 A. Get in the habit of keeping a straight edge (clear plastic rulers are excellent) to lay across the whole mount diagram so that the lettered guide lines in the right and left margins can be connected. This will reveal the exact level at which the cross section diagrams were made and will greatly facilitate the correlation of cross section and whole mount information.

The following comments are designed to direct attention to features of special interest in each section that is provided. It is quite probable that not all structures labelled in each sectional diagram will be mentioned specifically. The following paragraphs are numbered to correspond with the Figures 6-6 C–P.

Section C. In particular note that the **proamnion** underlying the head process is clearly composed of only extraembryonic ectoderm and endoderm; lateral mesoderm has not invaded this region nor will it ever do so. Note the very greatly expanded **optic vesicles** which are broadly in communication with the central lumen of the **forebrain.** The lumen is the **neurocoele. Head mesenchyme** can be seen in those regions where the neural tube or optic vesicles are not in contact with the epidermis. Lens induction occurs at about this time between the optic vesicles and the overlying epidermis. The **subcephalic pocket** is merely the space under the head.

Section D. The only special feature shown here that has not been noted in Section C is the **infundibulum** which here appears as a median ventral evagination from the floor of the foregut. The infundibulum will contribute to the formation of the pituitary gland and is a positive identification of that part of the forebrain that will form the diencephalon in later development. Some intraembryonic capillaries may be forming in the head mesenchyme; one, which is in the position that will be occupied by the **ventral aorta,** is indicated in this section.

Section E. Several additional structures should be noted. The anterior end of the **foregut** is shown; it, of course, is lined by endoderm. Note that in the ventral midline, the foregut endoderm and epidermis are in contact; this double membrane is the **oral membrane.** It marks the point at which the mouth will open at a considerably later time. Note under the neural tube the small, dense circular outline of the **notochord.** This will be present through the remainder of the sections back to Hensen's node, at the posterior end of the embryonal area. The presence of the notochord in this section is evidence that you have passed from the level of the forebrain into the level of the **midbrain.** The anterior tip of the notochord in vertebrates characteristically never extends farther anteriorly than the midbrain. Last, try to follow one or more of the head capillaries around the lateral walls of the foregut; if present, they can be reasonably interpreted as the capillary primordium of the **first aortic arch.**

Section F. All special structures described for Section E also are present and, to some degree, are even better formed in this section. In particular, attempt to identify the intraembryonic capillaries in the three special locations indicated in the diagram for the **ventral aorta, dorsal aorta,** and **anterior cardinal vein.** It is to be understood that all major vessels first arise from capillary beds, and that any one of the vessels in the primordial bed has an equal chance of being the precursor of the definitive vessel. Do not be surprised if there are several capillaries that appear to be equally good candidates as the primordium for these three major vessels.

Section G. This level is characteristically one of the most difficult ones for beginning students to master. The primary problems arise from the fact that the head fold is just making contact with the rest of the blastoderm. Go back and forth through this region several times. Each time follow a different structure so that the transitional area becomes thoroughly understood. Begin by tracing the fate of the **subcephalic pocket** from the point at which it is wide open, to the points at which it makes contact with one side, then the other side (this, of course, will trap a posterior part of the pocket inside the proamnion). Proceed to the level at which the pocket disappears. Then do the same for the ventral aortae. Begin where they are separate and clearly located beside the **oral membrane.** Proceed back until the oral membrane disappears and two vessels unite in the ventral midline. This fusion marks the anterior end of the **heart,** namely, the **conus arteriosus.** Note that just as the conus is forming from the fusion of ventral aortae, the lateral body folds are broadening, and extensions from the **extraembryonic coelom** enter the embryo and surround the conus; this is the anterior end of the **pericardial cavity.**

Section H. This section, through the midregion of the conus arteriosus, shows to advantage the early structure of the heart primordium. Note that the conus is a derivative of the **ventral mesentery of the foregut.** The wall of the conus can be seen to be composed of a very thin inner lining, the **endocardium,** which is continuous with the capillary lining of the ventral aorta. The external thicker layer is the **epimyocardium.** The latter is clearly an extension of the splanchnic mesoderm that surrounds the pharyngeal endoderm.

Section I. This section is much like Section H; however, the distinctive feature is the aggregate of **cranial neural crest** that will be easily seen on both sides of the neural tube and near the dorsal epidermis. These cells are the primordial beginnings of the neuroblasts for the 5th and 7th cranial nerves. As such, they provide a landmark in this section for the fact that, at last, you are surely at the level of the **hindbrain,** to which the 5th and 7th nerves will make neuronal contact. The other feature of importance is the displacement of the heart to one side of center. This is the **ventricle;** its growth is toward the right; therefore, you can identify the true right and left side of the embryo by the direction of ventricular displacement in the section. Note the dorsal connection of the ventricle to the floor of the foregut, the **dorsal mesocardium.** At the 33 hour stage, a dorsal mesocardium connects the heart along its entire length to the gut floor; later, most of the dorsal mesocardium will degenerate and permit the heart to "float" free in its pericardial cavity.

Section J. This section is essentially the same as the preceding one with the exception that the heart has again returned to a median and symmetrical location; this is the region of the **atrium.** In addition to this feature, note the greatly thickened epidermis dorsal and lateral to the brain. These are the **otic placodes,** the primordia of the inner ears. They have only recently been induced by the lateral walls of the myelencephalon.

Section K. The major feature to be noted here is the contact between the yolk sac and the heart wall. Note that the large capillaries of the yolk sac **(vitelline veins)** come in and unite with the endocardium of the heart so that the heart and capillary lumina are continuous. This region of contact between heart and vitelline veins is the **sinus venosus.**

Section L. This is a section through the **anterior intestinal portal,** namely, the point at which the closed foregut opens out into the midgut which still lacks a ventral endodermal floor. At this level and several sections following after it, favorable sections will be found through the developing **spinal neural tube, neural crest,** and **mesodermal somites.** It is of some interest to point out that the definitive head will extend back to the level including the first four somites. Refer to the whole mount embryos of 24 and 33 hours incubation and note the large percentage of the blastoderm that is destined to form the head.

Sections M and N. These are typical **open midgut** sections and show the basic axial organization of the vertebrate body to excellent advantage. Note the **neural tube** and **neural crest,** and the mesiolateral relationships of major mesodermal regions. Begin with the **notochord** in the midline. Find a median section through a **somite** and identify its central cavity or **myocoele.** This cavity will disappear without a trace but is of interest as paleomorphic evidence of primary enterocoelous somite formation which, in protochordates, is by evagination from the gut wall. **Nephrotome,** or **intermediate mesoderm,** will be identified as a small strand of mesoderm that usually can be seen connecting the somite regions with the more lateral **somatic** and **splanchnic** layers of **hypomere mesoderm.** As you proceed posteriorly through these sections, note that the **coelom** becomes less and less well formed and eventually disappears altogether.

Section O. This is a section through the posterior part of the embryonal area at the level of the **sinus rhomboidalis,** or **open neural plate.** At this level the simple 3-germ-layer composition of the embryo is at its clearest. Indeed, it was from sections such as this that the germ layer concept was born in the minds of early embryologists such as Wolff and von Baer.

In concluding this study, we must be impressed by the fact that, at the end of gastrulation and neurulation, whatever may have been the caenomorphic digressions imposed upon the embryos, be they bird, beast, or fish, they all eventually arrive at essentially the same basic structure as a beginning for molding the vertebrate body.

In the next two chapters further stages in the development of the chick will be studied. However, it is appropriate to emphasize that they are used as general examples of vertebrate morphogenesis; it is altogether incidental that the end product of their development will be a bird.

Legend for Figure 6-6. Structure of the 33 hour chick embryo.
Legend for the figures on the following pages.

Figure 6-6 A. Whole mount diagram with guide lines to show the levels of representative cross sections in 6-6 C through P.

Figure 6-6B. Diagrammatic reconstruction of the 33 hour chick (Adapted from B.M. Patten, *Early Embryology of the Chick,* W.B. Saunders, Co.).

Figure 6-6 C–P. Representative cross sections of the 33 hour chick embryo at levels indicated by guide lines on 6-6 A.

Legend and checklist of structures to be identified

Color code: blue — ectoderm, red — mesoderm, yellow — endoderm.

AI 1st aortic arch
AC anterior cardinal vein
AIP anterior intestinal portal
AMM anterior margin of invaginated lateral mesoderm
ANP anterior neuropore
AO area opaca
APL area pellucida
AT atrium of heart
BI blood islands
CA conus arteriosus
CAP capillaries
CC common cardinal vein
CL coelom
CNC cranial neural crest primordium of nerves 5, 7, 8.
DA dorsal aorta
DMC dorsal mesocardium
E ectoderm
EE epidermal ectoderm
EMB embryonal area
EMY epimyocardial layer of heart
EN endoderm
END endocardial lining of heart
FB forebrain, prosencephalon
FG foregut lumen
FGE foregut endoderm
HB hindbrain, rhombencephalon
HM head mesoderm
HN Hensen's node
IC internal carotid artery
IN infundibulum
LBF lateral body fold mesoderm
M mesoderm
MB midbrain, mesencephalon
MFH margin of free head
MG midgut lumen
MM mesomere, nephrotome, intermediate mesoderm
MYC myocoele of somite
NC neural crest
NCL neurocoele, neural canal
NF neural fold, primordium of neural crest
NM neuromeres of hindbrain
NO notochord mesoderm
NT neural tube, spinal cord
OM oral membrane
ONP open neural plate, sinus rhomboidalis
OP optic vesicle
OT otic placode
PA proamnion
PCV posterior cardinal vein
PE pericardial coelom
PF primitive fold
PG primitive groove
PS primitive streak
SCP subcephalic pocket
SI sinus terminalis (circular vein of yolk sac)
SM somite (epimere) mesoderm
SO somatic lateral plate (hypomere) mesoderm
SOP somatopleure
SP splanchnic lateral plate (hypomere) mesoderm
SPP splanchnopleure
SV sinus venosus
USM unsegmented somite mesoderm
V ventricle of heart
VA ventral aorta
VAS area vasculosa
VIT area vitellina
VV vitelline vein
YS yolk sac
ZJ zone of junction

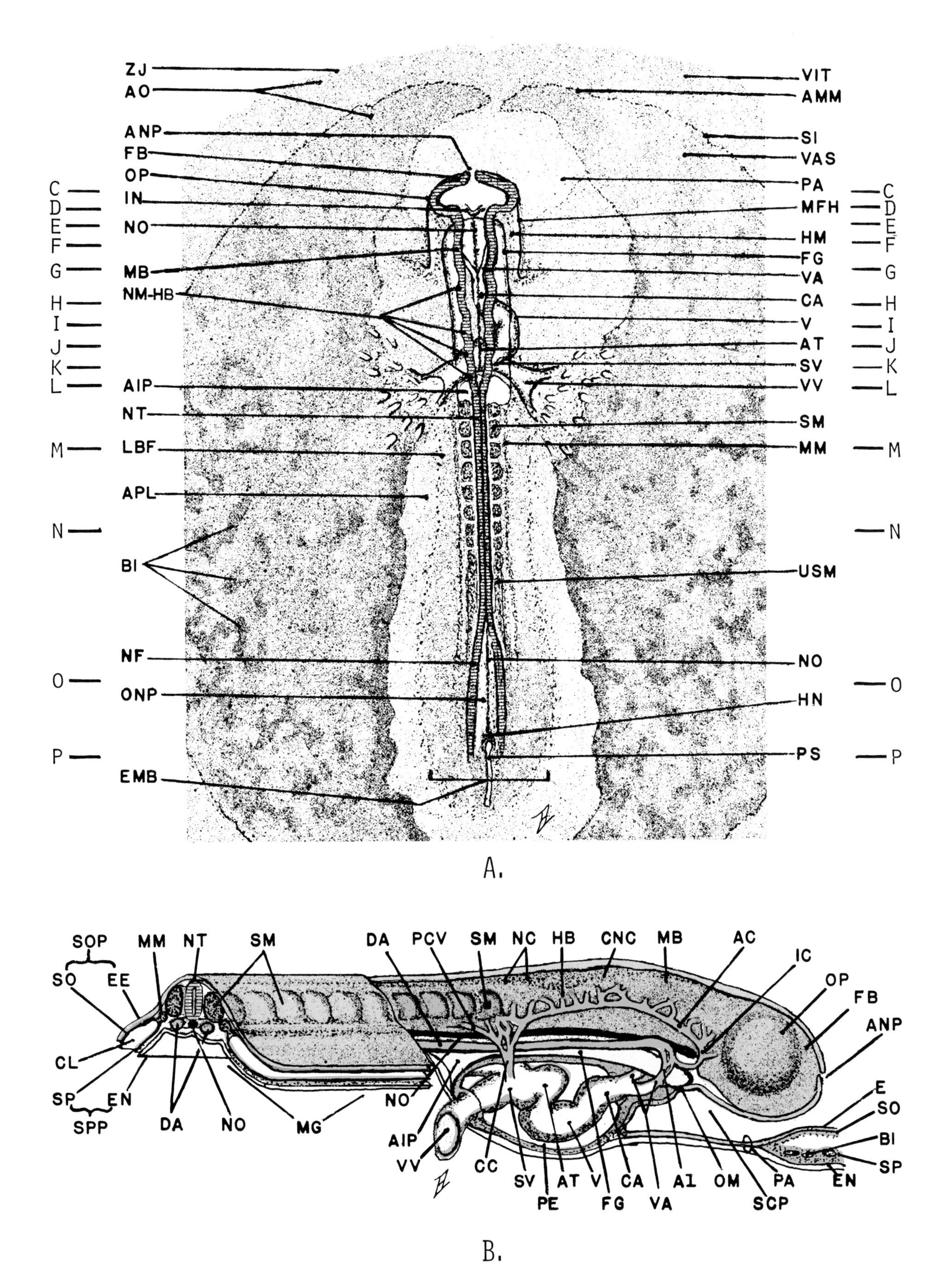

Figure 6-6. Structure of the 33 hour chick embryo. 6-6 A. Whole mount diagram with guide lines to show the levels of representative cross sections in 6-6 C–P. 6-6 B. Diagrammatic reconstruction of the 33 hour chick embryo. See legend.

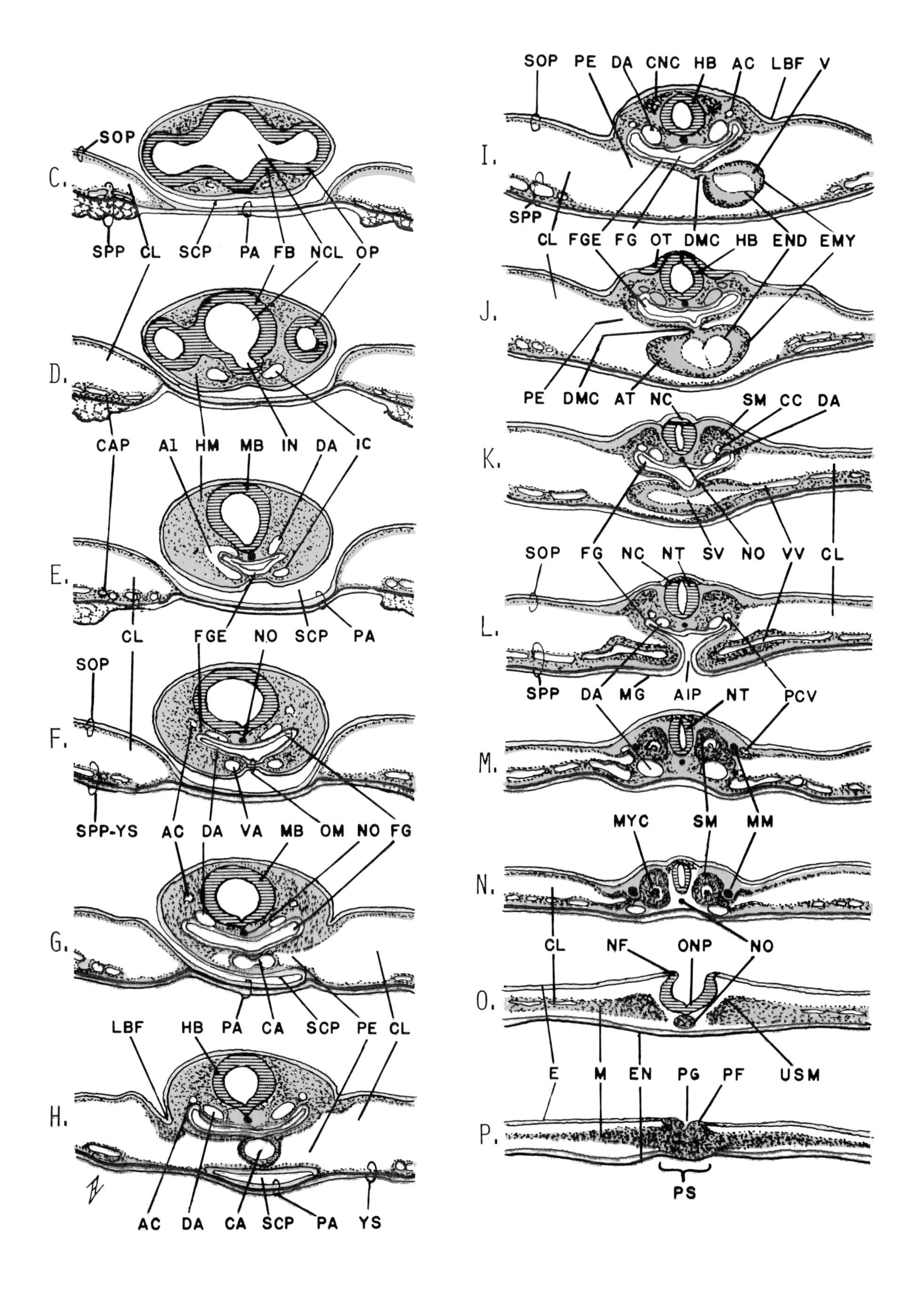

Figure 6-6 C–P. Structure of the 33 hour chick embryo, continued. Representative cross sections at levels indicated by guide lines in Figure 6-6 A. See legend.

D. RECONSTRUCTION OF THE 33 HOUR CHICK EMBRYO FROM SERIAL SECTIONS

The reconstruction of anatomical structures from serial sections is one of the standard tools of embryology. It is a particularly valuable technique to use with experimental cases in which as much information as possible must be extracted from a single specimen.

A serial reconstruction consists of making measurements of structures in every section through the entire series. These measurements are recorded section by section on graph paper in which each row of lines represents a single section. When all measurements are completed, the individual points are connected by a continuous line which then represents the contour of the structures measured at every level of the specimen. Two special precautions must be observed.

a. A constant axial point of reference must be provided as a base from which measurements are made. When it is expected that serial reconstructions are to be made, it is customary to embed with the embryo two animal hairs that lie in the AP axis and are sectioned along with the embryo to provide the reference guides.

b. Microscopic magnifications of measurements must be in agreement with the thickness of the tissue sections. Most embryonic materials are routinely sectioned at a thickness of 10 microns (0.01 mm). If there is to be an accurate reconstruction of the dorsoventral and mesiolateral proportions, their scales of magnification must be the same as that in the section thickness.

Reconstruct the 33 hour chick embryo from your serial sections from the level of the anterior neuropore at the tip of the head through the anterior intestinal portal. Proceed as follows.

1. Draw a line down the center of a page of millimeter graph paper. This line will correspond with the median axis of the embryo. Place points along this line at 2 mm intervals to represent the thickness of each section.

2. Use the 10 X ocular, 10 X objective, and the camera lucida arm length at 120 mm. These will give a magnification of approximately 200 X. Since the serial sections are cut at 0.01 mm, allow 2 mm for each slice in the AP axis. **Alternatively,** use an ocular micrometer with a known measured grid and calibrate it against a slide micrometer; then calculate the required linear dimension on the page to match the magnifications you need.

3. Count the number of sections in the series from the tip of the head fold back to the anterior intestinal portal. Mark off and number in sequence an equivalent number of 2 mm units across the width of the page; each unit will represent one section. Return to the first section in the series as the starting point for the reconstruction drawing.

4. Consider your guideline to be the notochord in the section. Make all measurements from the notochordal reference point, or the embryonic midline in regions lacking a notochord.

5. To make measurements with the camera lucida, align the notochord and guide line so that they are superimposed visually in the same axis. Then put marks to the right and left of the guide line in accord with the lateral displacement of the structure being measured. **To make measurements with an ocular micrometer,** align the middle line of the micrometer grid with the notochord and measure off directly the lateral displacement of the structure being measured.

6. Record measurements horizontally on the graph paper by a series of points to the right and left of the center guide line in linear agreement with actual observations.

7. Proceed to the next section and repeat Steps 5 and 6 at each 2 mm point on the reference line.

8. Connect all lateral points with a continuous line when all measurements have been completed and recorded. This will give an accurate picture of the anteroposterior dimensions of the structure you have measured through the series.

9. In this particular reconstruction it is suggested that separate colors be used in recording measurements of the following systems in the 33 hour chick embryo.

Use black for epidermal ectoderm
Use blue for neural tube
Use green for notochord
Use red for heart
Use yellow for foregut endoderm
Label all subdivisions of these structures that are revealed by the serial reconstruction.

E. EXPERIMENTS AND OBSERVATION ON THE LIVING 18, 24, AND 36 HOUR CHICK EMBRYOS

1. Incubation of fertile eggs. Fertile eggs can be obtained from local hatcheries. They can be kept for approximately 5 days before incubation is begun. However, once incubation is started, it may not be interrupted. **Eggs that are held for later incubation must be kept in a cool place but not refrigerated.** They are best held at about 18°C (= 65°F). Lower temperatures will kill blastoderms.

Incubate eggs at 37.5–38.5°C (= 99–100°F) in a saturated water atmosphere. This is accomplished by placing large bowls or pans of water in the incubator. Eggs should be rotated 180° twice a day during the period of incubation. This rotates the embryo within the shell and prevents adhesion of embryonic membranes to the inside of the shell. Rotation is not necessary prior to the third day of incubation, but is essential thereafter.

2. Observations on living blastoderms. Embryos incubated for 18, 24, and 36 hours will be available for each student. In the event that living materials are not available, models, charts, and films will be used to demonstrate the following features of normal development.

Materials needed for each student: Dissecting microscope or good hand lens, 500 ml 0.9% isotonic warm NaCl, 3 3–4 inch fingerbowls, 3 Petri dish halves, 1 wide-mouth waste jar, standard dissecting instruments (including 2 fine dissecting needles made from 000 stainless steel insect pins, small scissors, fine forceps, standard pipette), 1 plastic teaspoon, and dropping bottle of 0.01% Neutral Red and/or Nile Blue Sulfate. Other special materials will be indicated for particular techniques.

a. How to open incubated eggs. Eggs will be opened and embryos observed in fingerbowls of isotonic warm NaCl prepared by dissolving 9 g NaCl per liter of hot tap water. Rock salt or so-called ice cream salt is better for this purpose than chemically pure NaCl since it more closely approximates a balanced salt solution. The water should be appreciably warmer than body temperature before the fingerbowls are filled two-thirds full. The glass will cool the solution. The temperature should be warm, but not hot, to the touch before an egg is put into the bowl of saline.

Crack the egg in the usual "kitchen manner" by striking the egg sharply against the edge of the fingerbowl. Before opening the egg, submerge it completely in the warm saline solution. Carefully and slowly open the egg under water. Open the shell from the side or bottom (not the top). Allow the albumen and yolk to flow out into the bottom of the bowl.

Discard the two pieces of egg shell. Decant off half the saline into the waste bowl to a level that will just cover the yolk of the egg. Under ordinary circumstances, the yolk will orient by gravity, with the embryonic disc uppermost. The embryo is now ready to be observed on the dissecting microscope or with a hand lens.

Adjust the desk lamp close to the microscope to give maximum illumination from above. The heat of the lamp will also assist in warming the embryo. The embryo can be expected to remain alive under these conditions for from three to five hours. If there is any drying of the surface, pipette saline over the embryonal area from time to time.

b. Observation of early living embryos with the aid of vital dyes. At early stages before the blood begins to form, living embryonic tissues are very transparent. Most structures will be difficult or impossible to see without special assistance.

One of the simplest methods for making embryonic tissues visible involves the use of, so-called, "vital dyes." These are stains which, in low concentrations, may be absorbed by living cells without killing them. Vital dyes become bound in subcellular organelles and are permanently concentrated in them, thereby making the tissues visible. Two of the most rapidly bound and lowest toxicity dyes are **Neutral Red** and **Nile Blue Sulfate.**

Add a drop or two or 0.01% Neutral Red or Nile Blue Sulfate directly over the blastoderm after centering it in the field of vision on the microscope stage. Gradually during a period of five to ten minutes, the embryonic cells will concentrate the dye. In so doing the thicker parts of the embryo will become darker than the thinner cellular areas. Hensen's node, the primitive streak, neural folds, brain vesicles, etc., will gradually become visible. Observe each living stage over a period of half-an-hour to an hour during the period of progressive staining.

c. Removal of blastoderms from the yolk mass. The blastoderm is covered by a tough vitelline membrane which surrounds the entire yolk mass. With sharp scissors quickly snip around the blastoderm through the vitelline membrane and into the yolk. With a fine forceps, pick up the edge of the vitelline membrane and gently "peel" the blastoderm and membrane away from the yolk. Some yolk will usually adhere to the underside of the embryo.

With a spoon, transfer the blastoderm to a Petri dish of warm saline. Swirl the dish to wash off adhering yolk. Decant or pipette off the saline and replenish it with fresh warm saline. Embryos detached from the yolk and kept in warm saline will remain alive for several hours.

Note the level of development of the neural folds, the number of somites present on the embryos, the length of the head process, and other morphological characters of interest. Observe the living embryos over the duration of a laboratory period and record the status of morphogenesis at the end of the period. Remember that one somite per hour is usually added, and one hour will also make a substantial difference in the amount of fusion of the neural folds. Few laboratory preparations are so impressive in demonstrating the speed of the developmental processes as this one is.

Make quick sketches at the beginning and end of the laboratory period as a record of morphogenetic changes in living blastoderms. Record the time at which the drawings were made.

3. Experiments with early living blastoderms. Simple and basic experiments can be carried out on early embryos with little technical equipment.

Special materials needed: Plastic cone-shaped coffee cups, Saran Wrap to cover the cups, cup holders, incubator at 38.5°C = 99-100° F, glass needles or bent No. 000 stainless steel insect pins mounted in glass holders (see instructions for preparing micro-needles at the end of Chapter 5).

a. Experiments to demonstrate the totipotency of unincubated chick blastoderms to form whole embryos. The unincubated chicken egg has developed to the blastula stage. At this stage the blastoderm is a flat disc several cells thick which lies on the surface of the yolk mass. Before incubation the primary axes of the embryo are not entirely determined. Any half of the blastoderm has the capacity to organize a whole embryo. This totipotency can be demonstrated by the following simple experiment.

1. Open an unincubated egg into a **clean, dry plastic cup.**
2. Allow the yolk to orient spontaneously with the blastoderm at the upper surface.
3. With a clean, sharp, bent needle (glass needles are best but a #000 insect pin will serve) puncture the yolk a little to the side of the blastoderm.
4. Insert the needle point under the blastoderm until the point can be seen pressing against the vitelline membrane on the other side of the blastoderm. **Do not** puncture the vitelline membrane.
5. Then draw the needle back so that the point pressing against the vitelline membrane will bisect the blastoderm.
6. When the blastoderm has been cut into equal parts, carefully withdraw the needle.
7. Cover the container with Saran Wrap, and place it in the incubator for 24 hours. Embryos handled in this way will develop for approximately 1 to 2 days before bacterial infection kills them. This is sufficient time for each half of the divided blastoderm to develop an embryonic axis consisting of notochord, somites and neural tube.

b. Experimental demonstration of axial determination of 24 hour embryos. By the 24 hour stage the embryonic axes of the egg are determined. That is, anterior parts of the blastoderm will form only anterior structures; posterior parts will form only posterior structures. This can be demonstrated the following way:

1. Repeat the procedures described in the previous experiment on unincubated eggs.
2. Bisect the 24 hour blastoderm transversely so as to separate the primitive streak region (posterior) from the neural-fold-somite region (anterior).
3. Incubate the egg for 24 hours. Examine the egg and note the type of development that has taken place in the anterior and posterior halves of the blastoderm.

At the end of the laboratory period, clean up the work area and leave it in the condition that you found it at the beginning of the period. Discard egg shells in waste jars. Pour liquid waste and yolk into the sink. Wash out fingerbowls, waste jars, and Petri dishes thoroughly so that no traces of yolk remain in them. Rack the glassware to dry at your own desk area in readiness for the next laboratory section.

TABLE VI-3

COMPARATIVE MOVEMENTS IN CHORDATE GASTRULATION

AMPHIOXUS MOVEMENTS (See Figure 3-7 and accompanying explanation)	AMPHIBIAN MOVEMENTS (See Figure 3-10 and accompanying explanation)	AVIAN MOVEMENTS* (See Figure 6-4 and accompanying explanation)
	ENDODERMAL GASTRULA MOVEMENTS	
1. Embolic invagination of endoderm.	1. Polar ingression of 1/3 of endoderm.	1. Delamination of e.e. endoderm.
2. Abscission of endoderm from mesoderm.	2. Inversion(inrolling) of last 2/3 of endoderm.	-- Involution of embryonic endoderm.
3. Tubulation of archenteron by endoderm.	3. Tubulation and internal spreading of endoderm to form archenteron lining.	2. Internal spreading of e.e. endoderm
	MESODERMAL GASTRULA MOVEMENTS	
4. Involution of all mesoderm at blastopore lips.	4. Involution of all mesoderm around the blastopore lips.	3. Involution of all mesoderm around the primitive streak(blastopore lips).
6. Constriction and vesiculation of somite mesoderm & chorda from endoderm of archenteron wall.	5. Shearing of mesoderm from endoderm at lateral and ventral blastopore lips.	
-- Convergence of chorda mesoderm by contraction of the wide dorsal lip of the blastopore.	6. Convergence of chorda mesoderm and somite mesoderm at dorsal lip of blastopore.	4. Convergence of chorda (only) at Hensen's node (dorsal lip) of the primitive streak (blastopore).
5. Axial elongation of chorda mesoderm in AP-axis of body.	7. Axial elongation of chorda mesoderm in AP-axis of body.	6. Axial elongation of chorda mesoderm in AP-axis of body by regression of Hensen's node posteriorly.
7. Internal spreading of lateral mesoderm between ectoderm and endoderm of lateral and ventral body wall.	8. Internal spreading of lateral mesoderm between ectoderm and endoderm of lateral and ventral body wall.	5. Internal divergence of lateral mesoderm between surface ectoderm and yolk sac endoderm.
	ECTODERMAL GASTRULA MOVEMENTS	
9. Epiboly, general spreading of general ectoderm to cover mesoderm and endoderm.	9. Epiboly, general spreading of general ectoderm to cover mesoderm and endoderm.	7. Epiboly, general spreading of general ectoderm to cover embryonic and e.e. regions of the blastoderm.
	POST-GASTRULA, NEURULATION, MOVEMENTS	
8. Abscission (separation) of epidermal ectoderm from neural tube ectoderm.	11. Abscission (separation) of epidermal ectoderm from neural tube ectoderm.	-- Abscission (separation) of epidermal ectoderm from neurual tube ectoderm.
10. Tubulation of neural tube ectoderm.	10. Tubulation of neural tube ectoderm.	-- Tubulation of neural tube ectoderm.
-- Caudalization of the tail blastema to form the tail.	12. Caudalization of the tail blastema to form the tail.	-- Caudalization of the tail blastema to form the tail.

* Proliferation (mitotic cell replication) is very active in all germ layer regions of the chick blastoderm during gastrulation; however, proliferation is plays an insignificant role in the other two groups.

CHAPTER VII

THE PRIMORDIUM MOSAIC STAGE IN PRE-FUNCTIONAL DIFFERENTIATION

Illustrated with the 48 Hour Chick Embryo

TABLE OF CONTENTS

TABLES AND ILLUSTRATIONS

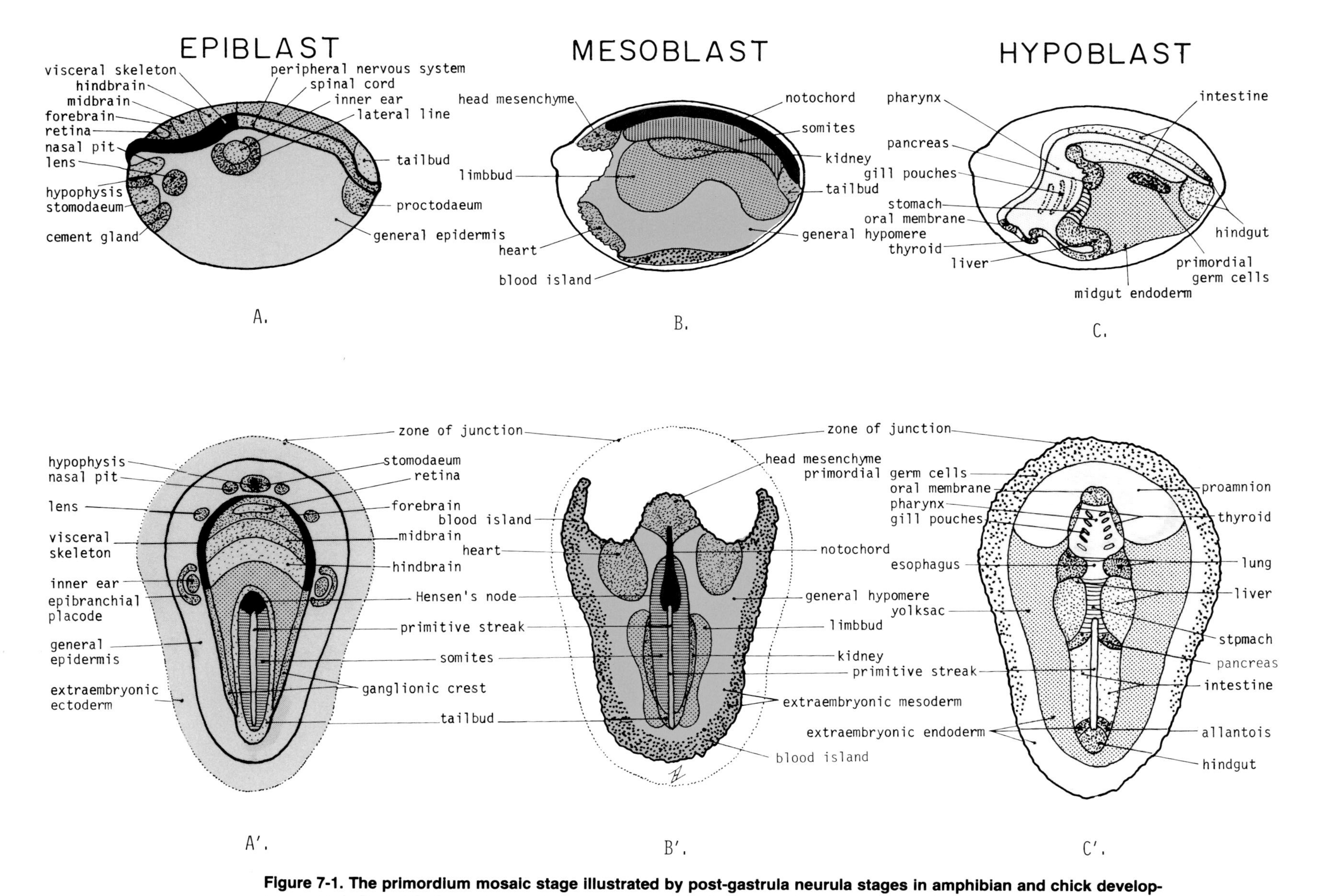

Figure 7-1. The primordium mosaic stage illustrated by post-gastrula neurula stages in amphibian and chick development. 7-1 A, B, and C. Organ forming fate maps of primordia in the amphibian epiblast (mainly ectoderm), mesoblast (mainly mesoderm), and hypoblast (mainly endoderm). 7-1 A', B', and C'. Organ-forming fate maps of primordia in the chick epiblast, mesoblast, and hypoblast. Color code: blue — ectoderm, red — mesoderm, yellow — endoderm.

CHAPTER VII

THE PRIMORDIUM MOSAIC STAGE IN PRE-FUNCTIONAL DIFFERENTIATION

Illustrated with the 48 Hour Chick Embryo

There is no generally agreed upon name that adequately describes the early post-neurula stage in embryonic development. In general, it is described by its morphological structure as the early tailbud stage. However, this in no way does justice to the dynamic properties that invisibly characterize the embryo at this early time.

I have chosen to characterize this period as the **primordium mosaic stage in prefunctional differentiation.** Each term is intended to convey special meaning which may not be self-evident from the words themselves; therefore, we will digress briefly to clarify what is implied.

First, a **primordium** (or organ rudiment, or *anlage;* plural, *anlagen*) in the accepted embryonic sense is a restricted part of the embryo which has been assigned a specific organ fate. That is, it has been embryonically determined by some epigenetic process involving ooplasmic segregation, induction, or field effect. As a consequence, the primordium will carry out its special developmental role in organ formation even though it may be isolated in tissue culture from all other parts of the embryo, or transplanted microsurgically to an altogether foreign tissue environment in the embryo. This compulsive capacity to self-differentiate according to its normal fate is, indeed, the way in which the time of determination in primordia is experimentally demonstrated. That is, before determination an isolated embryonic region in tissue culture will fail to undergo specialization, and after determination it will continue to differentiate independently.

Second, a **mosaic period** in development is achieved when virtually all parts of an embryo have been determined to form special primordia. Figure 7-1 shows prospective organ fate maps of primordial mosaic areas in amphibian and bird neurulae which are determined. The mosaic stage is achieved gradually with some primordia being determined very early in pre-gastrula stages (i.e., chorda, heart), others during gastrulation (i.e., brain, eye), and others after gastrulation (i.e., ear, lens). The boundary between primordia is not sharp. Instead there is some overlapping of primordial fields. This means that cells between two adjacent primordia may not yet be decisively assigned a specific fate. These cells are said to be in a state of **labile determination,** in as much as they are restricted in their developmental competence to only two (or at most a few) choices for differentiation without yet having been assigned a specific role to play in development.

Third, **prefunctional differentiation** is almost self-explanatory. It refers to the fact that each primordium in the mosaic, although embryonically determined to form a special part such as an eye, ear, heart, kidney, etc., as yet has not begun cellular specialization for a given physiological function such as nerve conduction, muscle contraction, glandular secretion, or sensory reception. The differentiation at this stage is invisible and at the molecular and nuclear level of commitment to a cell fate. In the absence of cell specialization, we find that cell types in primordia are exceedingly simple. Indeed, primordia are composed of only two basic cell types, epithelia and mesenchyme. Figure 7-2 shows the major embryonic variations in these two cell categories. Briefly, **epithelia** comprise sheets of cells attached to one another by a basement membrane so that there are no conspicuous intercellular spaces. If they have radial symmetry in movement, they will form either expanding or contracting sheets. However, if they have polarized, or axiate, symmetry, they will undergo elongation, invagination, or evagination in a given direction. **Mesenchyme** is composed of freely moving cells with no, or very few, intercellular attachments. They may have multiple filopodial extensions and they can move radially; their cell extensions and directions of movement are limited to one, or at most a few directions. It should be

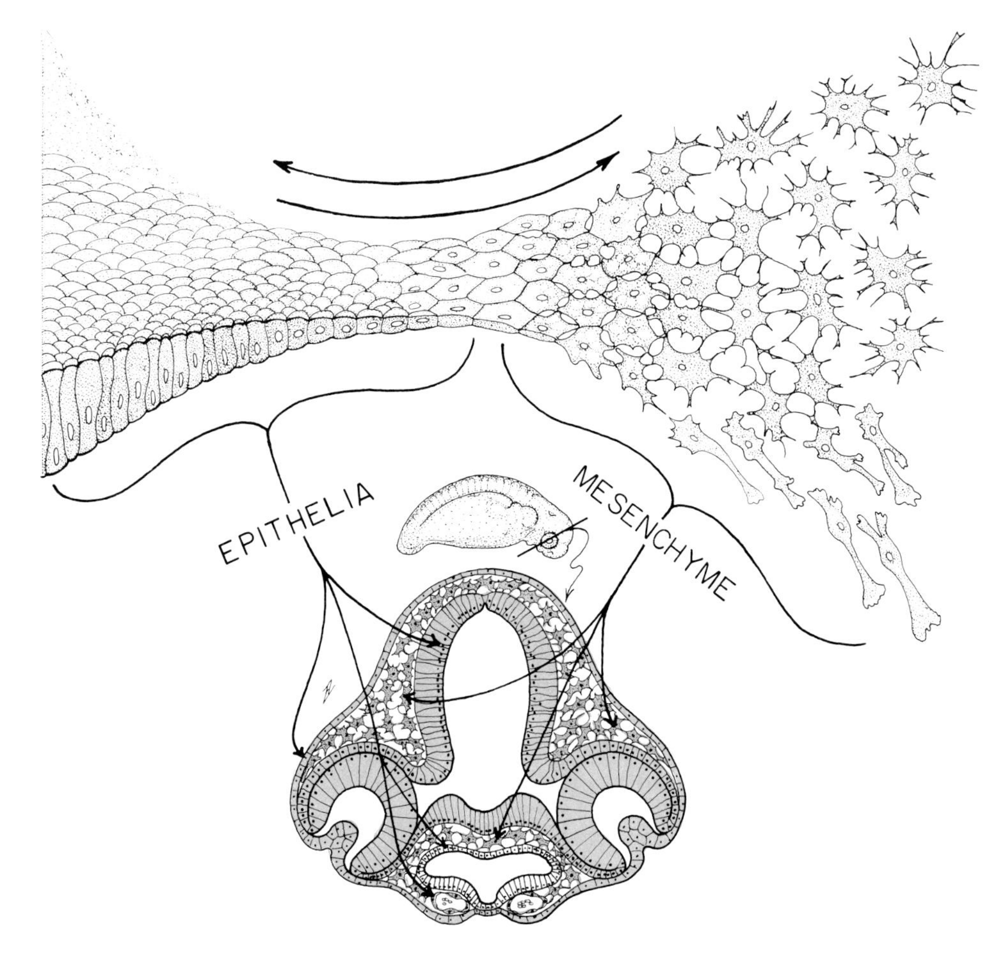

Figure 7-2. Graphic presentation of types of epithelia and mesenchyme in embryonic primordia.

emphasized that cells of a given primordium are not permanently restricted to being either epithelial or mesenchymal types. Mesenchyme can be converted into epithelium (i.e., vascular mesenchyme can form endothelia or capillary and heart linings), and epithelia can form mesenchyme (i.e., neuroepithelium can form neural crest mesenchyme). This is implied by the double arrows in Figure 7-2.

Finally, the **growth characteristics of primordia** should be pointed out. Following determination, a primordium usually undergoes an initial period of accelerated growth. Consequently, the simple sheets of cells that characterize the spherical or smoothly contoured late gastrula embryo rapidly develop many lumps and depressions during the neurula and tailbud stages. Each irregularity in the terrain marks some special primordial growth area. Growth is achieved by four primary means: **A. Mitotic activity,** as the primary step in growth, increases in primordia and adds to the local cell population. **B. Increase in individual cell size** then usually takes place. **C. Secretion of extracellular matrix,** usually of mucoprotein, results in a loose general spacing of cells in the primordium. **D. Water imbibition** by the matrix completes the early mechanisms for growth.

In the following study the 48 hour chick embryo has been selected to emphasize the character of primordia and the nature of their organization into a patterned mosaic within the vertebrate embryo. Constant reference will be made to invagination, evagination, proliferation, delamination, condensation, and aggregation as the structural aspects of various primordia. These active terms are used in an attempt to create an appreciation of the dynamic nature of developmental processes that are studied in static sections.

It is to be understood, in this study of the 48 hour stage, that the process of primordium formation did not begin with this stage, nor does it end here. The concept of primordium dynamics will be extended in our continuing study of progressive differentiation that is given in Chapters 8 and 9.

A. EARLY ORGANOGENESIS AND REGIONAL DIFFERENTIATION IN THE 48-56 HOUR CHICK EMBRYO

After two days of incubation, many primordia of special organs and organ regions are easily recognized in the chick embryo. This is particularly true of the anterior parts of the embryo which have a more rapid developmental rate than the more caudal regions. At this time, however, functional differentiation is almost exclusively restricted to the circulatory system. The heart has begun to beat by the 48 hour stage, and the basic pattern of both the embryonic and extraembryonic circulatory plans have been roughed in. Most organ primordia that were visible in the 33 hour embryo are present in a considerably more advanced morphological condition. The heart chambers and brain regions, identified as rather indistinct regions at the earlier stage, now have relatively clearcut morphological characteristics which permit them to be recognized with certainty. In addition, many new primordia have arisen. Special emphasis will be placed on these in the study of the 48 hour embryo. At this time note also the developmental advance in structures already identified at earlier stages.

The laboratory work will involve the study of whole mounts, cross sections, and demonstration whole mounts with the circulatory system injected. In studying these materials, follow the organization of the descriptive anatomy that is given below. This involves the study of organ systems rather than the study of regions. This will require repeated reference to the whole mount embryos while the serial cross sections are being studied. After completing the study of separate organ systems, review the serial sections and correlate all parts with the whole mount so that relationships of various primordia can be mentally visualized and understood.

1. General topography of the whole mount embryo (Fig. 7-6 A). Observe and identify the following general features concerning the embryo as a whole. The anterior end of the embryo is strongly bent posteroventrally in such a manner that the mesencephalic region of the brain is the most forward part of the embryo. This bend is the **cephalic flexure,** and consists of a 140° angle in the anterior-posterior axis. Structures anterior to the mesencephalon are located ventral and lateral to more posterior parts. The cephalic flexure initially causes the anterior end of the embryo to press down into the yolk. However, a lateral twisting of the body axis, called **thoracic torsion,** permits the anterior half of the embryo to lie on its left side. At the posterior end of the embryonal area, note that the **tail fold** is just beginning to form. Here the embryo is in the process of being undercut by the **body folds** in essentially the same manner as the head fold was formed in the 21 to 22 hour embryo. Eventually the body fold will completely undercut the embryo and separate it from the underlying yolk mass. The head and tail fold will meet in the midtrunk region at about the 5th day of incubation and, together, will completely close the midgut except for the narrow yolk-stalk which persists until the day before hatching. In embryos of 52 or more hours incubation, the **cervical flexure** may also be identified. This is a second ventral bend in the anterior-posterior axis in the region of the anterior (neck) somites. At this time the cervical flexure makes only a 15 to 25° angle in the AP axis. Later, at the 72 hour stage, it will be increased to 90°. The cephalic and cervical flexures are two of the four major flexures of the vertebrate body axis. The remaining 2 are the **pontine** and **lumbo sacral flexures** which appear in the region of the metencephalon and tail respectively at a somewhat later time. Identify the **amniotic folds** which, by the 48-hour stage, have progressed posteriorly to about the level of the 18th to 20th somite. The anterior two-thirds of the embryo is now covered by a double layer of somatopleure, namely the amnion and serosa, which will be clearly seen in sectioned material. Identify the remaining structures in the whole mount embryo as you study the organ systems in serial sections.

A brief digression will be made before beginning the serial sections to point out the amnion and its manner of formation in the 48 hour embryo. Note that almost half the entire embryonal area is enclosed within a double fold of somatopleure (Fig. 7-6 A). The diagrams in Figure 7-3 are provided to assist in understanding the relationship of the embryo to the developing amnion, amniotic cavity, serosa, seroamniotic cavity, yolk sac, and intra- and extraembryonic coeloms. Note that the serosa and amnion are both extensions of the embryonic lateral body wall. These two membranes consist of **epidermal ectoderm** and **somatic lateral plate mesoderm.** They are nonvascular and never independently develop any blood vessels. The amniotic folds are elevations of this **somatopleure** layer that begin to grow over the embryo at the anterior end of the area pellucida, just in front of the proamnion. The

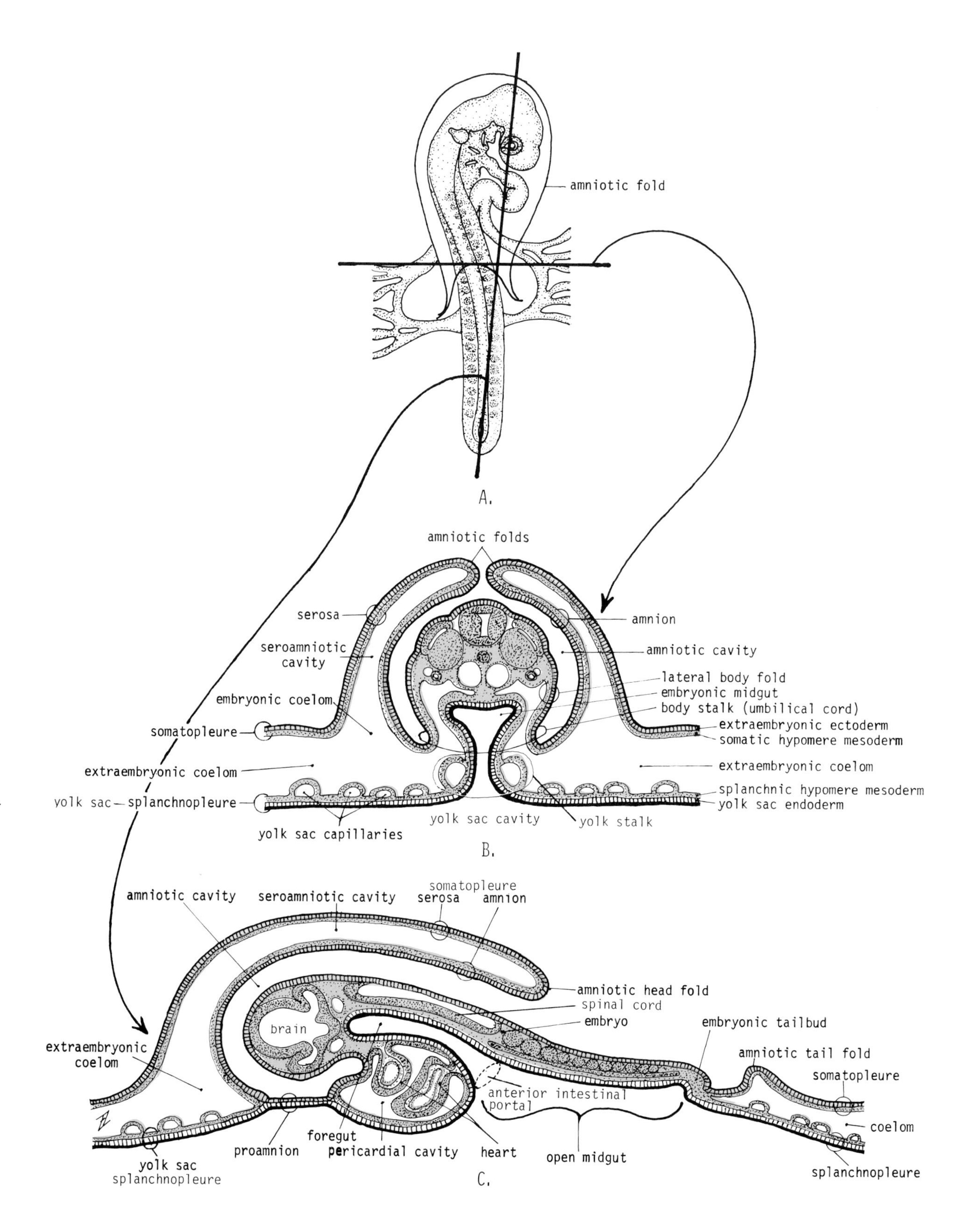

Figure 7-3. Extraembryonic relationships in the 48 hour chick embryo. 7-3 A. Diagram of the 48 hour embryo in its amniotic folds. 7-3 B. Diagrammatic cross section of embryo and amniotic folds. 7-3 C. Diagrammatic median sagittal section of the embryo in its amniotic folds. Color code: blue — ectoderm, red — mesoderm, yellow — endoderm.

amniotic folds grow back over the embryo, trapping some of the external space which is thereafter called the **amniotic cavity;** note that this cavity is lined by ectoderm. The lining of the amniotic cavity is the **amniotic membrane.** The external component of the amniotic fold is the **serosa.** It retains a connection with the internal amniotic membrane for a considerable period along a line of fusion between right and left amniotic folds which meet over the embryo. This fusion line is the **seroamniotic raphe.** The space between serosa and amnion is the **seroamniotic cavity;** it is merely an extension of the **extraembryonic coelom.** The yolk sac membrane at this time is the entire extraembryonic splanchnopleure. It is composed of endoderm and splanchnic lateral plate mesoderm. It can be distinguished from somatopleure by the blood vessels that develop in the mesodermal layer of the yolk sac. In Figure 7-3 C note that just under the head fold of the embryo is a persistent mesoderm-free region called the **proamnion** which will never acquire mesoderm. In the following study of serial sections it will be necessary to identify the extraembryonic membranes that invest the embryo. Additional details of extraembryonic membrane formation will be given in association with the descriptions of the three-day chick and ten millimeter pig in Chapters 8 and 9.

2. Instructions for the study of serial cross sections of the 48-56 hour chick (Fig. 7-6 B–P). It will be necessary to go through the series repeatedly, learning one system after another in an orderly manner, with the goal of achieving a thorough understanding of the position and manner of formation of the constellation of primordia that make up the 48-56 hour chick stage.

Keep in mind that each series will be sectioned at slightly different planes and, therefore, don't expect to find an exact correspondence between the diagrams provided and the sections you study. Age differences will also modify appearance of structures; young embryos can be expected to be less well differentiated than older ones. Laboratory specimens labelled "48 hour chick embryo" will vary in actual age (as determined by somite count + 20) from 47 to 55 hours of incubation (25 to 31 somites respectively). Older embryos will have better developed sensory placodes, aortic arches, and cranial nerves than younger ones; these are the special primordia that will show greatest variation.

As early as possible in the study of serial sections, **determine the plane of section of your own set of slides** by comparing the presence of several known structures in a section with their location on the whole mount diagram. Place a straight edge across the whole mount diagram, Figure 7-6 A, connecting these structures. This will give the approximate plane of section. Keep in mind that no one plane of section is inherently better than any other and, if study of the sections in one preparation is complete, then you should be able to identify comparable parts in any embryo sectioned at a different plane. After completing the study, exchange slides with a laboratory associate in order to gain greater familiarity and proficiency in interpreting the material.

It is also of importance to **distinguish between the embryonic right (serosal) and left (yolk sac) sides of the sections.** Remember that the embryo is lying with the left side down on the yolk sac. In cross sections near the anterior end, **identify the yolk sac by the presence of extraembryonic blood vessels** in its splanchnic mesoderm layer; this will mark the left side of the embryo. As an additional check, note that the serosa on the opposite side (i.e., right side of embryo) will lack blood vessels in its somatic mesoderm layer.

Ectodermal Derivatives

a. Brain (Fig. 7-4 and Table VII-1). Owing to the cephalic flexure, the brain stem is bent sharply on itself. The **midbrain** (mesencephalon) is the most forward brain region and will be encountered first in serial cross sections. The more anterior brain regions, namely, the **diencephalon** and **telencephalon** are bent ventrally under the **hindbrain** (rhombencephalon). The latter is differentiated into two parts, i.e., the **metencephalon** and **myelencephalon.** Figure 7-4 gives the common plan of the central nervous system.

Identify the five definitive brain vesicles on the whole mount. At this stage they appear as slight expansions of the neural tube. The only boundary that is indistinct is the one which separates the metencephalon from the myelencephalon. In whole mounts the anterior border of the posterior choroid plexus serves as an approximate boundary between these two regions. In cross sections, the more accurate landmark is the anterior border of the 5th cranial nerve and its ganglion (see discussion below).

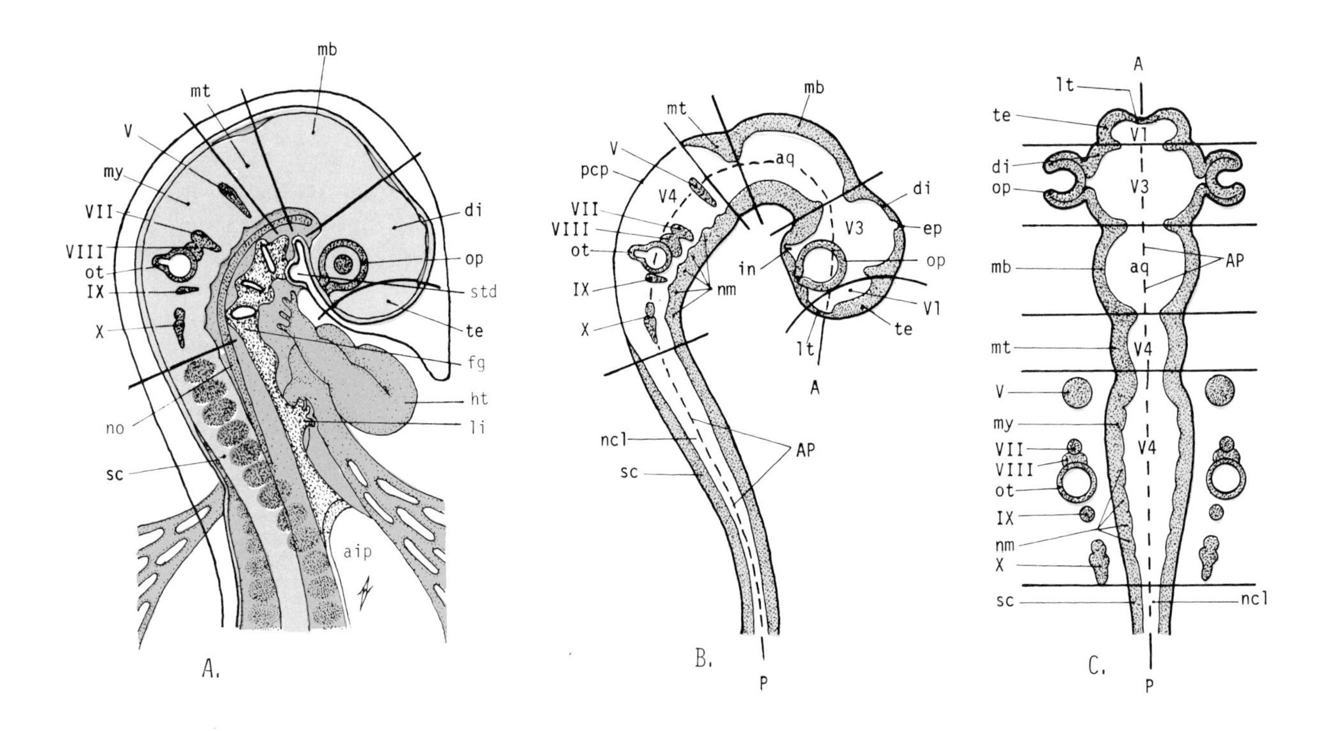

Figure 7-4. Diagram of the structural organization of the vertebrate brain as illustrated by the 48 hour chick embryo. 7-A. Topography of anterior whole embryo. 7-B. Major subdivisions of the central nervous system seen in lateral view. 7-C. Major subdivisions of the central nervous system as seen dorsally with the AP-axis straightened. Heavy lines transect the brain stem to mark boundaries between the five major brain regions. Color code: blue — ectoderm, red — mesoderm, yellow — endoderm.

Legend

V	5th cranial nerve	fg	foregut	P	posterior pole
VII	7th cranial nerve	ht	heart	pcp	posterior choroid plexus
VIII	8th cranial nerve	in	infundibulum		
IX	9th cranial nerve	li	liver diverticulum	op	optic cup
X	10th cranial nerve	lt	lamina terminalis	ot	otocyst
A	anterior pole	mb	midbrain, mesencephalon	sc	spinal cord
aip	anterior intestinal portal	mt	metencephalon	std	stomodaeum
AP	AP axis of embryo	my	myelencephalon	te	telencephalon
aq	cerebral aqueduct	ncl	neurocoele, spinal canal	V1	lateral ventricles
di	diencephalon	nm	neuromeres	V3	third ventricle
ep	epiphysis	no	notochord	V4	fourth ventricle

In serial cross sections, the mesencephalon will be the first brain region encountered. As you proceed posteriorly, you will encounter two sections through the brain (Fig. 7-6 B–F). Orient your slide so that the hindbrain sections are above the forebrain. The hindbrain at the level of the myelencephalon can be clearly distinguished from either of the two forebrain regions by the presence of a very wide and thin roof plate which forms the dorsal wall of the neural tube. This thin membrane is the **posterior choroid plexus.** The diencephalon and telencephalon walls are nearly uniform in thickness on all sides. In the diencephalon, two special regions can be identified which anticipate the formation of later structures. They are slight evaginations from the ventral and dorsal wall of this brain region. They are respectively the **infundibulum** (*anlage* of the posterior lobe of the pituitary gland) and **epiphysis** (*anlage* of the pineal gland Fig. 7-6 C). Note also that the **optic cup** is connected to the lateral wall of the diencephalon by means of the **optic stalks.** (For additional details concerning the eye, see the section on sense organs below.) General relationships of these parts are summarized in Table VII-1.

TABLE VII-1

STRUCTURAL RELATIONSHIPS OF NEUROSENSORY PRIMORDIA IN EARLY CHICK EMBRYOS

PRIMORDIA IN 30 TO 40 HOUR EMBRYOS	STRUCTURES RECOGNIZABLE IN 48 TO 72 HOUR EMBRYOS: SECONDARY DERIVATIVES	STRUCTURES RECOGNIZABLE IN 48 TO 72 HOUR EMBRYOS: TERTIARY DERIVATIVES	LATE EMBRYONIC AND DEFINITIVE SPECIAL STRUCTURES: DORSAL WALL	LATERAL WALLS	VENTRAL WALL	NEUROCOELE
I. NEURAL TUBE:						
PROSENCEPHALON (forebrain)	TELENCEPHALON & telocoele	Pallial plate & olfactory lobe	cerebral cortex or pallial plate	olfactory lobe	basal nucleus or corpus striatum	lateral ventricles
	DIENCEPHALON & diocoele	epiphysis, optic cup, thalamus infundibulum	epiphysis & anterior choroid plexus	epithalamus, thalamus, & hypothalamus	infundibulum, optic chiasma, retina, choroid, iris, & optic nerve	3rd ventricle
MESENCEPHALON (midbrain)	MESENCEPHALON & mesocoele	optic tectum & tegmentum	optic tectum	tegmentum	cerebral peduncles	cerebral aqueduct
RHOMBENCEPHALON (hindbrain)	METENCEPHALON & metacoele	cerebellum & pons	median lobe of cerebellum	lateral lobes of cerebellum	pons	4th ventricle (anterior end)
	MYELENCEPHALON & myelocoele	posterior choroid plexus & medulla	posterior choroid plexus	sensory column of medulla	motor column of medulla	4th ventricle
NEURAL TUBE & neurocoele	spinal cord & neural canal	alar sensory plate & basal motor plate	roof plate	sensory & motor column	floor plate	neural canal
II. NEURAL CREST:						
GANGLIONIC CREST	5th cranial ganglion	trigeminus nerve	(origin from 1st neuromere of medulla; innervates 1st visceral arch)			
	7th cranial ganglion	facial nerve	(origin from 3d neuromere of medulla; innervates 2nd visceral arch)			
	9th cranial ganglion	glossopharyngeal nerve	(origin from 5th neuromere of medulla; innervates 3d visceral arch)			
	10th cranial ganglion	vagus nerve	(origin from 7–9th neuromeres; innervates 4th, 5th, & 6th arches.			
	spinal ganglia & autonomic ganglia	somatic sensory and visceral sensory nerves post-ganglionic visceral motor nerves				
ECTOMESENCHYME	Visceral skeleton and other non-neuronal crest derivatives (see Table III in Introduction).					
III. SENSORY EPITHELIUM:						
SENSORY PLACODES	OLFACTORY PLACODE	olfactory pit	nasal epithelium and 1st cranial nerve (olfactory).			
	LENS PLACODE	lens vesicle	lens of eye			
	OTIC PLACODE	otic vesicle	inner ear and 8th cranial nerve (acoustic).			
	LATERAL LINE PLACODES	placodal neurons	vestigial somatic sensory neurons of cranial nerves V, VII, IX, & X; usually they degenerate in amniotes.			

b. Neural crest and cranial nerves. In posterior sections of the neural tube, at the level of the 21st to 23rd somite, the primitive condition of the neural crest is most clearly seen (Fig. 7-6 M, N). The crest cells at this level appear as small aggregates of mesenchyme located between the epidermis and dorsal half of the neural tube. These cells in anterior sections have already migrated away from the dorsal midline and can no longer be distinguished from the general mesenchyme of the embryo. The crest cells will later reaggregate to form the spinal and autonomic ganglia, and will give rise to the somatic sensory, visceral sensory, and postganglionic visceral motor neurons of the body. Primordia of spinal ganglia will be seen in the study of the 72 hour chick and 10 mm pig embryo. However, at the 48 hour stage, the only ganglionic *anlagen* that are identifiable are situated in the myelencephalon region of the brain. In this region the ganglia of the 5th, 7-8th, 9th and, rarely, the 10th nerves appear as distinct condensations of neural crest material. Use the **otic vesicle** as the landmark for identifying these ganglionic primordia. Immediately anterior to the otic vesicle is the common ganglionic aggregation of the 7th and 8th nerves, collectively called the **acousticofacialis** ganglion at this time. It should be pointed out that the ganglion of the 7th and 8th nerves, although indistinguishable in the appearance of their cells, come from altogether different sources of ectoderm, namely: the ganglion of the 7th is derived from neural crest ectoderm, and that of the 8th is derived from epidermal ectoderm of the otic vesicle which it will later innervate. The 7th nerve will supply muscles of the second visceral arch (Fig. 7-6 A–D). For the nature of visceral arches, see Figure 7-3 and the following section of the present chapter. Just anterior to the origin of the acousticofacialis ganglion is the large and well formed ganglion of the **5th (trigeminal) cranial nerve** which innervates the first visceral arch. Therefore it will be the nerve of the upper and lower jaws (Fig. 7-6 A–D). Just back of the otic vesicle the ganglionic condensation of the **9th (glossopharyngeal) nerve** can be identified and can be traced back into the region of the third visceral arch (Fig. 7-6 F).

c. Sense organs. Primordia of the three major sense organs of the head are present at various stages of differentiation in the 48 hour chick embryo. Begin at the anterior end of the telencephalon (Fig. 7-6 F) and work forward in the serial sections.

Nose. Note that the surface epidermal layer covering the head at the anterior extremity of the telencephalon is thickened on both sides of the head. These are the **nasal placodes.** They are approximately at the same stage of differentiation as were the otic placodes in the 33 hour embryos. In later development the nasal placodes will invaginate to form pits which are the primordia of the nasal canals, olfactory sensory epithelium, and the olfactory (first cranial) nerve.

Ear. The two components of the definitive ear present at the 48 hour stage are the **otic vesicle** and the **first visceral pouch** (Fig. 7-6 A–D). The former will form the membranous labyrinth of the inner ear, and the latter will form the cavity of the middle ear. The otic vesicle has arisen from an epidermal placode that was observed in the 33 hour embryo. At the 48 hour stage it has invaginated deeply but is still attached to the surface epidermis (Fig. 7-6 D, E). It will soon become detached from the surface layer but will retain a close association with its inductor, the myelencephalon. Note again the close association of the acousticofacialis ganglion and otic vesicle. The posterior half of this ganglion will later separate and form the sensory ganglion of the 8th (acoustic or auditory) cranial nerve. The development of the first pharyngeal or visceral pouch and its association with the developing middle ear cavity should be kept in mind; however, the discussion of the pouch, and the close association of the pouch to the otic vesicle, will be again considered in the discussion of the pharynx which will follow.

Eye. The two primary components of the eye, namely the **lens** and **retina,** are clearly formed in the 48 hour embryo. These structures can be identified in sections through the diencephalon (Fig. 7-6 C–E). Identify the **lens vesicle** which originated as an epidermal placode and has now invaginated to form a deep, thick-walled invagination from surface epidermis. The lens vesicle will soon become completely separated from the surface layer and will retain an intimate association with the optic cup. In the 33 hour embryo the primordia of the retina were present as the paired optic vesicles which formed as lateral evaginations from the wall of the forebrain. In the 48 hour embryo, the cells forming the lateral extremity of the optic vesicles have invaginated so as to form a double-walled **optic cup** (Fig. 7-6 C–E). The internal layer of this cup will form the sensory retina of the adult eye. The external wall of the optic cup will form the pigmented retina. In whole mount embryos note the median ventral cleft in the wall of

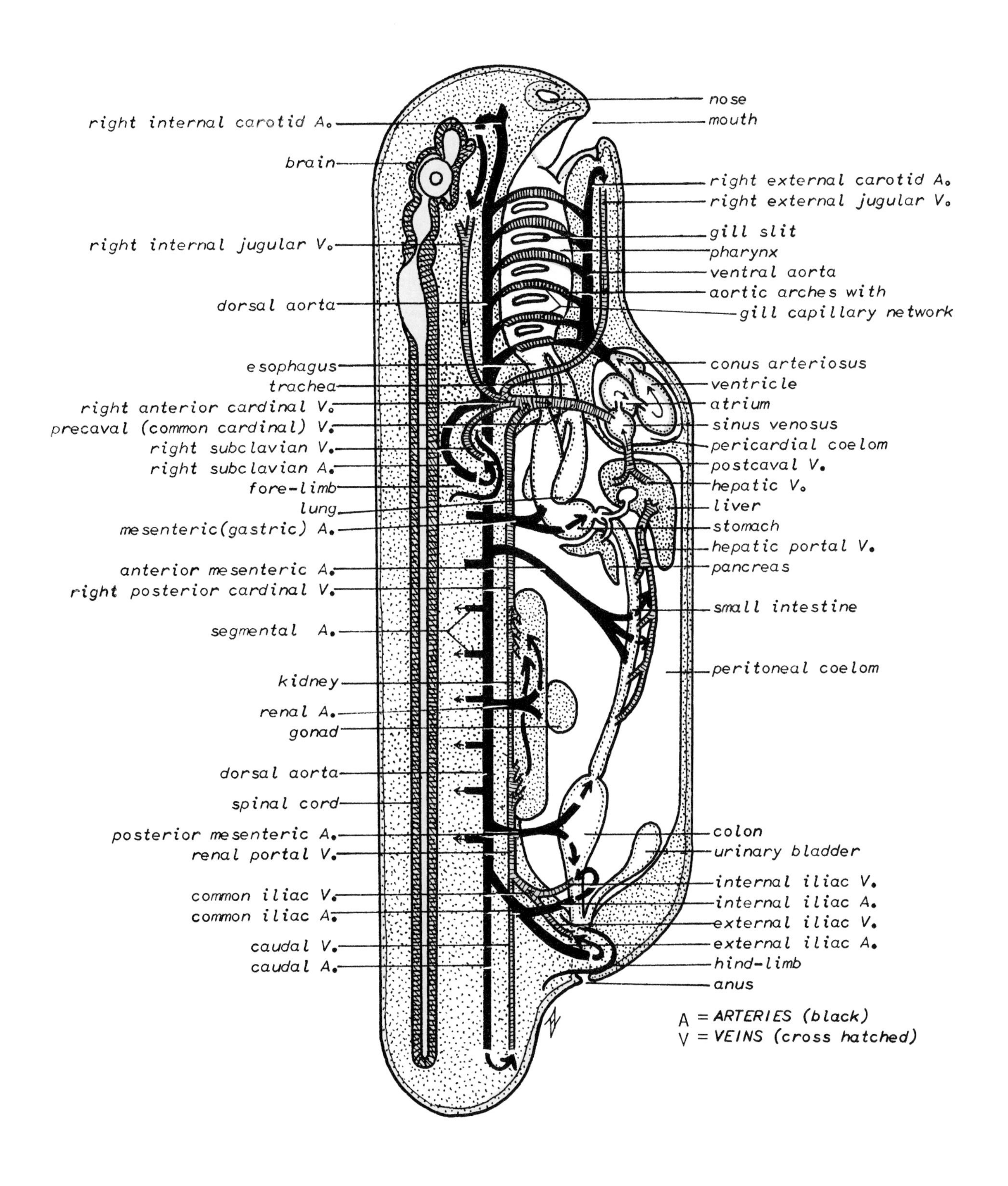

Figure 7-5. Diagrammatic common plan of the vertebrate circulatory system. (From H.E. Lehman, *Laboratory Studies in General Zoology,* Hunter Publishing Co., 1981).

the optic cup called the **choroid fissure.** Optic nerve fibers originating from the sensory retinal layers inside the optic cup will grow out of the cup through the choroid fissure back to the brain. After the emergence of the optic nerve fibers, the choroid fissure normally fuses and completes the wall of the optic cup. Go posteriorly through the sections of the eye until you reach the point at which the optic cup attaches to the lateral wall of the diencephalon. This connection is the **optic stalk.** It is merely the constricted attachment of the optic cup to the brain wall. It eventually will contribute to the connective tissue capsule (neurilemma) surrounding the optic nerve.

d. Stomodaeum and mouth. The stomodaeum is a midventral invagination of the epidermis which will form the lining of the anterior half of the mouth cavity. At the 48 hour stage it is located between the floor of the diencephalon and the anterior end of the foregut (Fig. 7-6 D). Its relationship can best be determined by tracing the head epidermis from a level anterior to the mouth where the head is not connected to the rest of the embryo (Fig. 7-6 E). Then proceed forward to the point at which the anterior head region fuses with the rest of the embryo (Fig. 7-6 D); you can see that a pocket of epidermis is "trapped" by lateral bridges of tissues. This ectoderm-lined chamber is the **stomodaeum.** The lateral points of fusion between the head and remainder of the body are the primordia of the jaws. Note that these masses of tissue lie just anteroventral to the first visceral pouch and therefore should be interpreted as the **1st visceral arch.** In the midline between the two first visceral (mandibular) arches, the stomodaeal epidermis is fused to the endoderm of the foregut. This fusion of ectoderm and endoderm is the **oral membrane.** It will persist until about the 72 hour stage before degenerating. After that time, the stomodaeum and foregut cavities are connected and the mouth opening is complete. Trace the stomodaeal cavity to its anterior extremity and note that it becomes greatly reduced in size and forms a small blind sack that is very closely applied to the floor of the diencephalon in the region of the infundibulum (Fig. 7-6 C). This evagination of the stomodaeum is **Rathke's pouch.** It is the primordium of the anterior and intermediate lobes of the pituitary gland of the adult. As mentioned earlier, the infundibulum will contribute the posterior lobe of this gland.

Endodermal Derivatives

a. Midgut. Begin the study of endodermal derivatives in the open midgut region. The midgut is that part of the embryonic digestive system which lacks a ventral floor and opens into the yolk sac. Choose any section in the posterior third of the embryo at a level comparable to that shown in Figures 7-6 M, N. Note that the embryonic endoderm is continuous with the endoderm of the yolk sac splanchnopleure that extends out from the lateral body folds.

b. Hindgut and posterior intestinal portal. Follow the sections posteriorly to the end of the embryo. Identify the **posterior intestinal portal,** the opening between the open midgut and the short hindgut. The latter possesses an endodermal floor which is formed by the median ventral fusion of splanchnopleure. This process of hindgut formation has just begun and sections of embryos of less than 50 hours incubation may lack a closed hindgut. The formation of the hindgut and tail fold is essentially similar to the ventral fusion of body folds which gave rise to the head fold at a much earlier time. With time, the fusion of lateral body folds, under the posterior end of the embryo, increases the length of the tail fold, and the posterior intestinal portal advances anteriorly. As stated previously, the anterior and posterior intestinal portals meet and form the base of the yolk stalk on the 5th day of incubation.

c. Foregut. Follow the sections anteriorly from the open midgut region to the anterior intestinal portal (Fig. 7-6 K) which marks the posterior boundary between the open midgut and closed foregut. The portal is located just posterior to the entrance of the vitelline veins into the sinus venosus. Note, in particular, the small anterior evaginations from the ventral wall of the foregut between the two viteline veins. These are the **liver diverticula** (Fig. 7-6 J). In later development the rapidly growing liver will completely block the main channel of the vitelline veins. The portion of the vitelline veins anterior to the liver will persist in the adult as the anterior end of the inferior vena cava; the segment of vitelline veins posterior to the liver will form the hepatic portal vein of the adult. In the 48 hour embryo the foregut anterior to the liver diverticula has not yet differentiated into a recognizable stomach, esophagus, trachea, or lungs. These will be seen in older embryos. At this time this region is merely designated as posterior foregut. In older embryos of 52 to 56 hours of incubation, there is a slight lateral expansion of posterior foregut which anticipates the formation of the trachea and lungs, but, aside from this, there are no other landmarks which permit the accurate designation of organ regions. Foregut derivatives will receive special consideration in the study of the 72 hour chick and 10 mm pig.

The anterior half of the foregut is the **pharynx.** By definition, it is that part of the digestive tube possessing **gill pouches** (Fig. 7-6 D, F, G). In the 48 hour embryo three of these lateral, paired evaginations are recognized. They are numbered consecutively from anterior to posterior. Eventually two additional pouches will form, making a total of five. The visceral pouches can be identified in sections as bilaterally symmetrical expansions of the foregut cavity which extend outward until the endodermal lining comes in contact with the superficial epidermis of the embryo. These points of contact are the **closing plates of the visceral clefts.** Proceed through the serial sections and identify the three visceral pouches which have formed (Fig. 7-6 D, F, G). In the midventral floor of the pharynx between the 1st and 2nd pouches, locate a thickening of endoderm which is the primordium of the thyroid gland (Fig. 7-6 E). This tissue will evaginate and form an endodermal vesicle that will completely separate from the pharynx. Anterior to the **thyroid diverticulum** at the level of the first visceral pouch, note again that the ventral wall of the pharynx is closely applied to the stomodaeal ectoderm and participates in the formation of the **oral membrane** described previously (Fig. 7-6 D).

Mesodermal Derivatives

a. General mesodermal structures. The simplest arrangement of the mesoderm can be seen in the posterior third of the embryo at approximately the level of Figure 7-6 N. At this level identify the following mesodermal structures and follow their progressive differentiation through anterior sections.

Notochord. The notochord at this stage differs little from that observed in the 33 hour chick and appears as a cylindrical rod of mesoderm located under the neural tube. It extends from the tailbud anteriorly to the level of the mesencephalon.

Somite or epimere. The somites, at almost all trunk levels, show differentiation into two primary subdivisions. One is a dorsolateral condensation of mesoderm, the **dermomyotome** (Fig. 7-6 H–N); it will give rise to **segmental muscles** (striated or voluntary) and a large part of the dermis of the skin. The mesioventral half of each somite forms a diffuse mass of mesenchyme, the **sclerotome,** which will form the axial skeleton (spinal column, ribs, and sternum) of the trunk region and tail. Even in very early stages of somite differentiation, as is found in the posterior fourth of the embryo, these two regions can be identified by the presence of a crescent-shaped cavity, the **myocoele** (Fig. 7-6 M, N). It forms a partial separation between the dorsolateral dermomyotome and scleratome elements of the primitive somites. The myocoele will disappear without a trace in later stages, as can be seen by following the myotomes anteriorly. However, the myocoeles are of evolutionary, paleomorphic importance because they are part of what might be called the "racial memory" of the way in which mesodermal somites first formed (in protochordates such as Amphioxus, or cyclostomes) as enterocoelic evaginations once directly connected with the gut wall and lumen. Trace the somite region posteriorly into the **unsegmented mesoderm** region that is entirely comparable to that observed in much younger embryos. At the extreme posterior end of the body, all three primary germ layers fuse into an amorphous mass of undifferentiated cells. This is the **tailbud blastema.**

b. Nephrotome or mesomere. The intermediate mass of mesoderm that lies lateral to the somites and mesial to the lateral plate is the primordium of the pro-, meso-, and metanephric kidney which will develop at different levels along the axis of the trunk. At the 48 hour stage it will be possible to identify three derivatives of the nephrotome in the region of the anterior half of the midgut (Fig. 7-6 L–M). The paired pronephric tubules are at the levels of the 5th through 16th somites. The pronephric ducts grow posteriorly along the dorsolateral margin of the nephrotome. They eventually will terminate in the dorsolateral wall of the hindgut. Last, primordia of **mesonephric tubules** can be seen as solid condensations of mesomere located mesioventral to the pronephric duct at somite levels 16 through 30 (Fig. 7-6 L, M). These condensations will later hollow out and open into the pronephric duct. When this fusion occurs, the pronephric duct is usually called the **mesonephric** or **Wolffian duct.** It is the functional duct of the embryonic mesonephros and forms the vas deferens of the adult male. Note in particular the close relationship between the pronephric duct and the post-cardinal vein (Fig. 7-6 L, M) which lies dorsolateral to the duct. This vein develops primarily as a venous supply and drainage channel for the pro- and mesonephric kidney. The further development of nephrotome will be considered in detail in the study of the 72 hour chick and 10 mm pig.

c. Lateral plate mesoderm and coelom. The general relationships and derivatives of the lateral

Figure 7-6. Structure of the 48 hour chick embryo. Legend for figures on following pages.

Figure 7-6 A. Whole mount diagram of the 48 hour chick embryo (Stage 14-16). Guide lines to the right and left of the figure refer to levels of correspondingly lettered cross section drawings. See legend.

Figure 7-6 B-P. Representative cross sections of the 48 hour chick embryo at levels of corresponding guide lines in 7-6 A.

Legend and checklist of structures to be identified

Color code: blue — ectoderm, red — mesoderm, yellow — endoderm.

Cranial nerves and ganglia:

5 V, trigeminus
5g semilunar ganglion of V
7-8 VII & VIII, facial-auditory
7-8g acousticofacialis ganglion
9 IX, glossopharyngeal
10 X, vagus nerve & ganglion

AA1–3 aortic arches 1, 2, 3
AC anterior cardinal vein
AIP anterior intestinal portal
ALV allantoic vein (capillaries)
AM amniotic membrane
AMC amniotic cavity
AMH head fold of amnion
AP alar plate
AT atrium of heart
AVV anterior vitelline vein
BP basal plate
CA conus arteriosus
CC common cardinal vein
CEF cephalic flexure
CH chroid fissure
CL general embryonic coelom
CP1 1st closing plate, eardrum
DA dorsal aorta
DI diencephalon
DMC dorsal mesocardium
DMY dermomyotome of somite
E ectoderm
EE epidermal ectoderm
EMY epimyocardium of heart
EN endoderm
END endocardial heart lining
EP epiphysis, pineal body
FG foregut lumen
FGE foregut endoderm
FP floor plate of neural tube
HG hindgut lumen
IC internal carotid artery
IN infundibulum
ISA intersegmental arteries
LBF lateral body folds
LE lens vesicle
LI liver diverticulum
LMC lateral mesocardium
MB midbrain, mesencephalon
MG midgut lumen
MN mesonephric tubules
MT metencephalon
MY myelencephalon
MYC myocoele of somite
NC neural crest ectoderm
NCL neurocoele of neural tube
NO notochord, chordamesoderm
NP nasal placode
NT neural tube, spinal cord
OM oral membrane
OP optic cup
OPS optic stalk
OT otic vesicle, inner ear
P pharynx
P1-3 pharyngeal pouches 1-3
PA proamnion
PCP posterior choroid plexus
PCV posterior cardinal vein
PE pericardial coelom
PIP posterior intestinal portal
PND pronephric (Wolffian) duct
PR pigmented retina of eye
RF roof plate
RP Rathke's pouch
RT sensory retina
SA seroamniotic cavity
SAR seroamniotic raphe
SC scleratome
SL sulcus limitans
SM somite, epimere
SO somatic hypomere
SOP somatopleure
SP splanchnic hypomere
SPP splanchnopleure
SR serosal membrane
STD stomodaeal lumen
SV sinus venosus
TB tailbud
TE telencephalon
TF tail fold
TH thyroid diverticulum
TT thoracic torsion
USM unsegmented somite mesoderm
V ventricle of heart
VA ventral aorta
V3 3rd ventricle of diencephalon
V4 4th ventricle of myelencephalon
VS1,2,3 visceral pouch 1,2,3
VTA vitelline artery
VV vitelline vein
XCL extraembryonic coelom
YS yolk sac
YSC yolk sac capillaries

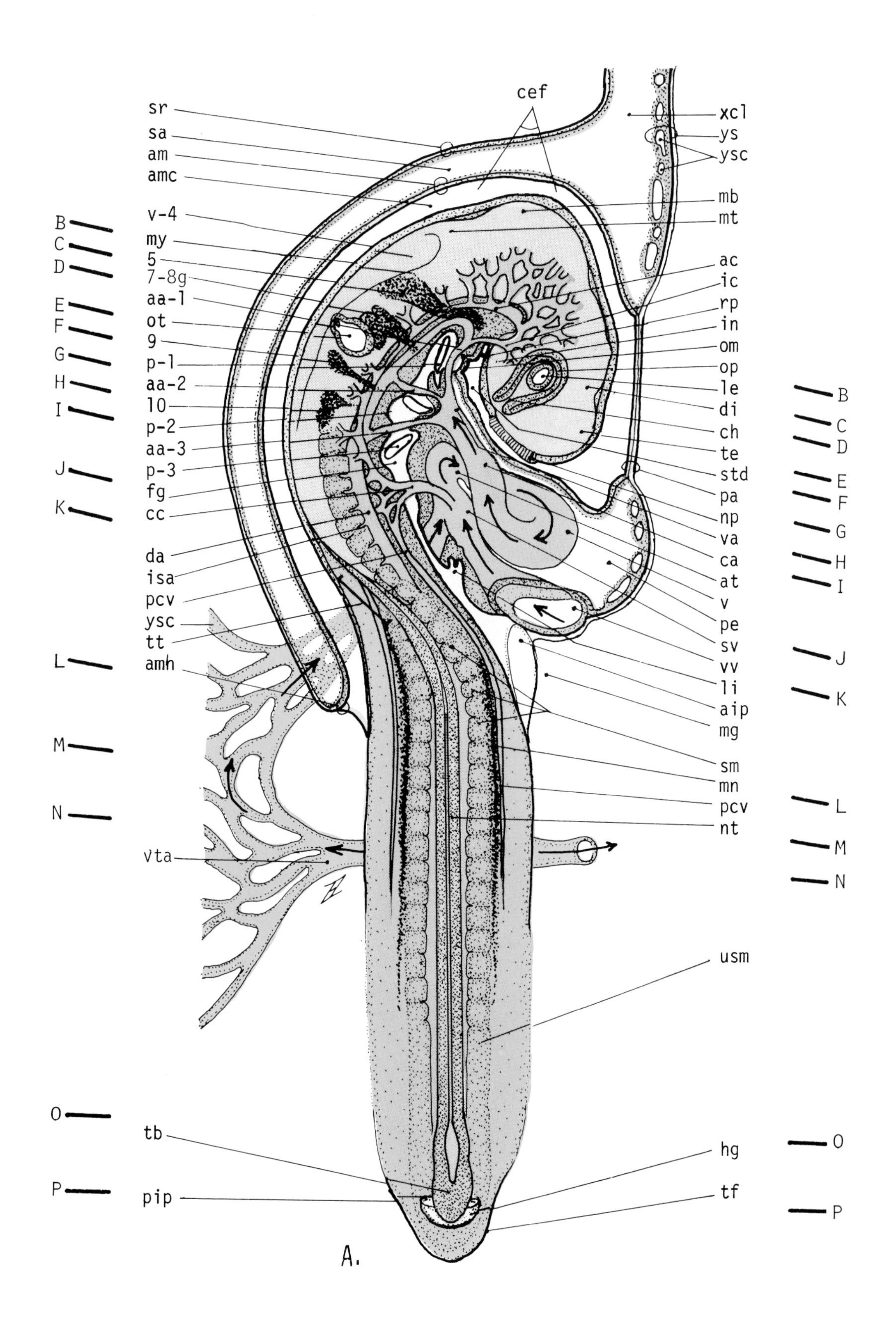

Figure 7-6 A. Whole mount diagram of the 48 hour chick embryo (Stage 14-16). Guide lines to the right and left of the figure refer to levels of correspondingly lettered cross section drawings. See legend.

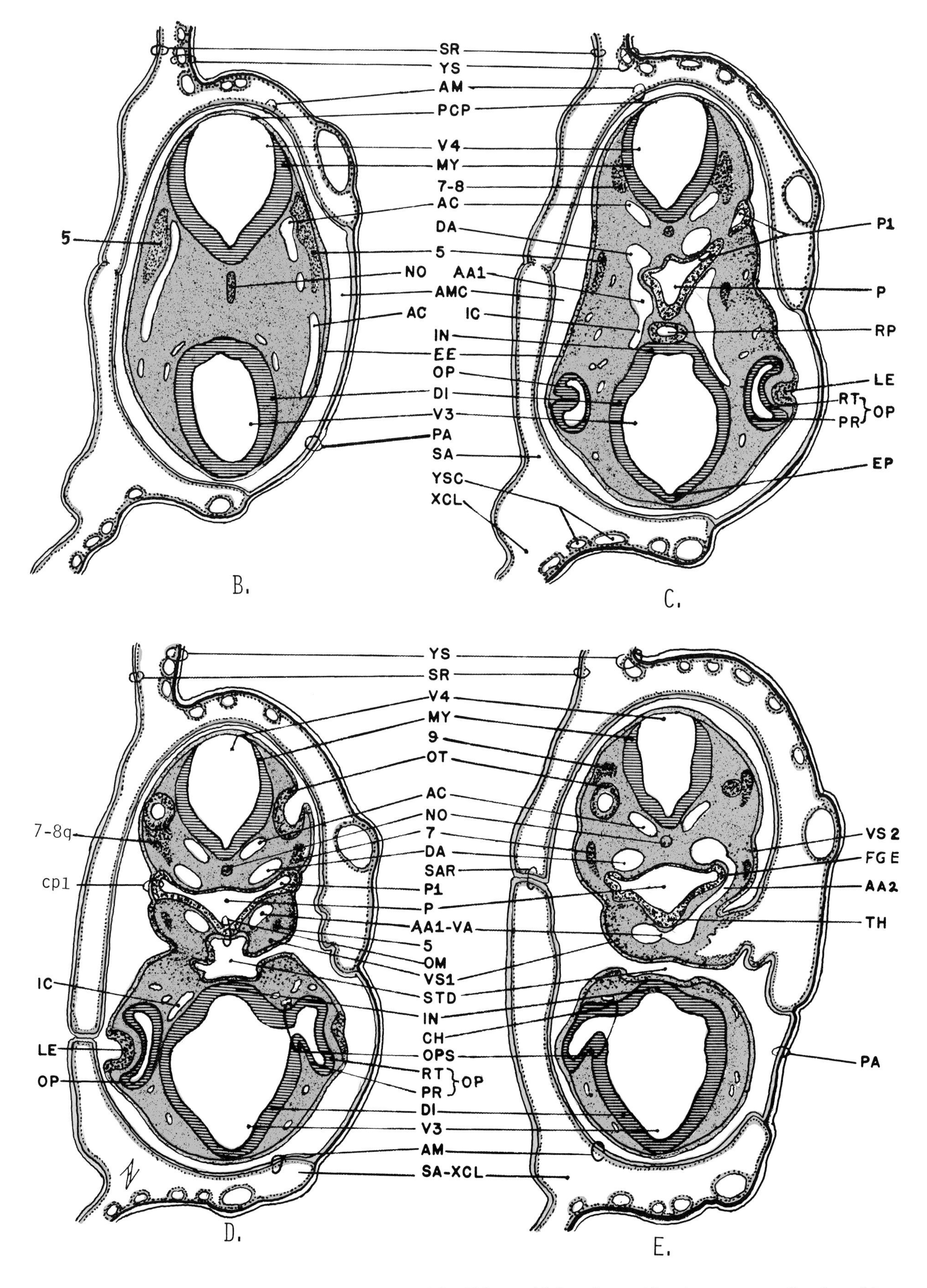

Figure 7-6 B–E. Representative cross sections of the 48 hour chick embryo at levels corresponding to guide lines in 7-6A. See legend.

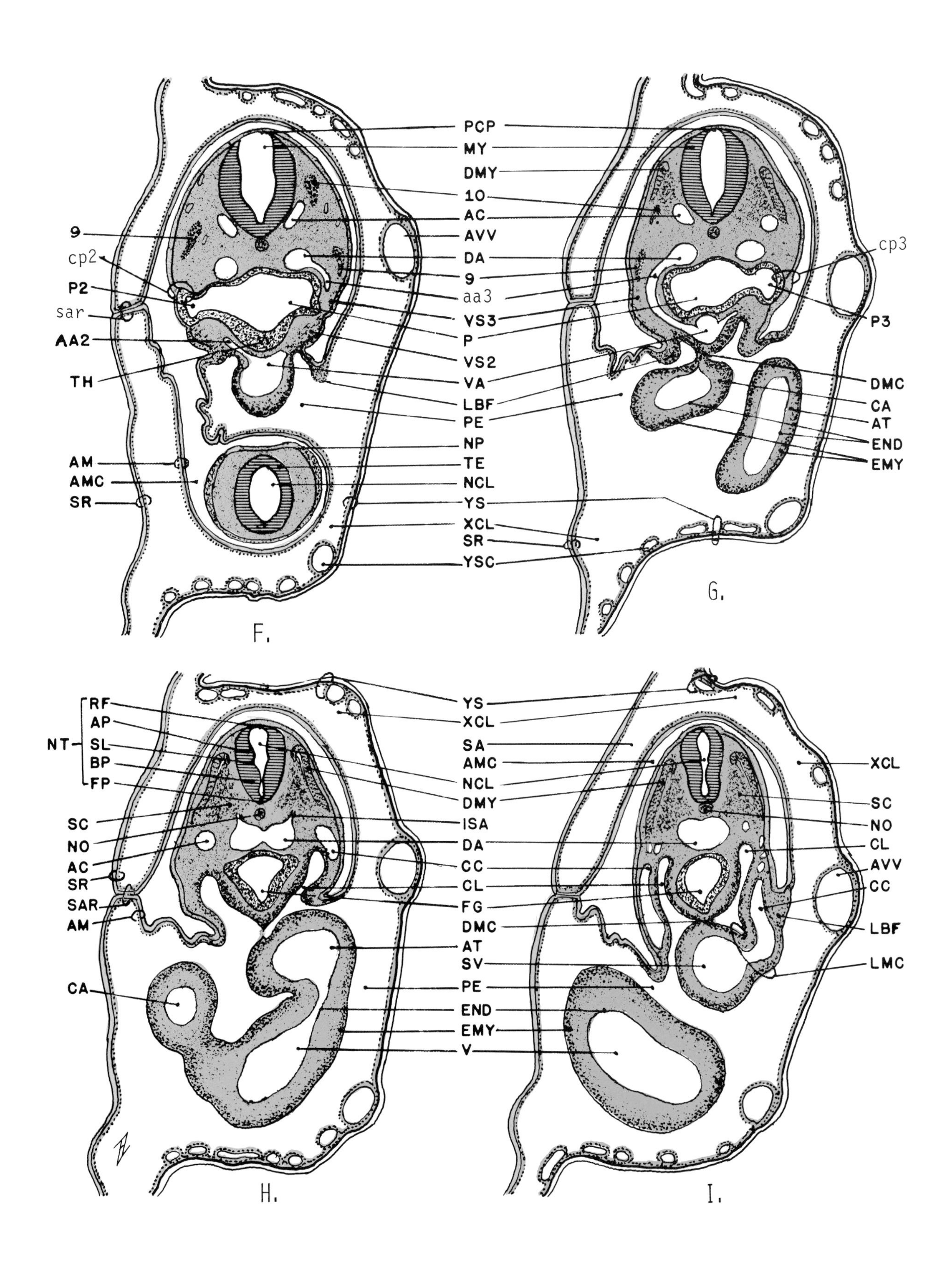

Figure 7-6 F–I. Representative cross sections of the 48 hour chick embryo at levels corresponding to guide lines in 7-6 A. See legend.

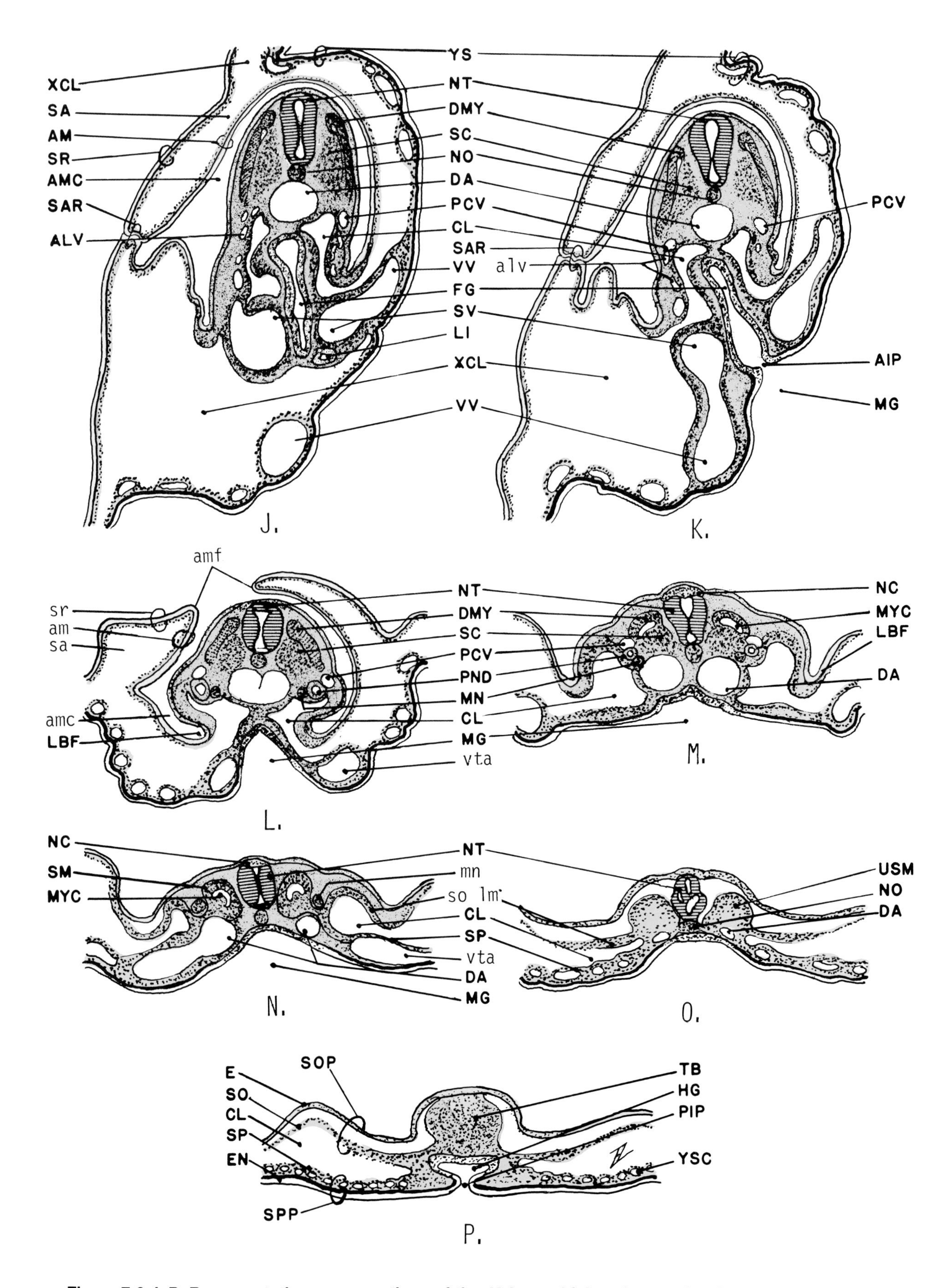

Figure 7-6 J–P. Representative cross sections of the 48 hour chick embryo at levels corresponding with guide lines in 7-7 A. See legend.

plate (hypomere) mesoderm are essentially the same as have been described for earlier stages in connection with the development of the coelom and formation of somatopleure and splanchnopleure in the 33 hour chick. The major derivatives of lateral mesoderm in the embryo are associated with formation of the heart (discussed separately in the next section), and the intraembryonic coelom which will be most clearly interpreted in posterior sections where the **amniotic folds** are not fused over the embryo (Fig. 7-6 L–N).

Identify the **somatopleure** and **splanchnopleure** and note the constriction in the coelom at the lateral edges of the embryonal area where the lateral body folds are in the process of undercutting the embryo. These constrictions in the coelom mark the approximate boundary between the **intra-** and **extraembryonic coeloms.** Trace the sections anteriorly and note the manner in which the **amniotic folds** arch upwards and eventually fuse over the embryo. The point of fusion of the right and left amniotic folds is the **seroamniotic raphe,** which can be identified as a slender connection between the **serosa** and **amnion** in most anterior sections (Fig. 7-6 B–K). The space between the amnion and embryo is the **amniotic cavity.** The space between amnion and serosa is the **seroamniotic cavity.** It is an extension of the extraembryonic coelom. Trace both of these cavities and the two membranes anteriorly. In the region of the cephalic flexure, be particularly attentive to the course of the amnion around the front end of the head. Try to visualize how the torsion and flexure of the head will, of necessity, carry the amnion with it as the axis of the embryo changes. The **yolk sac splanchnopleure** should also be followed forward from the region of the open midgut. It can be easily identified by the large number of **vitelline blood vessels** in the walls. At the region of the anterior intestinal portal, note the entrance of the **vitelline veins** from the yolk sac into the embryonic region. They empty into the sinus venosus (see following section for the discussion of the heart and circulatory system). Trace the yolk sac forward and find the region where the two amniotic folds become attached to it. The membrane between these points of attachment is the **proamnion** Figs. 7-3, 7-6 B–E). It is composed of one layer of extraembryonic ectoderm and an inner layer of yolk sac endoderm. Begin at the open midgut region and also trace the **intraembryonic coelom** through the anterior boundary of the heart. The part of this cavity which surrounds the heart is the pericardial coelom, and the part which lies lateral to the region where the lungbuds are beginning to form (Fig. 7-6 H–K) is the **pleuroperitoneal coelom.** At this stage all subdivisions of the coelom communicate with one another and there are no distinct boundaries which separate them.

d. The circulatory system. The major components of the circulatory system are the blood, heart, and vessels. We have discussed the origin of blood islands in the area vasculosa of the 24 and 33 hour chick embryos and will not give further consideration to blood forming tissues at this time.

The heart is a derivative of the ventral mesentery. At the 33 hour stage this was clearly evident by the presence of a **dorsal mesocardium** along the entire length of the heart, which connected the epimyocardium directly to the spanchnic mesoderm surrounding the foregut. By the 48 hour stage, this relationship is not so clearly evident since the dorsal mesocardium has degenerated in the region of the ventricle and atrium. It persists only as a vestige at the anterior and posterior extremities of the heart, in the regions of the conus and sinus venosus. The disappearance of the dorsal mesocardium is apparently associated with the enormous growth and twisting of the atrial and ventricular regions and may be a functional adaptation that frees the atrium and ventricle from restrictions to their movements in contraction essential for their roles in moving the blood. The heart tube is still single and gives little indication of its later separation into right and left halves.

Identify the four primary chambers of the heart in injected whole mount preparations and familiarize yourself with the general shape and torsions of the heart tube before studying the serial cross sections. Begin the study of sections at the posterior end of the heart in the region of the anterior intestinal portal. At this level trace the entrance of the **vitelline veins** into the embryo where they empty into the posterior end of the **sinus venosus.** In tracing the course of the heart, it will be necessary to go first anteriorly, then posteriorly, and then anteriorly again, to follow the S-shaped curvature of the heart tube. The heart begins to beat at about 40 to 44 hours of incubation, and complete circulation is well established by 48 hours. Depending upon the state of contraction of the heart chambers at the time of fixation, and upon the age and degree of torsion of the heart, you can expect to find considerable minor variations in the diameter of the heart regions in different serial sections. Therefore, it is essential to learn to identify the chambers of the heart on the basis of constant relationships which do not vary.

The following landmarks can be used to identify the heart chambers. The **atrium** always occupies a dorsal position in the pericardial cavity and is displaced slightly to the left of the midline; that is, toward the yolk sac side of the embryo (Fig. 7-6 H). The **ventricle** is U-shaped and occupies a median ventral position in the pericardial cavity. It is also the only one of the three anterior heart divisions which extends posteriorly beyond the level of the sinus venosus, as will be seen in the whole mount diagram (Fig. 7-6 A). The **conus arteriosus** is the most anterior part of the heart and is the smallest of the four chambers. It is always located in a right-dorsal position in the pericardial cavity, that is, on the amnion-serosa side of the embryo (Fig. 7-6 G, H). The posterior boundary of the **sinus venosus** is marked by the mid-ventral fusion of the two vitelline veins which enter the heart from the yolk sac (Fig. 7-6 I, J). The dorsal wall of the sinus venosus is attached to the floor of the foregut by the **dorsal mesocardium.** The lateral walls of the sinus venosus are fused to the **lateral body folds;** these are the **lateral mesocardia.** Identify the branches of the **posterior cardinal veins** in the lateral body folds. These branches fuse to form a large venous channel that passes through each lateral mesocardium into the sinus. These veins are the **common cardinal veins.** The paired common cardinal veins (or ducts of Cuvier, Fig. 7-6 I) empty into the lateral walls of the sinus venosus. Anterior to this point, the heart wall loses contact with the lateral body folds and the dorsal mesocardium also disappears (Fig. 7-6 H). This point is usually taken as the landmark for the boundary between the sinus venosus and the atrium. Before progressing to more anterior regions of the heart, it should be pointed out, in connection with the future divisions of the embryonic coelom, that the fusion of the lateral body folds to the sides of the sinus venosus cuts off a small pocket of coelom dorsally (Fig. 7-6 I, J). Note that the lungbuds expand into this cavity. This region of the coelom is destined to form the **pleural** region of the **peritoneal cavity.** The lateral mesocardia themselves are components of the transverse septum which will eventually play an important part in the separation of the **pericardial cavity** from the intraembryonic coelom. In mammals the lateral mesocardia will also be important components of the diaphragm which will separate the pleural coeloms from the peritoneal cavity.

Aortic and visceral arches. Before proceeding to the study of aortic arches, review the discussion of visceral arches as given in Figure 5-6, and review the common plan of the vertebrate circulatory system that is given in Figure 7-5.

At the anterior end of the **pericardial cavity,** the **conus arteriosus** opens through the **dorsal mesocardium** into the single median ventral aorta (aortic sac). The ventral aorta lies immediately under the **pharynx.** It gives rise to three pairs of arteries which pass laterally around the pharynx and reunite dorsal to the pharynx and form the paired **dorsal aortae.** The three pairs of arteries arising from the ventral aorta are the first 3 **aortic arches** (Fig. 7-6 C, E, G). In lower vertebrates the aortic arches provide the respiratory circulation of the gills. Each gill, or **visceral arch,** is paired and consists of the total tissue mass that lies between two consecutive **pharyngeal pouches.** The latter are evaginations of the pharyngeal wall which fuse with the superficial epidermis. Each visceral arch normally contains an aortic arch and musculature (both of which are derived from head mesoderm), a branch of a cranial nerve, and a visceral skeletal bar of cartilage (both of cranial neural crest origin). In the 48 hour embryo only the nervous and arterial components of the first three visceral arches are differentiated sufficiently to be identified in sectioned material. The visceral arches were given detailed consideration in the study of the salamander tadpole. In the 48 hour embryo, learn to identify the first three arches by the following landmarks.

(1). First aortic arch (Fig. 7-6 C, D.). Trace the ventral aorta anteriorly to the point at which it divides completely into right and left halves. Each of these two tubules represents the ventral base of the first pair of aortic arches. The condensations of mesoderm in the lateral body wall, in which each first aortic arch lies, are the paired first visceral arches. The latter will form the jaws of the older embryo. Note the relationship of the first visceral arch to the **stomodaeum** and **oral membrane.** These relationships provide the best landmarks for the identification of the first aortic and visceral arch. Anteriorly, trace the first aortic arches around the sides of the pharynx. When they reach the dorsal side of the pharynx, they form the anterior end of the paired **dorsal aortae** (Fig. 7-6 C). A second reliable landmark for the identification of the first visceral arch is its relation to the ventral extension of the **5th cranial nerve.** Note that a branch from this ganglionic aggregation is in the process of growing down into the first aortic arch region.

(2). Second aortic arch (Fig. 7-6 E). Return to the ventral aorta and note that just posterior to the point at which it divides into the two first arches, a small pair of branches is given off anterior to the **thyroid diverticulum.** These branches extend around the pharynx and enter the dorsal aortae. These are the **second aortic arches.** The mass of tissue in the lateral body wall that surrounds each second aortic arch is the **second visceral arch.** The most reliable landmarks for the second aortic and visceral arch are their intermediate position between the **first** and **second pharyngeal pouches** and their relation to the 7th cranial nerve dorsally and the thyroid diverticulum ventrally.

(3). Third aortic arch (Fig. 7-6 F, G). This arch is very small at this time and may not be fully formed in some specimens. It leaves the ventral aorta at a posterodorsal angle so that only rarely can you get a section that shows the full length of this vessel connecting the ventral and dorsal aortae around the lateral walls of the pharynx. The most reliable landmarks for the third aortic and visceral arches are their location just posterior to the second visceral pouch and the entrance of the third arch into the **ventral aorta** or **aortic sac** posterior to the point at which the conus arteriosus connects with the ventral aorta. If the 9th cranial nerve is sufficiently well developed, it will be clearly aligned with the dorsal margin of the third arch.

Intraembryonic circulation. Review the circulatory plan of the embryo as seen in the whole mount diagram (Fig. 7-6 A), in demonstration-injected specimens, and in the diagram of the vertebrate circulatory plan (Fig. 7-5).

The **anterior intraembryonic circulation** is simple; it consists of the paired **internal carotid arteries** and the **anterior cardinal veins** along with capillaries that connect them. The former arise from the first aortic arches and carry blood to the forebrain where they break up into a capillary network that drains into the paired anterior cardinal veins. These relations can be seen in sections comparable to those shown in Figure 7-6 B–D. The anterior cardinal veins return blood to the sinus venosus and, in their course posteriorly, they lie close to the ventrolateral wall of the neural tube (Fig. 7-6 B–F). At the level of the third visceral arch, the anterior cardinal veins move ventrally and unite with the posterior cardinal veins to form the **common cardinal veins** which pass through the lateral mesocardia into the lateral walls of the sinus venosus (Fig. 7-6 H–J).

The middle and posterior intraembryonic circulation consists of the **dorsal aorta** and its **intersegmental arteries** which are given off between each pair of somites (Fig. 7-6 H). The capillary drainage from these vessels is collected by the paired **posterior cardinal veins** (Fig. 7-6 J–M); these return blood to the sinus venosus by way of the common cardinal veins. Review the course of these vessels in serial sections and note that the dorsal aorta is paired anterior to the level of the third aortic arch. Behind this region the two dorsal aortae fuse in the midline and form a single vessel at the level of the 18th to 20th somites. Posterior to the 20th somite, the dorsal aortae remain separate throughout the remainder of their extent into the tailbud.

The last paired intraembryonic vein to be mentioned is one that is usually recognizable in whole mount injected specimens but is rarely identifiable in sectioned material. They are the **allantoic veins.** It is generally considered to be homologous with the umbilical veins of mammals and with the lateral abdominal veins of lower vertebrates. As the latter name implies, their course is along the sides of the body. In the chick, they run forward from the hindgut region and will empty directly into the posterior wall of the common cardinal veins. At the 48 hour stage, any capillaries identified in cross sections of the lateral body folds, posterior to the heart, can be considered a part of the developing allantoic vein capillary beds which are just beginning to form at this time. They will become the major extraembryonic veins to the chorion.

Extraembryonic circulation. The system of vessels external to the embryo at the 48 hour stage consists entirely of the yolk sac, or vitelline, circulation. The **vitelline arteries** arise on the right and left sides of the body as lateral branches from the dorsal aortae at about the level of the 22nd to 24th somites (Fig. 7-6 A, M). In the yolk sac splanchnopleure, the vitelline arteries break up into a rich capillary network that eventually returns blood to the sinus venosus by means of a large pair of **vitelline veins.** These have already been identified as entering the heart just anterior to the anterior intestinal portal.

After completing the foregoing study of primordia in the 48 hour chick embryo, it will be instructive to reread the introductory paragraphs of this chapter. Refresh your memory on the nature of **primordia, mosaic stages,** and **prefunctional levels of differentiation,** and mentally fill in examples of these from the study of the 48 hour chick.

By way of review, also compare the 48 hour chick with comparable early stages in the development of other vertebrates that have been discussed. In particular compare the vertebrate common plan illustrated in Figures Int.-3, 1-1, 1-2, 4-3 A, 4-4 A, 5-2, and 7-6 A, with the purpose of seeing more clearly the examples of paleomorphosis that have been presented up to this point.

B. EXPERIMENTS AND OBSERVATIONS ON LIVING 2 TO 3 DAY CHICK EMBRYOS

Incubated living two and three day chick embryos will be used. They have been treated in the manner described for incubating eggs at the end of Chapter 6. The basic materials and solutions described in Chapter 6 for the handling of chick blastoderms will also be needed, as well as a few additional items.

Basic set of materials needed for each student: Dissecting microscope, focusable microscope lamp, 500 ml 9% isotonic warm NaCl, 3 4-inch fingerbowls, 3 Petri dish halves, 1 widemouth waste jar, standard dissecting instruments (including 2 dissecting needles made from No. 000 insect pins, 2 pairs of fine forceps, several standard medicine dropper pipettes), 1 plastic teaspoon, 1 dropping bottle of 0.1% Nile Blue Sulfate, 3 2-inch low, wide-mouth collecting jars, and other special materials required for specific techniques which will be given in association with each procedure described below.

1. Observation of the living 2 and 3 day chick embryos and removal of the blastoderm. The directions given for opening and handling 18, 24, and 33 hour eggs in Chapter 6 will be followed in all details. Open the eggs under warm saline, then decant off the excess water so that the embryo on the yolk is immediately under the surface film of water. Place the fingerbowl, with the embryo still on the yolk, on the microscope stage and focus the light on the blastoderm.

Use the whole mount diagram (Fig. 7-6 A), the normal series for the chick in Figure 6-1, and Table VI-1 of developmental stages for the chick, as aids in diagnosing stages of development. Note the circulatory pattern of yolk sac blood vessels and the rapidly pulsating heart which dominates the center of the embryo in size and scarlet color.

Identify the major elements of the extra- and intraembryonic circulatory channels. The yolk sac circulation is very precisely limited peripherally by a circular vein called the **sinus terminalis.** Note the extensive network of **yolk sac capillaries** and the connections between these capillaries and the embryo. The **vitelline arteries** are slightly posterior to the **vitelline veins,** which enter the embryo just anterior to the anterior intestinal portal. Note the S-shaped heart with its four chambers in their typical positions that have already been described in association with the study of serial sections.

Intraembryonic vessels can easily be distinguished by color and position. Identify the three or four pairs of aortic arches, dorsal and ventral aortae, internal carotid artery, capillary network of the head, and the anterior cardinal vein that drains the blood back to the sinus venosus via the common cardinal vein. The primary embryonic vein posterior to the heart is the posterior cardinal vein. Its intersegmental branches connect with the dorsal aorta.

Add, dropwise, sufficient Nile Blue Sulfate to give a pale blue tint to the surrounding saline. Continue to observe the embryo while the vital stain is being incorporated into the living tissues. As this happens, special primordia will become much more readily visible. Identify the amniotic folds, extraembryonic areas including the zone of junction, and area vasculosa, although the latter is readily visible by virtue of its rich vascularization. In the embryo, the stain will make the lens, nasal placodes, visceral arches, and somites stand out with particular clarity.

2. Fixation and storage of embryos. After observations on living embryos are complete, the embryos should be preserved for possible later use in making whole mount or serial section preparations. These are standard embryological techniques to virtually all developmental materials.

a. Embryological fixing fluids. The formulae in Table VII-2 include some of the widely-used general fixatives for embryological materials.

b. Make embryo mounting rings. Framing is a convenient way to fix chick embryos for ease in handling and for recording data on individual cases. Frames are cut out of non-glossy, heavy, construction paper. Frames are most easily prepared by folding the paper with two right angle folds so the tip can be cut off to give a square hole and the outer edge can be trimmed to give a border about half an inch wide around the central hole. The central hole should be large enough to "frame" the embryo and whatever extraembryonic parts that are desired for a particular preparation. Make several frames so that enough will be on hand when they are needed. With pencil or India

TABLE VII-2
STANDARD EMBRYO FIXING SOLUTIONS

BOUIN'S PICRO-FORMOL-ACETIC FLUID

75 ml	Saturated picric acid
25 ml	Commercial formalin
5 ml	Glacial acetic acid

KAHLE'S ALCOHOL-FORMOL-ACETIC FLUID

32 ml	95% Ethanol
16 ml	Commercial formalin
2 ml	Glacial acetic acid
60 ml	Distilled water

4% FORMALDEHYDE SOLUTION: 10 ml Commercial formalin + 90 ml Distilled water

Most of the above solutions were developed as general fixatives for both adult and embryonic tissues. My own preference is to use less strong solutions on the soft tissues of embryos and recommend the following variations as being specially good for eggs and embryos.

CALCIUM-PICRO-FORMOL-ACETIC (CaPFA)

0.5 g	$CaCl_2 \cdot H_2O$
0.8 g	Picric acid crystals
5.0 ml	Commercial formalin
5.0 ml	Glacial acetic acid
90.0 ml	Distilled water

***CALCIUM-ALCOHOL-FORMOL-ACETIC (CaAFA)**

0.5 g	$CaCl_2 \cdot H_2O$
20.0 ml	95% Ethanol
5.0 ml	Commercial formalin
5.0 ml	Glacial acetic acid
70.0 ml	Distilled water

ALKALINE GLYCERIN-FORMOL (AGF)**

0.2 g	Sodium bicarbonate ($NaHCO_3$)
5.0 ml	Commercial formalin
10.0 ml	Glycerin
85.0 ml	Distilled water

*The solution will often be slightly cloudy due to the formation of calcium formate which will fall out of solution as a fine precipitate.

**4 drops 0.1% Phenol Red indicator can be optionally added to indicate pH; solution should remain a lavender-pink; yellow indicates acid range.

ink, write on the paper any data needed, such as stage, date, experimental case number, and name of preparator.

c. Mounting living embryos in paper rings. It is very important that living embryos be inverted in Petri dishes of warm saline so that the yolk sac side is up and the serosa–vitelline membrane side is down. This means that the general curvature of the head will be to the left (not the right) when viewed from above (i.e., it should be the mirror image of embryos shown in whole mount diagrams in Chapter 6, 7, and 8). Usually it is necessary, therefore, to refloat the embryo with the addition of more saline so that it can be turned over, with the embryo spoon, with a minimum of damage. Center and flatten the inverted embryo again in the center of the Petri dish from which most of the water has been decanted or pipetted away. **There should be no free standing water covering the bottom of the dish before proceeding with framing.**

Remove the dish from the microscope stage to a region that will permit you to look directly down on the embryo in the dish. Then, holding a paper frame in a forceps, center the frame over the embryo and see that the frame is large enough to include the entire embryo, but is still too small to permit all extraembryonic membranes to be included in the opening. Select a suitable frame and lower it carefully, with the embryo centered in the opening. The blastoderm should be spread like a drum head across the opening in the frame with the embryo in the middle.

Proceed to **Step d,** or directly to **Step e,** which follow.

d. Microinjection of the embryonic circulatory system. If microinjections are to be attempted, it will be necessary to make all preparations in advance of mounting an embryo in a paper ring. Therefore, read the following directions and make sure that all materials are at hand in readiness for immediate use when required.

First make several micropipettes. These must be hand-pulled from Pasteur pipettes or medicine droppers. The pipette tip must be heated in a micro-Bunsen flame so that the tip can be pulled out quickly to give a rapid taper to an almost invisible diameter. Break off the tip so that it will end about three-quarters of an inch from the beginning of the taper. The tip can be no larger in diameter than one-half the diameter of the vitelline veins of the embryo. With this in mind, you can judge when suitable pipettes have been constructed.

Next prepare a suspension of carbon black in saline. This can be done by mixing dry carbon or lamp black powder with physiological saline. Or you can use India ink, China black, or poster paint black, which should be diluted with an equal volume of saline. In all cases it is essential to filter the solutions to remove all but the colloidal particles. Otherwise, injection pipettes will become hopelessly clogged. The pipette tip will be too small to load in the usual way. Therefore, fill the pipette with a few drops of saline-ink mixture **from the rear** with a Pasteur pipette. Blow on the pipette opposite the point and force out a small drop of fluid. Then cap the blunt end of the pipette with a small cork or with a rubber pipette bulb. It is recommended that the pipette tips, when not in use, be submerged in saline up to the level of their ink. A wide-mouth collecting jar, with a cotton wool pad flooded with 0.9% saline, is convenient to use. Unless this is done, the ink will dry very quickly and clog the tip. Make sure that a small dropping bottle of fixing solution is on hand; CaAFA is recommended (Calcium-Alcohol-Formol-Acetic; see Table VII-2).

Prepare an embryo for injection by inverting it and mounting it in a carefully constructed paper ring in the manner described in **Step c** above. If extraembryonic circulation is to be part of the preparation, the window must be carefully trimmed so as to include the area vasculosa, but the frame must overlay the area vitellina.

Making injections requires infinite patience, steady hands, and muscle coordination. Do not be discouraged if you are not successful on the first try. **Proceed as follows.** Put the framed embryo, yolk sac up, on the microscope slide and focus the light so that the anterior intestinal portal is clearly illuminated and in sharp focus. Pick up a loaded micropipette near its slender bill. As much as possible **avoid unnecessary finger contact with the large bore part of the pipette** since warming it will cause the ink to be forced out again. Puncture the vitelline vein at its junction with the sinus venosus. Warm the glass with a finger to cause the ink to be forced gently into the posterior chamber of the heart. In a successful injection, there will be an immediate perfusion of the circulatory system with carbon black, and the capillary networks throughout the blastoderm and embryo will become visible almost instantly. When optimum injection has been reached, the embryo must be killed immediately with the addition of fixative to prevent the leakage of carbon from the point of injection.

You can expect several failures before this technique is perfected. You also can expect that there will be some leakage of carbon in and around the point of puncture. This should be washed away with strong squirts of saline immediately after the fixative has stopped heart contractions.

With successful and less perfect attempts, proceed with the method for fixation and preservation that follows.

e. Fixation and preservation of chick embryos. Dropping bottles of one or more of the above fixatives should be on hand for use. The Calcium-Alcohol-Formol-Acetic fixture is specially recommended for use with chick tissues.

Whole embryos and blastoderms can be dropped into low, wide-mouth collecting jars of fixing fluids. However, **it is strongly recommended** that specimens of experimental value, and ones that are planned for later sectioning or for the preparation of whole mounts, be in paper frames in the manner described in **Steps b** and **c.**

Add fixative dropwise to the center of the embryo. Observe the fixation process under the microscope and note the progressive change in opacity of the embryonic tissues as fixation goes to completion. Surface layers become opalescent first and reveal with remarkable clarity the amniotic folds, otic capsules, limb buds, gill slits, visceral arches, and allantoic diverticulum. Gradually the tissues become completely opaque, and the blood loses its crimson color; but it remains slightly browner than the rest of the tissues which will become milky white.

When penetration of fixative is complete, squirt fixative under the embryo to loosen it from the bottom of the dish. Do not attempt to do this in less than about 20 to 30 minutes, which will allow time for some hardening of tissues to take place. Flood the dish with fixing fluid to a level of one-quarter inch, pick up the card, and turn the embryo over. It is desirable to remove the vitelline membrane at some point before final preparations are complete. This can be done by pulling it off slowly with the aid of fine forceps. However, removal of the vitelline membrane should not be done until the embryo has hardened for an hour or more in fixative.

The frame with embryo can then be transferred to a wide-mouth collecting jar for storage. As a general rule, store embryonic or histological materials in a volume ratio of fixing fluid to tissue of at least 20:1.

f. In toto staining of embryos for whole mounts. Transfer fixed specimens to low wide-mouth staining jars. In general it is best to have one embryo per jar, or at most not more than a few, so that progressive staining can be observed and controlled. Proceed as follows.

Wash specimens in two 10 minute changes of distilled water, or in sufficient changes to remove picric acid stain, if picro-fixatives have been used.

Add stain dropwise to the jar, which should be about half full of tap water. Add either 0.1% aqueous Rose Bengal, or 0.1% aqueous Nigrosin, or 0.1% Naphthol Blue Black until the water is tinted a clear transparent color. Add only one color to each jar. These are progressive permanent stains with a strong affinity for protein, and the initial concentration should be quite dilute. Embryos will concentrate the dyes. Staining can continue for hours or even days until the desired intensity is reached. Tissues are difficult to destain!

3. Dehydrating, clearing, and mounting stained embryos. Three methods for making whole mount preparations are summarized below. Each requires separate handling and solutions. They will be described individually.

For all three methods, it is necessary to wash-unbound stain out of embryos with successive changes of tap water until all traces of unbound dye are removed.

General materials needed for all mounting methods. Standard microscope slides, preferably round No. 2 thickness cover glasses, embroidery rings (these come in sizes ranging from half-inch to one inch diameter and can be obtained at fabric, department, or ten-cent stores; various sizes should be on hand to meet the needs of different size embryos). In anticipation of use, embroidery rings should be glued onto microscope slides with transparent cement such as balsam, permount, or plastics. This can be done most easily by placing a small amount of cement in a low Syracuse dish or bottle cap. Then after touching one side of the ring to the cement, place it on a microscope slide. The cement should be just enough to seal the ring completely to the glass with no air spaces between the ring and the glass.

a. Glycerin clearing and glycerin-jelly mounts. This is the simplest but the least permanent of the three methods for preparing whole mounts. It has the advantage of easily permitting an embryo to be recovered for sectioning, but among the disadvantages are sensitivity to heat above usual room temperatures, since gelatin will melt at about 30°C/85°F.

Special materials needed for glycerin-jelly mounts. Stoppered bottle of 25% aqueous glycerin, clear fingernail polish with brush, and a dropping bottle of glycerin-jelly. **(Formula for glycerin-jelly:** Dissolve 20 g Knox unflavored gelatin in 50 ml warm water. Filter and add an equal volume of 100% glycerin and a crystal of phenol or thymol to serve as a mold deterrant. Store in dropping bottles; it will keep indefinitely.)

Dehydrate and clear the specimen in glycerin by the following method. Fill a specimen jar half-full of 25% aqueous glycerin and mark the fluid level. Transfer embryos that have been stained and washed to the jar. Place the jar, with an aluminum foil dust cover, in an incubator, or on a warm table, or under a desk lamp for several days until the fluid level is reduced to one-quarter the starting value. At this point clearing and dehydration are complete.

Trim the embryo by transferring the specimen to a clean glass plate. Cut off the card and extraembryonic membranes to a size that will fit conveniently in the area of a mounting ring on a slide. This can be done easily with a sharp razor blade.

Mount the specimen in a mounting ring on a microscope slide. Mounting is best done under a desk lamp or on a warm stage. Melt the sample of glycerin-jelly in its dropping bottle by placing it under the desk lamp or by holding the bottle tightly in the palm of the hand. Add several drops of glycerin-jelly to the ring mount and place the specimen in the ring. Add sufficient glycerin-jelly to fill the ring to brimming full. Remove any air bubbles that may have been trapped. This can be done best with a hot dissecting needle heated in an alcohol or Bunsen flame.

Cover with a round coverslip and press the cover down until it is seated firmly on all sides against the mounting ring. Set the slide in a cool place until the overflow of jelly has set firmly.

Brush the external jelly away and carefully clean the area with an alcohol swab. Allow the preparation to air dry. Paint a ring of clear fingernail polish around the slide as a sealant. The seal should be built up with several coats to help stabilize the preparation against mechanical damage.

Special precautions. Glycerin-jelly mounts never get really hard or dry. They must be kept cool to prevent melting, and they must be cleaned very gently to avoid removing the coverslip. With proper care these preparations remain useful for many years.

b. Alcohol dehydration, toluene or xylene clearing, and resin mounting. This is without question the most widely used and most generally satisfactory method for making whole mount preparations. Its disadvantages are associated primarily with the relatively long time required to complete the drying of preparations.

Special Materials needed for resin mounts. Stoppered bottles for alcohol series at concentrations of 50%, 70%, 80%, 95%, and 100% toluene or xylene, dropping bottle of resin (i.e., commercially available under trade names including Canada Balsam, Damar, Permount, Eukitt, etc.).

Dehydrate washed and stained specimens in successive half hour changes of either methyl, ethyl, or isopropyl alcohol at concentrations incrasing from 50% to 70%, 80%, 95%, and100%. It is advisable to repeat the final 100% change two or three times to assure complete dehydration before clearing.

Trim the preparation to a suitable size for mounting in a ring and be particularly careful to avoid any drying, which may occur very quickly owing to the high volatility of alcohol. Metal rings should be used since many plastics are soluble in toluene or xylene.

Clear the embryos in two, half-hour changes of either toluene or xylene (varsol or charcoal lighter fluid may be substituted).

Mount a trimmed, cleared embryo on a slide with a suitable mounting ring. Add several drops of thick resin dissolved in toluene or xylene. Remove any trapped air bubbles by touching them with a dissecting needle heated in an alcohol or Bunsen flame. Add sufficient resin to fill the ring completely. Place the slide on a warming table under a dust cover. Each day examine the preparation and add additional resin until the ring is filled with firm resin. Last, add a final drop of resin and a cover glass. If the cover glass is added too soon, additional evaporation of solvent will cause large air bubbles to form under the glass and it will be necessary to dissolve off the cover with xylene and start over again.

c. Acetone dehydration and plastic clearing and mounting. This method has been developed specially for embryological materials by this author and has not been previously published. It is simple, inexpensive, quick, and permanent.

Special materials needed for plastic mounts. Capped bottles of 50% acetone, 100% acetone, 50% monomere plastic in acetone, dropping bottle of 100% monomere plastic, dropping bottle of 50% catalyst in acetone, toothpicks, Silicone (i.e., Dow-Corning Slipicone-silicone lubricant spray, obtainable at hardware stores), monomere plastic (i.e., clear plastic for embedding, obtainable from hobby shops under the trade name Deep-Flex Plastic, Titan Plastic, Bio-Plastic, Hobby-Plastic, etc.), 1 inch Nalgene or Teflon squares (cut from flexible plastic, deep-freeze food containers).

Dehydrate stained, washed, and trimmed embryos in acetone as follows. Each change should be for at least 15 minutes but can be of indefinite duration without harm. Decant off three-quarters of the fluid in each change and fill up to unit volume with the next solution in the dehydration and clearing schedule. Make two changes in 50% acetone, followed by two changes in 100% acetone; the last acetone change should be at least of an hour duration, or overnight in a capped jar, or until the tissue has settled to the bottom of the container.

Clear in a monomere plastic. This is best done by pouring off the second 100% acetone change and replacing it with a one-quarter volume of fresh 100% acetone. To this add 3 parts 50% monomere plastic diluted in acetone. This mixture will settle to the bottom and the embryo will float in pure acetone on top. Cap the jar tightly. Gradually the acetone will go into solution in the acetone-monomere mixture and, as this occurs, the embryo will become infiltrated with plastic and will settle to the bottom of the jar. When the embryo has settled, which will usually take several hours to overnight, the jar should be gently shaken to make sure that the acetone-plastic mixture is uniform.

Infiltration is accomplished merely by removing the cap from the jar and placing it near a desk lamp to hasten the evaporation of acetone. When the fluid level has reached one-quarter the starting volume, the acetone has evaporated and pure monomere remains.

Catalysis and mounting are carried out individually in slides with mounting rings already cemented in place. First, sufficient monomere is added to the mounting ring to nearly fill it. Then one drop of catalyst diluted with an equal volume of acetone is added to the ring mount and is carefully stirred into the monomere with a toothpick or blunt dissecting needle until a smooth mixture is achieved. Usually some air bubbles are trapped in the process. They should be avoided and must be removed with a glass needle or by adding a drop of 50% monomere-acetone mixture, which has a lower surface tension and causes air bubbles to break. A hot needle can also be used.

Transfer an embryo to the mounting ring and gently push it down into the catalyzed plastic. Remove any additional bubbles that may be formed. Add a final drop of 50% plastic-acetone to the preparation to flatten the mixture and break any air bubbles. **The rim should now be brimming full.** Add a small square of Teflon to the top of the ring and press it down to flatten and give a reasonably smooth surface to the preparation.

Polymerization will begin almost immediately after adding catalyst to the monomere. In 15 minutes polymerization is quite noticeable, and the mixture will usually be hard within 2 hours. Polymerization can be hastened by placing slides under desk lamps to warm them. The overflow plastic should be removed when the plastic has undergone preliminary hardening into a firm gel. It can be flicked off easily and cleanly at this time. Later, when it is completely hardened, it is almost impossible to remove, and earlier attempts to remove it may result in a smeared mess. The cover square of Teflon should remain in place throughout a full day of polymerization, after which it can be removed.

The plastic mount will undergo some contraction during polymerization; as a result, some air spaces will usually form between the plastic ring and the plastic mount itself. This can largely be prevented, however, if special precautions are taken to evaporate all acetone from the monomere, and if special care is taken to mix the catalyst thoroughly with monomere in the mounting ring.

The above procedure will give a reasonably flat surface. However, if a glass cover is desired, it can be added after polymerization and contraction of plastic are complete, that is, about two or more days after catalysis. A cover glass should be coated with a very thin film of Silicone and wiped smooth, (Silicone serves to bind the cover to the plastic and prevent bubbles from appearing under the glass.) Add a drop of catalyzed plastic to the top of the preparation and add the coverslip in the usual way as in ordinary resin preparations. Allow polymerization to take place overnight before cleaning the cover glass. In the event that embedding plastics for electron microscopy and the equipment necessary for handling and polymerizing them are available, these can also be used for the preparation of embryo whole mount preparations of the type described above.

Label all slide preparations with information relating to type of material, developmental stage, stain, mountant, date, and name of preparator.

In concluding this chapter it will be pointed out that further consideration of the so-called mosaic stage in embryonic development will be continued without digression in Chapter 8 which follows.

CHAPTER VIII

THE PRINCIPLE OF CORRESPONDENCE OF EMBRYONIC STAGES

Illustrated with the 72 Hour Chick

TABLE OF CONTENTS

TABLES AND ILLUSTRATIONS

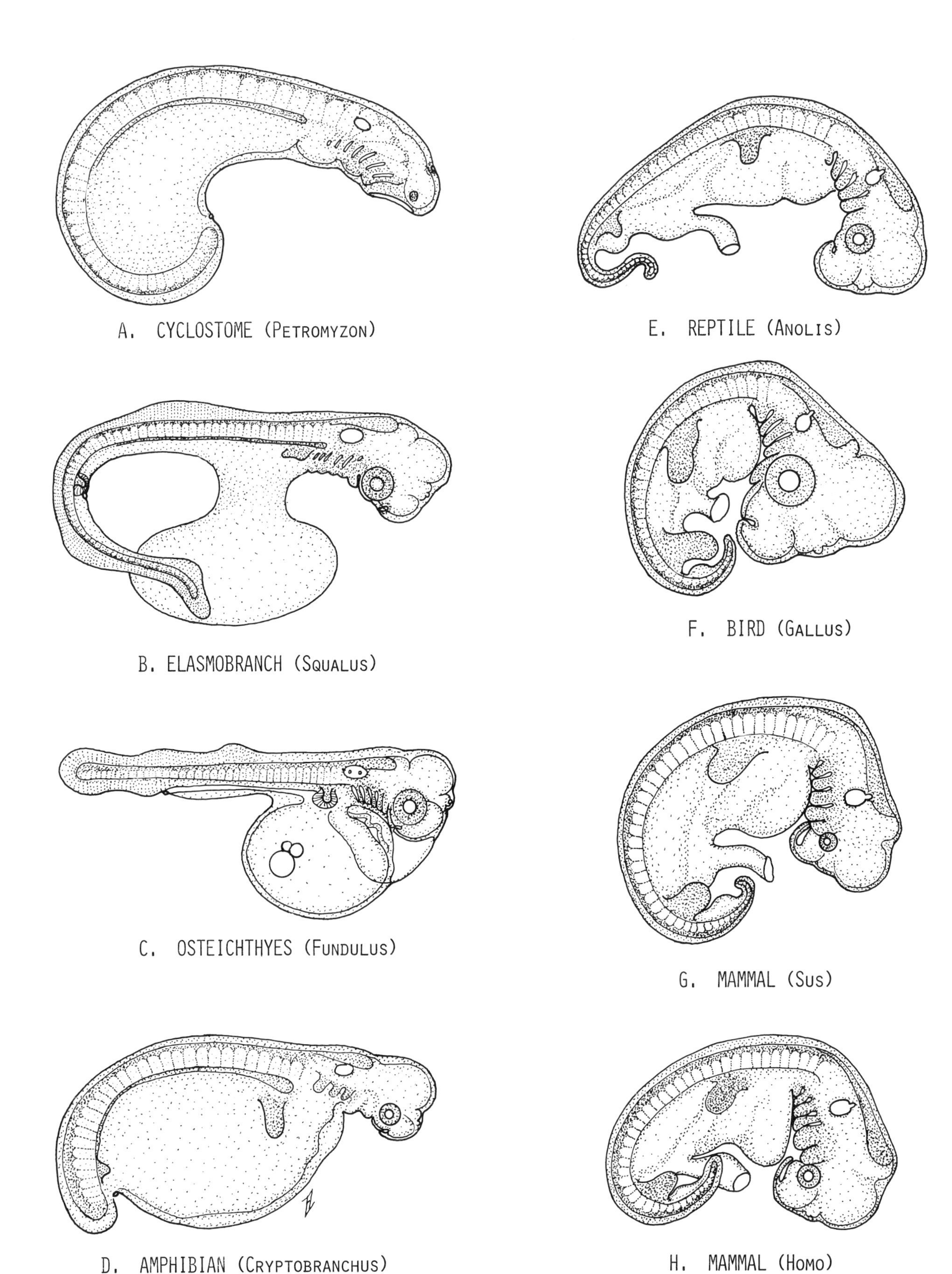

Figure 8-1. Embryos of representative vertebrate classes to illustrate von Baer's fourth law, the principle of correspondence of embryonic stages. See Table VIII-1 and explanation in text.

CHAPTER VIII

THE PRINCIPLE OF CORRESPONDENCE OF EMBRYONIC STAGES

Illustrated with the 72 Hour Chick

Shortly after the publication of Darwin's great book *The Origin of Species by Natural Selection,* in 1859, attention was directed toward the importance of embryology as a source of evolutionary evidence for relationships among species. The most dramatic interpretation of embryonic similarities was given by Ernst Haeckel's *Recapitulation Theory* (1867) which has been summarized in some detail in Chapter 5 along with the comparison of structures in the larval cyclostome and salamander. The central theme of the theory is captured in the phrase, **ontogeny recapitulates phylogeny;** that is, life histories (and particularly embryonic stages) summarize evolutionary histories. This theory, in the context of its time, attracted much support from evolutionists, and provided the major impetus that led to the description of embryology in virtually all major taxonomic groups before the end of the last century.

A brilliant theory sometimes serves most by stimulating research, even though the theory itself is unable to survive in the light of facts it has helped to engender. So it was with the fate of the recapitulation theory which is not highly regarded at present. The diametrically opposite theory to account for embryonic similarities had been proposed by von Baer in 1837 in the second volume of his pioneer studies in comparative embryology. The four major conclusions from this study are summarized in Table VIII-1; they are now known as von Baer's Laws, although he referred to them as the "fundamental corollaries" of embryonic development.

TABLE VIII-1
VON BAER'S LAWS OR COROLLARIES OF DEVELOPMENT

The following statements of von Baer's corollaries, are free-translations from his great study, *UeberEntwicklungesgeschichte der Thiere (On the Developmental History of Animals),* Part II, 1837, pages 219-231. The explanatory notes are this author's commentary and are not quotations from von Baer. The last "law" given below relating to germ layer derivations was stated by von Baer but was not included by him among his four corollaries. (See also Fig. 9-1 for an illustration of these laws.)

I. **LAW OF CONVERGENT DEVELOPMENT: Development proceeds from the general to the specific.** This means that in embryonic development general (phylum) characters of an animal group will appear before special (specific) ones. That is, there will be a developmental common plan expressed by all members of a phylum early in organogenesis.

II. **LAW OF INCREASING SPECIALIZATION: General characters gradually give rise to more special ones.** This means that phylum characters of an animal group will appear earlier in development than the distinguishing characters for classes. Class characters will in turn appear before ordinal traits. Order characters will appear before familial ones. Family characters appear before generic ones, Genus characters appear before specific ones. Species diagnostic characters will only become manifest at the end of development.

III. **LAW OF PROGESSIVE DIVERGENCE: Embryos of different species will become more and more different as development proceeds.** This means that the embryos of different species in a phylum or other taxonomic group will resemble each other most early in their development, but, as development proceeds, difference between them will gradually make them depart along increasingly more unique paths of development.

IV. **LAW OF CORRESPONDING STAGES: Embryos of (evolutionarily) advanced groups will pass through early stages that resemble embryos of (evolutionarily) primitive groups, but late embryos of advanced and primitive groups will not resemble one another closely.*** This means that for the recognition of homologies in development, comparisons must be made at corresponding stages in development; i.e., blastulae must be compared with blastulae, gastrulae with gastrulae, neurulae with neurulae, etc. (One should note that this corollary is in direct opposition to Haeckel's Recapitulation Theory which specifically makes comparisons of adults of lower species to embryonic stages of higher species.).

LAW OF UNIFORMITY IN GERM LAYER DERIVATIONS OF ADULT STRUCTURES. This means that in normal development similar or homologous structures in different species belonging to the same taxonomic group will be formed from the same embryonic germ layer or layers. The general truth of this law is overwhelmingly supported by the facts of descriptive embryology as summarized in Tables Int.-5, Int.-6, and Int. 7 respectively for the derivatives of ectoderm, mesoderm and endoderm in vertebrates. Exceptions do exist but they are more remarkable for their rarity than as refutations of the general truth of this great law which is probably the most useful generalization of normal development in all descriptive embryology. That experimental treatments can modify germ layer fate is totally irrelevant to this law since it deals only with the concept of prospective fate in normal development and not with developmental competence under experimental conditions.

---

* Von Baer did not himself accept the theory of organic evolution. This is the more surprising since embryological evidences are now held to be among the most convincing ones underlying principles of homology and evolution.

The principle generally known as von Baer's fourth law states that **homologies between species must be drawn from corresponding stages in the life cycle.** Followers of von Baer have usually chosen to refer to this principle as the **"biogenetic law."** However, the term "biogenetic law" in recent years has often mistakenly been applied to Haeckel's "Recapitulation Theory." It will be seen in Table VIII-1 that von Baer's fourth law and Haeckel's recapitulation theory are in diametric contradiction to one another. In order to bypass this unhappy state of confusion, I have chosen to drop entirely the much used and confused term, "biogenetic law." In its place I propose to use von Baer's own terminology and refer to the fourth law as the **Principle of corresponding embryonic stages.** In brief, this law, or principle, or corollary, maintains that valid homologies can only be made by comparing equivalent developmental stages among species. That is, comparisons should be made from blastula to blastula, or gastrula to gastrula, or tailbud to tailbud. However, homologies become progressively less certain when we compare early developmental stages with later ones, as is done in the recapitulation theory where the embryonic stages of one species are compared with adults of another. According to the principle of correspondence of embryonic stages, we can expect that the early embryos of evolutionarily advanced species in a phylum will resemble embryonic stages in primitive members of the phylum, but that the wider the evolutionary separation between the species, the greater will be the differences in later developmental stages. The following statements attempt to show the basic differences between the two theories that attempt to explain relationships between ontogeny and phylogeny:

The recapitulation theory of Haeckel maintains that the cause of differences in both life histories and evolutionary change is **the addition of adult stages to the life cycle** of ancestral forms so that former adult stages in primitive animals become larval or embryonic stages in evolutionarily advanced forms.

The correspondence of embryonic stages principle of von Baer suggests that the cause of phylogenetic or evolutionary change is due to the **insertion of changes into the life cycle** with the result that later stages will diverge from the ancestral plan.

Time does not permit a review of the relevant arguments supporting these two theories. Notwithstanding the dramatic appeal of the recapitulation theory, modern views of developmental genetics and evolution support von Baer. At present it seems most reasonable to explain evolution of embryonic differences by an extension of standard mutation theory. The simple provision is that some genetic effects act early rather than at the end, or adult stage, of the life cycle. By this it can be seen that genetic novelty is **inserted** into an existing life cycle. The earlier in development that a gene begins to exert its effects, the greater is the likelihood that it will have far-reaching effects of major importance on later development.

A. DEVELOPMENT OF THE AMNIOTE EGG ILLUSTRATED WITH THE EXTRAEMBRYONIC MEMBRANES OF BIRDS

Two classic examples of the insertion of developmental novelty into life cycles are found in the evolution of higher insects and vertebrates. Both of them evolved at about the same time at the end of the Paleozoic Era when the warm wet period in the Carboniferous came to an end and the worldwide climate became relatively cold and dry in the Permian Period. Holometabolous insects evolved the "pupa" stage as an adaptation for surviving by dormancy through times of cold and drying. Vertebrates met the same problems by the reptilian evolution of the shelled amniote egg in which the embryo is enclosed by membranes and a shell in its own small lake, the amniotic cavity. In both cases the imago and embryo respectively are entirely isolated from the exterior by special investing structures, and these stages can in no way be considered to represent ancestral conditions of adults. These caenomorphic adaptations must have been inserted into the life histories of more primitive members in their respective phyla that previously did not possess these characters.

The development of the reptilian amniote egg was one of the most important evolutionary advances by reptiles over amphibians. It freed the reptiles and their descendants, the mammals and birds, from the necessity of returning to water for the completion of the life cycle, as is the case for nearly all fish and amphibians. All descendants of reptiles have retained the amnion and amniotic cavity. Therefore, the reptiles, birds, and mammals are known as amniotes.

The origin of **somatopleure** and **splanchnopleure** as the primordia that give rise to extraembryonic membranes in amniotes was described very briefly in Chapter 7 (Fig. 7-3); they will now be examined in more detail.

The five extraembryonic membranes of reptiles and birds (**serosa** and **amnion** of somatopleure derivation, **yolk sac** and **allantois** of splanchnopleure origin, and the **chorion** formed by fusion of serosa and allantois) collectively are adaptations for water retention, respiration, nutrition, and excretion in the exceedingly confined space of a shelled egg. Graphic illustrations of the essential features of the reptile-bird egg are given in Figure 8-2. Details of the mammalian membranes and placentae are deferred to Chapter 9.

In birds the **amnion** and **serosa** are formed from extraembryonic **somatopleure** which folds over the embryo in such a manner as to enclose entirely the embryo by a double fold of somatopleure. The inner layer is the amnion and the outer is the serosa. The cavity between the embryo and amnion is the **amniotic cavity.** The cavity between the amnion and serosa is the **seroamniotic cavity,** which is a part of the extraembryonic coelom. Neither the amnion nor serosa is vascular; however, they develop smooth muscle. The contractions of the amnion are believed to prevent adhesions of the embryo to the extraembryonic membranes. The fluid of the amniotic cavity probably functions as a shock-absorbing buoyant chamber which reduces mechanical damage to the embryo within the egg. The serosa functions primarily in respiration and is closely applied to the inner surface of the shell membrane.

The **yolk sac** is formed from the extraembryonic splanchnopleure. It spreads radially over the yolk mass and eventually entirely encloses it. The yolk sac acts as a digestive and absorptive membrane during the development of the embryo. The **allantois** is homologous with the amphibian urinary bladder and develops as a median ventral evagination from the hindgut. It protrudes from the body of the embryo and expands until it almost entirely displaces the seroamniotic cavity and extraembryonic coelom. The allantois is highly vascular and fuses with the superficial serosa. The fusion forms the **chorion,** the primary respiratory membrane of the embryo. The allantoic cavity serves as a storage reservoir for excretory wastes produced by the embryo during its development.

In the following laboratory study, each pair of students will be issued embryos that have been incubated at 48 hour intervals over a period of from three to eight days, that is, they will be available as 3-4, 5-6, and 7-8 day embryos.

Materials needed for each pair of students. 1 liter warm 0.9% NaCl 4 4-inch fingerbowls, 1 wide-mouth waste jar, standard set of dissecting instruments, small wide-mouth preserving jar with fixative (i.e., either glycerin-formol, calcium-alcohol-formol-acetic, or calcium-picro-formol-acetic as described at the end of Chapter 7), 2 plastic embryo spoons, and 3 Petri dish halves.

Proceed to open the eggs as described in Chapter 6 and 7. However, **observe the following additional precautions.**

1. A number has been placed on the side of each egg which corresponds with its days of incubation. The position of the number also marks the position of the embryo in the egg. **Do not turn the eggs in handling them.** This will keep the embryo under the mark.
2. Crack the egg **opposite** the position marked on the shell.
3. Submerge the egg in the usual manner and **open the shell carefully and slowly from the bottom** so as to allow the extraembryonic membranes sufficient time to separate from the inner surface of the shell membrane and slide out into the dish.
4. Special care should be taken to avoid injury to the extraembryonic membranes and to assure the correct upright orientation of the embryo as it emerges from the shell; it may no longer be able to reorient itself by gravity alone.
5. If the embryo is not properly oriented, try to move it back into a position on top of the yolk by use of a plastic embryo spoon and dull forceps. **Do not use any sharp instruments** because if membranes are torn, extensive hemorrhaging will occur resulting in an early death of the embryo as well as a loss of definition in all extraembryonic vessels.

Open all stages that are available and compare the levels of development in the embryos. Most details can readily be determined with the naked eye, or with a low power hand lens. Figure 8-3 A–I can be used to assist in identifying the following features in membrane and embryo formation. Use models and charts as additional information on membrane structure.

In the event that living materials are not available, a film of extraembryonic membrane formation and function in the bird will be shown.

1. Living 3 to 4 day chick embryos (Fig. 8-3 A–D). Note the general shape of the embryo and the appearance and distribution of the area vasculosa on the yolk sac. In the embryo itself, identify the **brain, sense organs, limb buds, tail, somites, yolk stalk,** and **heart.** Observe, in particular, the beating of the heart and the direction blood flows in the embryo and yolk sac. The heart at this time is a simple twisted tube Note that the beat originates in the posteriodorsal (sinus venosus) region and continues in a wave anteriorly. Observe the large vitelline arteries and veins which leave the embryo and branch out over the **yolk sac.** Try to trace the capillary connection between the **vitelline arteries,** which carry blood from the embryo, and the **vitelline veins,** which return the blood to the embryo. Note the **sinus terminalis,** a circular blood vessel that collects blood along the peripheral border of the yolk sac. At the posterior end of the embryo, note the small vascularized sac which projects out of the embryo posteriorly. This is the **allantois,** which has just begun to expand. Note that the embryo is enclosed in an almost completely transparent membrane lacking blood vessels; this is the **amnion.** The entire blastoderm is covered by the non-vascular **serosa.** The seroamniotic raphe still connects the serosa and amnion and marks the point at which the amniotic folds have closed. This structure is generally not visible in living preparations.

Make a quick topographic sketch of the entire embryo and its extraembryonic membranes. Show any details that you are able to distinguish.

2. Living 5 to 6 day chick embryos (Fig. 8-3 E, F). Most embryonic and extraembryonic structures and regions which were observed in the earlier stage can be identified in 5-6 day embryos. Note, in particular, the increased area of the yolk mass now covered by the **yolk sac** and the great expansion of the **allantois** into the extraembryonic coelom. The latter now appears as a distinct balloon-like vesicle at the posterior end of the embryo. Observe the allantoic or umbilical arteries and veins running through the allantoic stalk. Watch for periodic contractions of the amnion which rock the embryo suspended in the amniotic cavity. In the embryo, note the greater curvature of the major body axis and the relatively great increase in the size of the head and eyes. Identify the structures shown in Figure 8-3 E, F and pay particular attention to the developing **gill slits** and **limb buds.**

Make a quick topographic sketch of the entire embryo within its embryonic membranes. Add any special details that you are able to distinguish.

3. Living 7 to 8 day chick embryos (Fig. 8-3 F–G). By this time, the embryo is approaching the definitive condition with respect to extraembryonic membrane formation. Figure 8-3 H is a diagrammatic presentation of the arrangement of extraembryonic membranes as they might appear in an optical section of the embryo within the shell. The major feature of late membrane formation is associated with the **allantois** which fuses peripherally with the **serosa** to form the **chorion.** The latter is strongly vascularized by vessels contributed from the allantois. The chorion will be applied to the interior of the shell membrane and will serve as the "lung" or respiratory membrane for the developing embryo.

After completing the observations on living embryos within their extraembryonic membranes, use a pair of fine forceps to tear the chorion so as to expose the embryo within it. Start with the youngest stages and progress to the older ones.

Observe the transparent **amniotic membrane** that surrounds the embryo and the **amniotic cavity** that is contained within the membrane. Note the **umbilical cord** or **body stalk** through which pass the allantoic stalk, yolk stalk, and the great veins and arteries on their ways to the **allantoic** and **yolk sac membranes,** respectively.

4. Study of embryos separated from their extraembryonic membranes (Fig. 8-3 C–H). With sharp scissors cut the umbilical cord and free the embryos at all stages from their extraembryonic membranes. With an embryo spoon, transfer the embryos to a Petri dish of clean warm saline. Observe the superficial structures shown in Figure 8-3 B–H. Note the striking amount of morphogenetic change that takes place in the 48 hours of incubation that separate each stage available for study. Note the gradual emergence of distinctive avian characters that become very distinct by 7 to 8 days of incubation. The 3-4 day embryos give no evidence of their class characters which would distinguish them from reptiles or mammals at equivalent developmental stages (see Fig. 8-1). Note the transformation of the **first gill cleft** into the **external ear canal** and the gradual development of **digits** on the **limbs** from primitive paddle-like limb buds. In the oldest embryos, observe the appearance of black pigment in the outer layer of the eye cup and the development of distinct upper and lower jaws (bills). In the skin, **feather germs** appear as dermal papilla which mark the major feather tracts of the skin. Although not visible externally, the heart is now completely divided into right and left atria and ventricles. Most other internal organs have also achieved adult avian characters by this time. Functional differentiation of

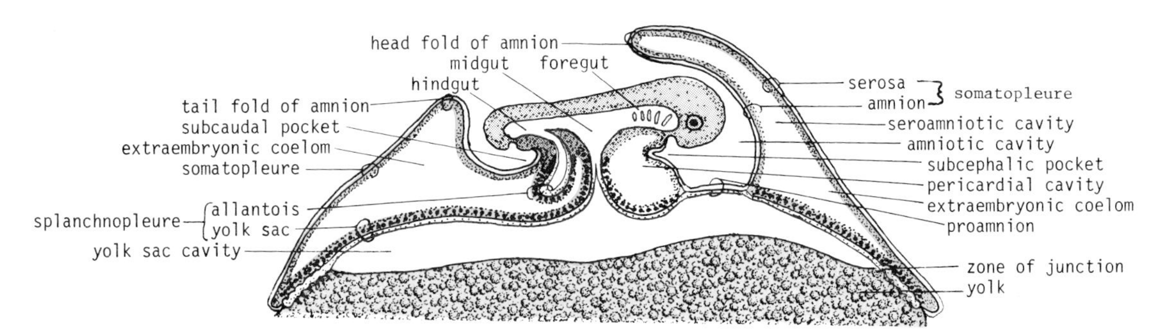

A. SCHEMATIC EXTRAEMBRYONIC MEMBRANES IN A 3 TO 4 DAY CHICK

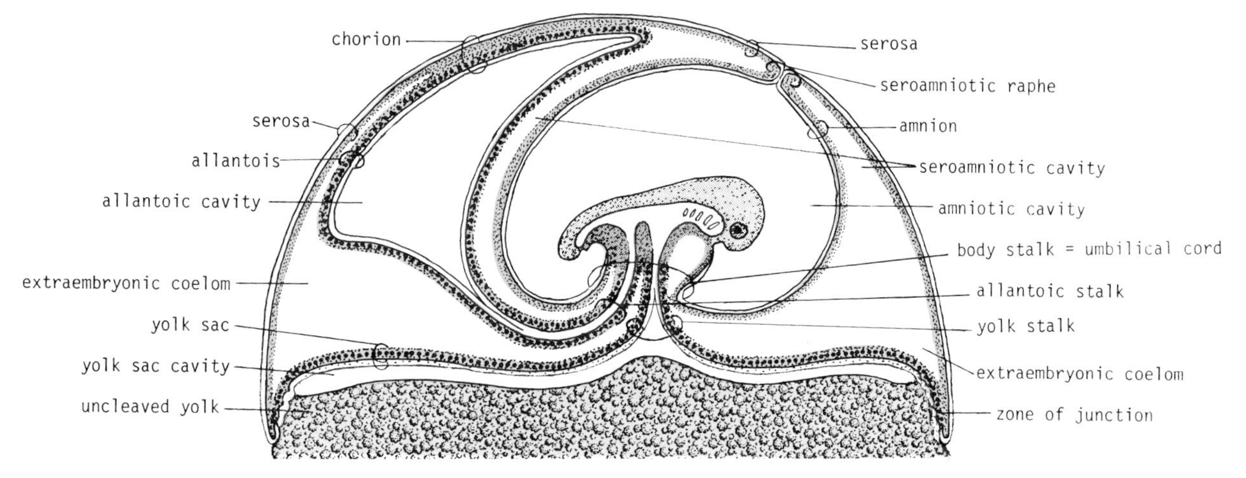

B. SCHEMATIC EXTRAEMBRYONIC MEMBRANES IN A 5 TO 7-DAY CHICK

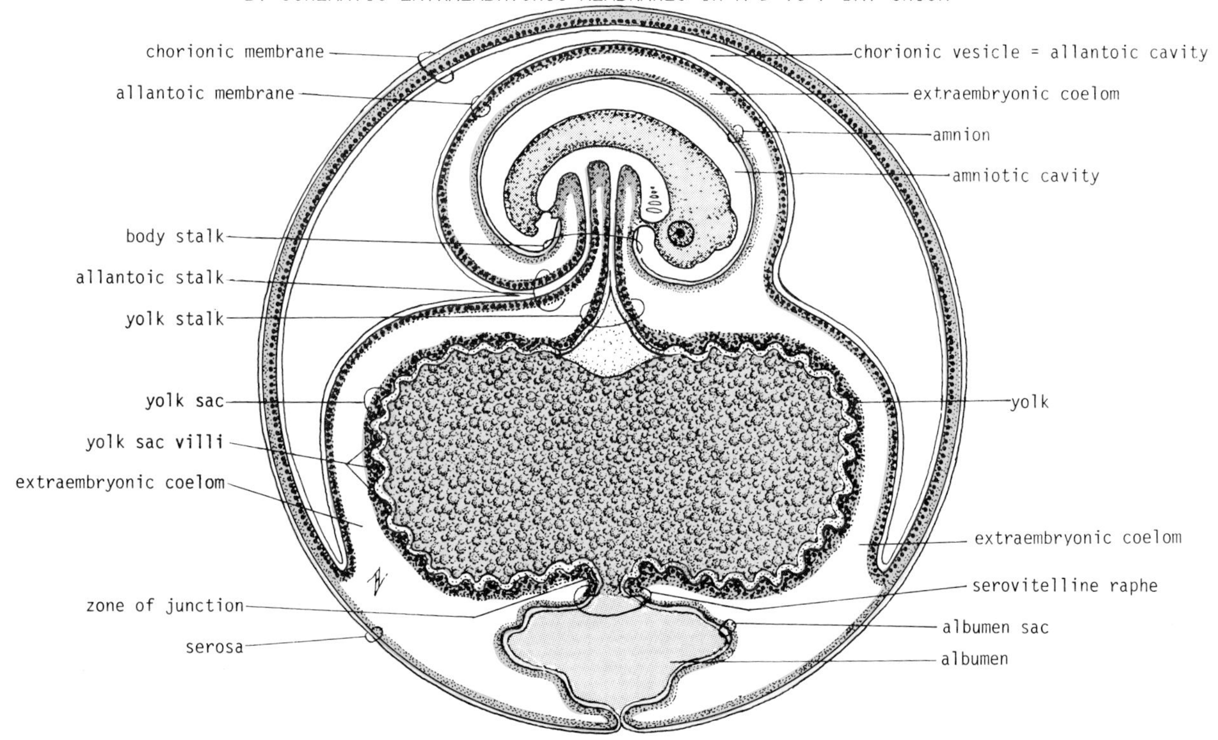

C. SCHEMATIC DEFINITIVE EXTRAEMBRYONIC MEMBRANES OF THE REPTILE--BIRD EGG

Figure 8-2. Diagrams of relationships of generalized reptile–bird extraembryonic membranes. See explanation in text. Color code: blue — ectoderm, red — mesoderm, yellow — endoderm, green — uncleaved yolk.

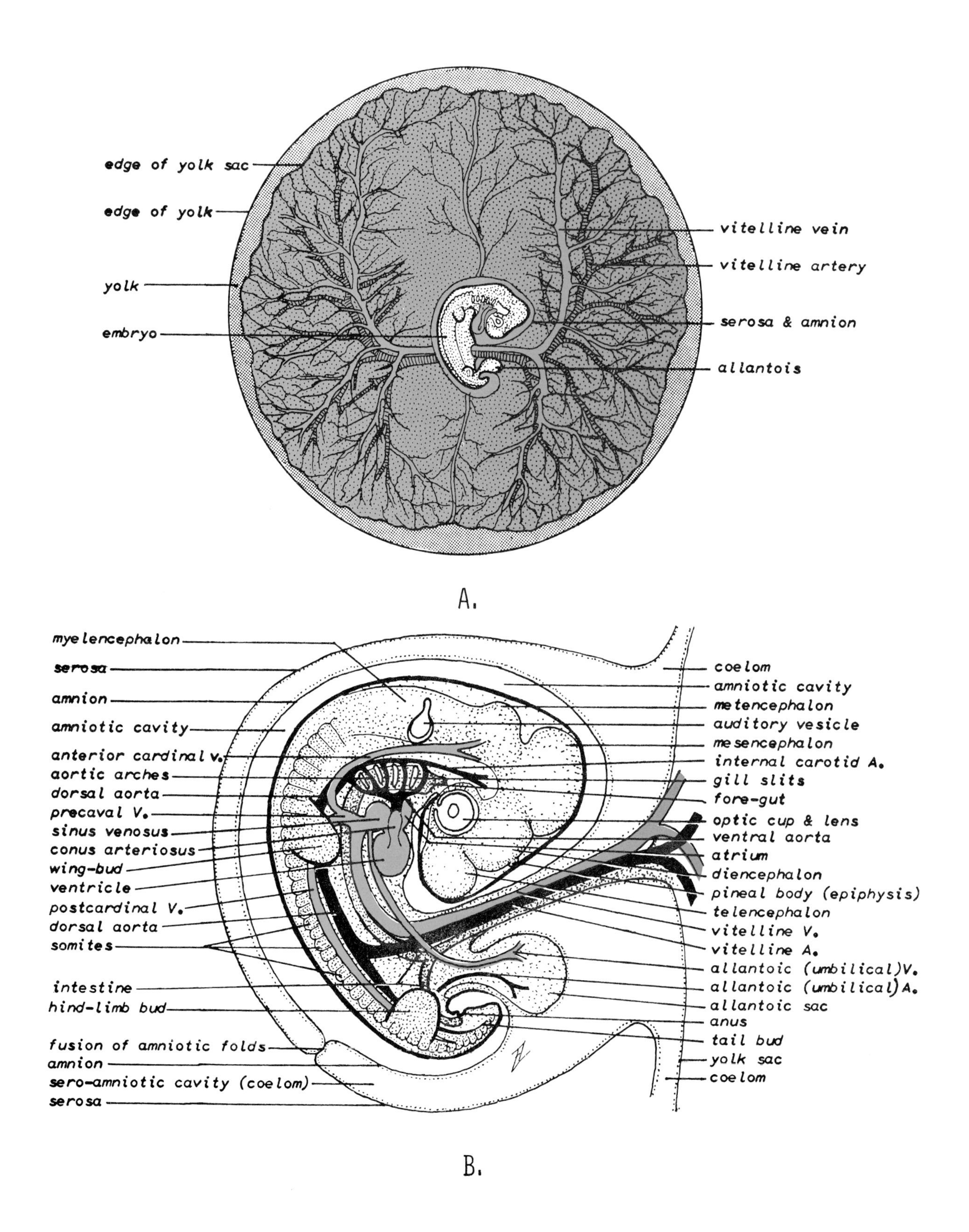

Figure 8-3. Diagrams of late chick embryos and their extraembryonic membranes. 8-3 A and B. 4 day chick embryos. (From H.E. Lehman, *Laboratory Studies in General Zoology,* Hunter Publishing Co., 1981).

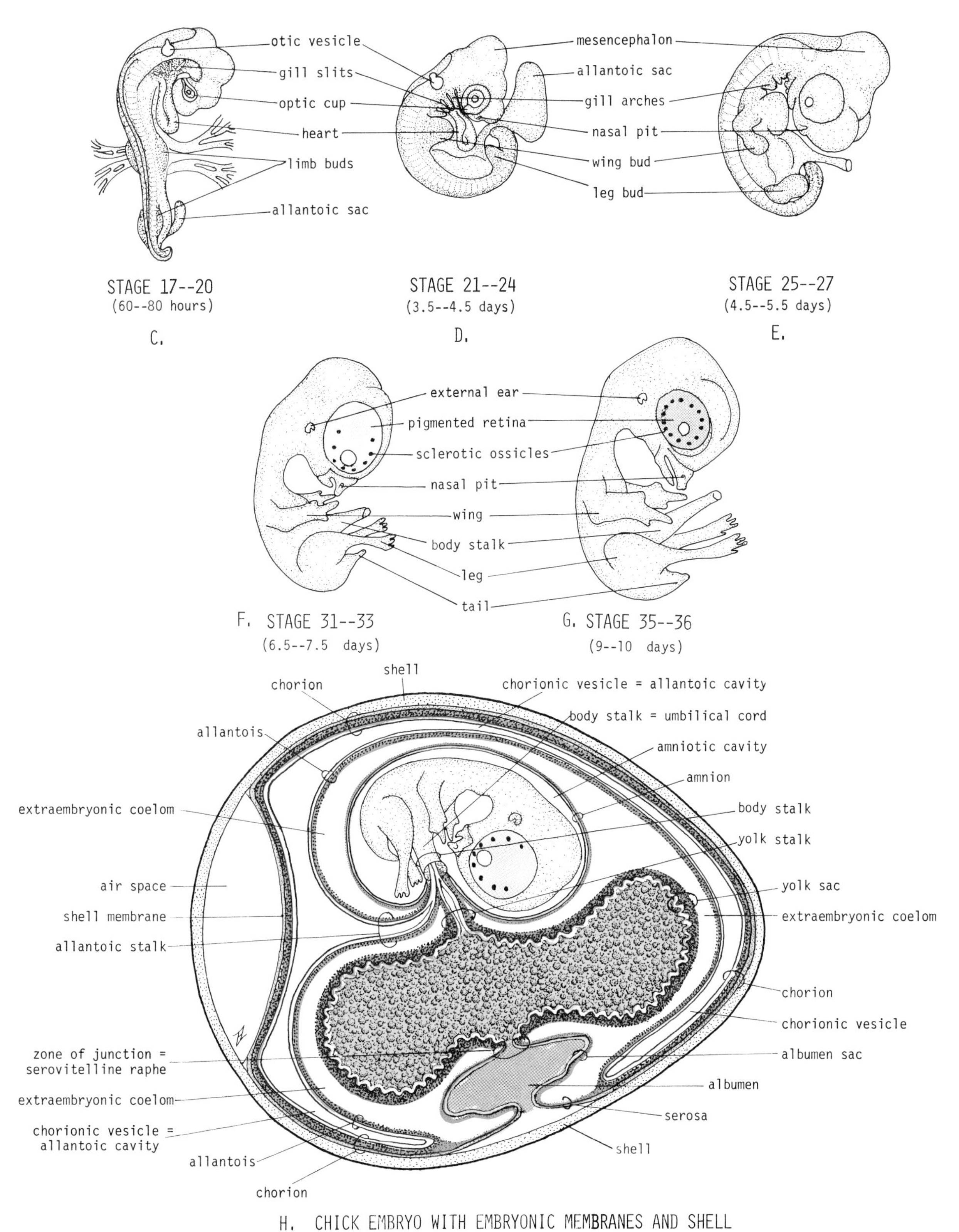

Figure 8-3. Diagrams of late chick embryos and their extraembryonic membranes, continued. See also Figures 8-2 and 6-1. Color code: blue — ectoderm, red — mesoderm, yellow — endoderm, green — uncleaved yolk.

8-3C 3-day embryo, Stage 17 — 20	8-3F 6-day embryo, Stage 31 — 33
8-3C 4-day embryo, Stage 21 — 24	8-3G 7-day embryo, Stage 35 — 36
8-3C 5-day embryo, Stage 25 — 27	8-3H embryo in its membranes

These stages are adapted from photographs in Hamilton, *Lillie's Development of the Chick.* H. Holt and Co. 1952.

muscles has advanced sufficiently for spontaneous body movements which occur as slow dorsoventral bendings of the trunk and neck. The limbs can also move independently to a slight degree.

Make quick topographic sketches of embryos for comparison with earlier stages shown in diagrammatic form elsewhere.

After completion of observations on living embryos, **add fixative** to the Petri dishes and observe the fixation process, which will reveal details of surface structures with exceptional clarity unmatched either in the living or in the completely fixed tissues. In particular, note the progressive steps in fixation of the sense organs, gill slits, limb buds, and feather germ patterns. CaAFA is recommended as the fixative (See Table VII-2).

Transfer embryos to wide-mouth jars for storage in fixative. They will be used for comparative study of late vertebrate stages in Chapter 10. Label the jar with name, embryonic stages, fixative, laboratory section, and desk number.

B. THE BASIC VERTEBRATE DEVELOPMENTAL PLAN ILLUSTRATED BY THE 72 HOUR CHICK EMBRYO

The fundamental plan of vertebrate development can be shown by any representative embryo at the tailbud stage, such as those shown in Figure 7-1. For reasons of tradition and ready accessibility, the 72 hour chick tailbud stage is selected as the "type" of the amniotes. It is significant that directions given for the chick could serve almost equally well for the study of the 6 mm human embryo, or for the comparable stage in reptilian development, so similar is the basic organization and manner of organ formation in all parts of the body.

In the study of the 3 day chick embryo, attention will be given to the progressive differentiation of structures identified in the 48 hour embryo. In addition, the appearance of structures which have become morphologically distinct during the intervening day of incubation will be stressed. The individual preparations for study can be expected to vary from 68 to 80 hours of incubation (i.e., 33-39 somites). Accordingly, some structures which are just making their appearance at this time may not be clearly distinct if the individual specimen studied is relatively young. Before beginning the study of the 72 hour embryo, review the morphological development of the 48 hour embryo and compare the whole mount preparations at these two stages. Note the differences in proportions of head structures and changes in flexures of the body axis. The 48 hour embryo is the point of departure for the study of the 72 hour embryo and a complete understanding of the earlier stage is prerequisite to an understanding of the latter.

1. General topography of the 72 hour whole mount embryo (Fig. 8-7 A). Identify the primary flexures in the anteroposterior axis of the body. The **cephalic flexure** was fully formed at the 48 hour stage. It bends the anterior brain region ventrally so that the diencephalon and telencephalon lie at a 140° angle to the rest of the brain. The **pontine flexure** has just begun to appear as a sharp construction near the anterior end of the metencephalon. The **cervical flexure** is now fully formed and has bent the entire head region of the embryo at a 90° angle to the trunk region. Owing to this flexure, the serial cross sections will pass through the hindbrain structures first and will section this part of the brain and the pharynx underlying it in the frontal plane. The **lumbosacral flexure** is located posterior to the hind-limb buds. This flexure bends the posterior part of the embryo anteroventrally. When all flexures have completely formed, the embryo will present the typical C-shape that is characteristic of vertebrate embryos generally. At this time only the open midgut region remains relatively flat on the yolk mass. Accordingly, the **thoracic** and **sacral torsions** are still present and represent lateral twisting of the body axis to permit the anterior and posterior ends of the embryo to lie on the left side.

Identify the **amniotic folds;** they now entirely enclose the embryo, except for a very small region at the level of the hind-limb buds where the tailfold of the amnion is progressing forward to meet the headfold of the amnion. The most conspicuous change in the embryo is in the posterior end where the **tailfold** is now distinctly formed and a rudimentary **allantoic evagination** of the hindgut will be seen protruding from the ventral floor of the hindgut. Identify also the paired **limb buds** which are just making their appearance as lateral thickenings at the level of the thoracic and sacral torsions.

With the aid of the whole mount diagram of the 72 hour embryo (Fig. 8-7 A), identify as many structures as possible. Many fine details of the circulatory system will be visible in injected whole mount embryos on demonstration.

2. Instructions for the study of serial cross sections and sagittal sections (Fig. 8-7 B–W, 8-8A–F). In the study of organogenesis in the 72 hour chick embryo, initial reference should be made to whole mount preparations. Then study the serial cross sections, and finally the serial sagittal sections. Each type of preparation will show different features to advantage and comparisons of all three aspects will increase comprehension of the relationships of the various organs.

As early as possible in the study of serial sections, determine the plane of section by comparing the presence of two or more known structural landmarks in a given section to their location as indicated in the whole mount diagram. Connect these landmarks with a straight edge; this line will give the plane at which the embryo was cut. Each set of slides will vary from the others in age or plane of section. No plane of section is inherently better than any other. When you understand the organization of the embryo, you will be able to interpret any series. If time permits, exchange serial sections with others in the laboratory. The greater range of material studied, the more complete will be your understanding of the structure and arrangement of organs within the embryo.

Ectodermal derivatives:

a. Spinal cord and brain. The diagram of a generalized vertebrate reflex arc is given in Figure 8-4 along with a table of functions in the so-called columns of the central nervous system. These will serve as a short review of the functional morphology of the spinal cord before beginning the study of development in the central nervous system. Begin the study of serial sections in the open midgut region as represented by sections at levels equivalent with Figure 8-7 F–G. Identıfy the thin dorsal **roof plate** and **ventral floor plate.** The lateral walls of the tube are much thicker and it is possible to distinguish two regions in them. The dorsolateral component of the neural tube is the **alar plate** or **sensory column** and the ventrolateral component is the **basal plate** or **motor column.** A groove in the inner wall of the neural tube called the **sulcus limitans** is a visible landmark that permits the boundary between these two functional regions to be identified.

Spinal cord. Begin at the open midgut region, at approximately the level of Figure 8-7 Q and proceed anteriorly. Trace the spinal cord from very simple to progressively more advanced levels of differentiation. However, at no level will the degree of regional specialization diagrammed in Figure 8-4 be achieved; this diagram is designed to indicate the significance of these structures, which in actural fact are only emerging as recognizable structural areas.

Myelencephalon (Fig. 8-7 B, C and 8-8 A). As a consequence of the cervical flexure, the myelencephalon is the first part of the brain encountered in cross sections. This part of the brain will be cut in a frontal plane. Sagittal sections are particularly favorable for showing the relationships of the thin-walled roof plate **(posterior choroid plexus)** to the 4th ventricle and brain floor of this region. The cross sections do, however, show the most distinctive feature of the myelencephalon, namely, the presence of **neuromeres** which appear as a series of from 5 to 7 lateral enlargements in its floor. Neuromeres are restricted to this brain region and can be taken as reliable landmarks for the myelencephalon.

Metencephalon (Fig. 8-7 B, C, and 8-8 B, C). This region in the 3 day embryo possesses clear landmarks which permit it to be distinguished from the adjacent brain regions. It differs from the myelencephalon by the absence of neuromeres, and is set off from the mesencephalon anterior by the initial formation of the **pontine flexure.** The latter is a deep indentation of the brain roof at the level of the posterior metencephalon. In addition, the anterior end of the metencephalon is strongly constricted in its lateral walls to form the **"isthmus,"** the narrowest and most constricted part of the brain stem.

Mesencephalon (Fig. 8-7 B–H). This region, owing to the cephalic flexure, is somewhat curved upon itself, as can be seen in whole mounts and sagittal sections. In most cross sections it will present a circular contour. The cavity of the mesencephalon is still very greatly expanded and forms the **aqueduct** which, in later development, will become greatly reduced and form a narrow connection between the third and fourth ventricles.

Diencephalon (Fig. 8-7 J–M). This brain region characteristically appears as a laterally compressed thick-walled portion of the neural tube which has the appearance of a "key hole" in cross section. The cavity of the diencephalon is the **third ventricle.** The wall of the diencephalon shows the following regional specializations. The **epiphysis** (Fig. 8-7 P), a bulb-like evagination from the roof of the diencephalon, is the primordium of the pineal gland. The ventral floor of the diencephalon forms a narrow trough, the walls of which make up the **infundibulum** (Fig. 8-7 I, J). The ventral floor of the infundibulum is in close contact with **Rathke's pocket** or stomodaeal hypophysis and, as mentioned earlier, the infundibulum will form the posterior lobe (pars nervosa) of the pituitary gland.

TABLE VIII-2
MAJOR FUNCTIONAL COLUMNS IN THE CENTRAL NERVOUS SYSTEM OF VERTEBRATES

GENERAL FUNCTIONS OF COLUMNS	SPECIAL FUNCTIONAL CATEGORIES
SSS = SPECIAL SOMATIC SENSORY	Senses of smell, vision, motion, equilibrium, hearing, & lateral line senses of water pressure and vibrations.
GSS = GENERAL SOMATIC SENSORY	Senses of temperature, touch, muscle fatigue & general proprioception relating to skeleton & skeletal muscle.
SVS = SPECIAL VISCERAL SENSORY	Senses of taste, thirst, and touch as related to derivatives of the pharynx, stomodaeum (mouth & nose) and proctodaeum (anal & external genital structures).
GVS = GENERAL VISCERAL SENSORY	Senses of hunger — satiation, visceral well-being or malaise, proprioception of glands, & visceral organs.
GVM = GENERAL VISCERAL MOTOR	Motor control of smooth muscles of gut, heart rate, glandular secretion of blood pressure in arteries.
SVM = SPECIAL VISCERAL MOTOR	Motor control of striated muscles of visceral arches to the jaws, tongue, larynx, neck, back, and middle ear.
GSM = GENERAL SOMATIC MOTOR	Motor control of striated muscles of myotome origin to limbs, neck, trunk, body wall, and tail.
SSM = SPECIAL SOMATIC MOTOR	Motor control of striated muscles of extrinsic muscles of the eye ball and occipital myotomes to the tongue.

Rathke's pouch forms the anterior and intermediate lobes. Anterior to the infundibulum in the ventral floor of the brain, at the boundary between the diencephalon and telencephalon, is a thin-walled, non-nervous region which marks the point of closure of the anterior neuropore (Fig. 8-7 O). This, the **lamina terminalis,** is most clearly seen in median sagittal sections. The lamina terminalis persists as a permanent landmark of the primitive anterior end of the brain marking the place at which the anterior neuropore closed. Last, a pair of lateral evaginations in the floor of the diencephalon called the **optic recesses** (Fig. 8-7 L) mark the attachment of the optic stalks to the brain wall. These are permanent landmarks which identify the points at which the optic vesicles originally were attached to the brain.

Telencephalon (Fig. 8-7 O–Q). At this time the telencephalon has just begun to enlarge. Therefore, there is more structural variation with age in this region than in any other part of the brain. The telencephalon is clearly distinguished from the diencephalon as a pair of lateral evaginations of the brain wall which project dorsolaterally from the anterior end of the neural tube. These will form the cerebral hemispheres.

b. Cranial nerves and neural crest. In addition to the 5th, 7-8th, and 9th cranial nerves which were identified in the 48 hour embryos, several more cranial nerves are present in the 3 day chick. Those present, listed in order from anterior to posterior, are as follows:

I. Olfactory (Fig. 8-7 O). This nerve is just beginning to differentiate. In favorable sections it can be seen as an ingrowth of fibers originating from the inner face of the nasal pit and running to the floor of the telencephalon. As a derivative of the olfactory epithelium, this nerve is of epidermal ectoderm origin.

II. Optic (Fig. 8-7 L). Optic nerve fibers have not yet begun to grow out from the inner layer of the optic cup. At an appreciably later time they will grow through the choroid fissure of the eye cup, will follow the optic stalk back to the brain floor into the optic chiasma, and will continue on around the side of the brain wall into the roof of the midbrain, or optic tectum.

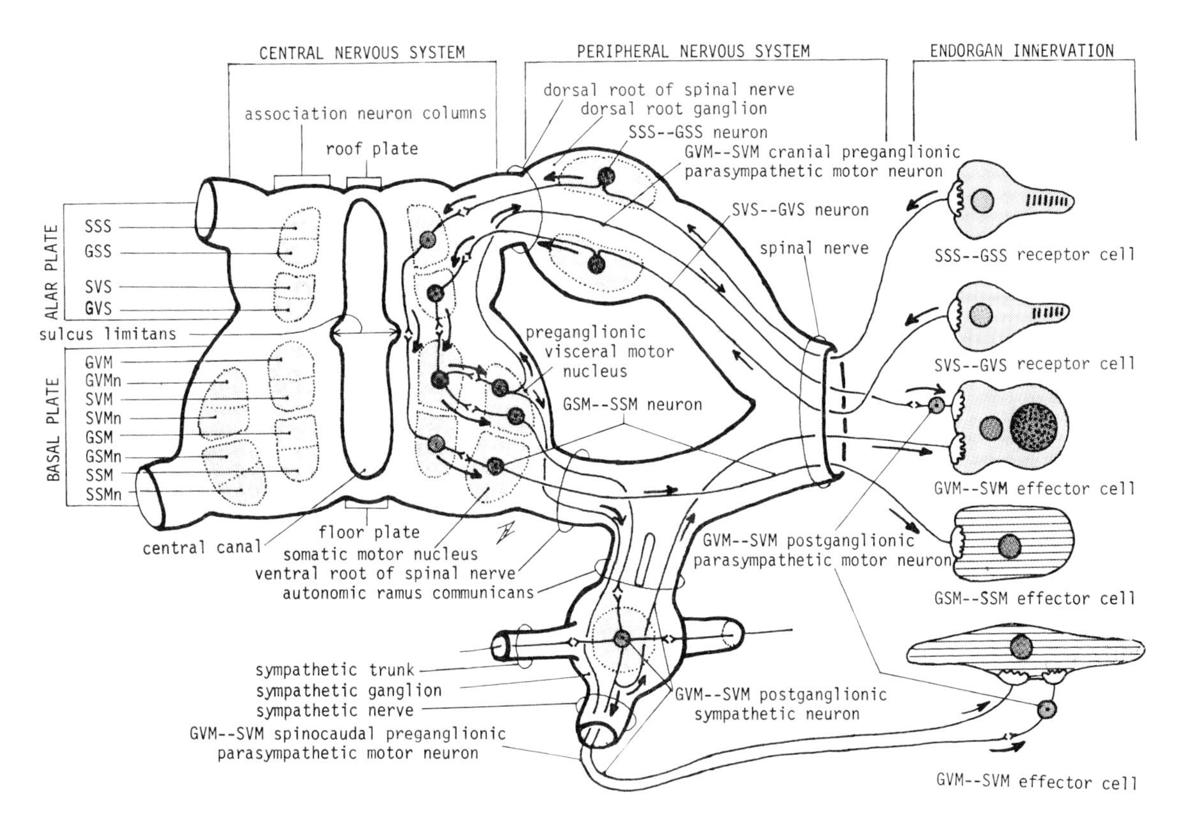

Figure 8-4. Diagram of general relationships of somatic and visceral reflex arcs in vertebrate central and peripheral nervous systems. Legend of abbreviations:

SSS	special somatic sensory	SVM	special visceral motor
GSS	general somatic sensory	SVMn	special visceral motor nucleus
SVS	special visceral sensory	GSM	general somatic motor
GVS	general visceral sensory	GSMn	general somatic motor nucleus
GVM	general visceral motor	SSM	special somatic motor
GVMn	general visceral motor nucleus	SSMn	special somatic motor nucleus

III. Oculomotor (Fig. 8-7 G). This pure somatic motor nerve arises from the ventral wall of the mesencephalon. It is directed antero-ventrally toward the eye where it will eventually innervate 4 of the 6 eye muscles (namely, the superior-, inferior-, and internal rectus, and inferior oblique muscles).

V. Trigeminus (Fig. 8-7 C–F). This nerve, situated at the level of the first neuromere of the myelencephalon, can serve as an anterior landmark of this brain region. Identify the **semilunar ganglion** of this nerve. Trace the course of this nerve ventrally and note that it gives rise to three branches. These branches are: (1) The **mandibular ramus,** a mixed somatic sensory-visceral motor nerve which supplies the lower jaw; (2) the **maxillary ramus,** a somatic sensory nerve to the upper jaw; (3) the **ophthalmic ramus,** or **profundus nerve,** a somatic sensory nerve to the nasal and oral derivatives of the stomodaeum.

VII. Facialis (Fig. 8-7 B–G). The geniculate ganglion of this nerve is located just anterior to the otic vesicle at the level of the third neuromere of the myelencephalon. It bears the same close relation to the 8th nerve as was seen in the 48 hour embryo. The facial nerve can be readily traced ventrally into the 2nd visceral arch.

VIII. Acoustic (Fig. 8-7 C). Although the **acoustic ganglion** of the 8th nerve is closely associated with that of the 7th nerve, a clear distinction between the two ganglionic masses is evident. The ganglion of the 8th nerve is applied directly to the anterior lateroventral wall of the otic vesicle from which it is derived and which it will also innervate after the semicircular canals have developed. The acoustic nerve is derived from the epidermal ectoderm of the otic vesicle.

IX. Glossopharyngeal (Fig. 8-7 C—F). This nerve is still quite small and is located posterior to the otic vesicle at the level of the 5-6th neuromeres of the myelencephalon. By tracing it ventrally, one can follow it for a short distance into the dorsal portion of the 3rd visceral arch.

X-XI. Vagus-Accessory (Fig. 8-7 D–G). This nerve complex arises from several condensations of ganglionic material which extend to the posterior end of the myelencephalon. The compound nerve can be followed along the anterior cardinal vein into the 4th visceral arch. Posterior branches of the vagus-accessory nerve into the regions of the 5th and 6th visceral arches are very poorly formed at this stage.

Neural crest, spinal, and autonomic ganglia. The primitive condition of the neural crest, as described for the 33 and 48 hour stages, can still be seen in the posterior third of the 72 hour embryo. In more anterior trunk and neck regions, the neural crest is represented by ganglionic aggregations of cells located segmentally between the dermomyotomes and neural tube. These are the primordia of the **spinal ganglia** (Fig. 8-7 F, G). Of special interest are the first 4 spinal ganglia at the level of the occipital somites. These ganglia will later degenerate, but they represent vestigial somatic sensory and visceral sensory components of the **IX accessory** (visceral motor) and **XII hypoglossal** (somatic motor) cranial nerves which are of spinal origin. The occipital ganglia degenerate in later development, and in mammals; the last pair is referred to as the **ganglia of Froriep** (Fig. 8-7 C–E).

Autonomic ganglia (Fig. 8-7 D–J). These are visceral motor derivatives of the neural crest. Early primordia of autonomic ganglia can occasionally be identified at the 72 hour stage. They will appear as condensations of mesenchyme near the dorsolateral margin of the dorsal aorta. These components of the peripheral nervous system will be much more clearly evident in the 10 mm pig embryo which will be described in Chapter 9.

c. Epidermal ectoderm derivatives and sense organs. The development of embryonic epidermal ectoderm has undergone only moderate amounts of change since the 48 hour stage. Note the following advances.

Nasal pit (Fig. 8-7 O). The nasal epithelium, which at the 48 hour stage formed a pair of differentiated placodes, has, by the 3rd day of incubation, formed a pair of clearly defined, thick-walled nasal pits. In favorable sections occasionally a few nerve processes are seen beginning to grow from the nasal epithelium toward the ventral floor of the telencephalon.

Eye (Fig. 8-K–M). The major changes in the development of the eye involve the complete separation of the lens pit from the overlying epidermis so that a completed **lens vesicle** is formed. The **choroid fissure** (Fig. 8-7 L) is a ventral cleft in the wall of the optic cup. Look for the development of optic nerve fibers from the sensory retina back through the choroid fissure and along the optic stalk. This particular feature will not be evident in the majority of specimens. Last the epidermis overlying the eye in the region of the lens vesicle can be designated as the earliest primordium of the **cornea** or conjunctiva.

Ear (Fig. 8-7 B–D). The otic vesicle, which will form the membranous labyrinth of the inner ear, is almost completely separated from the overlying epidermis. The connection is the **endolymphatic duct.** It will eventually lose its connection with the head epidermis, but will remain as a vestigial dorsal evagination from the inner ear vesicle. The relation of the otic vesicle to the 8th cranial nerve has already been indicated. No significant advance in the differentiation of the first visceral pouch into a middle ear has taken place since the 48 hour stage.

Stomodaeum and oral cavity (Fig. 8-7 F–1). The relationships of the stomodaeum, Rathke's pouch, pharynx, and jaws are not greatly modified from those observed in the 48 hour chick embryo. These relationships can be seen in median sagittal section (Fig. 8-8 A). It will be noted that **Rathke's pouch** projects as a blind evagination from the roof of the stomodaeum posterodorsally under the infundibulum (Fig. 8-7 H, I). The **oral membrane** which separates the stomodaeum from the pharynx has, in most instances, ruptured. However, parts of this membrane are usually present to mark the boundary between the ectodermally and endodermally lined parts of the oral cavity.

Proctodaeum and cloacal membrane (Fig. 8-7 W, 8-8 A). The formation of the anal opening into the hindgut involves an ectodermal invagination at the locus of the primitive streak and Hensen's node. Its formation is somewhat analogous to the formation of the stomodaeal plate in the frog (Fig. 4-3). At the posterior end of the embryo, the hindgut endoderm evaginates ventrally and becomes applied to a solid cord of invaginated epidermis. The fusion of ectoderm and hindgut endoderm is called the **cloacal membrane** or anal plate. In the 3 day chick embryo, the formation of the proctodaeum has just begun, and the cloacal membrane is still incompletely formed.

Endodermal derivatives (See Fig. 8-5).

a. Foregut derivatives. Few endodermal derivatives of the foregut, aside from the visceral pouches, can be seen in whole mount preparations. Accordingly, the following description is designed primarily to accompany study of the serial sections. Trace the foregut forward from the anterior intestinal portal.

Stomach, liver, and pancreas, (Fig. 8-7 M–Q). The **anterior intestinal portal,** at the 72 hour stage, is situated approximately at the boundary between the future stomach and duodenum. At this level, identify the solid proliferation of endoderm cells into the **dorsal mesentery of the gut.** This endodermal evagination is the primordium of the **dorsal pancreas.** Slightly anterior to the portal, note that a ventral evagination of the foregut endoderm can be found which projects into the thick walls of the vitelline veins. This is one of several **liver diverticula.** A few sections forward, a second liver diverticulum will be found situated dorsal to the fused vitelline veins. As stated in the discussion of the 48 hour embryo, the liver primordia will proliferate very extensively forming solid cords of cells that invade and eventually subdivide the main channel of the vitelline veins into a capillary network of liver sinusoids. The posterior part of the vitelline veins will form the hepatic portal vein. The anterior part of the vitelline veins will form the hepatic veins and the anterior end of the inferior vena cava. The expanded part of the foregut, just anterior to the region in which the liver diverticula are given off, is the primordium of the **stomach** (Fig 8-7 M—P). Note that the stomach is already beginning to expand toward the left in anticipation of the asymmetry in growth which will place the stomach predominantly on the left side of the body.

Esophagus, trachea, and lung buds (Fig. 8-7 J–L). At the 48 hour stage, no morphological landmarks were present to permit a clear distinction between these three regions. By the 72 hour stage, these primordia are sufficiently differentiated to permit the identification of each. Refer to the whole mount diagram and review the relationships of these parts. Then proceed anteriorly through the cross sections of the foregut. Anteriorly from the stomach, the lumen of the gut becomes greatly constricted. This narrow anterior continuation of the foregut is the *anlage* of the **esophagus.** Trace the esophagus anteriorly and observe that its ventral floor expands and gives off a deep evagination, the **laryngotracheal groove,** which is beginning to expand laterally as **lung buds** into the coelom. The wall of the lung buds will undergo enormous growth and branching and will form the bronchial tubes, bronchioles, and alveolar sacs of the definitive lung. Anterior to the origin of the lung buds, the esophagus, in cross section, has the appearance of a laterally compressed slit (Fig. 8-7 J). The dorsal, wide part of this slit is the anterior end of the esophagus, and the narrow ventral part is the **laryngotracheal groove** which will separate from the esophagus along most of its length and form the trachea and larynx. The esophagus and trachea share a common anterior opening into the greatly expanded lumen of the pharynx.

Pharynx and visceral pouches (Fig. 8-7 E–I). The pharynx is an expanded cavity possessing 4 or 5 pairs of visceral pouches which appear as paired lateral evaginations of the foregut wall. They make contact with the superficial epidermis of the corresponding visceral grooves. Each pouch carries the same number as the visceral arch immediately anterior to it. In some instances the cavity of the pharynx opens to the outside through the visceral pouch, but usually a **closing plate** consisting of a layer of visceral pouch endoderm and a layer of visceral groove epidermis prevents the formation of complete visceral clefts (i.e., visceral pouch + visceral groove). Owing to the cervical flexure, the anteroposterior axis of the pharynx is parallel to the plane of cross section. As a result, most series will possess a few sections in which all pharyngeal pouches and the stomodaeum will appear. Such sections are particularly favorable for the identification of the individual pouches in their linear sequence. It should be kept in mind that the first visceral arch (or the mandibular process of the arch at least) is bounded anteriorly by the stomodaeum; the latter should not be mistaken for the first visceral pouch.

The relationships between the pharynx, stomodaeum, and oral membrane were described earlier; these can be seen to best advantage in median sagittal sections (Fig. 8-7 A). Sagittal sections also show the **thyroid diverticulum** more clearly than do cross sections. The thyroid diverticulum is a spherical, mid-ventral evagination of pharyngeal endoderm at the level of the second visceral arch. This structure is one of the most useful landmarks for the identification of this particular region of the pharynx, and the ventral aorta, or aortic sac, which lies immediately ventral to it.

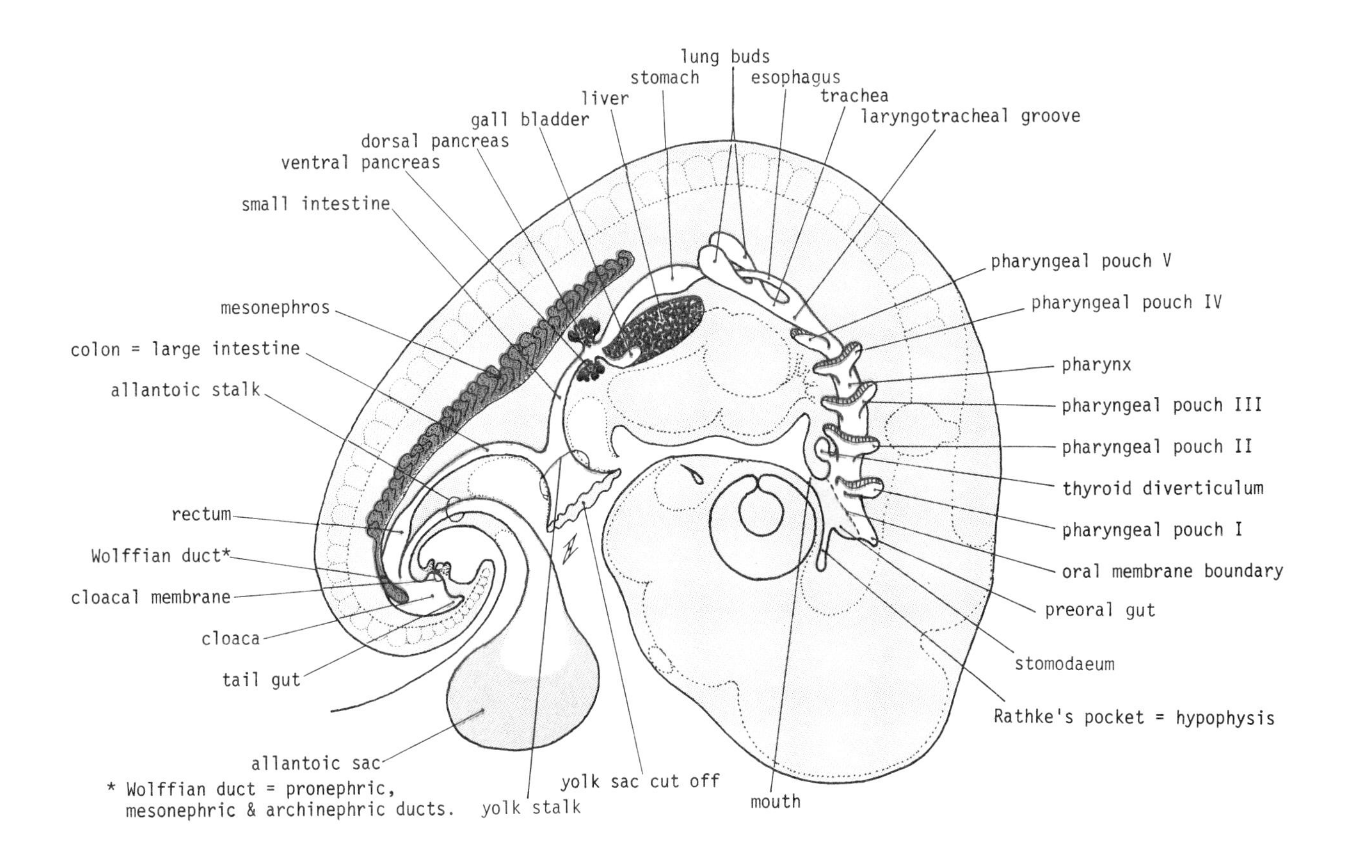

Figure 8-5. Diagram of primary derivatives of embryonic endoderm. The 4 day chick embryo showing associations of the gut with the mesonephros and stomodaeum. Color code: blue — stomodaeal and proctodaeal ectoderm, red — mesonephric mesoderm, yellow — endoderm.

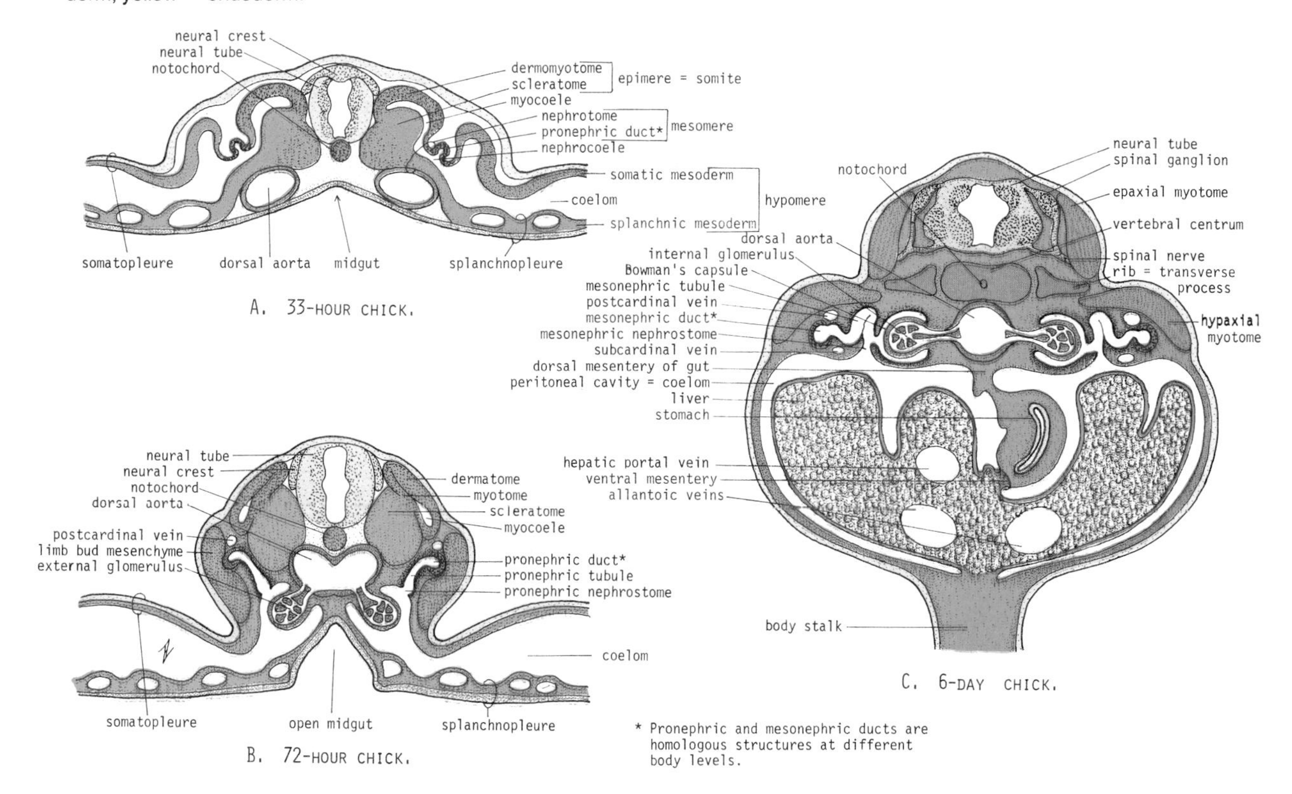

Figure 8-6. Pronephric and mesonephric vascular and coelomic relationships. 8-6 A. Primordial nephrotome and nephrocoele relationships. 8-6 B. Embryonic pronephric tubule general relationships. 8-6 C. Embryonic mesonephric tubule general relationships. Color code: blue — ectoderm, red — mesoderm, yellow — endoderm.

b. Midgut (Fig. 8-7R). The open midgut region of the 3 day embryo is restricted to the mid-region of the embryo from the approximate level of the 23rd to 30th somites. Anterior and posterior to these levels, the head and tail folds of the embryo have undercut the embryonal area and have provided an entire endodermal lining to the fore- and hindgut. There are no special features visible in the open midgut region that have not been discussed in previous sections. The only addition of significance is the presence of the well-formed **posterior intestinal portal** (Fig. 8-7 S), which clearly delimits the posterior extent of the region.

c. Hindgut derivatives (Fig. 8-7 S—W). The anterior end of the hindgut is marked by the **posterior intestinal portal.** It can be identified by the midventral fusion of the splanchnopleure layers from the two sides of the embryo which have advanced anteriorly approximately to the level of the 30th somite. Trace the closed hindgut posteriorly and identify the midventral evagination of the hindgut floor; it is the primordium of the **allantois** and **allantoic stalk.** Note that the endodermal evagination carries with it a layer of splanchnic mesoderm from the floor of the hindgut in which capillaries of the allantoic circulation can be seen. The allantoic stalk appears as a wide ventral evagination of the hindgut; the allantois tends to bend posterad and protrude into the seroamniotic cavity (extraembryonic coelom). In later development, the allantoic cavity becomes greatly expanded and will function as the embryonic urinary bladder. Posterior to the opening of the allantoic stalk, the hindgut is called the **cloaca.** This portion serves in adult birds as a common receptacle for digestive, excretory, and reproductive ducts. Note that the lateral walls of the cloaca are evaginated slightly and meet the paired **pronephric ducts** (also called mesonephric or Wolffian ducts), thus establishing the urogenital connection with the cloaca. The posterior end of the hindgut is continued for a short distance into the tailbud beyond the cloaca and **cloacal membrane.** This posterior end of the hindgut is the **postanal** or **tailgut.** It forms a special structure in birds called the **bursa of Fabricius** which for a time functions, as the mammalian thymus, as a "glandular" reservoir of primary cells for the immune system of the blood; that is, stem-cells for lymphocytes and other white cells of the blood are located in this organ before the spleen and other lymphoid tissues are formed. It will later be resorbed into the wall of the cloaca and has no special structural significance in the adult bird.

Mesodermal derivatives

Almost all comments relative to the appearance of the **notochord, somite, mesomere** and **lateral plate mesoderms** that were given for the 48 hour chick embryo can be applied directly to the 72 hour chick. The following additions to the preceding account should also be noted.

a. Somite derivatives. Throughout most of the body length posterior to the hindbrain, **dermomyotome** and **sclerotome derivatives** of the somites can be identified. In anterior regions, the muscle rudiments can be seen growing ventrolaterally under the skin. The sclerotomes can be identified as aggregations of mesenchyme around the notochord in anticipation of the formation of vertebral centra. Of particular interest are the first 4 somites since their development fate is quite different from that of the remaining segmental mesoderm. These are the **occipital somites** (Fig. 8-7 C–E). Occipital dermomyotomes migrate into the tongue and form the hypobranchial muscles which are innervated by the 12th cranial nerve. Occipital sclerotome is also unique in contributing to the posterior chondrocranium, namely, to the formation of the basi-and exoccipital bones.

b. Nephrotome or mesomere derivatives (Fig. 8-7 Q–V). The primary developmental advances of the nephrotome concern the posterior migration of the **pronephric ducts** (also, mesonephric or Wolfian ducts), and their fusion with the lateral walls of the cloaca (Fig. 8-7 V). Along the course of the pronephric duct, at the levels of the 16th through 30th somites, the mesonephric mesomere has become organized into discrete tubules with cavities that open into the pronephric duct (Figs. 8-7 F,G and 8-8 A, C). These segmental tubules are the mesonephric tubules. They are better developed in the more anterior segments. They can be seen particularly well in parasagittal sections through the mesonephros. Of particular interest are the **nephrostomes** whereby the mesonephric tubles communicate with the coelom (Fig 8-7 Q, P). A general consideration of the developmental and evolutionary fate of the nephrostome is provided in Table IX-7 in the next chapter. Here it will suffice to say that the nephric development is at an exceedingly primitive level. It corresponds structurally with the pronephric kidneys of lower forms such as the cyclostome Ammocetes and amphibian larvae (see Figs 1-6 and 5-7) in which body excretory wastes are first excreted into the coelom and then the coelomic fluid is taken directly into the nephric tubules by the ciliary action of nephrostomes. The coelomic fluid is then processed for excretory waste extraction by the kidney tubules. It should be emphasized that in the bird and mammal, although pronephric and/or mesonephric nephrostomes will form on some kidney tubules, they are truly vestigial and never function in the excretory process. They are thus paleomorphic characters that reveal a "racial memory" of ancient evolutionary steps in the development of the vertebrate excretory system. Figure 8-6 diagrams the structural relationships of vertebrate nephric tubules and shows their relations to the coelom and vascular system.

c. Lateral plate of hypomere mesoderm (Figs. 8-2, 8-3, 8-7, 8-8). The lateral plate contributions to the extraembryonic membranes remain essentially unchanged as compared with those seen in the 48 hour chick. However, note that the **tail fold of the amnion** has made its appearance and that, except for a small area in the region of the hind limb buds, the embryo is almost completely enclosed within the serosa and amnion. The splanchnopleure of the yolk sac merits no special attention. However, it is greatly thickened around the **allantoic stalk** and in the latter the beginnings of vascularization can be seen (Fig. 8-7 U, V).

Intraembryonic hypomere. Both the **fore-** and **hind limb buds** are massive enlargements of somatic lateral mesoderm. They will form the skeleton and intrinsic muscles of the limbs (Fig. 8-7 R–U). Intraembryonic splanchnic mesoderm should be noted in the region of the stomach. Here it can be seen to form the dorsal and ventral mesenteries of the gut (Fig. 8-7 N, O). The latter is continuous anteriorly with the **dorsal mesocardium** which supports the atrium and sinus venosus of the heart. The thick mesodermal wall surrounding the stomach endoderm is splanchnic and is destined to form the muscular layers of the gut as well as the visceral peritoneum.

Intraembryonic coelom and coelomic septa. Relationships and the manner of formation of mammalian coelomic cavities will be given special attention in Chapter 9. For the present it will be sufficient to indicate that the embryonic coelom of mammals is divided into four cavities, namely, 1 pericardial, 2 pleural, and 1 peritoneal cavity, whereas in birds there will be only 2 definitive coelomic cavities: i.e., 1 pericardial and 1 pleuroperitoneal cavity. These subdivisions are continuous in the three day chick embryo, but they are nevertheless sufficiently set off from one another to permit them to be identified with certainty. Primordia of the septa which will eventually divide these cavities are in the process of forming. These septa are: (1) The **transverse septum** will form from a fusion between the **ventral mesentery** and the right and left **lateral mesocardia** (Fig. 8-7 N). The transverse septum when fully developed will contain the liver and will form a ventrolateral wall between the **pericardial cavity** (below) and **peritoneal cavity** (above). A second septum which has begun to form in the 72 hour embryo is the **pleuroperitoneal septum.** As the name implies, it separates the **pleural cavity** from the general **peritoneal cavity.** Pleuroperitoneal septa can be seen in the sections shown in Figures 8-7 M, N. The lung buds have not, as yet, grown back into these small cavities which anticipate the formation of the **pleural coeloms.** Remember that the pleuroperitoneal septa never completely separate the pleural and peritoneal cavity in birds. However, in mammals, these septa will contribute, along with the transverse septum and lateral mesocardia, to the formation of the mammalian diaphragm which will be given special consideration in Chapter 9.

d. Circulatory system. Review the general circulatory paths in injected whole mount embryos and whole mount diagram (Fig. 8-7 A) and circulatory common plan (Fig. 7-5). In the study of these preparations it is strongly recommended that you trace the various arterial and venous pathways in an orderly manner through the embryo, comparing the 72 hour injected embryos with the whole mount diagram. Note that the **intersegmental branches** of the dorsal aorta have been omitted. The diagram also presents the allantoic circulation as consisting of discrete arteries and veins, whereas, in fact, injected whole mount specimens show that it consists of a diffuse capillary bed with no clearly defined, main arterial and venous channels. Trace the remaining structures of the circulatory system in the following order.

Heart (Fig. 8-7 J–G). The landmarks for the identifications of the heart chambers are essentially as described for the 48 hour chick. However, note that the **atrium** (Fig. 8-7 L, M) has begun to expand toward the left in anticipation of the division of this region into right and left atrial chambers. Note that the communication between the **sinus venosus** and atrium is through the right side of the atrium. This relationship will persist throughout development, and the right atrium will continue to receive all systemic blood returning to the heart. The left side will develop new venous connections and receive blood from the pulmonary veins after the lungs develop.

Aortic and visceral arches (Fig. 8-7 E–I). Four well developed aortic and visceral arches will be found in the 3 day embryo. The first aortic arch will soon degenerate, and in older specimens, may be reduced in size. Each aortic arch arises from the **aortic sac (ventral aorta)** and passes around the lateral wall of the pharynx. They then unite to form the **dorsal aortae.** The latter are paired anterior to the 3rd aortic arch, but fuse to form a single vessel posterior to the 3rd aortic arch. The plane of section, owing to the cervical flexure, is usually in the frontal plane through the pharynx and visceral arches. Accordingly there are usually several sections in which all, or most, aortic arches can be seen in a single section. The following landmarks can be used as reliable identifications of the specific aortic and visceral arches: **(1) Cranial nerves** 5, 7, 9 and 10 innervate, respectively, the 1st, 2nd, 3rd, and 4th-5th-6th visceral arches. **(2)** The **thyroid diverticulum,** in the mid-ventral floor of the pharynx, marks the level of the 2nd visceral arch. **(3) Pharyngeal pouches** and grooves 1, 2, 3, and 4 are located posterior to the aortic and visceral arches of corresponding number. If any one of these landmarks is known, the remaining structures can be determined with certainty.

Intraembryonic circulation. The circulatory plan of the 72 hour embryo differs from that of the 48 hour chick primarily in having more clearly defined blood channels than were previously noted. However, a few new vessels are also present either as distinct channels, or as incipient capillary beds.

Anterior to the sinus venosus, the primary vascular plan involves the transport of blood anteriorly from the dorsal aorta through the paired **internal carotid arteries.** They are branches of the first aortic arches. They break up into the extensive capillary bed on the surface of the fore- and midbrain. This blood is carried back to the heart by way of the **anterior cardinal veins** and **common cardinal veins** along the same course as seen in the 48 hour embryo. A new, unpaired, median artery is also beginning to form immediately ventral to the hindbrain floor; this is the **basilar artery.** This vessel is not functional at this time. Its relationships to the internal carotid and vertebral arteries will be discussed in connection with the description of the 10 mm pig embryo. Last, note, in particular, the paired right and left **external carotid arteries** which are beginning to appear as anteroventral outgrowths from the aortic sac. They enter the mandibular processes of the first visceral arches. These vessels will ultimately replace the first aortic arches as the primary vascular source for the lower jaw.

Posterior to the level of the sinus venosus, the **intersegmental arteries** are given off dorsally between successive pairs of somites. They provide the primary source of blood for organs of the trunk and tail. The return flow of blood is accomplished by the **posterior cardinal veins** as has been described for the 48 hour embryo. In addition, the following vessels are beginning to take form from the general capillary beds of the trunk and tail. The **subcardinal veins** are secondary venous channels that parallel the postcardinal veins and lie mesioventral to the mesonephric tubules. They will later replace the posterior cardinal veins as the major veins of the functional mesonephroi. At the level of the forelimb buds, one pair of intersegmental arteries will become converted into the **subclavian arteries** to the wings. However, at the 72 hour stage there is no clear indication of the main arterial channel within the capillary bed of the limb bud. The same situation applies to the hind limb bud. The **femoral artery** will form from one of the intersegmental arteries in the sacral region. The paired **allantoic arteries,** although not clearly formed at this time, will arise as anteroventral outgrowths from the paired femoral arteries. (Allantoic arteries are homologous with the hypogastric arteries which supply the urinary bladder, genitalia, and groin of the adult). These vessels will all be clearly differentiated in the 10 mm pig embryo and, if not seen now, will be found in the more advanced embryo.

Extraembryonic circulation. In the 3 day chick embryo the extraembryonic circulation is primarily vitelline. The only modification of the vitelline system concerns the manner in which the **vitelline veins** enter the **sinus venosus** through the newly formed **ventral mesentery of the stomach.** In the younger stages previously studied, the right and left vitelline veins entered the sinus venosus by separate lateral openings. Note in injected specimens that the two vitelline veins of the 72 hour embryo enter the sinus venosus posteriorly by a common opening. This change is accomplished by the posterior regression of the head fold

Legend and checklist of structures to be identified in Figures 8-7 and 8-8.

Color code: blue — ectoderm, red — mesoderm, yellow — endoderm.

Cranial nerves and ganglia:

1	olfactory (sss)*
3	oculomotor (ssm)*
5	trigeminus (ss & vm)*
5G	semilunar ganglion of trigeminus
5GP	placodal ganglion of trigeminus
5M	mandibular ramus of trigeminus
5O	opthalmic ramus of trigeminus
5X	maxillary ramus of trigeminus
7	facialis (vs & vm)*
7G	geniculate ganglion of facialis
7GP	placodal ganglion of facialis
8	acousticus nerve and ganglion
9	glossopharyngeus
9G	superior-petrosal ganglion
9GP	placodal ganglion of glossopharyngeus
10	vagus (vs & mv)*
10G	jugular-nodos ganglion of vagus
11-12G	vestigial ganglia of occipital nerves (i.e., Froriep's ganglia)

Abbreviations for nerve* functions are as given in Table VIII-4.

General abbreviations:

AA	aortic arch
AA 1, 2, 3, & 4	aortic arches 1 through 4
AC	anterior cardinal vein
AG	autonomic ganglia
AIP	anterior intestinal portal
AL	allantoic membrane
ALA	allantoic artery
ALC	allantoic cavity
ALS	allantoic stalk
ALV	allantoic vein
AM	amniotic membrane
AMC	amniotic cavity
AMH	head fold of amnion
AMT	tail fold of amnion
AP	alar (sensory) plate
AQ	aqueduct of Sylvius, or mesocoele
AT	atrium of heart
BA	basilar artery
BP	basal (motor) plate
CA	conus arteriosus
CAP	capillaries
CC	common cardinal vein, or duct of Cuvier
CEF	cephalic flexure
CER	cervical flexure
CF, CH	choroid fissure
CL	general coelom
CLO	cloaca
CM	cloacal membrane
CO	cornea, conjunctiva
CP	gill closing plate
CP 1, 2, & 3	closing plates of gill slits 1 to 3
DA	dorsal aorta
DI	diencephalon
DM	dorsal mesentery
DMC	dorsal mesocardium
DMY	dermomyotome
EC	external carotid artery
ED	endolymphatic duct
EE	epidermal ectoderm
EMY	epimyocardial layer of heart
END	endocardial layer of heart
EP	epiphysis, primordium of pineal body
FL	fore limb (wing) bud
FP	floor plate
GB	gall bladder
HG	hind gut
HL	hind limb (leg) bud
IC	internal carotid artery
IN	infundibulum
IS	isthmus of metencephalon
ISA	intersegmental arteries
IT	intestine
LB	lung bud
LBF	lateral body fold
LE	lens vesicle
LI	liver diverticulum
LMC	lateral mesocardium
LSF	lumbosacral flexure
LT	lamina terminalis
LTG	laryngotracheal groove
MB	midbrain, mesencephalon
MG	midgut
MN	mesonephric tubule
MND	mesonephric (Wolffian) duct
MT	metencephalon
MY	myelencephalon
MYC	myocoele of somite
NC	neural crest
NCL	neurocoele
NE	nephrotome
NES	nephrostome
NM	neuromere of myelencephalon
NO	notochord
NP	nasal pit
NT	neural tube spinal cord
OC	occipital myotome
OE	esophagus
OM	oral membrane
OP	optic cup
OPR	optic recess
OPS	optic stalk
OT	otic vesicle
P	pharynx
P 1, 2, 3, & 4	pharyngeal pouches 1 through 4
PAN	pancreas
PC	peritoneal coelom
PCP	posterior choroid plexus
PCV	posterior cardinal vein
PE	pericardial coelom
PF	pontine flexure
PIP	posterior intestinal portal
PL	pleural coelom
PPM	pleuroperitoneal membrane
PR	pigmented retina
RF	roof plate
RP	Rathke's pouch
RT	sensory retina
SA, SAC	seroamniotic cavity
SAR	seroamniotic raphe
SB	subcardinal vein
SC	scleratome of somite
SG	spinal ganglion
SL	sulcus limitans
SM	somite mesoderm
SO	somatic hypomere
SOM	somatopleure
SP	splanchnic hypomere
SPP	splanchnopleure
SR	serosa
ST	stomach
STD	stomodaeum
STR	sacral torsion
SV	sinus venosus
TB	tailbud
TE	telencephalon
TG	tailgut
TH	thyroid diverticulum
TM	tail mesoderm
TT	thoracic torsion
V	ventricle of heart
V 1	lateral ventricle of telencephalon
V 3	3rd ventricle of diencephalon
V 4	4th ventricle of myelencephalon
VA	ventral aorta or aortic sac
VG	visceral grooves
VG 1, 2, 3, & 4	visceral grooves 1 through 4
VM	ventral mesentery
VS	visceral arches
VS 1, 2, 3, & 4	visceral arches 1 through 4
VS 1'	mandibular process of arch 1
VS 1''	maxillary process of arch 1
VT	vitelline artery
VV	vitelline vein
VVA	vitelline anastomosis of vitelline veins
XCL	extraembryonic coelom
YS	yolk sac

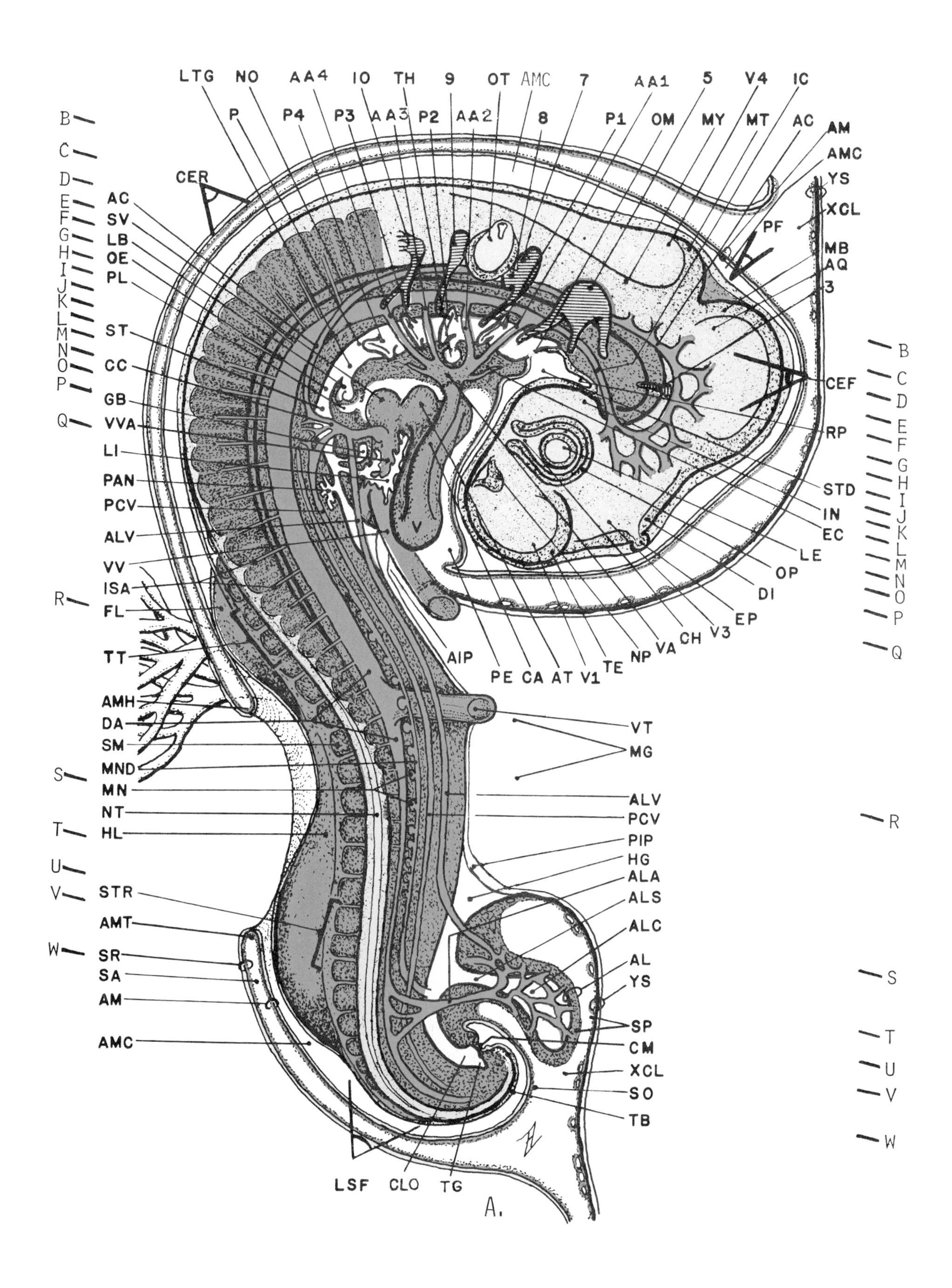

Figure 8-7. Structure of the 72 hour chick embryo. 8-7 A. Whole mount diagram of the 72 hour chick embryo (Stage 17–20 hours). Guide lines to the right and left of the figure refer to levels of correspondingly lettered cross section drawings. See legend.

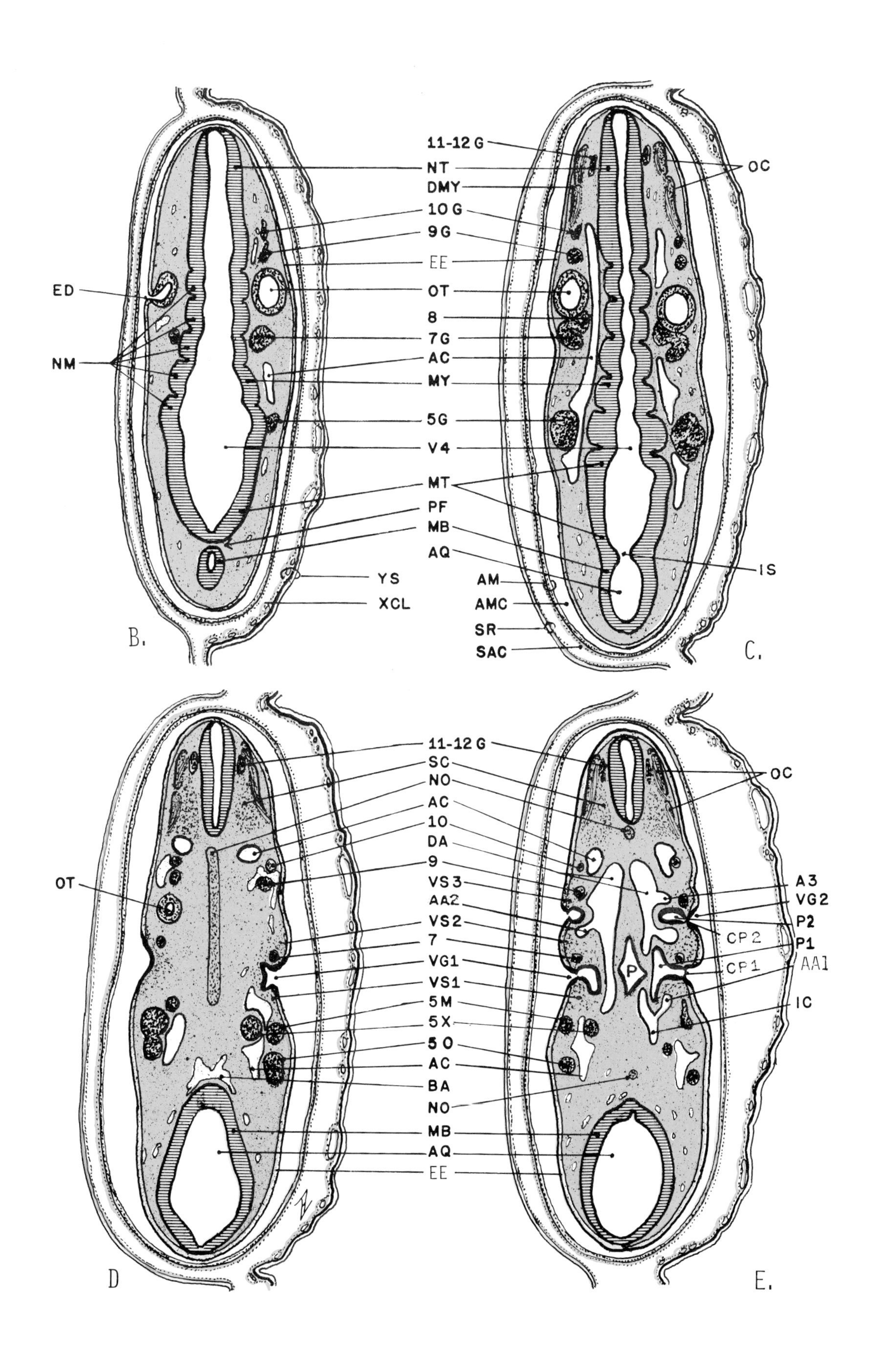

Figure 8-7 B–E. Representative cross sections of the 72 hour chick embryo. See legend.

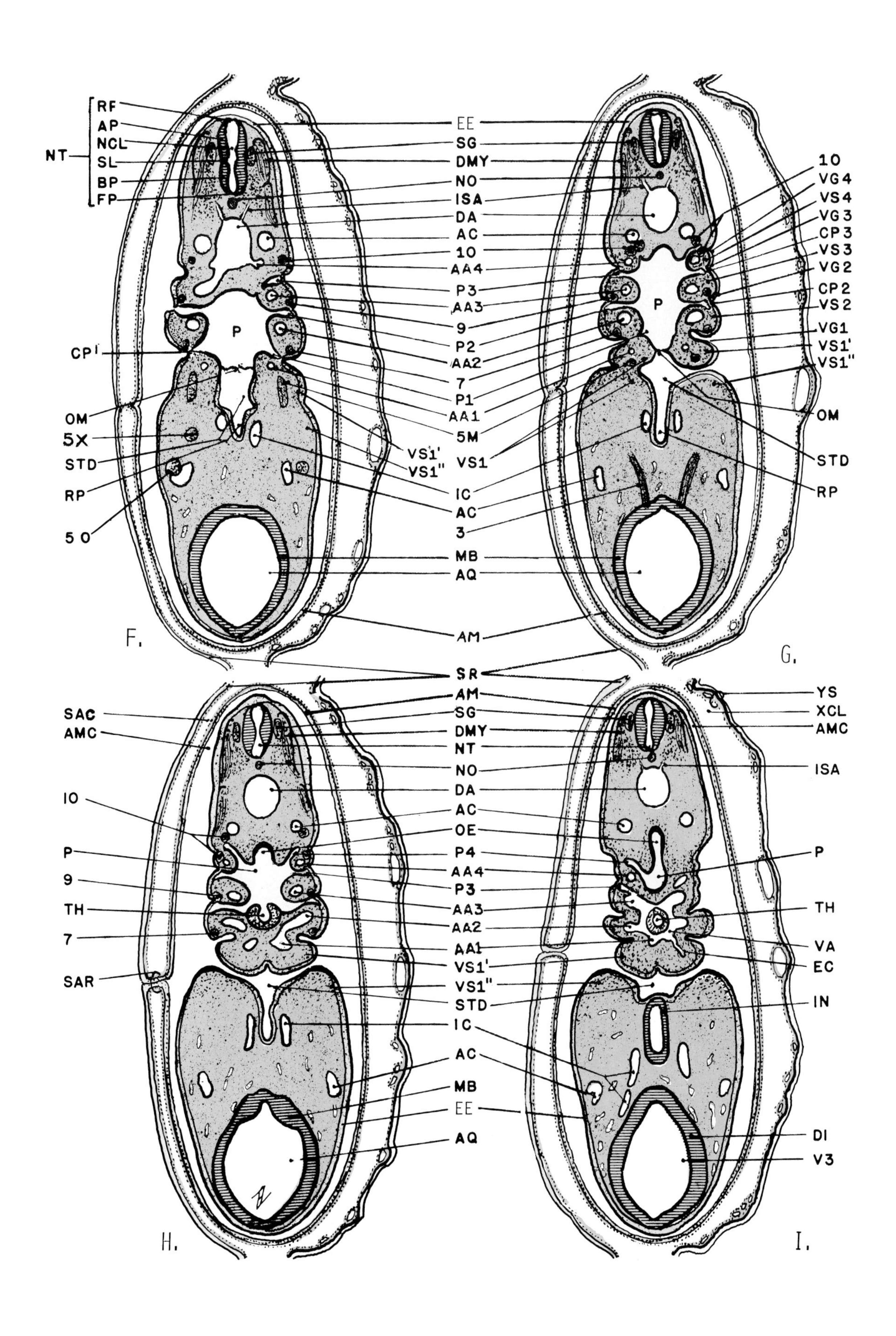

Figure 8-7 F–I. Representative cross sections of the 72 hour chick embryo. See legend.

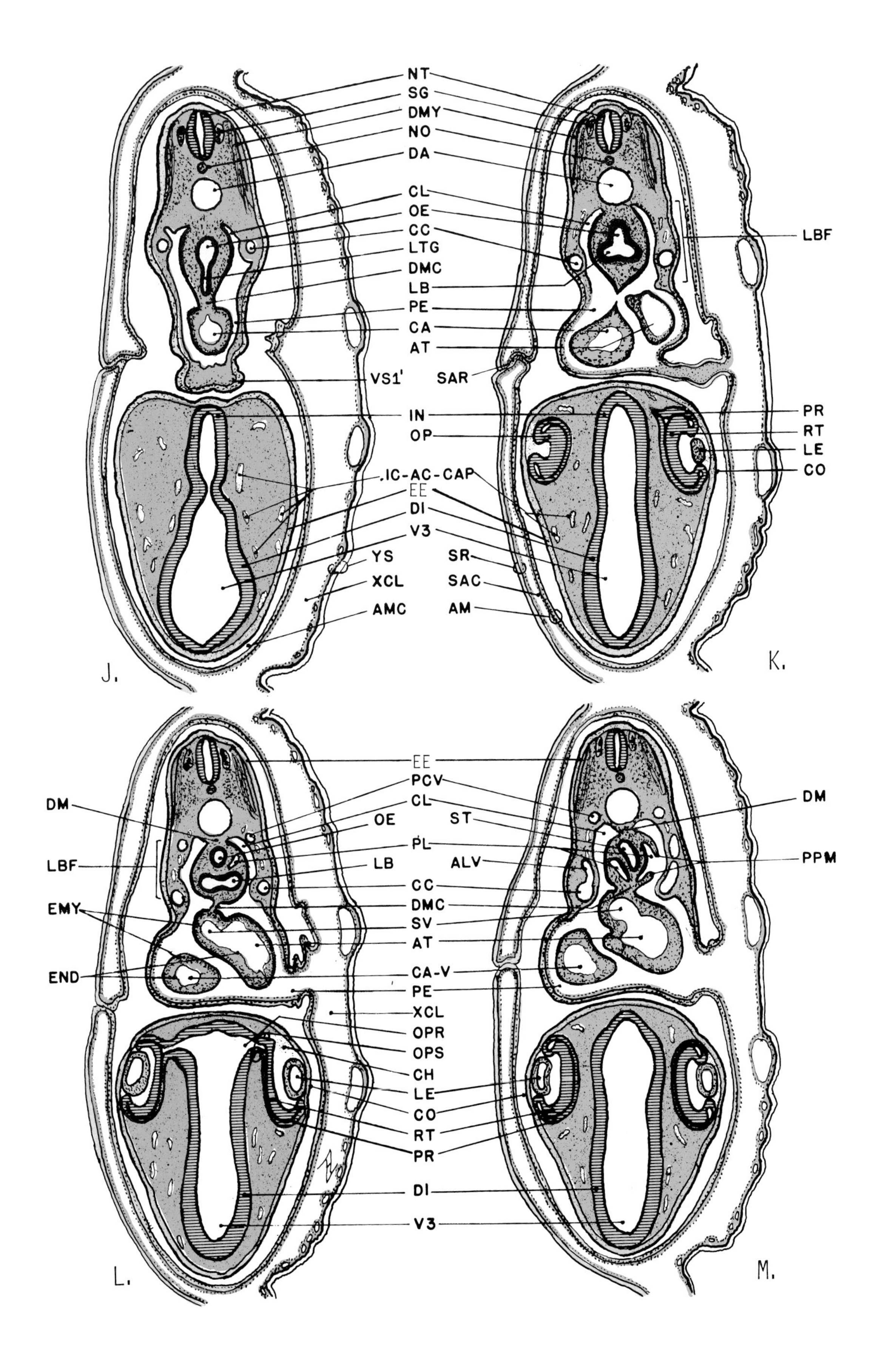

Figure 8-7 J–M. Representative cross sections of the 72 hour chick embryo. See legend.

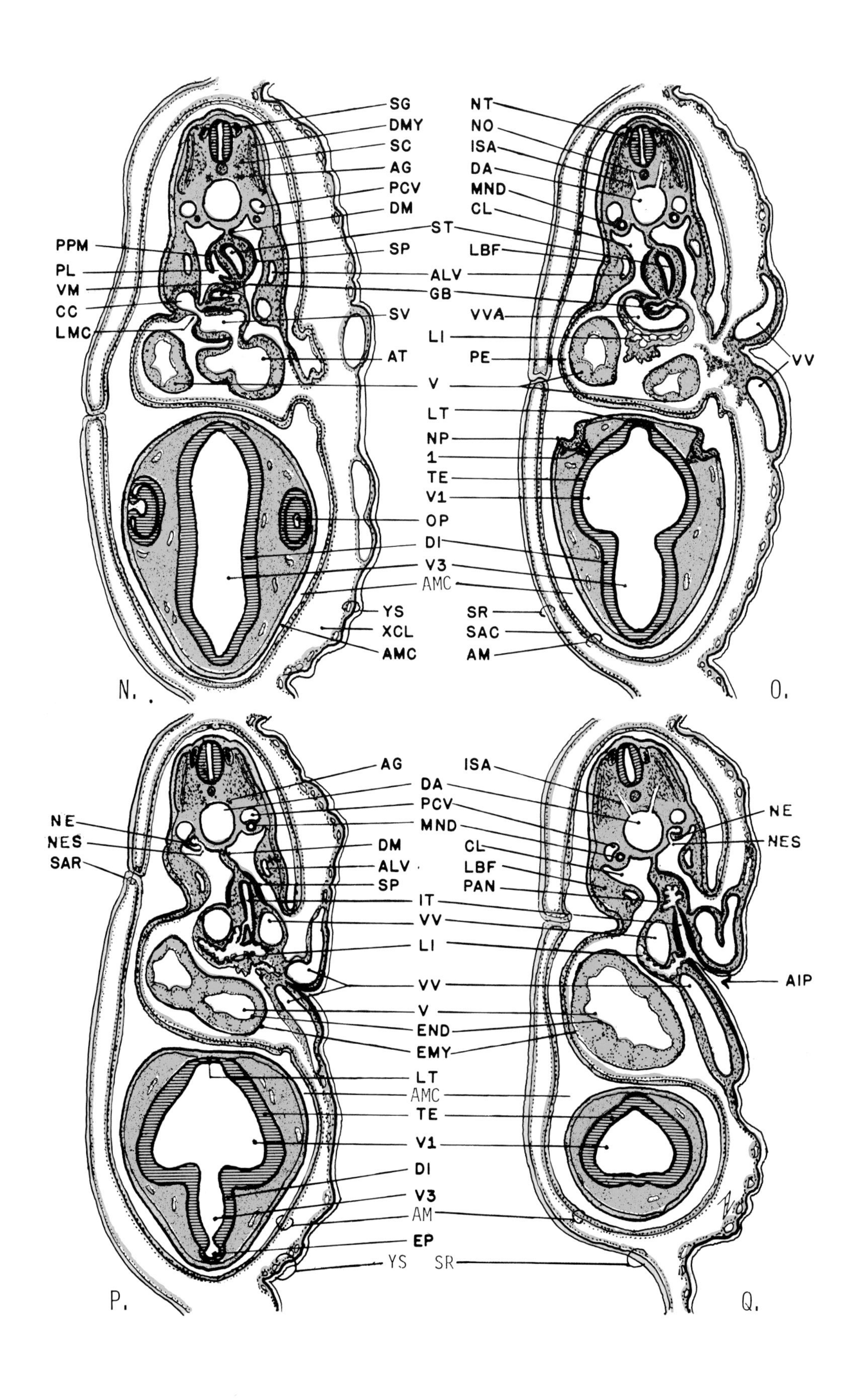

Figure 8-7 N–Q. Representative cross sections of the 72 hour chick embryo. See legend.

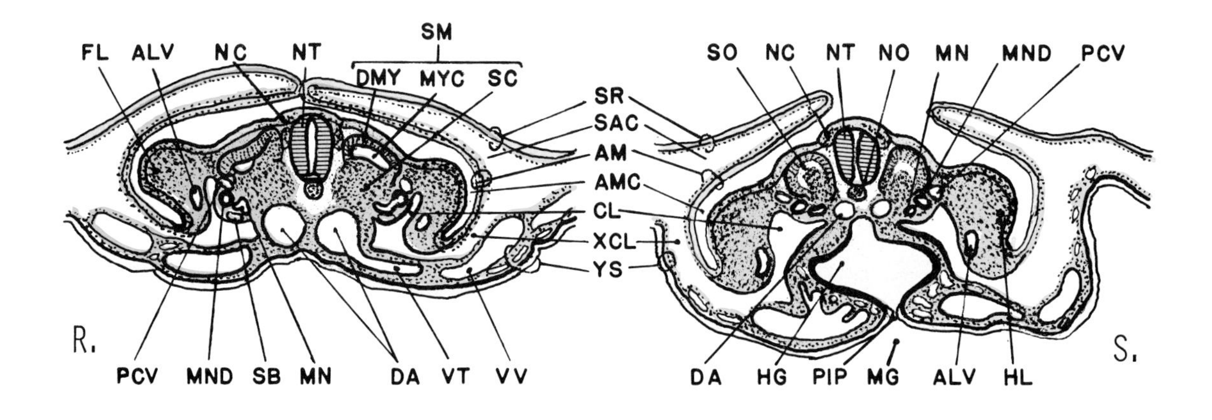

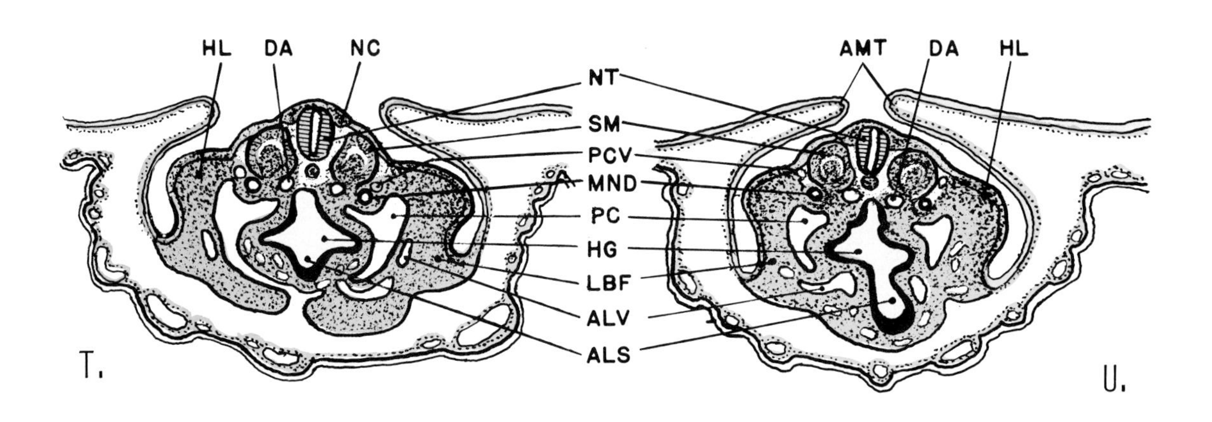

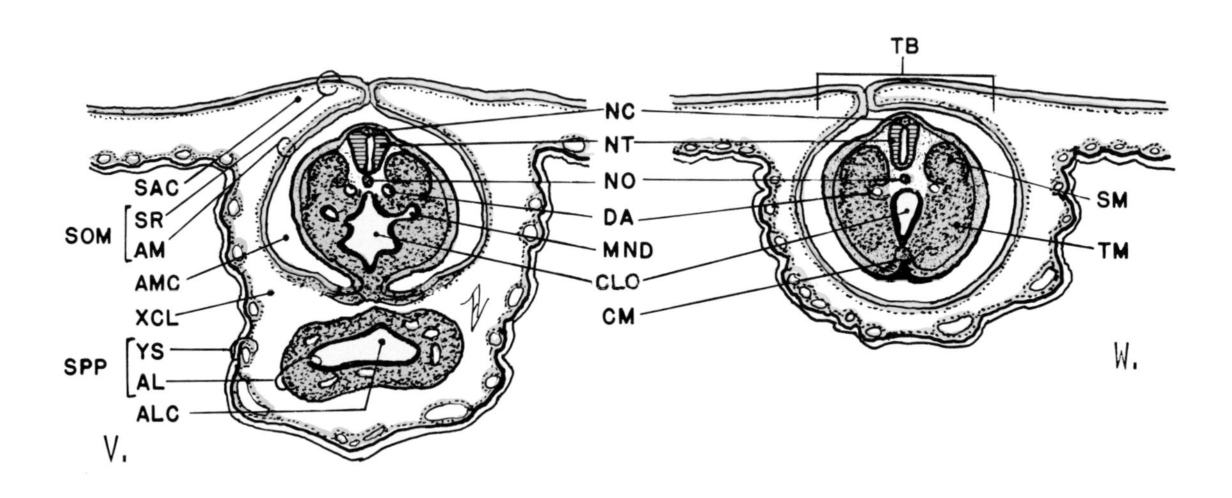

Figure 8-7 R–W. Representative cross sections of the 72 hour chick embryo. See legend.

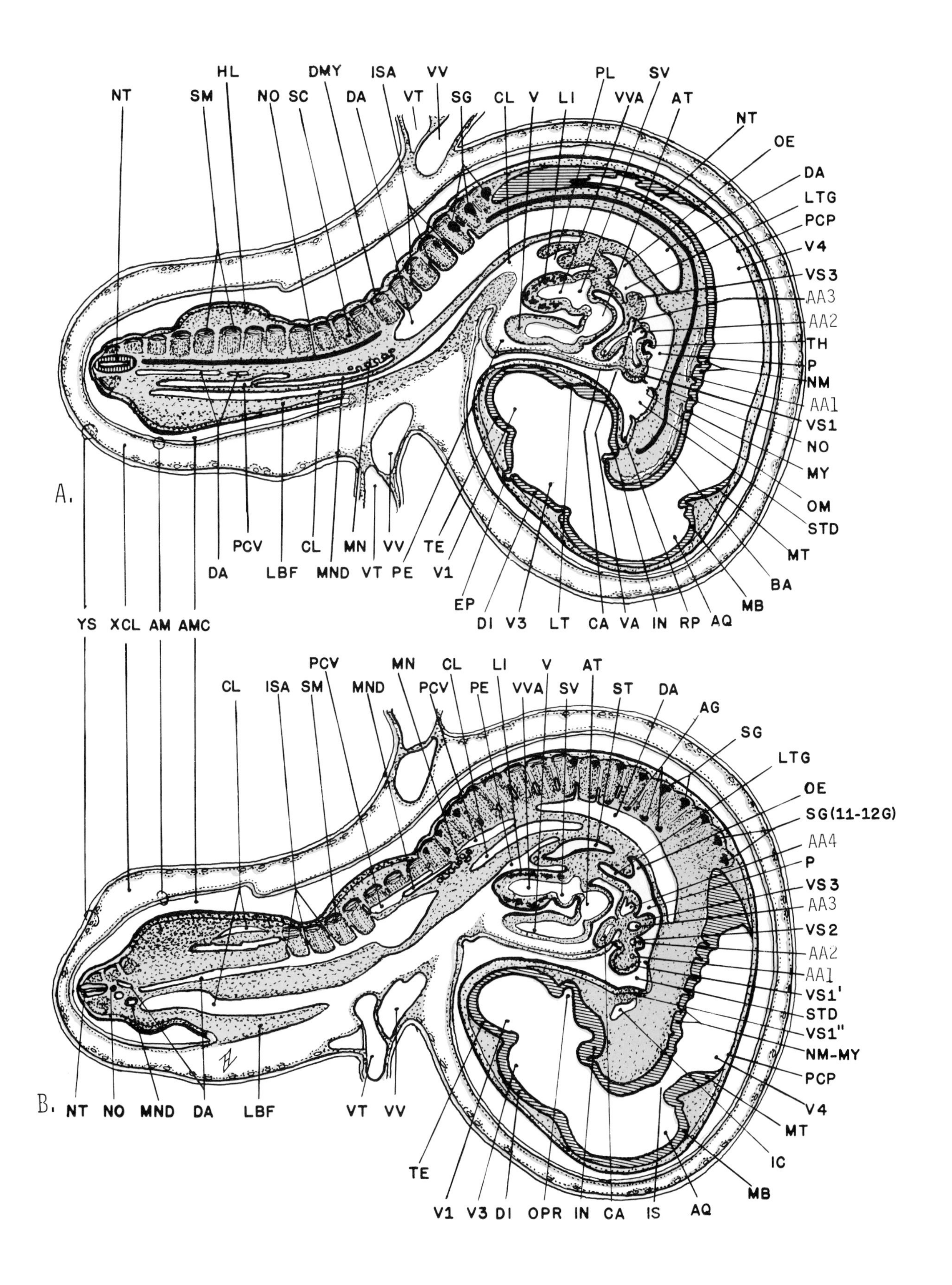

Figure 8-8 A–B. Representative sagittal sections of the 72 hour chick embryo. See legend.

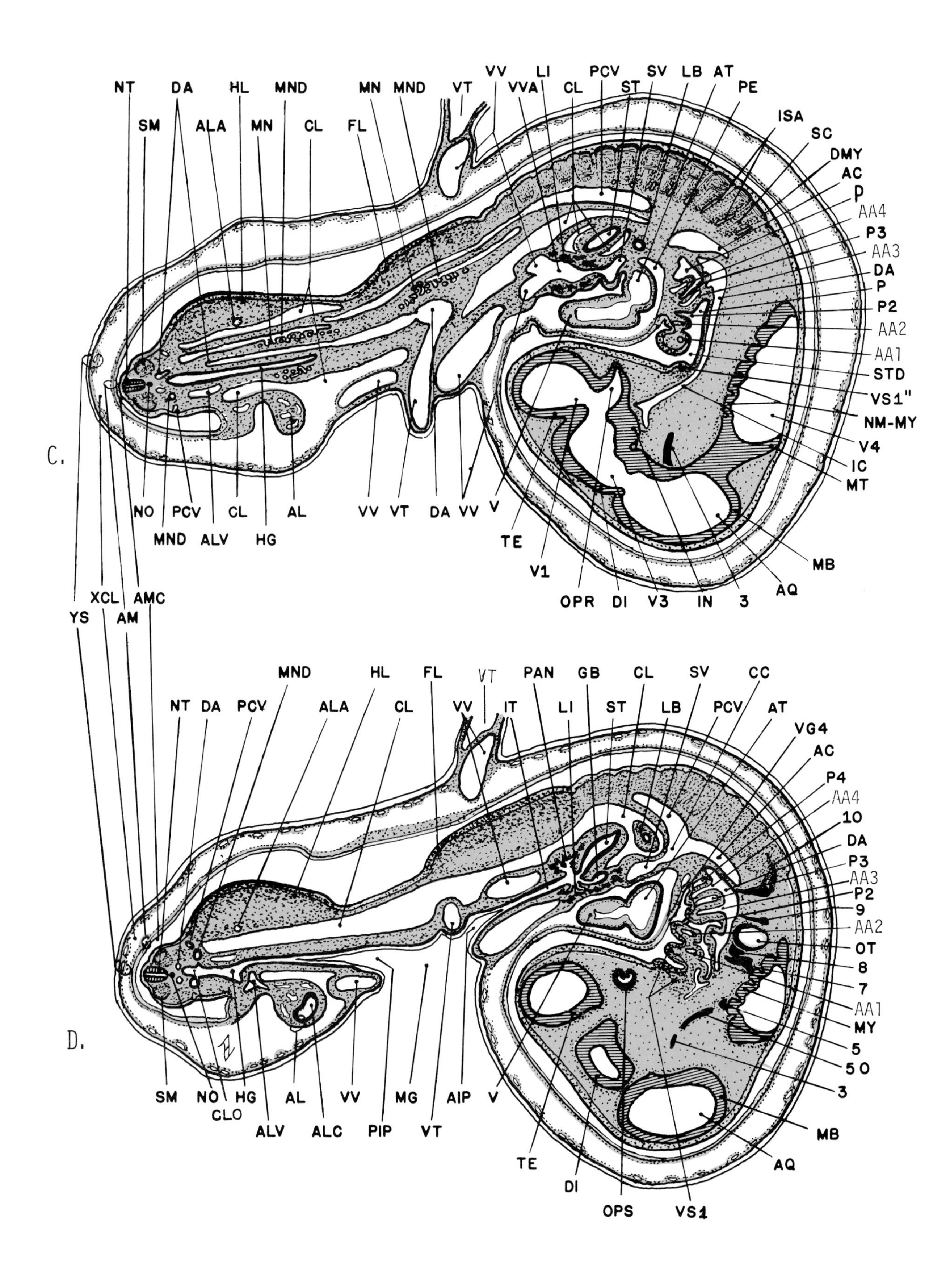

Figure 8-8 C–D. Representative sagittal sections of the 72 hour chick embryo. See legend.

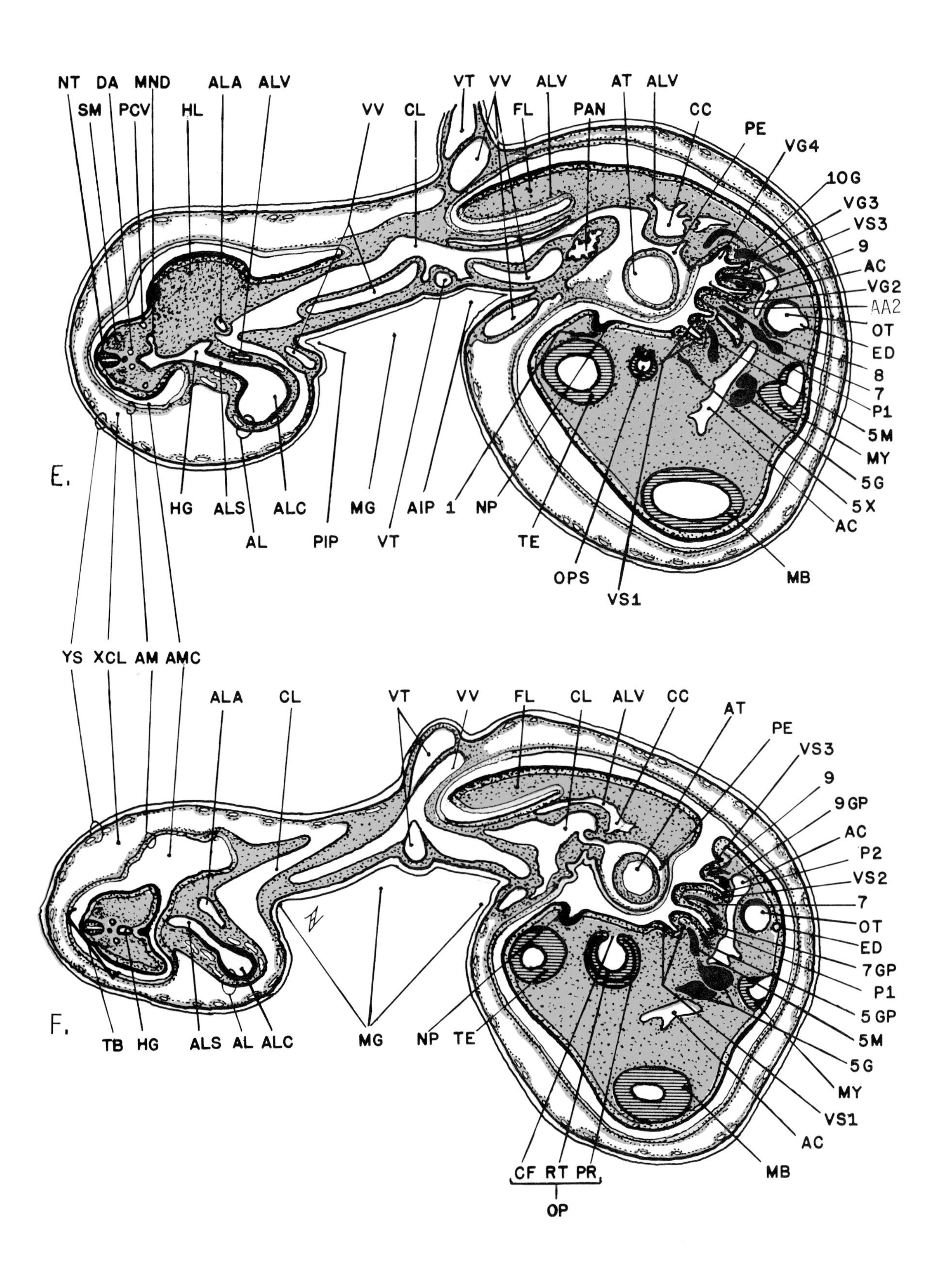

Figure 8-8 E–F. Representative sagittal sections of the 72 hour chick embryo. See legend.

of the body wall which undercuts the embryo and separates it from the yolk sac. In so doing, the roots of the vitelline veins are forced posterad. At the same time, the lateral body folds undercut the embryo from the sides and force the vitelline veins toward the midline. As a consequence, the two vitelline veins are brought together behind the heart. The walls between them degenerate and a single channel, the **vitelline anastomosis,** is formed (Fig. 8-7 O) as a posterior extension of the sinus venosus. Unfortunately the fused vitelline vein at this time is sometimes called the ductus venosus. The latter term, however, should be reserved for a new venous channel which will appear somewhat later as an embryonic connection between the left allantoic vein and the inferior vena cava (see discussion of the 10 mm pig which follows in Chapter 9).

The paired **allantoic veins** are in the process of forming a rudimentary capillary bed situated in the lateral body folds posterior to the sinus venosus. They can be traced through the lateral body folds from the allantoic capillaries forward to the common cardinal veins which enter the sinus venosus through the lateral mesocardia. As mentioned above, the allantoic arteries are much less well formed at this time; they will eventually arise from intersegmental branches of the dorsal aorta at the level of the hind-limb buds.

After the study of the 72 hour chick has been completed, it is particularly rewarding to refer to comparable stages in the development of embryos or larvae from other vertebrate classes. For example, compare the whole mount diagram of the 72 hour chick (Fig. 8-7 A) with the diagrams in Chapters 1 and 5 of Amphioxus (Fig. 1-4), Ammocetes (Fig. 1-6), salamander larva (Fig. 5-7 A, B) and the vertebrate developmental common plan (Fig. 1-2) that were provided as the basic reference point for the studies that have since been made. The ground plan of vertebrate development, hopefully, has been made abundantly clear. It is now appropriate that we proceed to a more advanced developmental stage in which characteristics that distinguish among classes of vertebrates begin to appear. We will conclude our study with some examples of mammalian development in Chapters 9 and 10.

C. OPERATIVE EXPERIMENTS WITH CHICK EMBRYOS TO DEMONSTRATE SELF-DIFFERENTIATING CAPACITIES OF ORGAN PRIMORDIA

Chapters 7 and 8 have presented the chick embryo as representative of early amniote development, at a time in organogenesis when most body parts are already programmed as a mosaic of self-differentiating primordia. In Chapter 5, a series of experiments was outlined to demonstrate the self-differentiating capacities of organ primordia in frog and salamander embryos. The basic kinds of experiments described for them included **extirpation** (removal of parts), **transplantation** (relocation of parts), and **tissue culture** (isolation of parts). These are also the major experimental tools for the study of embryology in the chick and other amniote embryos.

In many respects, experiments with chick embryos are simpler than those described for the amphibians. This is certainly true with regard to obtaining eggs at predetermined stages for year-round use. Many of the instruments and culture conditions are also more easily improvised. On the other hand, the requirements for sterile precautions and temperature regulations are much more demanding in the chick.

General materials needed at each station. Dissecting microscope lamp, 3 fingerbowls, alcohol lamp, beaker of 70% alcohol, plastic embryo spoon, 4 medicine droppers, cotton swabs, 2 metal instrument racks (see Fig. 5-8), standard dissecting kit (including fine scissors, fine forceps, dissecting needles), 2 No. 000 stainless steel insect pins mounted in glass holders and suitable for flame sterilization, wide-mouth specimen jars with fixative (see formulae in Chapter 7, Table VII-2; CaAFA recommended), and 200 ml sterile Tyrode or Hanks solution as given in Table VIII-3.

Three different types of experiments will be described in the following pages. They are arranged in order of increasing complexity. However, all are sufficiently simple to assure some levels of success in classroom situations. They have been adapted from research methods outlined in the short but excellent volume by D.A.T. New, *The Culture of Vertebrate Embryos,* Academic Press, 1966.

Before beginning any one of the following experiments, **read through the entire set of directions** relating to it. Make sure that all materials are ready and on hand before starting. **Pay particular attention to all directions relating to sterile technique.**

1. Sterile injections of biologically active chemicals into 72-hour incubated eggs. Many experiments can be improvised involving the injection of living eggs with various substances to see their possible effect on embryonic development. Indeed, this is one of the standard industrial tests for the biological activity and permissible dosage of pharmaceuticals before they are released

TABLE VIII-3
FORMULAE FOR AVIAN PHYSIOLOGICAL SALT SOLUTIONS

TYRODE SOLUTION:[1]
8.0 g NaCl
0.2 g KCl
0.2 g $CaCl_2$
0.1 g $MgCl_2 \cdot 6\,H_2O$
0.05 g $NaHPO_4 \cdot 2\,H_2O$
1.0 g $NaHCO_3$
1.0 g Glucose
H_2O to make 1 liter

HANKS SOLUTION:[2]
8.0 g NaCl
0.4 g KCl
0.14 g $CaCl_2$
0.2 g $MgSO_4 \cdot 7\,H_2O$
0.06 g KH_2PO_4
0.06 g $Na_2HPO_4 \cdot 2\,H_2O$
0.35 g $NaHCO_3$
1.0 g Glucose
0.02 g Phenol Red indicator
H_2O to make 1 liter

Use only distilled water; for tissue cultures double pyrex distilled water is recommended. **Adjust pH to 7.6–7.8** with 0.1 N NaOH or 0.1 N HCl. Optionally 10 drops 0.1% Phenol Red indicator dye can be added to Tyrode or Hanks solutions; correct pH can be visualized by a tangerine pink (not purple pink = too alkaline; not orange yellow = too acid).

1. M.V. Tyrode (1910) *Arch. Int. Pharm., 20:* 205–223. (Waters of hydration not included in original reference.)
2. J.H. Hanks and R.E. Wallace (1949) Proc. Soc. Exp. Biol. Med., 71: 196-200.

for human use. The procedure is exceedingly simple.

Special materials needed. In addition to the general materials, each student will need four 72 hour incubated eggs, adhesive plastic tape, one 5 ml hypodermic syringe and needle, and sterile test solutions (see below). For general use an egg candling box will be needed.

A **candling box** is used to determine whether or not an embryo is present in an egg before opening it. A candling box can be prepared most simply by cutting an oval hole two-thirds the diameter of an egg in the top of a paper box of sufficient size to hold an ordinary desk lamp inside. The lamp should have the shade removed or turned so light can shine up. Since the effectiveness of the candling box is dependent upon having a maximum amount of light pass through the egg against a black background, the hole in the box should be lined with soft black felt against which the egg can be pressed snugly. The box should be placed in a shaded corner or dark room for maximum effective use. Proceed as follows.

a. Candle each of the four 72 hour eggs by placing it over the hole in the light box. Look for the shadow of the embryo, the vitelline veins, and the sinus terminalis. The general appearance of a candled, normal, three-day embryo will appear somewhat like a plus in a circle + . Use only candled eggs with a proven embryo in them.

b. Sterilize the egg shell, instrument tray, syringe, scissor blades and fingers with 70% alcohol.

c. Hold the egg with one hand and hold one point of the scissors near its tip in the other. With a twisting movement, drill a small hole through the blunt end of the egg. The hole should be just large enough to admit a hypodermic needle.

d. Place the egg in a paper egg crate with the hole uppermost and put a small drop of 70% alcohol on the hole to keep it sterile while preparing the remaining eggs and solutions for injection.

e. With a felt tip marking pen write your seat number on each egg and add the letters A, B, C, and D consecutively to each. These numbers and letters will be part of the experimental coding required for data recording (see below).

f. Each student will inject a different test material in a series of dilutions to be carried out as follows. Stock solutions will all be made up at 1% concentrations by weight or volume in Tyrode solution and sterilized by boiling or autoclaving before use. Dilutions of stock solutions will be made with sterile Tyrode solution. The average volume of a chicken egg is approximately 50 ml. With this volume in mind, it is apparent that a 0.5 ml injection will be equivalent, in final dilution in the egg, to 1/100th the stock concentration, or 0.01% in the egg. Make dilutions, injections, and calculate concentrations in accord with the method given in Table VIII-4.

TABLE VIII-4
SCHEDULE FOR DILUTION AND INJECTION IN CHICK EGGS

Egg Number	For dilution use: Stock Solution		Tyrode Solution	Injection Volume	Final Dilution in Eggs
A	0.5 ml	+	0	0.5 ml	0.01%
B	0.5 ml	+	4.5 ml	0.5 ml	0.001%
C	0.05 ml*	+	4.95 ml*	0.5 ml	0.0001%
D (control)	0	+	0.5 ml	0.5 ml	0%

*This volume is most easily measured with accuracy by using at the start 0.5 ml of the solution for egg B, of which an excess has been prepared, and then diluting it up to 5 ml with sterile Tyrode solution.

Suggested Classes of 1% Stock Solutions for Testing

Inorganic salts:	calcium nitrate, ammonium sulfate, sodium fluoride.
Buffer salts:	sodium carbonate, sodium phosphate, glycine, or others.
Heavy metal salts:	lead chloride, mercurous chloride, silver nitrate, or others.
Alcohols:	methyl, ethyl, propyl alcohol.
Amino acids:	cysteine, glutamic acid, tryptophane, arganine, or others.
Common drugs:	aspirin, caffeine, nicotine, chloretone, MS-222 or others.
Antimetabolites:	inhibitors of nucleic acids, vitamins, respiration.
Vital dyes:	Toluidine Blue, Neutral Red, Nile Blue, or others.

g. After completing the dilutions and injections, seal the hole in the shell with a piece of adhesive tape. Incubate the eggs for 5 to 7 days and then open the eggs and examine the embryo. Evaluate the experimental results by recording them on the blackboard table for class results.

Record the chemical name of the compound tested and evaluate the effects at all dilutions on a 5-point scale as follows:

+ + Development strongly enhanced as compared with control egg D.
+ Development slightly enchanced as compared with comtrol egg D.
0 No detectable effect
– Development slightly impaired as compared with control egg D.
– – Development stopped or embryo killed at time of injection.

Experimental cases of general interest will be placed on demonstration for class observation. Where advisable, **preserve the embryo in fixative** for later observation and study.

2. Operations on egge in opened shells. This technique entails the use of host egg shells as the culture chamber.

Special materials needed. In addition to the general materials, each student will need four 72 hour incubated eggs. They should be in a cardboard egg crate with the pointed ends down and blunt ends up. Also needed are a vaseline gun made from a hypodermic syringe, 4 one and a half inch squares of glass or stiff plastic (the latter can be made from plastic cover sheets for photographs and are available at camera stores), or small squares of Saran Wrap, and an alcohol lamp or gas flame.

a. Sterilize the egg, fine forceps, scissors, dissecting needles, instrument tray, and hands with 70% alcohol.

b. Place the egg in the egg crate with the pointed end down and blunt end up. Hold the egg securely with one hand and tap a hole in the blunt end of the shell with the point of a sterile scissors or forceps. This will expose the air space inside the shell.

c. With sterile scissors and forceps, trim the shell down to about 2mm above the level of the shell membrane which makes the floor of the air space. Add one squirt (i.e., about 1 ml) of sterile Tyrode solution to the surface of the membrane. With sharp fine forceps tear around the edge of the membrane so that it can be removed. The host embryo will lie immediately beneath the shell membrane and will be clearly visible.

d. Primordia in the embryo will be removed by heat cautery. To do this, heat a fine dissecting needle in an alcohol flame and burn a single selected primordium. You will need to move quickly and with precision from flame to embryo to be effective. **Choose from the following list of primordia:** tailbud, leg bud, wing bud, otic vesicle, lens, mandibular arch, or any other primordium. However, **do not damage the heart** since all primordia are dependent upon heart function for continued development.

e. Seal the egg after experimental cautery is complete. Place a ring of vaseline around the rim of the cut shell. With 70% alcohol sterilize and air dry a square of glass or plastic. Then lower it over the open end of the egg and press it down into the vaseline. Additional vaseline can be added around the outside edge to make sure that there are no air passages still open.

f. With a felt tip marking pen, number the egg. Record on the egg (and in your notes) the primordium that was cauterized.

g. Return the egg to the incubator in its upright position for 5 to 7 days. Then the egg should be opened and the embryo examined. Note the effects of the experiment. For cases of special interest, embryos should be fixed and placed on demonstration for class observation.

3. Transplantation of organ primordia onto the chorionic membrane of 8 day embryos. Experience has shown that this kind of experiment is best done with primordia from embryos of three to four days incubation transplanted onto host embryos of eight to ten days incubation. The experiments must be terminated by the 18th day of host incubation because, at that time, the chorion begins to regress as part of the preparation for hatching.

Special materials needed. In addition to the general materials, each student will need one or two 72 to 96 hour eggs, and four 8-day eggs. Also needed are 2 sterile Petri dishes with covers, embryo spoon, vaseline gun, alcohol lamp, and egg crate to hold the eggs in an upright position.

Preparation of Donor Embryos:

a. Sterilize the shell of a 3-4 day egg, the scissors, dissecting needles, forceps, embryo spoon, instrument tray and hands with 70% alcohol. Place the egg in the egg crate with the pointed end down and blunt end up.

b. With points of scissors tap a hole into the blunt end of the egg. Trim the shell down and remove the shell membrane at the bottom of the air space in the manner described for the previous experiment.

c. Wet the top of the shell membrane with several drops of sterile Tyrode solution and remove the membrane with forceps. This will expose the embryo under the membrane.

d. With sterile forceps and embryo spoon remove the embryo from its extraembryonic membranes and transfer the embryo to a sterile covered Petri dish with about 5 ml sterile Tyrode solution.

e. With sterile fine dissecting needles crossed against each other, make scissor-like cuts that will isolate primordia for chorionic grafting. The following are suggested: tail bud, leg bud, wing bud, eye, mandibular arch, pieces of brain or heart, or any other primordia. **Cover the dish with isolated primordia until host eggs are prepared.**

Preparation of Host Embryos:

f. Host embryos of 8 to 10 days incubation should be sterilized along with instruments and hands, with 70% alcohol. The blunt end of the egg should be opened and trimmed in the manner described previously for cautery experiments. When the shell membrane is removed, the host chorion and its rich vascularization will be exposed at the surface.

g. With a blunt, hot needle cauterize a point on the surface of the chorion near the branching of a large vessel. This will cause a blood clot which will be used as the site of a tissue transplant.

h. With sterile, fine forceps transfer one piece of donor primordium to the clot. Push it gently into the clot. Two or three grafts can be placed on a single host at separate clot locations on the chorion.

i. Prepare a vaseline seal around the edge of the shell and cover the egg with a sterilized cover square or Saran Wrap. Complete the sealing with additional vaseline to close all air spaces between the shell and cover square.

j. Return the egg to the incubator in an upright position and incubate it for 5 to 7 days. Then open the egg and search the chorionic membrane for successful grafts. Observe the levels of development achieved by the grafts.

Grafts of general interest should be preserved in fixatives and placed on demonstration for class observation.

Many other types of experiments can be carried out on chick eggs that are beyond the scope of the present volume. Regrettably, time and space do not permit a review of even the more common of them. Refer to Appendix I for full bibliographic references on basic operative techniques by Hamburger, Rugh, and Wilt and Wessels; or for more detailed tissue culture methods by Cameron, New, and Parker for simple and useful summaries of practical techniques.

It will be desirable to make slides for histological staining of many experimental embryos. In the event that you plan to take a formal course in microtechnique following the present laboratory experience, it is suggested that experimental materials be sectioned in association with such a course. If that opportunity does not present itself, refer to any of a number of good references on microtechnique. Of these, Appendix I gives complete citations for the ones by Gray, Bancroft, Humason, and Preece.

CHAPTER IX

THE PRINCIPLES OF PROGRESSIVE SPECIALIZATION AND DIVERGENCE IN EMBRYONIC DEVELOPMENT

Illustrated by Early Development in the Mouse and Pig

TABLE OF CONTENTS

TABLES AND ILLUSTRATIONS

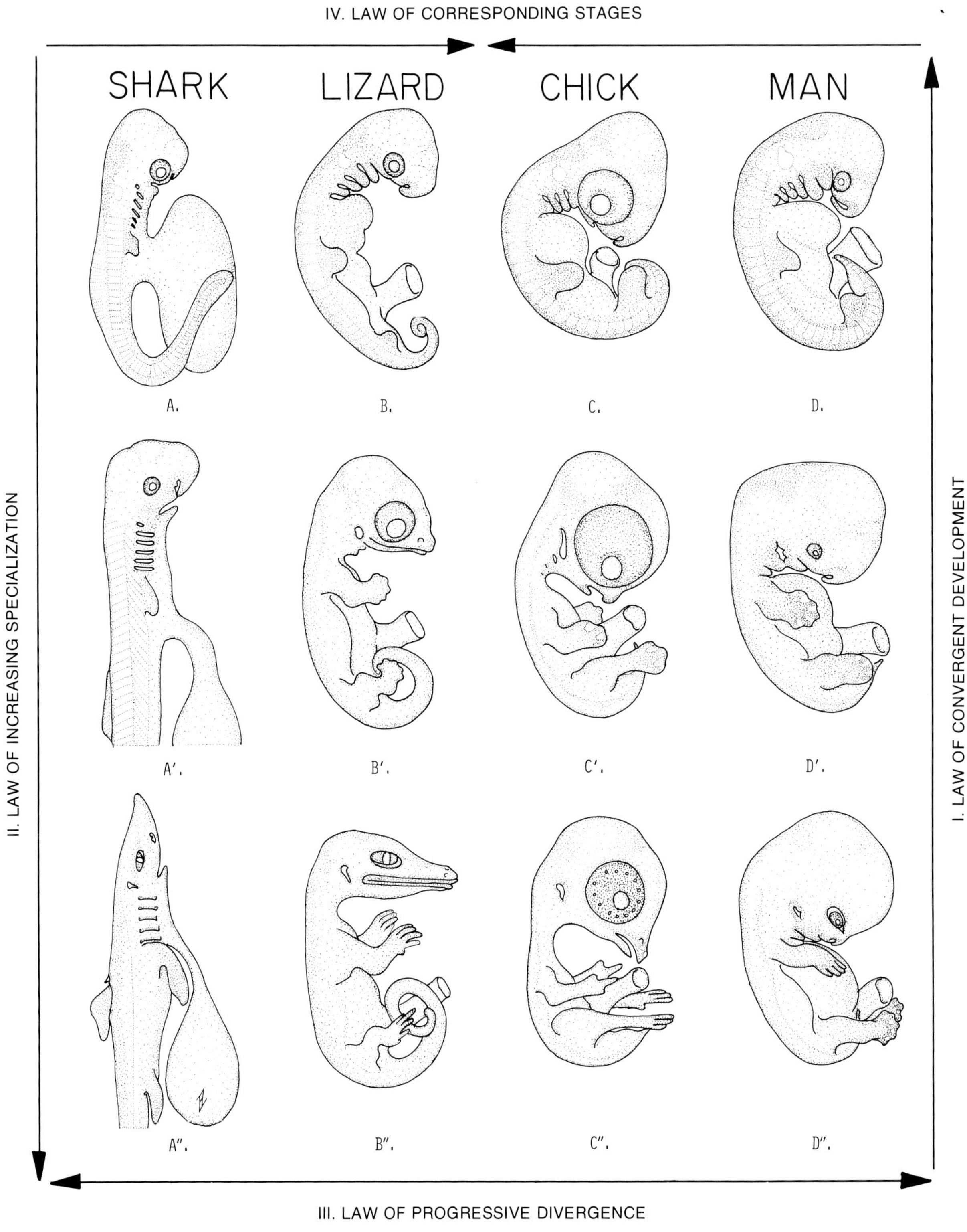

IV. LAW OF CORRESPONDING STAGES

SHARK LIZARD CHICK MAN

A. B. C. D.

A′. B′. C′. D′.

A″. B″. C″. D″.

II. LAW OF INCREASING SPECIALIZATION

I. LAW OF CONVERGENT DEVELOPMENT

III. LAW OF PROGRESSIVE DIVERGENCE

Figure 9-1. Comparison of vertebrate embryos to illustrate von Baer's Laws. Arrows framing the figure indicate the direction in which comparisons are to be made among embryonic stages to illustrate the respective Laws of von Baer named on each side of the figure. For full statements of these Laws, see Table VIII-1 and the explanation in the text.

9-1 A, A', A''.	Early, middle and late embryonic stages of the shark, *Squalus acanthius.*
9-1 B, B', B''.	Early, middle, and late embryonic stages of the lizard, *Anolis carolinensis.*
9-1 C, C', C''.	Early, middle, and late embryonic stages of the chick, *Galus domestica.*
9-1 D, D', D''.	Early, middle, and late embryonic stages of man, *Homo sapiens.*

CHAPTER IX

THE PRINCIPLE OF PROGRESSIVE SPECIALIZATION AND DIVERGENCE IN EMBRYONIC DEVELOPMENT

Illustrated by Early Development in the Mouse and Pig

Previous laboratory examples have emphasized general aspects of development that apply, for the most part, across the breadth of chordate development. Common plans of cleavage, gastrulation, germ layer derivation, and primordium formation for the phylum at large have consistently been stressed. Differences have been pointed out for each group but, for the most part, they have been explained as the consequence of special adaptations for embryo survival in their respective habitats (as, for example, differences related to relative amounts of stored yolk, and/or adaptations for survival in shelled eggs or intra utero).

A study of early embryo formation in mammals would provide additional examples of the common plan in vertebrate development that would differ in no fundamental way from that already described in the 18, 24, 33, 48, and 72 hour chick embryos. There are, of course, differences in the rate of development and in the relative size and growth rate of specific primordia. However, these differences between the bird and mammal early tailbud embryo are relatively trivial as compared with the overwhelming and remarkable similarities that can be seen in comparing equivalent stages in the development of birds (see Fig. 8-1) and mammals (Fig. 9-2). Although, from the beginning, the genes of different embryos are programmed to produce a bird in one case and a human or a mouse in the others, the beginning steps in embryo formation show point-by-point homologies in the arrangement, structure, and manner of development of all their parts. Viewed in a somewhat different way, if these tailbud embryos are taken for their own morphological values without considering what their final goals will be (bird, pig, or mouse, etc.), then all of them would simply qualify as "vertebrate embryos" which as yet carry no visible imprint of the essential characters that will distinguish the classes as adults. Any fish, amphibian, reptilian, avian, or mammalian embryo at the tailbud stage could serve equally well in illustrating the early primary vertebrate body plan of development.

Chapters 1, 4, 5, 6, 7, and 8 provide many examples that illustrate von Baer's first and fourth laws, namely: **I. the law of convergent development,** and **IV. the law of corresponding stages.** Chapters 8, 9, and 10 are more directly concerned with his second and third laws; i.e., **II. the law of increasing specialization,** and **III. the law of progressive divergence.** A more complete statement of these principles is given in Table VIII-1. The interrelationships of von Baer's laws are diagrammatically presented in Figure 9-1. Arrows on the four sides of the figure indicate the direction in which the figures of comparative stages are to be "read" in order to illustrate each of the four laws. It is of interest to point out that the data summarized in this figure have often been used to support both the views of von Baer and the opposing views by Haeckel. Note that von Baer's views are clearly in agreement with the direct embryological data. Embryological stages **do not** repeat adult stages as Haeckel proposed in his recapitulation theory. Instead, as von Baer stated: (I) as one goes back in development, embryonic stages converge on a common plan of development; (II) as development in a single line proceeds, there is increasing specialization of parts; (III) when one compares older stages of embryos of different species, one sees them diverge from the common plan and become more and more different; and (IV) as development continues in different groups, the most primitive forms share fewer corresponding stages with advanced forms. The genius of von Baer lies in the acuteness of his observations of undisputable fact and the perceptiveness of his interpretation of the common principles to be derived from them. His concepts of development wear remarkably well with the passage of time primarily because they state observed fact and do not indulge in premature flights of theoretical fiction.

A. EARLY DEVELOPMENT AND PLACENTAL RELATIONSHIPS IN MAMMALIAN EMBRYOS

1. Early mammalian development illustrated by the pig (Fig. 9-2). The major aspects of early mammalian development are well illustrated by the embryology of the pig as summarized in Figure 9-2. Eggs of placental mammals are isolecithal and, following fertilization, the egg divides holoblastically and almost equally. The overall pattern of cleavage following the 4-cell stage, however, is generally irregular, and cleavage furrows are not directly referable to any axial planes of symmetry in the future embryo. This is related to the fact that by the 4-cell stage, there is already ooplasmic segregation in which two of the original blastomeres are determined as so-called **trophoblast cells** which are destined to give rise to all extraembryonic ectoderm. The other two cells (called **embryo cells**) will give rise to the entire embryo and to all extraembryonic endoderm and mesoderm as well. Trophoblast and embryo cells divide at very different rates; embryo cells are the slower of the two.

As is shown in Figure 9-2, late cleavage results in the formation of a hollow blastula which, in mammals, is usually called a **blastocyst.** The embryo cells form an eccentric thickening on one side, called the **embryonic disc,** or **inner cell mass;** the surface of the embryonic disc represents the future dorsal surface of the embryo. Gastrulation begins by precocious delamination of **extraembryonic endoderm** from the inner surface of the embryonic disc much as is the case already described for birds. Extraembryonic endoderm cells quickly spread out and line the inner surface of the trophoblast layer. The central cavity of the blastocyst is homologous with the archenteron and **yolk sac cavity** of bird eggs even though no yolk is present to fill the cavity. The next phase of gastrulation is the formation of a **primitive streak** at the posterior end of the embryonic disc. The formation of the primitive streak is the first morphological evidence of anteroposterior axial differences. After this time the embryo has visible bilateral symmetry. From the primitive streak, **embryonic endoderm and all mesoderm** will invaginate and proliferate to the interior in a manner similar to that previously described for birds in Chapter 6. Embryonic endoderm will replace the extraembryonic endoderm in the region of the embryonic disc. **Extraembryonic mesoderm** will migrate outward from the region of the disc in all directions and become interposed between extraembryonic trophoblast ectoderm and yolk sac endoderm.

Legend and Checklist for Figure 9-2.

Color code: blue — ectoderm, red — mesoderm, yellow — endoderm.

als allantoic stalk
am amniotic membrane
amc amniotic cavity
amf amniotic folds
bc blastocoele
cl coelom
di diencephalon
e eye
ea ear
ecl extraembryonic coelom
een extraembryonic endoderm
emc embryo cells
emd embryonic disc
en embryonic endoderm
flb forelimb bud
ga' first gill arch
gs' first gill slit
hlb hind limb bud
hp head process, notochord
ht heart
mes general mesoderm
mg midgut
mm margin of mesoderm
mn mesonephros
mr mammary ridge
ms mesencephalon
mt metencephalon
my myelencephalon
nf neural folds & neural plate
no notochord
np neural plate
op olfactory pit, nose
ot otocyst, inner ear
pe pericardial coelom
pn primitive knot = Hensen's node
pp primitive pit
ps primitive streak
sac seroamniotic cavity
sc spinal chord
so somite or myotome
sop somatopleure
spp splanchnopleure
sr serosa (trophoblast + extraembryonic mesoderm)
tb tailbud
te telencephalon
tr trophoblast, extraembryonic ectoderm
trc trophoblast cells
umb umbilical cord
ys yolk sac membrane
ysc yolk sac cavity
yse yolk sac endoderm
yst yolk stalk
zr zona radiata, chorion or fertilization membrane

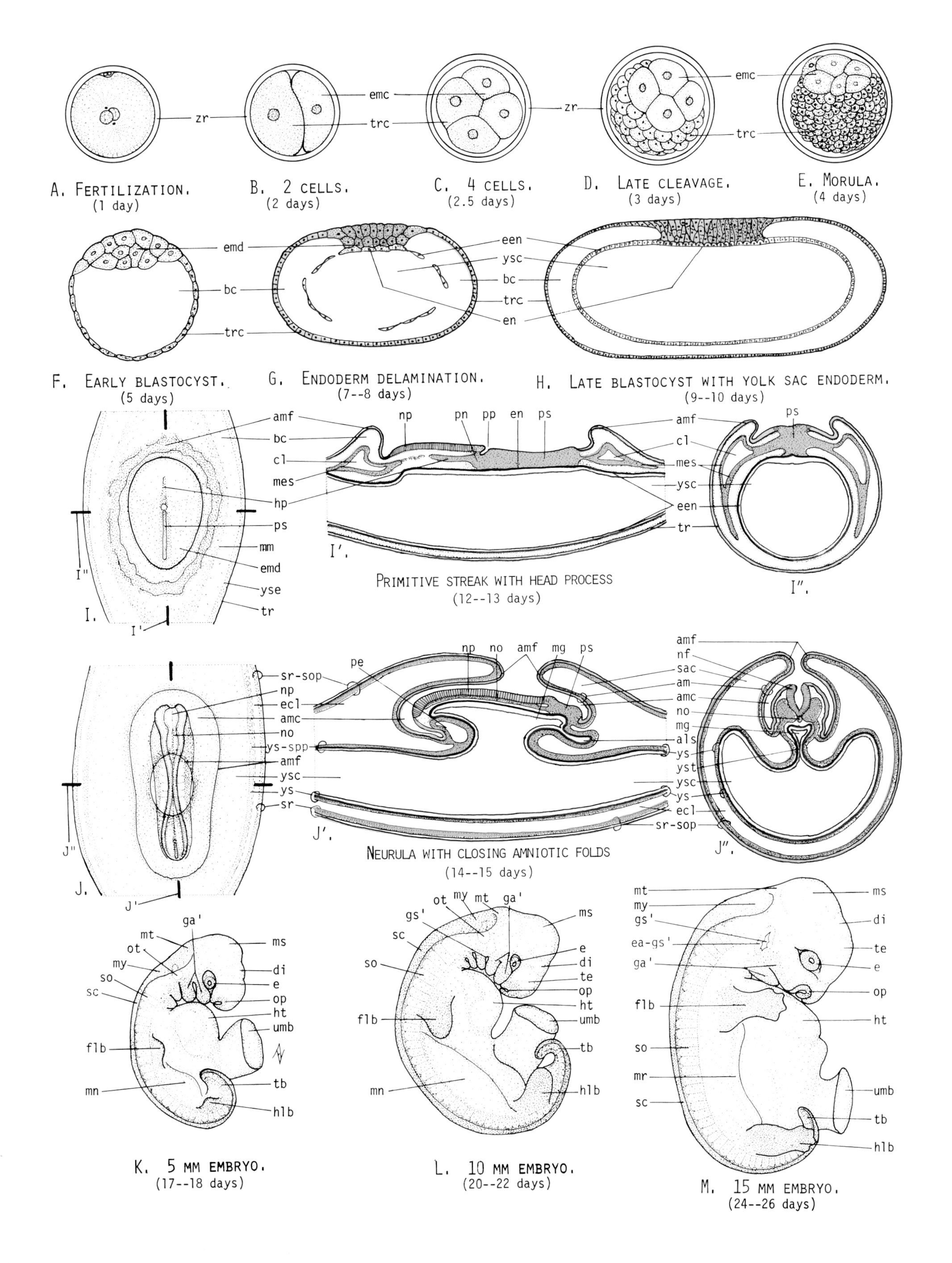

Figure 9-2. Early development in placental mammals illustrated by cleavage, blastulation, gastrulation, neurulation, and early organogenesis in the pig embryo. See legend.

TABLE IX-1. COMPARISON OF ORIGIN, STRUCTURE, AND FUNCTION IN EXTRAEMBRYONIC MEMBRANES OF REPTILES, BIRDS, AND MAMMALS

EXTRAEMBRYONIC MEMBRANES		REPTILES, BIRDS & MONOTREMES	MARSUPIALS & LOWER PLACENTALS	ADVANCED RODENTS & PRIMATES
1. AMNION:	Germ layer composition	Extraembryonic ectoderm & somatic mesoderm (mesoderm involutes at primitive streak).	Same as in reptiles & birds.	Ectoderm is derived from embryonic disc & mesoderm is delaminated from trophoblast.
	Morphogenesis	Formed by folding of extraembryonic somatopleure which forms the amniotic folds.	Same as in reptiles & birds.	Formed directly by delamination from the embryonic disc (i.e., from inner cell mass).
	Membrane function	Secretes amniotic fluid, muscle contractions rock the embryo & prevent tissue adhesions.	Same as in reptiles & birds.	Same as in reptiles, birds & lower mammals.
	Cavity formation	The embryonic exterior is trapped by fusion of the amniotic folds over the embryo.	Same as in reptiles & birds	Formed directly by delamination & splitting of embryonic disc (i.e., from inner cell mass).
	Cavity fluid function	Fluid prevents drying and acts as a flotation and shock absorbing support for the embryo.	Same as in reptiles & birds.	Same as in reptiles, birds & lower mammals.
2. SEROSA:	Germ layer composition	Extraembryonic ectoderm & somatic mesoderm (mesoderm involutes at the primitive streak).	Same as in reptiles & birds.	Both ectoderm & mesoderm are formed from the trophoblast (i.e., mesoderm is delaminated from the inner wall of the blastocyst).
	Morphogenesis	Formed by folding of extraembryonic somatopleure which forms the amniotic folds.	Same as in reptiles & birds.	Formed directly from trophoblast without formation of amniotic folds.
	Membrane function	Functions in respiratory exchange as the outer component of the chorion.	As the outer foetal component of the yolk sac & chorionic placentae it functions in respiratory, excretory & nutritive exchanges.	Formed as in lower placentals; however, no functional yolk sac placenta is formed.
	Seroamniotic cavity	Part of the extraembryonic coelom.	Same as in reptiles & birds	Same as in reptiles, birds & lower mammals.
3. YOLK SAC:	Germ layer composition	Extraembryonic endoderm and splanchnic mesoderm (mesoderm involutes at primitive streak).	Endoderm from embryonic disc & splanchnic mesoderm involuted from primitive streak.	Embryonic midgut endoderm & splanchnic mesoderm delaminated from trophoblast (i.e., not by involution from the primitive streak).
	Morphogenesis	Formed by overgrowth of the yolk mass at the zone of junction by yolk sac splanchnopleure.	Endoderm is delaminated from the embryonic disc and organizes a yolk sac vesicle directly (i.e., there is no zone of junction or yolk).	Same as in marsupials and lower placentals except for differences in the origin of splanchnic mesoderm indicated above.
	Membrane function	Nutrition by digestion & absorption of yolk.	As part of the early yolk sac placenta it functions briefly in respiratory, excretory & nutritive exchanges.	Completely vestigial with no function.
	Yolk sac cavity function	Storage of nutrient yolk.	Vestigial cavity lacking yolk.	Same as in marsupials & lower placentals.
4. ALLANTOIS:	Germ layer composition	Hindgut endoderm & splanchnic mesoderm (mesoderm involutes at the primitive streak).	Same as in reptiles & birds.	Same as in reptiles, birds & lower mammals.
	Morphogenesis	Arises from floor of hindgut, then expands into the extraembryonic coelom & fuses with serosa.	Same as in reptiles & birds.	Vestigial & restricted to an allantoic stalk which contributes only allantoic splanchnic (vascular) mesoderm to the chorion.
	Membrane function	Allantoic mesoderm forms vascular tissue of allantois, and chorionic placenta. endoderm aids in concentrating excretory wastes.	Same vascular function by allantoic mesoderm as in reptiles & birds; however, there is minimal waste concentration by endoderm.	There is no allantoic membrane; however, allantoic vascular mesoderm enters the serosa & forms vessels in chorion & chorionic placenta
	Allantoic cavity function	Storage of excretory waste until hatching (homologous with amphibian urinary bladder).	Temporary storage of dilute excretory waste which will be removed by placental exchange.	Allantoic cavity is completely lacking except for the lumen of the allantoic stalk.
5. CHORION:	Germ layer composition	Ectoderm & somatic mesoderm of serosa & endoderm & splanchnic mesoderm of allantois.	Same as in reptiles & birds	Same as in reptiles, birds & lower mammals except that it lacks allantoic endoderm.
	Morphogenesis	Derivation by fusion of serosa somatopleure and allantoic splanchnopleure.	Same as in reptiles & birds.	Serosa ectoderm & somatic mesoderm are derived from the trophoblast & chorionic vascular mesoderm is derived from allantoic stalk.
	Membrane function	Respiratory exchange of O_2 & CO_2.	As the foetal component of the chorionic placenta it functions in respiratory, excretory & nutritive exchange.	Same as in marsupials and lower placentals.
	Chorionic vesicle cavity	Homologous with allantoic cavity (i.e., it is lined by allantoic endoderm.	Same as in reptiles & birds.	Homologous with the extraembryonic coelom (i.e., it is lined by mesoderm not endoderm).

The formation of extraembryonic somatopleure and splanchnopleure, and the extraembryonic membranes derived from them, occurs in several steps which bear strong homologies with those events in birds; however, in the absence of an intrinsic nutritive yolk reserve, the mammalian embryo greatly shortens the steps in membrane formation and hastens the process of establishing food lines with the maternal uterine circulation. The sequence of steps in membrane formation is shown in Figure 9-2. **First, extraembryonic coelom** is formed by a splitting of extraembryonic mesoderm into somatic and splanchnic mesoderm layers; the former then attaches to the trophoblast ectoderm and forms **somatopleure** externally, and the latter fuses to the yolk sac endoderm to form the **splanchnopleure** internally. **Second,** the **amniotic folds** form simultaneously around the entire embryonic disc which they then overgrow completely. A seroamniotic raphe is formed as the amniotic folds fuse to create a closed **amniotic cavity** over the embryonic disc. The **amniotic** and **trophoblastic (serosa) membranes** bear precisely the same topographic relationships to each other and to the amniotic cavity and extraembryonic coelom as was observed in the corresponding chick membranes and cavities. **Third,** the head, tail, and lateral **body folds** form simultaneously around the embryonic disc and undercut it, forming a very broad umbilical cord which thereafter will connect embryonic and extraembryonic regions. The embryonic disc at this time is no longer circular but is an elongated, shield-shaped region with an open neural plate at one end and a primitive streak at the other; in all respects the embryonic disc is comparable to the 24 hour chick embryo in basic topography. **Fourth,** the **yolk sac** soon acquires an abundant vitelline circulation and undergoes a rapid expansion. It presses against the serosa and fuses with it. Together the serosa and yolk sac press against the uterine wall and form a temporary **yolk sac placenta** through which metabolic exchange between maternal and embryonic circulations takes place. **Fifth,** a posteroventral evagination from the embryonic hindgut grows ventrally through the umbilical cord pushing splanchnic mesoderm before it. This is the primordium of the **allantoic stalk** and **allantoic membrane;** the lumen of the outgrowth is the **allantoic sac.** This evagination expands with great speed and soon displaces the extraembryonic coelom and the yolk sac. The allantoic membrane, with its vascularized layer of splanchnic mesoderm, will fuse with the overlying trophoblastic serosa and somatic mesoderm to form the **chorionic membrane. Last,** the chorion will send out finger-like **chorionic villi** into the endometrial lining of the uterine cavity. Maternal and fetal circulatory systems come into intimate contact with one another in this location; this association of chorion and uterine tissues is called a **chorionic placenta.** The placenta is a physiological adaptation for improved efficiency in the exchange of nutritive, respiratory, and excretory products between fetal and maternal circulations. The type of placenta found in pigs is the most primitive type of chorionic placenta found in placental mammals. More advanced types have evolved in carnivores, primates, and rodents in which a progressive reduction in the tissue barriers to osmotic exchange of metabolites has evolved as a further adaptation for placental efficiency. Placental adaptations in mammals are shown in Figure 9-3, which attempts to show homologies and differences that have evolved in the reptile–bird, marsupial, lower placentals and higher primates including monkeys, apes and man.

Observe the demonstrations of preserved materials, models, and the film on early mammalian development and placental relationships.

2. Progressive divergence, specialization, and placental relationships in mammals illustrated by living mouse embryos. Early mammalian development within placental membranes and embryonic specialization will be demonstrated with living mouse fetuses ranging in age from 8 day neurulae to 16 day pre-partum fetuses; full term is 21 days. Students will work in groups of four for the dissection of a freshly killed pregnant mouse. Since there are usually from 8 to 14 conceptions in each female, an exchange of stages between dissection teams will be made so that fetuses at varied ages will be available for all to examine and preserve.

Materials for each team of four will include: 1 enamel dissecting pan with a liner of three layers of paper towels, 500 ml warm 0.9% NaCl, small scissors; 2 4-inch fingerbowls, 4 Petri dishes or low dissecting dishes, 4 wide-mouth collecting jars with plastic screw caps, 100 ml 5% commercial formalin in 0.9% NaCl or 100 ml CaAFA fixing solution (see formula in Table VII-2). Each student will need fine forceps and dissecting needles.

To skin the mouse, place it belly-side up on the paper liner and orient it with the head away from the operator. Make a small "V-shaped" cut in the posterior belly skin with the scissors as follows. Holding the skin up with the fingers of one hand or

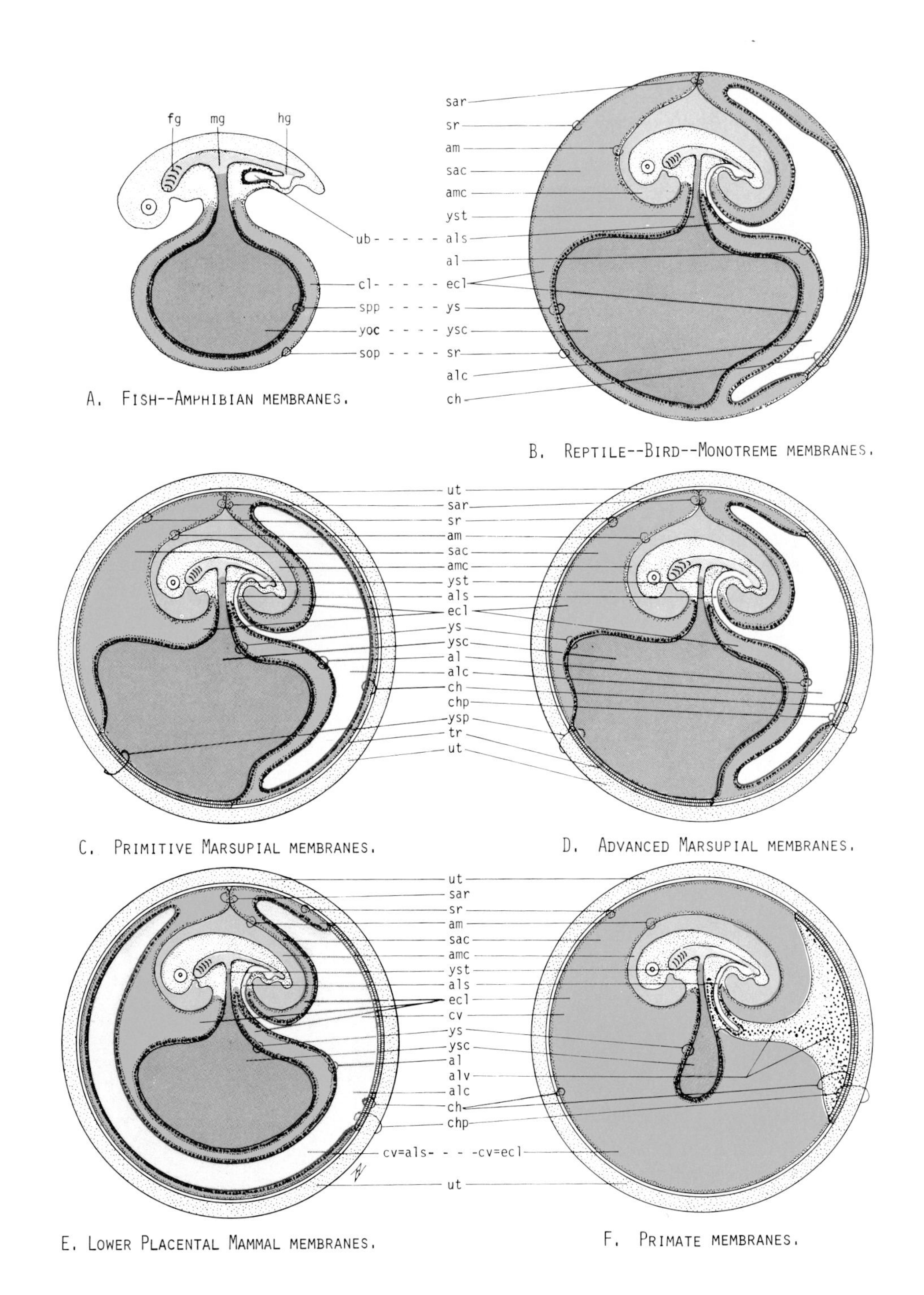

Figure 9-3. Diagrams of comparative evolution of mammalian placentae. See Table IX-1 and explanation in text. Color code: blue — amniotic cavity, red — extraembryonic coelom, yellow — allantoic cavity, green — yolk sac cavity. Legend of abbreviations:

al	allantoic membrane	cv	chorionic vesicle	tr	trophoblast ectoderm
alc	allantoic cavity	ecl	extraembryonic coelom	ub	urinary bladder
als	allantoic stalk	fg	foregut	ut	maternal uterine wall
alv	allantoic splanchnic (vascular) mesoderm	hg	hindgut	yoc	yolky endoderm cells
		mg	midgut	ys	yolk sac
am	amniotic membrane	sac	seroamniotic cavity	ysc	yolk sac cavity
amc	amniotic cavity	sar	seroamniotic raphe	ysp	yolk sac placenta
ch	chorionic membrane	sop	somatopleure = body wall	yst	yolk stalk
chp	chorionic placenta	spp	splanchnopleure = gut wall		
cl	coelom	sr	serosa		

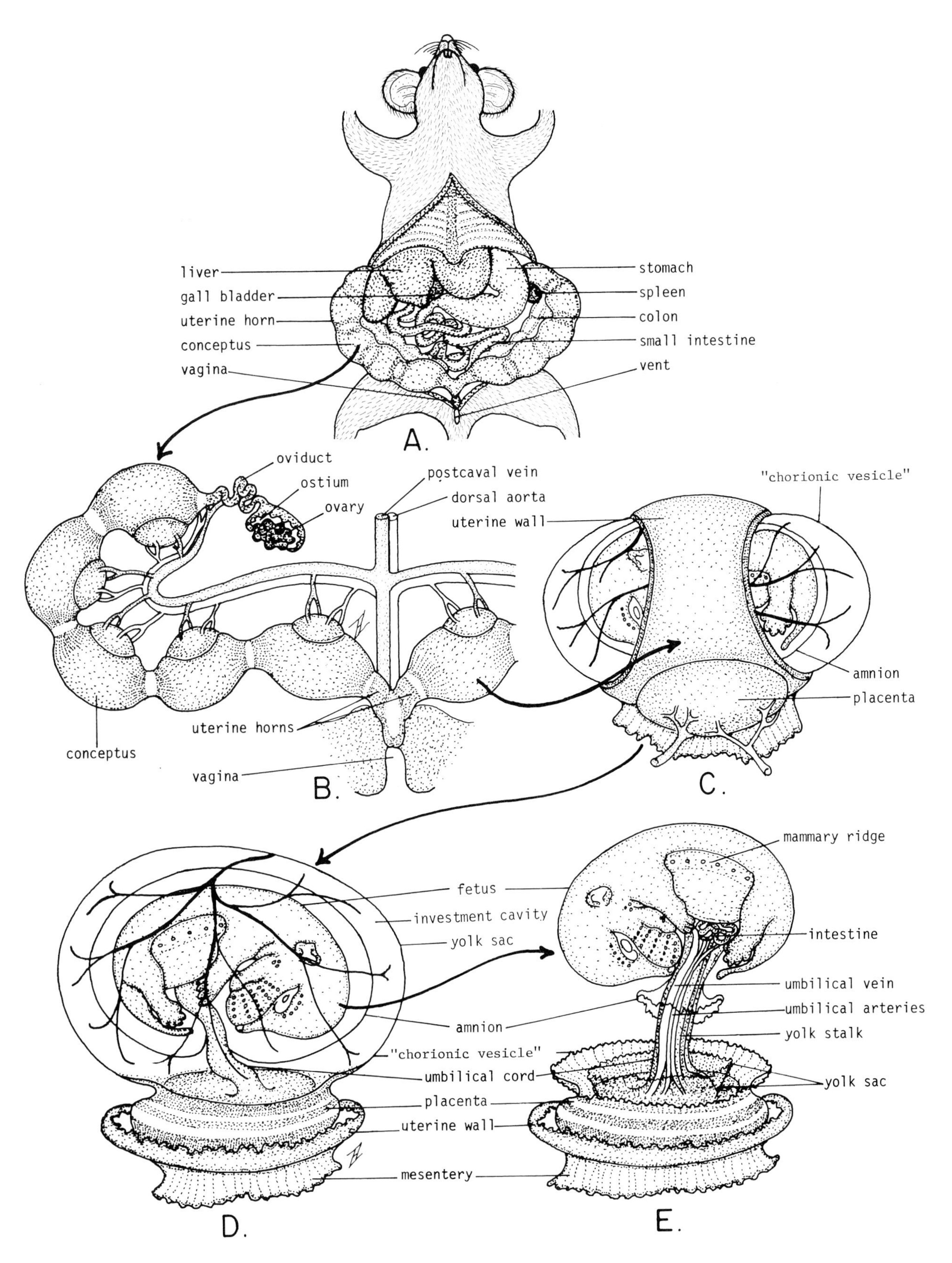

Figure 9-4. Maternal and fetal relationships in the 14 day mouse embryo. 9-4A. The general topography of the pregnant female reproductive system. 9-4B. Gravid uterus removed from the body cavity. 9-4C. The appearance of a conceptus after cutting across the uterus between individual implantations. 9-4D. A fetus in its chorionic membrane after the uterine wall has been cut away. 9-4E. The fetus after chorionic and amniotic membranes have been removed. See *Note,* page 246, for a discussion of peculiarities in formation of the "chorionic vesicle" of rodents.

with forceps, make a single clean cut about one-quarter inch long through the fold of skin. Mouse skin is very thin and soft and will tear easily. Hold the mouse by the tail and hind legs and, with the other hand, pinch the belly skin in front of the cut. Then pull the skin forward and it will tear in a widening "V" to expose the entire ventral surface. With the fingers quickly widen the gap and pull the skin up over the head. This will expose the entire belly and chest region. All this can be done in less time than it takes to read these directions.

The belly muscles are thin and it will be possible to see the outlines of the intestines, ribs, and sternum clearly through them. Also note the outlines of the greatly enlarged uterine horns which extend along each side of the abdomen. If the female is near full-term, it may even be possible to see periodic swellings in the uterine horns which mark the positions of embryos or fetuses within them.

Open the abdominal cavity by making a median incision with scissors that extends from the pubis to the sternum (breast bone). Make a pair of lateral cuts through the muscles that follow the posterior contour of the ribs. Lift back the abdominal muscles and expose the visceral organs. Note the extensive peristaltic movements of the intestines and stomach. Press them anteriorly and identify the right and left **uterine horns** which converge toward the posteroventral midline. The uterine horns may be partially covered by a loosely applied layer of grey fat. Lift this fat away and then pour a small amount of warm 0.9% saline over the visceral organs in sufficient amount to wet the paper around the mouse generously. The uterine horns are dark red in color owing to the abundant vascularization of the entire area. Lift both uterine horns out of the body cavity and extend them laterally so that the full extent of each horn can be seen. Note the large blood vessels which course through the mesentery attaching the horns to the dorsal body wall. Each swelling in the uterus represents one **conceptus** (embryo or fetus). Before proceeding with the next step in the separation and dissection of embryos and uterus, take time to open the thoracic cavity to expose the heart and lungs. Cut through the lateral wall of the rib case and through the diaphragm which lies just anterior to the liver and stomach. The mediastinal septum which connects the pericardial sac to the inner surface of the sternum will also need to be cut to free the rib case from the underlying visceral organs. For general interest, quickly review the internal anatomy of the mouse as an example of an adult mammal.

Remove each uterine horn by cutting its mesentery with a scissors. Cut across the posterior end of the uterus at its junction with the vagina (Fig. 9-4 A). Lift out the uterine horn so that it will lie on the wet paper liner. Return to the anterior end of the uterus and trace it to its origin in the dorsal surface of the abdominal cavity behind the liver or stomach. Identify its small anterior termination, the **oviduct,** and the small lobular **ovary** that is more or less enveloped by the **ostium** of the oviduct. Remove all these parts of the female reproductive system by cutting through their mesenteries. As quickly as possible complete the above dissection and transfer each uterine horn to warm 0.9% saline in a fingerbowl. There will be extensive and unavoidable hemorrhaging at this time. Allow all of the blood to drain from the uterine tissue. Then transfer them to a fresh bowl of warm saline and make the following observations with bright illumination and with a hand lens if one is available.

On the surface of the uterus note the **longitudinal** muscle bundles which give the uterine wall a striped appearance. Observe the slow peristaltic contractions of the uterus. With a sharp scissors make a **single clean cut through the entire uterus** between two of the enclosed embryos. Repeat the process between each successive embryo in the horn, thereby isolating the embryos for individual study. Cutting of the uterus will stimulate the smooth muscle to contract and force the embryo within its chorionic vesicle to extrude from one end of the short section of uterus. If this does not occur, use forceps to tear or strip back the uterine tissue and expose the chorionic vesicle. The chorionic vesicle of the mouse is analogous, but not identical, with corresponding structures in pig and human development as shown in Figures 9-3 E & F.

Note: A pecularity of rodent development is retention of the yolk sac as a very conspicuous membrane which fuses with the trophoblast (the serosal ectoderm) and forms a second fluid-filled sac surrounding the amnion and embryo. This is the "chorionic vesicle" and is an endodermally lined yolk sac cavity. The term "chorionic vesicle" is used in this text to identify analogous structures with similar function in placental mammal development. That is, the chorionic vesicle in the pig and many lower placental mammals is an **"allantoic chorionic vesicle"** (Fig. 9-3E); in primates it is a **"serosal or trophoblastic chorionic vesicle."** (Fig. 9-3F); but in mice and other rodents it is a **"yolk sac chorionic vesicle."** In all cases the function of the chorionic vesicle is the same—it supports the embryo and isolates it within the uterus and these are therefore analogous membrane bound cavities. They are not homologous, however, because their embryonic germ layer and tissue compositions are different. The above remarks refer only to non-placental functions of metabolite exchange between fetal and maternal circulations in rodents in a chorionic and not a yolk sac placenta (as in the case with all euplacental mammals). For additional details on variations and evolution of mammalian placentae, see Jenkinson, *Vertebrate Embryology,* 1913.

Do not tear the chorionic vesicle. Under low magnification, note that the chorionic vesicle is more or less transparent and contains a few large blood vessels (Fig. 9-4 B). The embryo can be clearly seen through the chorion. Note that the chorionic vesicle is firmly attached at one point to the uterine wall by a disc-shaped thickening of the extraembryonic membranes. This attachment to the uterine wall is the **placenta.** It is through this intimate vascular contact that nutritive, respiratory and excretory exchanges take place between fetal and maternal circulations.

With needles and forceps tease the placenta free from the uterine wall. Some uterine tissue will, of necessity, remain attached to the placenta. Note that, owing to the loss of blood, the uterine tissue is quite pale whereas the embryonic part of the placenta remains a dark red. With the chorionic vesicle fully exposed, use Figure 9-4 to aid in the identification of **chorionic vessels, placenta, umbilical cord,** and the **large umbilical arteries** and **vein.** Note that the force of the pulse rocks the embryo within its membranes. Try to identify the almost transparent amnion which surrounds the embryo very closely.

Open the **chorionic vesicle** by tearing it with forceps. As you do this, you will see the very thin amnion bulge out of the opening first and then the embryo will follow. Usually the amnion is broken during this process of liberation. The amnion is not vascular and is very thin; if it is not seen when the chorion is ruptured, it probably will not be seen at all. With forceps, tear the chorionic membrane away from its attachment to the placenta and expose the embryo completely. Note that the embryo is attached to the placenta by the **umbilical cord** through which the **umbilical arteries and vein** carry blood to and from the placenta.

Arrange the embryos in a developmental series and estimate the ages by comparing them with the series provided in Figure 9-5. Table IX-2 gives comparative data on developmental stages in the mouse and human.

On the embryo, identify the nose, eyes, and ears. The **external ear** is in various stages of being molded from the first gill cleft. In older stages the mouth will have a well formed **tongue,** and **lips.** Observe the pattern of **hair follicles** that are especially prominent around the nose and eyes. Several branches of the **anterior cardinal vein** and **external jugular vein** will be found draining blood from the surface of the head and entering the skull back of the ear.

On the trunk, identify the **limb buds** with rudimentary **digits** forming on them. The **milk ridge** is the primordium of the mammary glands. This is a small elevated line that runs between the bases of the fore and hind limbs. Note that the **small intestine** in early fetal stages projects out into the base of the umbilical cord. This is a normal condition which results from the fact that the **liver** is very greatly over-developed at this time and occupies most of the coelomic cavity. The excessive size of the liver is correlated with its role as the primary blood forming organ of the embryo.

At the posterior end of the body, note the **external genitalia** between the hind limb buds. At this stage the external genital organs are at an indifferent period in development and it is not possible to identify the sex of the fetus.

Preserve at least one embryo of each stage; however, additional embryos are useful. Any embryonic fixative will serve (see formula in Table VII-1); however, CaAFA is recommended. These embryos, along with the preserved stages in chick development, will be used in Chapter 10.

At the end of the dissection, wrap all waste materials in paper towels and dispose of them in the waste container provided. Wash all glassware and rack it at the sink area for air drying.

TABLE IX-2
DEVELOPMENTAL TIME SCHEDULES FOR MOUSE AND HUMAN FETUSES

DEVELOPMENTAL STAGES	MOUSE AGE	MOUSE CROWN-RUMP LENGTH	HUMAN AGE	HUMAN CROWN-RUMP LENGTH
Fertilization	0	–	0	–
2 cell stage	0.5–1.0 days	–	1.0–1.5 days	–
4 cell stage	1.0–1.5 days	–	1.5–2.0 days	–
8 cell stage	1.5–2.0 days	–	2–3 days	–
16 cell stage	2.0–2.5 days	–	3–4 days	–
Morula	2.5–3.0 days	–	5–6 days	–
Blastocyst	3–4 days	–	1 weeks	–
Embryonic disc	4–5 days	–	1.5 weeks	–
Primitive streak	5–6 days	0.5 mm	2 weeks	1.0 mm
Neurula	7 days	1 mm	3 weeks	2–3 mm
Early tailbud	8 days	2 mm	4 weeks	3–4 mm
Mid-tailbud	9 days	3 mm	5 weeks	5–7 mm
Late tailbud	10 days	4 mm	6 weeks	8–12 mm
Fetus	11 days	5–6 mm	7 weeks	15–17 mm
Fetus	12 days	6–7 mm	8 weeks	1 inches
Fetus	13 days	8–9 mm	10 weeks	1.5 inches
Fetus	14 days	10–11.5 mm	3 months	2–2.5 inches
Fetus	15 days	12–14 mm	4 months	4–5 inches
Fetus	16 days	14–16 mm	5 months	5.5–6.5 inches
Fetus	17 days	16–17.5 mm	6 months	7–9 inches
Fetus	18 days	18–19 mm	7 months	9–10.5 inches
Fetus	19 days	19–20 mm	8 months	11–12 inches
Fetus	20 days	20–21 mm	9 months	12–13 inches
Birth	21 days	21–22 mm	9.5 months	13–14 inches

Data from various sources including: L.B. Arey (1974) *Developmental Anatomy,* Saunders; W.J. Hamilton, J.D. Boyd and H.W. Mossman (1945) *Human Embryology,* Williams & Wilkins; and R. Rugh (1964) *Vertebrate Embryology,* Harcourt, Brace & World.

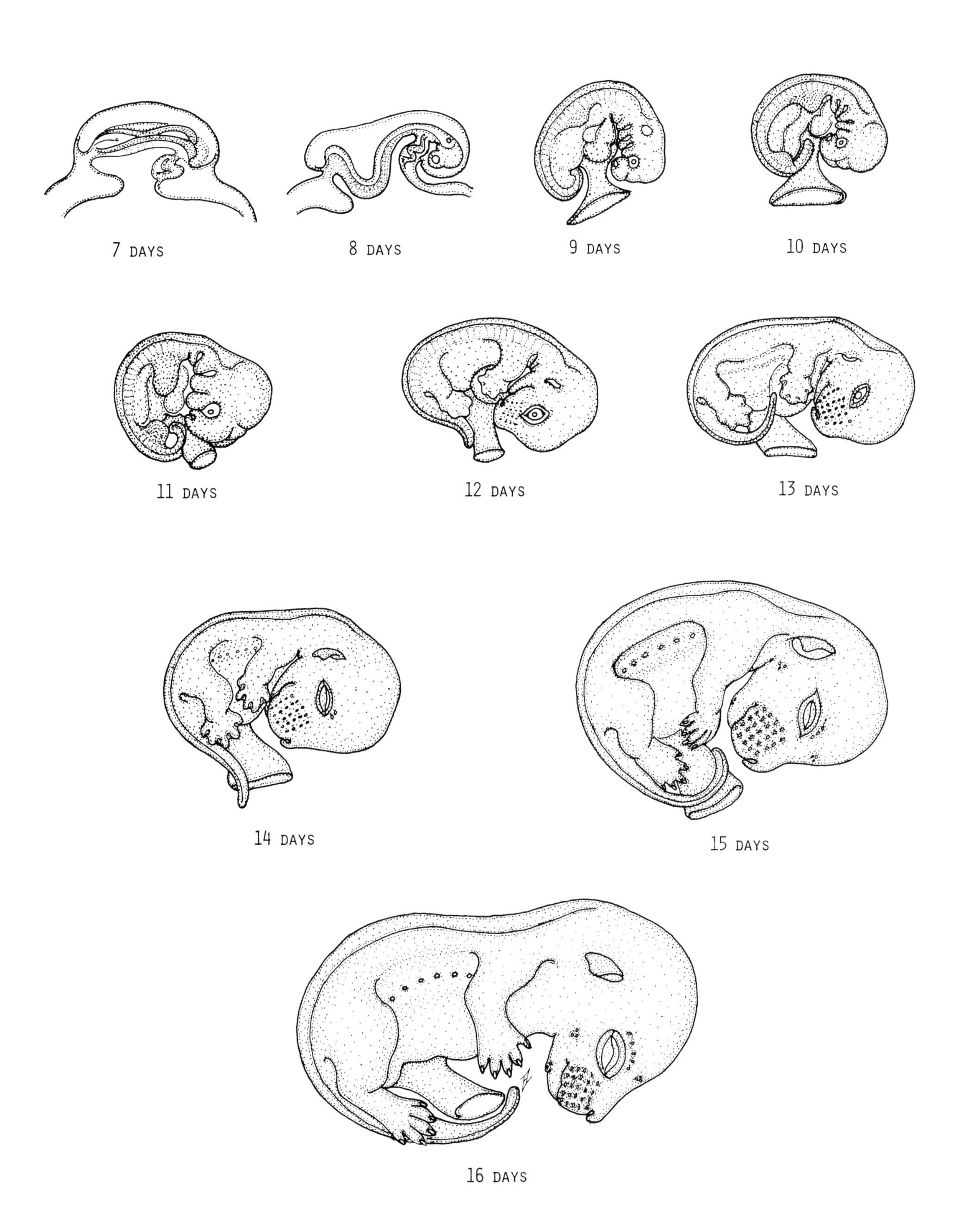

Figure 9-5. Mouse development from 7 to 16 days gestation. Compare with stages given in Table IX-2.

B. TRANSITION BETWEEN EMBRYONIC AND ADULT MAMMALIAN ORGANIZATION AND STRUCTURE ILLUSTRATED BY THE 10 MM PIG EMBRYO

The anatomical structure of the 10 mm pig embryo is particularly instructive in providing a link between the primordial tailbud stage and older fetal stages possessing the diagnostic characters distinctive of the Class Mammalia. Many organ rudiments are still in a primordial condition (i.e., limb buds, nasal pits, eyes, metanephros, genital ridge, brain vesicles, etc.), but others such as the heart, mesonephros, muscles, and nerves are in the process of acquiring physiological functions. The 10 mm pig may reasonably be taken as a near mid-point in the progressive prefunctional morphogenesis of mammals.

At this stage it is possible to visualize most structures in the light of their simple origins, as studied under parallel conditions in the development of Amphibian and chick. The 10 mm pig embryo does not show specific characteristics which permit it to be distinguished from any other mammal and therefore can be taken as representative of the entire Class Mammalia. More interesting, however, is the fact that most structures that will be identified have not yet reached the "mammalian level" in structural specializations, if we consider the homologous organs from the point of view of diagnostic adult mammalian traits. Indeed, only one primordium exists at the 10 mm stage that is exclusively of mammalian type, this is the mammary ridge which is restricted to this class alone.

1. General topography of the 10 mm pig. A large measure of the general information given in the discussion and study of the 48 and 72 hour chick embryos is directly applicable to the 10 mm pig. The understanding of these simpler stages is a necessary background for the following study of the 10 mm pig. To be sure, there are some differences in general proportion in different organs, but it will be found that the structural relationships of the sense organs, brain regions, cranial nerves, pharyngeal pouches, and arches are quite comparable to those observed in the 72 hour chick. In other organs and regions, however, the 10 mm pig shows numerous advances which represent transitions between the early fish-like organization of the young embryo and the specialized modifications of the adult mammalian body plan. It will be found, for example, that the early tubular condition of the heart is in the process of being remolded into separate right and left halves. Many early blood channels are in the process of degenerating and new definitive vessels are appearing. The derivatives of the mid- and hindgut are more advanced than those seen heretofore. Particularly impressive is the degree of differentiation to be found in the development of the meso- and metanephroi.

TABLE IX-3
APPROXIMATE TAXONOMIC LEVELS OF DIFFERENTIATION IN ORGAN PRIMORDIA AT THE 10 mm STAGE IN THE DEVELOPMENT OF THE PIG

CYCLOSTOME	ELASMOBRANCH	OSTEICHTHYES	AMPHIBIA	REPTILIA	MAMMALIA
forebrain, midbrain, otocyst	nasal pits, gill slits, venous system	limb buds, kidney	arterial system, conus arteriosus	ventricle, coelom, cranial nerves	mammary ridge

It is clear from the foregoing partial list of primordia that the 10 mm pig embryo can in no way be said to represent a single adult vertebrate Class in the evolutionary history of mammals; however, it is equally clear that different organs and organ systems do pass through characteristic ancestral steps in their acquisition of definitive adult characteristics. These gradual transformations of the common plans for the various systems underscore the validity of von Baer's Laws. In a manner of speaking, one can say that Haeckel's "recapitulation" although not applying to entire developmental stages, does seem to apply to the development of organs and organ systems. However, different parts of the same embryo are "out of step" with others in the time sequence of development.

The study of the 10 mm pig will follow the same general procedure that has been used in the study of chick embryo. Individual series of cross and sagittal sections are supplied for study. In addition, a number of whole mounts of pig embryos of graded age are on demonstration. A whole mount diagram and representative serial sections are given in Figures 9-2, 9-12, 9-13 A–Z, and 9-14 A–H as aids for identification of special structures and to provide a permanent record of your observations. As early as possible in the study of your cross sections, identify the plane of section of your own series in order to orient yourself with regard to the arrangement of parts seen in the organization of the entire embryo.

Identify the **mesencephalic, pontine, cervical,** and **lumbo-sacral flexures** which have been defined earlier in studies of the 48 and 72 hour chick embryos. As in the chick, these flexures will bend the anteroposterior axis of the pig into the shape of a "C". Identify the thick **umbilical cord** which protrudes obliquely anterad from the ventral body wall. The base of the umbilical cord represents the location at which the head and tail folds of the embryo have completed the separation of the embryonal area from the extraembryonic regions. In these preparations and in the serial sections, the extraembryonic membranes have been removed; it will not be possible to study the relationships of the amnion, serosa, allantois, and yolk sac. Note that there is no open midgut remaining and that, with the complete undercutting of the embryonal area by the head and tail folds, there are no thoracic or sacral torsions which characterized the 72 hour chick. Use the whole mount diagram to identify as many general organ regions as is possible (i.e., **brain, sense organs, visceral grooves, heart, liver, mesonephros, limb buds,** and **tail**). Also note the **mammary ridge** extending between the **fore** and **hind limb buds.**

2. Study of serial cross sections and sagittal sections (Fig. 9-13 A–Z' and 9-14 A–H). The remaining study of the 10 mm pig will be based on the study of sections, since the embryo is too large for use as whole mounts. Most median structures will be seen to advantage in the sagittal plane, but cross sections will merit primary attention in the detailed study of organ structure. Many structures to be found in the 10 mm pig are quite similar to, and are readily interpretable by, a thorough understanding of the 72 hour chick. These similarities will be pointed out, but **major emphasis will be placed on developmental advances which anticipate the adult mammalian body plan.**

Ectodermal derivatives

a. Brain and spinal cord. The **cervical flexure** bends the entire head region sharply ventrad. As was the case in the 72 hour chick, the posterior choroid plexus, covering the 4th ventricle of the myelencephalon, is the first part of the nervous system to be encountered in serial cross sections. Begin the study of the nervous system with the myelencephalon and proceed anteriorly through the head to the telencephalon. Most directions given previously for the study of the brain in the 72 hour chick are directly applicable to the 10 mm pig and should be consulted in your study. Note the following additions.

Myelencephalon (Fig. 9-13 A–E). Under high power examine the wall of the myelencephalon and note that it is now differentiated into clearly recognizable nucleated and fibrous layers. The nucleated inner two-thirds of the wall will form the gray matter of the brain. It is now called the **mantle layer.** The innermost layer of cells of the mantle layer which lines the neural canal is the **ependymal layer.** It functions as the germinal epithelium of the neural tube. By repeated mitotic divisions in this layer, new neuroblasts are added to the mantle layer. The outer layer of the brain wall stains poorly and is composed primarily of nerve fibers. This layer is the primordium of the white matter of the brain. At this time it is called the **marginal layer.** These three layers of the neural tube can be identified in all levels of the brain and most of the spinal cord. Proceed ventrally through the myelencephalon and note that the neuromeres which characterize the wall of the myelencephalon are primarily associated with the nucleated mantle layer and do not appear in the outer marginal layer of nerve fibers. For the relationships of the myelencephalon to cranial nerves V through XII, refer to the following section on cranial nerves.

Metencephalon (Fig. 9-13 A, B). This region has undergone little developmental advance over that seen in the 3 day chick embryo. The landmarks of its boundaries are the laterally compressed **isthmus** regions anteriorly, and the first neuromere and **pontine flexure** posteriorly. No nerves leave this brain region.

Mesencephalon (Fig. 9-13 A–D). As in the 3 day chick, this brain region is typically circular in cross section; this is its most reliable identification key. No other features of this brain region merit

TABLE IX-4
STRUCTURE AND FUNCTION OF THE MAMMALIAN CRANIAL NERVES

NUMBER AND NAMES OF NERVES	ORIGIN OF NERVE & GANGLION**	RELATIONSHIP TO BRAIN	ENDORGANS AND FUNCTIONS*	STATUS IN THE 10 mm PIG EMBRYO
I. OLFACTORY NERVE	**(EPID) no ganglion	enters the floor of the telencephalon	*(SSS) origin from the nasal epithelium	not differentiated
II. OPTIC NERVE	(NP) no ganglion	enters the floor of the diencephalon forming the optic chiasma	(SSS) origin from the nervous layer of the optic cup	not differentiated
III. OCULOMOTOR NERVE	(NP) no ganglion	arises from the floor of the mesencephalon	(SM) to superior-,inferior-, and internal rectus, and inferior oblique muscles of eye	well differentiated
IV. TROCHLEAR NERVE	(NP) no ganglion	arises from the roof of the mesencephalon at the isthmus	(SM) to superior oblique muscle of eye	not differentiated or very poorly developed
V. TRIGEMINAL NERVE	(NP, CNC, PL1) semilunar ganglion	enters myelencephalon at the level of the first neuromere	nerve of first (mandibular) arch:	well developed nerve and ganglion
opthalmic (profundus) ramus	- - -	- - -	(SS) from skin of nose and anterior scalp	well developed
maxillary ramus	- - -	- - -	(SS) from upper jaw, lips and teeth	well developed
mandibular ramus	- - -	- - -	(SS,VM) to lower jaw, teeth, and muscles of mastication	well developed
VI. ABDUCENT NERVE	(NP) no ganglion	arises from the floor of myelencephalon at level of 3rd & 4th neuromeres	(SM) to external rectus muscle of eye	poorly developed or rarely absent
VII. FACIAL NERVE	(NP, CNC,PL2) geniculate ganglion	enters lateral wall of myelencephalon at level of 3rd neuromere anterior to otic vesicle	nerve of 2nd (hyoid) arch; (VS) from anterior mouth & tongue; (VM) to facial muscles & most salivary gland tissue	well developed nerve and ganglion
VIII. AUDITORY NERVE	(EPID) acoustic ganglion	enters lateral wall of myelencephalon at 3rd neuromere with 7th nerve	(SSS) from sacculus, utriculus and cochlea of inner ear	ganglion present but nerve fibers absent
IX. GLOSSOPHARYNGEAL NERVE	(NP, CNC, PL3) superior & petrosal ganglion	enters lateral wall of myelencephalon at level of 5th neuromere posterior to the otic vesicle	nerve of 3rd visceral arch; (VS) from lining of posterior mouth and anterior pharynx; (VM) to some muscles of mouth tongue and anterior pharynx	well developed nerve and ganglia
X. VAGUS NERVE	(NP, CNC, PL4) jugular & nodos ganglion	enters the lateral wall of the myelencephalon at the level of neuromeres 6 & 7	nerve of 4th, 5th, & 6th arches; (SS) from posterior scalp; (VS) from posterior pharynx, lung, trachea, gut, lungs, heart, etc; (VM) to the above visceral organs	well developed nerve and ganglia
XI. SPINAL ACCESSORY NERVE		arises from dorsolateral wall of anterior spinal chord and follows path of vagus into shoulder	(VM) to visceral muscles of neck and shoulder of post-6th arch origin (trapezius and sternomastoid)	well developed
XII. HYPOGLOSSAL NERVE		arises by several roots from posterior floor of myelencephalon	(SM) to hypoglossal muscles of tongue of occipital myotome origin	well developed

* (SSS) = special somatic sensory; (SS) = somatic sensory; (VS) visceral sensory; (SM = somatic motor; (VM) = visceral motor.

** (CNC) - cranial neural crest; (EPID) = epidermis of sensory placode of nose or ear; (NP) = neural plate ectoderm; (PL 1--4) = epidermal placodes of the first through fourth visceral grooves.

attention except its relationship to the origins of the III and IV cranial nerves which will be considered later.

Diencephalon (Fig. 9-13 E–K). In cross section the diencephalon is compressed laterally and the **infundibulum** in the ventral floor of the diencephalon is relatively much smaller than in the 3 day chick. Its relation to Rathke's pouch (stomodaeal hypophysis) is, however, quite comparable. There is no clearly recognizable epiphysis evaginating from the roof of the diencephalon; in this respect the 10 mm pig brain is less advanced than that seen in the 72 hour chick. The **optic recesses** are well formed, and their continuation from the **third ventricle** through the **optic stalk** to the **optic cup** can be clearly traced. No optic nerve fibers connect the sensory retinal layer of the optic cup with the floor of the diencephalon.

Telencephalon (Fig. 9-13 K–O). The bilaterally paired primordia of the cerebral hemispheres are particularly well shown in anterior sections through the brain. However, aside from minor differences in proportion and in thickness of the brain, there are no special features present which were not identified in the 3 day chick embryo. The **lamina terminalis** marks the ventral boundary between the telencephalon and diencephalon. It is best seen in median sagittal section, although the difference in thickness of this region as compared with the remainder of the brain wall is not as distinct as that seen in the 3 day chick.

Spinal cord (Fig. 9-13 E–Z). Almost any section posterior to the brain will show the typical structure of the spinal cord. In addition to the **ependymal, mantle,** and **marginal layers** described above in connection with the myelencephalon, note the beginnings of functional organization in the wall of the neural tube. Identify the **roof plate** (thin and non-nervous), **alar plate** (dorsolateral sensory column), **basal plate** (ventro-lateral motor column), **floor plate** (thin and non-nervous). The **sulcus limitans** appears as a depression in the lining of the neural canal and marks the boundary between sensory and motor areas in the lateral wall of the neural tube (Fig. 9-13 R). Note that the ventral half of the spinal cord is relatively thicker than the dorsal half. This is primarily due to the rapid increase in motor neuroblasts in the **mantle layer** of the basal plate. Nerve fibers from these cells will leave the spinal cord segmentally and will form the ventral (motor) roots of the spinal nerves. Choose a section which shows a well formed spinal ganglion and note the connection between the nerve fibers of the **sensory neuroblasts** of the spinal ganglion with the dorsal wall of the spinal cord. Also trace the sensory and motor nerve fibers of the dorsal and ventral roots of a spinal nerve to the adjacent myotome. These are the **pioneer nerve fibers** which make contact with the neighboring myotomes and establish the pathways for contact guidance of nerve fibers which develop later. Compare the structures seen in spinal cord sections with the diagram of the central and peripheral nervous system in Figure 8-4.

b. Cranial nerves. In general, all remarks made previously in the discussion of the 48 and 72 hour chick embryos concerning the topography of cranial nerves and their relationships to the brain wall or respective end-organs also will apply to the 10 mm pig. In addition, the cranial nerves of the pig show several advances. These include the presence of additional ganglia and a few new nerves. For economy of space the cranial nerves will not be discussed individually. Instead, refer to Table IX-3 which tabulates the state of differentiation of the cranial nerves in the 10 mm pig embryo. Figure 9-13 A—I shows the locations of the nerves as they appear in cross sections. A final word should be added concerning the appearance of the nerves in sectioned preparations. In almost all cases a nuclear stain has been used in these slides, and cytoplasmic structures such as nerve fibers will stain poorly. Frequently the fibrous parts of the cranial nerves (i.e., the 3rd, 6th, 11th, & 12th, in particular) will appear in sections as transparent rather than as stained structures.

c. Spinal and autonomic nerves (Fig. 9-13 C–Z). The spinal nerves have already been noted in connection with the study of the spinal cord. The segmental arrangement of these structures can best be seen in sagittal sections (Fig. 9-13C) just lateral to the neural tube, or in posterior cross sections which pass through the spinal cord in a frontal plane at the level of the lumbo-sacral flexure (Fig. 9-13 Z'). In sections such as these, the linear segmental arrangement of the large dorsal root ganglia (composed of somatic sensory and visceral sensory neuroblasts) and their relationship to the myotomes and spinal cord are clearly seen. In cross sections the relationship of the dorsal and ventral roots to the spinal nerves and neural tube can be seen at the level of the forelimb bud where the **brachial plexus** is rapidly differentiating (Fig. 9-13 Q). **Autonomic ganglia** are also differentiated and segmentally arranged. They are located dorsolateral to the dorsal aortae and can be seen in Figure 9-13 K–M. They are distributed along most of the length of the trunk. Nerve fibers can occasionally be seen connecting the autonomic ganglia to the spinal nerves by a **ramus communicans** (Fig. 9-13 O, P). Refer to Figure 8-4 for a review of the common plan for the reflex arc in the vertebrate nervous system. The relationships of various types of functional neurons to their respective association centers in the central nervous system, and to their end organs, as given there, also applies to mammals.

d. Sense organs. Review the description of development of the nose, eye, and ear primordia in the 72 hour chick embryo. Very few additional comments need be made concerning these organs in the 10 mm pig. The **nasal pit** (Fig. 9-13 L, M) of the pig embryo is deeply invaginated but, as yet, only rarely are olfactory nerve fibers from the nasal epithelium traceable to the floor of the telencephalon. The **optic cup, choroid fissure, lens vesicle, cornea,** and **optic stalk** are less well developed in the 10 mm pig than in the 3 day chick, but the various components, mentioned earlier for the chick, and shown in Figure 9-13 A–E, illustrate these structures of the developing eye quite clearly.

The **otic vesicle** (Fig. 9-13 A—E) exhibits one clear advance over that of the 3 day chick in that the **semicircular canals** can be distinguished as short evaginations in its lateral, anterior and posterior walls. The mid-regions of the canal primordia are still open and communicate directly with the cavity of the otic vesicle which is the *anlage* of the **sacculus** and **utriculus.** The **endolymphatic duct** is also more clearly defined than in the 3 day chick. It arises as a median dorsal evagination from the mesial wall of the otic vesicle. Note again the close association between the **acoustic ganglion** of the VIII cranial nerve and the anteroventral wall of the otic vesicle.

e. Stomodaeum and **oral cavity** (Fig. 9-13 F–L). No vestige of the oral membrane remains to provide a landmark for the boundary between the stomodaeum and anterior foregut. The relationships of the mouth to **Rathke's pouch, jaws,** and **pharynx** are essentially similar to those seen in the 3 day chick and can best be seen in median sagittal sections. The close contact between Rathke's pouch and the **infundibulum** is also seen in this plane. The relationship of the jaws to the mouth cavity shows to best advantage in cross sections; they can be traced by beginning in front of the head at the level shown in Figure 9-13 L and proceeding forward through the cross sections until contact between the oral and pharyngeal cavities is established. In doing this, note, first, that the **nasal pit** (Fig. 9-13 L) has not opened internally into the stomodaeum. This will occur somewhat later in development. Going forward through the oral cavity, identify the **mandibular** and **maxillary processes** of the first visceral arch which form the lateral margins of the oral cavity. They consist of dense aggregations of mesenchyme in which branches of the 5th cranial nerve can be identified (Fig. 9-13 G–J). In addition, branches of the **external carotid artery** will be found in the lower jaw, but no complete first aortic arch persists at this time. In the upper jaw primordium (maxillary process), note that, although the roof of the mouth cavity is not greatly arched, there are nevertheless aggregations of mesenchyme in the roof that will later form the palate in mammals and will separate the oral and nasal cavities. These are the **palatine processes** (Fig. 9-13 H–J). As you progress forward to the point at which the maxillary and mandibular processes fuse to form the angle of the jaws, the connection between Rathke's pouch (stomodaeal hypophysis) and the stomodaeum can be seen. At approximately this level, the mouth cavity begins to expand greatly between the two mandibular processes. In a few sections the stomodaeum will open into the anterior end of the pharynx. In the absence of an oral membrane, the division between the pharynx and stomodaeum can be identified by the presence of the **first pharyngeal pouch** (see Table IX-5). The latter lies between the mandibular process and the hyoid arch. At this level (Fig. 9-13 G), the **tongue** can be seen as a mid-ventral evagination at the median fusion of the right and left **hyoid arches,** in the floor of the oral cavity. Follow the sections a short distance forward until the posterior part of the pharynx opens into the oral cavity. This will occur in cross sections at the level of the 3rd and 4th visceral arches (Fig. 9-13 F–H).

The identification of various structures in the **visceral arches** will not be given special consideration in these directions for the study of the 10 mm pig embryo. However, their identification and relationships are implied in the study of the cranial nerves (see above), the stomodaeum, pharynx, and aortic arches. For a review of the pertinent facts concerning the visceral arches, refer to Figures 5-5 and 5-6 and the discussion accompanying them in the directions for the study of the salamander larva. Use this opportunity to review the structural relationships of the visceral arches and the fate of these arches in later development. Recall that each visceral arch in the primitive condition contains a nerve, visceral cartilage, an aortic arch, and visceral muscles. The first two of these arch components are of neural crest origin and the latter two are of head mesenchyme origin.

TABLE IX-5
DERIVATIVES AND FATE OF MAMMALIAN PHARYNGEAL POUCH ENDODERM

1st Pouch: The dorsal and ventral horns contribute to the tympanic cavity of the middle ear with the three ear bones developing between them. The final expansion of both horns will suspend the ear bones across the tympanic cavity by small endodermal "mesenteries." Closing plate I is the primordium of the eardrum; most of the membrane, **tympani tensor,** representats that part of the closing plate associated with the ventral horn of pouch I. The much smaller dorsal, **tympani flaccida,** corresponds with closing plate associated with the dorsal horn of pouch I. The base of Pouch 1 forms the Eustachian tube which in the adult connects the middle ear cavity to the pharynx. The first visceral groove external to closing plate I will form the **external auditory canal.**

2nd Pouch: The dorsal and ventral horns degenerate. The base of the pouch remains as the crypt of the palatine tonsil (non-functional).

***3rd Pouch:** The dorsal horn will form the **posterior parathyroid glands.** The ventral horn will form the **posterior (thoracic) thymus glands** exclusive of lumphoid cells.

***4th Pouch:** The dorsal horn will form the **anterior parathyroid glands.** The ventral horn will form the anterior (cervical) thymus glands exclusive of lumphoid cells.

5th Pouch: The dorsal horn usually fails to form in mammals at all; the ventral horn forms the **ultimobranchail body** which is the source of the recently discovered hormone, thyrocalcitonin, involved in regulation of the parathyroid glands.

***Note:** In the above Table there is an apparent inconsistency in the anterior and posterior relationships of the embryonic 3rd and 4th pouches and the final positions of their derivatives. This reversal of anterior — posterior polarity is a secondary consequence of posterior migration by the heart from its embryonic location in the upper neck and thoat, to its final mammalian position deep in the thoracic cavity. A discussion of this phenomenon will be given in association with the development of heart and aortic arches in a following sction. Here it is sufficient to indicate that the anterior attachment of the conus arteriosus and ventral aorta is directly under arch 3 in the early embryo; as the neck elongates and the heart regresses posteriorly, associated structures related to arch 3, and pouch 3 which lies immediately behind, tend to be "dragged along." As a consequence the derivatives of pouch 3 will come to lie posterior to those of pouch 4 (which, it might also be added, are also displaced posteriorly to some degree). In view of the above events, it is not surprising that there is a very high incidence of "anomalies" in the positions often encountered in the location of parathyroid and thymus glands of adult humans and other mammals.

Pay particular attention to the appearance and location of the dorsal and ventral ganglia of the 9th and 10th cranial nerves since they serve as useful landmarks for the identification viscera and aortic arches (i.e., nerve 9 — arch 3, nerve 10 — arches 4-5-6).

Endodermal derivatives

a. Pharynx (Fig. 9-13 F—K). Continue the study and identification of pharyngeal derivatives by going back through the sections just covered in the study of the stomodaeum and oral cavity. However, now pay attention to the fate of the pharyngeal part of the oral cavity. Identify the 5 pharyngeal pouches. Note the **closing plates** of the first three and the visceral grooves associated with them (Fig. 9-13 F—I). The 4th and 5th pharyngeal pouches are atypical in that they do not approach the surface epidermis and form closing plates. Nevertheless, the surface epidermis is depressed in this general region and forms a deep groove called the **cervical sinus** which is homologous to a combined 4th and 5th visceral groove (Fig. 9-13 J). The 5th pharyngeal pouch is greatly modified and actually seems to arise as a slender ventrolateral extension of the 4th pouch. Each 5th pouch ends in a small, expanded vesicle termed the **ultimobranchial body.** The fates of the endodermal linings of the five pharyngeal pouches, along with some of their closely associated closing plates and epidermal grooves, are summarized in Table IX-5. Refer to Figure 8-5, which diagrams derivatives of the foregut, and note that, primitively, each pharyngeal pouch except the last, is bilobed and possesses so-called **dorsal** and **ventral horns.**

The **thyroid diverticulum** of the 10 mm pig is no longer attached to the pharyngeal floor, nor does it retain the vesicular shape that characterized this organ in the 3 day chick embryo. Thyroid endoderm will be found deep in the mesenchyme of the 2nd visceral arch in close contact with the anterior border of the aortic sac, or ventral aorta (see Fig. 9-13 H, I).

b. Esophagus, trachea, and lungs (Fig. 9-13 I–Q). The **laryngotracheal groove** begins to evaginate from the mid-ventral floor of the pharynx at the level of the fourth pharyngeal pouch. At its anterior end, the groove is surrounded by an exceedingly thick and compact mass of mesenchyme, the primordium of the **glottis** and **larynx.** It is composed of an indistinguishable mixture of neural crest and splanchnic mesoderm cells which will contribute respectively to the formation of visceral skeleton and visceral muscles of the larynx. Posteriorly, the **esophageal** and **tracheal** regions of the foregut become entirely separated from one another. The trachea lies in the ventral mesentery of the esophagus (also dorsal mesocardium of the heart). The **lung buds** are just forming and expanding laterally from the posterior end of the trachea. They show little advance over the level of differentiation seen in the three day chick.

c. Stomach, liver, pancreas, and omentum (Fig. 9-13 R–U). Follow the esophagus posteriorly and note that the gut lumen rapidly enlarges and becomes obliquely oriented toward the left. This part of the gut is the **stomach.** The dorsal mesentery of the stomach will give rise to the **greater omentum,** a unique mammalian structure consisting of an extensive fold of mesentery that will extend over the ventral surface of the abdominal viscera. The greater omentum if primarily a fat-storing, insulating structure that protects the abdominal viscera; it also forms the "bay windows" of successful or indolent fat persons. At the 10 mm stage in the development of the pig, a small portion of the coelom will be seen between the right side of the stomach and the dorsal mesentery. This cavity is the **omental bursa** (lesser peritoneal cavity). The opening between the omental bursa and the remainder of the peritoneal cavity (Fig. 9-8 F, 9-13 T) is at the left between the **caval mesentery** and a liver lobe. This opening is elegantly named the **epiploic foramen of Winslow.** The ventral mesentery of the stomach, or **lesser omentum,** connects the stomach to the **transverse septum** which contains the liver. Follow the sections posteriorly beyond the expanded stomach region to the point at which the gut lumen becomes greatly reduced in size. This is the **duodenal** part of the small intestine. Trace the course of the **common bile duct** from the intestine through the ventral mesentery into the **gall bladder** and **liver** (Fig. 9-13 T). Posterior to this region the gut possesses only a dorsal mesentery; no vestige of a ventral mesentery persists, except at the posterior end of the body cavity in the region of the urogenital sinus where the ventral mesentery reappears as the median ligament of the urinary bladder. This structure is not yet evident but will form as an expansion of the allantoic stalk near its origin from the hindgut. At the level at which the common bile duct and gall bladder appear (Fig. 9-13 T), identify the dense glandular mass situated in the dorsal mesentery of the duodenum. This is the primordium of the **dorsal pancreas.** At this time there is no lumen, and it is generally difficult to trace its origin as an evagination of the endodermal gut wall. Examine the **liver** and note that it is very large and spongy in appearance. The spaces in the liver tissue are **blood sinusoids;** they represent divisions of the fused vitelline veins which were invaded by rapidly growing liver cords. At this period in development, the liver is the most important blood-forming organ of the embryo, and its apparent excessive size is associated with this circulatory function.

d. Intestine and cloaca (midgut and hindgut, Fig. 9-13 U–X). On the whole mount diagram (Fig. 9-12), review the general course of the intestine posterior to the stomach. Note that the intestine forms a loop that extends deep into the umbilical cord and that it is attached to a vestigial yolk sac at quite a distance from the body. Most sections for study do not include the entire **umbilical loop of the intestine** because the umbilical cord was severed close to the body. In the study of serial sections, follow the course of the anterior and posterior arms of the intestinal loop as far as possible even though it may not be possible to find a point at which they unite. Note that along most of its length, the intestinal lumen is very greatly reduced, or entirely absent. In the latter event the endodermal cells form a solid cord that will form a new lumen during later developmental stages.

Begin the study of the hindgut at its posterior end. Start with the tail fold at a level showing the **cloacal membrane,** a solid plate of invaginated epidermal ectoderm (Fig. 9-13 U). At this level, identify the postanal gut, or tailgut. Trace the postanal gut forward and observe that it expands into a cavity of considerable size, namely, the **cloaca** (Fig. 9-13 V). Continue anteriorly through the cloaca and note that it becomes divided into a dorsal **rectum** and a ventral **urogenital sinus.** The tissue separating these subdivisions of the cloaca is the **urorectal septum** (Fig. 9-13 W). Trace the rectum to its connection with the posterior arm of the umbilical loop of the intestine. Return to the urogenital sinus and trace it forward to

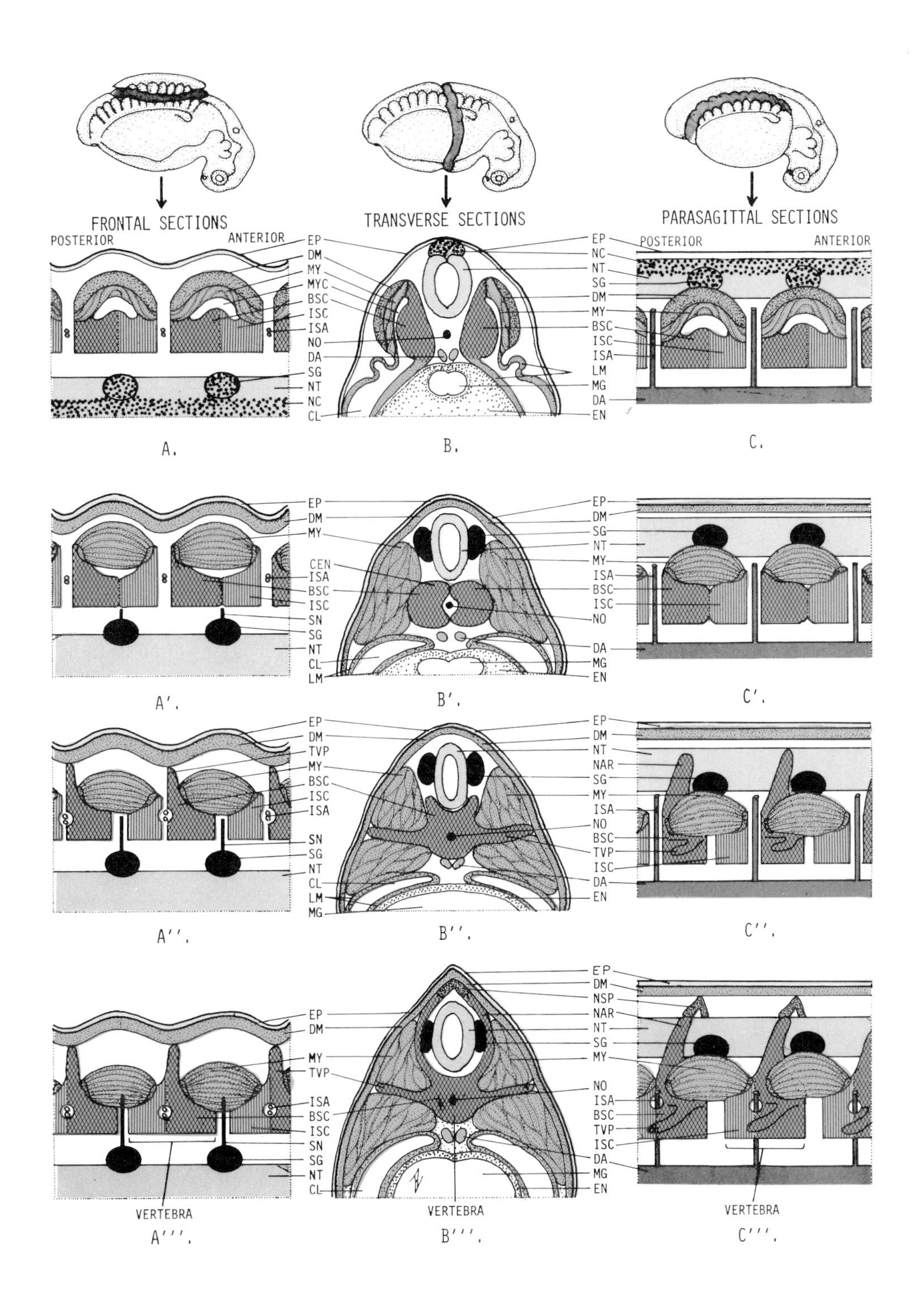

Figure 9-6. Development of intersegmental vertebrae from primary segemental sclerotomes. Diagrams of four progressive developmental stages are shown in each of the major planes of body symmetry, i.e.: 9-6 A, A', A'', A''' = frontal sections; 9-6 B, B', B'', B''' = transverse sections; 9-6 C, C', C'', C''' = parasagittal sections. Note that the sclerotomes of early stages are a major component of the embryonic somites and are therefore initially segmental in arrangement. However, at the end of development each sclerotome becomes divided and the halves of the original sclerotome will realign so that the anterior half of each sclerotome will fuse with the posterior half of the one in front of it. As a consequence the definitive vertebral centrum (composed of two half-sclerotomes of different segments) will be morphologically intersegmental with the myotomes or muscle components of embryonic comites.

Legend:

BSC	basisclerotome	ISA	intersegmental artery	NC	neural crest
CEN	centrum	ISC	intersclerotome	NO	notochord
CL	coelom	LM	lateral plate mesoderm	NSP	vertebral neural spine
DA	dorsal aorta	MG	midgut	NT	neural tube
DM	dermatome or dermis	MY	myotome	SG	spinal ganglion
EN	endoderm	MYC	myocoele	SN	spinal nerve
EP	epidermis	NAR	vertebral neural arch	TVP	vertebral transverse process

TABLE IX-6
ORIGIN AND FATE OF MYOTOME AND VISCERAL MUSCLES I MAMMALS

HUMAN MYOTOME NUMBER	MYOTOME REGION	GENERAL MYOTOME FATE	SPECIAL MYOTOME FATE
A, B, C	OPTIC or ORBITALS (3) (a single mass in tetrapods)	Extrinsic eye muscles (innervated by cranial nerves III, IV, & VI)	A = superior rectus, inferior rectus internal rectus, inferior oblique B = superior oblique C = external rectus
---	OTICS (2)	Present only in Amphi-oxus, Ammocetes larva, and fish embryos	Disappears without a trace in all tetrapod embryos and adults
1 to 4	OCCIPITALS (4)	Hypobranchial muscles to tongue	Tongue muscles innervated by cranial nerve XII; i.e., sternohyoid, sternothyroid, thyrohyoid, etc.
5 to 12	CERVICALS (8)	Muscles of neck	C3 to C5 also form diaphragm, C4 to C8 form extrinsic arm muscles
13 to 24	THORACICS (12)	Muscles of upper back, belly & rib case	T1 also forms extrinsic arm muscles
25 to 29	LUMBARS (5)	Muscles of lower back and belly wall	L3 to L5 form extrinsic thigh muscles
30 to 34	SACRALS (5)	Muscles of hip	S1 and S2 for extrinsic thigh muscles S3 to S5 form anal and genital muscles
35 to 39+	CAUDALS (5+)	Mucles of tail (vestigial in man)	Cd1 also contributes to anal muscles
VISCERAL MUSCLES	CRANIAL NERVE	GENERAL VISCERAL FATE	SPECIAL VISCERAL FATE
Arch I	V	Muscles of jaw and ear	Masseter, anterior half of diagastric, tensor, tympany, mylohyoid, temporalis, etc.
Arch II	VII	Muscles, of the jaws, ears, face and scalp	Buccinator, posterior half of digastric, auricularis, platysma, stapedius, etc.
Arch III	IX	Muscles of anterior larynx and pharynx	Stylopharyngeus
Arches IV, V	X	Posterior laryngeal and pharyngeal muscles	
Arches VI, Post VI	XI	Some back and neck muscles	Sternomastoid and trapezius, etc.
POST-PHARYNGEAL	X	Smooth muscles of gut and many visceral organs	

the point at which it gives off the **allantoic stalk** as a ventral evagination into the umbilical cord (Fig. 9-13 W). At approximately this level, also observe that the posterior ends of the paired **mesonephric ducts** enter the dorsolateral walls of the urogenital sinus. The allantoic stalk and urogenital sinus will contribute to the formation of the urinary bladder, urethra, uterus, and vagina in later development.

Mesodermal derivatives

a. Somite (epimere) derivatives. Posterior to the level of the myelencephalon, almost any section through the body will show some stage in the differentiation of the 3 primary somite derivatives. They are the **dermatome, myotome,** and **sclerotome.** In the tail region the somites are at an early stage of differentiation and are very similar to those already seen in the study of the 48 and 72 hour chick embryos. **Dermomyotome,** separated from the **sclerotome** by a **myocoele,** can still be rocognized in the caudal somites. Anteriorly these three derivatives show more advanced levels of differentiation.

The **dermatome** disappears as a recognizable unit in the trunk region. Its fate is not thoroughly known, but the general consensus is that its cells migrate out under the skin and contribute to the connective tissue of the dermis, the underlying muscle sheaths, and certain bones evolutionarily derived from scale plates including the roof bones, the brain case, neural spines of vertebrae, and possibly the clavicle in the shoulder girdle. The **myotomes** of each segment are in the process of differentiating into clusters of cells that are the primordia of the segmental muscles. These structures are perhaps best seen in terminal cross sections where the embryo is cut in the frontal plane through the lumbosacral flexure (Fig. 9-13 Z').

In sections such as these, the neural tube is cut frontally in the long axis and the dorsal ganglia of the spinal nerves can readily be identified on both sides of the neural tube. Lateral to each ganglion will be found a myotome appearing as a thick spindle-shaped mass which stains relatively poorly. In

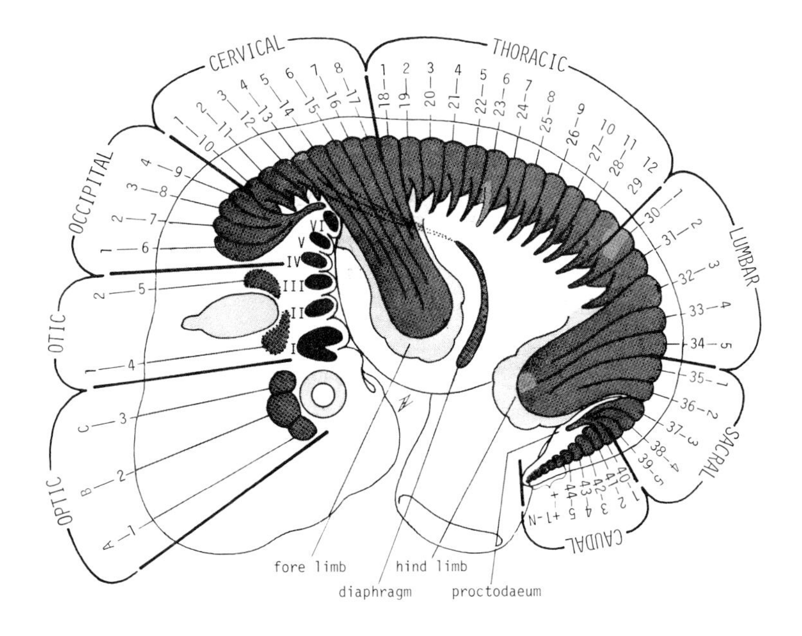

Figure 9-7. Summary of derivatives of myotome and visceral muscles in mammals. Arabic numerals refer to somite and myotome numbers. Roman numerals indicate muscle masses for visceral arches I, II, III, IV, V, and VI. See Table IX-6 for special derivatives of muscles in different body regions. In mammals the optic myotomes are all fused into a single "orbital muscle mass" and cannot be distinguished individually.

progressing forward through the ganglia, identify the **sclerotome** component of the somite. It is in the process of forming the **vertebral centra.** Figure 9-6 diagrammatically summarizes the major steps involved in the transformation of primordial sclerotomes into vertebral centra and neural arches. The transformation is most readily understood by recognizing that the early sclerotomes and myotomes are segmental in their derivation from somites. However, in order for muscles and bones to alternate in positions with vertebrae joints between muscle attachments, the final functional requirements of muscle and skeletal action demand that one of them must become intersegmental. **Myotomes and muscles derived from them retain the primitive segmental position. Sclerotomes, although primitively segmental, give rise to intersegmental vertebral centra.** This transition is accomplished by the division of each sclerotome into an anterior, **intersclera,** and a posterior, **basisclera,** half. The joint and articular surfaces of definitive vertebral centra will correspond with the space between the inter- and basiscleral components of a single embryonic sclerotome. Thus, the final articulations will be intersegmental. In primitive vertebrates, there are often two vertebrae per embryonic segment; these correspond to interscleral and basiscleral components respectively. In mammals, however the definitive vertebral centra are formed by the fusion of the basiscleral element of one sclerotome with the intersclera of the segment immediately posterior to it. In so doing the intersegmental arteries and veins become trapped inside the vertebral centrum but remain as landmarks for embryonic boundaries between adjacent sclerotomes. Myotomes remain connected, across the newly formed vertebral joint, to their original attachments with the inter- and basiscleral elements of their own somite level. The initial steps in scleral and central differentiation can be recognized in the 10 mm pig by using the myotomes as landmarks for each primitive somite level (Fig. 9-13 Z'). The presence of **intersegmental arteries** from the dorsal aorta, which pass between the somites, can also be used as points of reference to determine the original segmentation of the embryo. Note that the mesenchyme of each sclerotome shows two centers of condensation; the posterior half is the most conspicuous of the two. The segmental and intersegmental relationships of spinal ganglia, myotomes, sclerotomes and intersegmental arteries can also be seen to advantage in parasagittal sections just lateral to the neural tube (Fig. 9-14 C, D).

Cross sections through the body at any mid-trunk level will show the mesiolateral relation of the myotomes to the spinal nerves (Fig. 9-13 T). They will also show these muscle elements in the process of growing into the ventral body wall where they will form the hypaxial muscles. In cross sections through the trunk region, the sclerotomes are not conspicuous but are represented by uniformly stained, massive aggregations of mesenchyme ventral to the neural tube and surrounding the **notochord.** The latter is in the process of degenerating. The sclerotome mass is the primordium of the vertebral centra. At appropriate intersegmental levels, condensations of sclerotomes project dorsally between the ganglia. These will form the **neural arches** of the vertebrae and will, eventually, entirely surround the neural tube.

b. Nephrotome (mesomere) derivatives. Refer to the whole mount diagram of the 10 mm pig and observe the extent of the **mesonephros** and the position of the **metanephros.** Note the relationships of the anterior end of the mesonephros to the **postcardinal** and **subcardinal veins.** Observe that the postcardinal vein is absorbed into the mesonephros but that it reappears again at the posterior end of this organ. Note the relation of the **postcaval vein** to the mid-ventral surface of the mesonephros. In particular, familiarize yourself with the connections between the **mesonephric duct, metanephric duct (ureter), urogenital sinus,** and **cloaca.**

With these relationships in mind, trace the structure of the mesonephros through the serial cross sections beginning at a level comparable to that shown in Figure 9-13 P. The most anterior mesonephric tubules appear and separate the **posterior cardinal vein** into two parts. The dorsal half is the posterior cardinal and the ventral half is now termed the **subcardinal vein.** Trace the mesonephros posteriorly and observe that the main channels of both these veins become subdivided into the general capillary bed of the mesonephros. Continue back through the sections to a level comparable to that shown in Figure 9-13 S, T. Several **renal arterial branches** of the dorsal aorta supply the mesonephroi with arterial blood. Trace one of these renal arteries into the **mesonephric glomeruli** which lie on the mesial face of this organ. The glomeruli are capillary knots which project into the expanded ends of mesonephric tubules (Bowman's capsules). The mesonephric arteries will, for the most part, degenerate with the atrophy of the mesonephros that occurs somewhat later. Two or more branches will remain as the arteries of the gonad (i.e, spermatic or ovarian arteries) in the adult. In sections through the mesonephros at approximately the level at which the common bile duct enters the intestine (Fig. 9-13 T), it is possible again to identify the **subcardinal veins.** They are situated just under the peritoneum lining of the mesial surface of the mesonephroi. Trace the sections back until the connection between the right subcardinal and postcaval vein is established. Several sections posteriorly, the right and left subcardinal veins will unite in the midline at the base of the dorsal mesentery of the intestine (Fig. 9-13 V). At this level, where the subcardinals unite to form the **subcardinal anastomosis,** two other structures in the mesonephros should also be noted. First, identify the **genital ridge,** which at this time is slightly more than a thickening of the peritoneum on the ventromesial face of the mesonephroi. The genital ridge is the primordium of the germinal epithelium of the gonad (Fig. 2-2 C). Under higher magnification, **primordial germ cells** can be seen among the smaller interstitial cells derived from the genital germinal epithelium. Sex cannot yet be determined. Identify the paired **mesonephric** or **Wolffian ducts.** In cross sections they will appear as the largest cavity at the ventral apex of each mesonephros. The connection between the duct and the mesonephric tubules can be observed in many sections. The mesonephric duct, as mentioned earlier, is a part of the pronephric duct and it will form the vas deferens of the adult male. Table IX-7 will summarize these relationships.

Trace the mesonephros posteriorly through the series, keeping in mind that, owing to the lumbosacral flexure, the terminal sections will be in an oblique frontal plane, as can be seen in Figure 9-12. Be particularly attentive to the course of the mesonephric duct. At the level of the hind-limb buds, it will leave the mesonephros and extend posteriorly into the lateral wall of the **urogenital sinus** (Fig. 9-13 W,X).

Review again the relationships of the mesonephric and metanephric ducts in the whole mount diagram. Then, in cross sections, trace the mesonephric duct from its entrance into the urogenital sinus back to the level at which the **metanephric duct** or **ureteric bud** is given off. Follow the metanephric duct until it expands into a laterally compressed cavity of considerable size (Fig. 9-13 Y, Z). This cavity is the primordium of the **metanephric renal pelvis.** The lining of the renal pelvis is a thick epithelial expansion of the metanephric duct. This lining will form the **renal medulla,** and the collecting tubules of the definitive kidney. The metanephric ducts will persist as the **ureters** which connect the kidney to the bladder. The primordium of the **renal cortex** can be recognized as a dense condensation of mesenchyme surrounding the renal medulla. This is **metanephric mesomere** that has been induced by the metanephric duct (ureter). The renal cortex will later differentiate into the functional metanephric tubules of the adult kidney. **It should be emphasized that only the cortex of the definitive kidney is of metanephric mesomere origin. The medullary portion derived from the ureteric bud is of pronephric origin.**

One final relationship should be noted in terminal sections through the mesonephros and metanephros. This relates to the reappearance of the **posterior cardinal veins** along the dorsolateral margins of the mesonephroi (Fig. 9-13 Y, Z). In this region these veins are homologous with **renal**

TABLE IX-7
DIFFERENTIATION OF MAMMALIAN MESOMERE AND NEPHROGENIC CORD

CHARACTERISTICS	PRONEPHRIC MESOMERE	MESONEPHRIC MESOMERE	METANEPHRIC MESOMERE
1. Located at body level	Somite level 7-14	Somite level 15-25	Somite level 26-31
2. Functional in excretion in:	Pronephric kidney in larval Cyclostomes, in embryos of fish, & in larvae of amphibia	Mesonephric kidney in adult fish and amphibia; in embryos of amniotes	Metanephric kidney in adult birds reptiles, & mammals
3. Nephrostome	Present & functional	Present & vestigial	Absent
4. Arterial glomerulus	External (projects into coelom)	Internal (projects into Bowman's capsule)	Internal (projects into Bowman's capsule)
5. Venous drainage	Post-cardinal Vein	Sub-cardinal Vein	Supra-cardinal Vein
6. Duct	Wolffian duct = pronephric duct	Wolffian duct = mesonephric duct	Ureter (derived from Wolffian duct)
7. Inductive interrelationships	Pronephric mesoderm is determined as early as the blastula stage in the mesodermal field	Mesonephric tubules are induced by the action of the Pronephric duct as it grows back to the cloaca	Metanephric tubules are induced by action of the ureteric bud of the pronephric duct on metanephric mesomere
8. Definitive mammalian derivatives	Wolffian duct gives rise to: metanephric renal medulla & collecting ducts, renal pelvis, & ureter; **In Males** also to the vas deferens & seminal vesicle; **In Females** also to vestigial Gartner's canal	Adrenal cortex, interstitial cells of testis & ovary; **In Males** also the epididymis with its vasa efferential & rete tubules, and vestigial ducts of the paradidymis; **In Females** also the vestigial ducts of epoophoron & paroophoron	Metanephric renal cortex and nephrons (nephron = Bowman's capsule, proximal convoluted tubule, loop of Henle, and distal convoluted tubule of kidney)
9. Mullerian Duct	a. It is induced from parietal (somatic) peritoneum by the Wolffian duct. b. **Fate in females:** it gives rise to the ostium, oviducts, and uterus. Functions in females: The ciliated ostium picks up eggs shed into the coelom; oviducts transport eggs to the uterus and often they secrete jelly, albumen, and shells around eggs; the uterus is the final location of eggs before deposition, or, as in mammals and so-called placental sharks, it is the locus for implantation of the fetus or embryo. c. **Fate in males:** It degenerates except for the two ends of the duct which persist as vestigial remnants. The ostium forms the appendix testis which is without function, and the uterus persists as the utriculus masculinum (see Figure 2-7). The latter is embedded in the prostate gland and opens into the urethra just below the sphincter of the urinary bladder. It has been suggested that the utriculus masculinum may be the embryonic inductor of the prostate gland from endoderm of the urogenital sinus.		

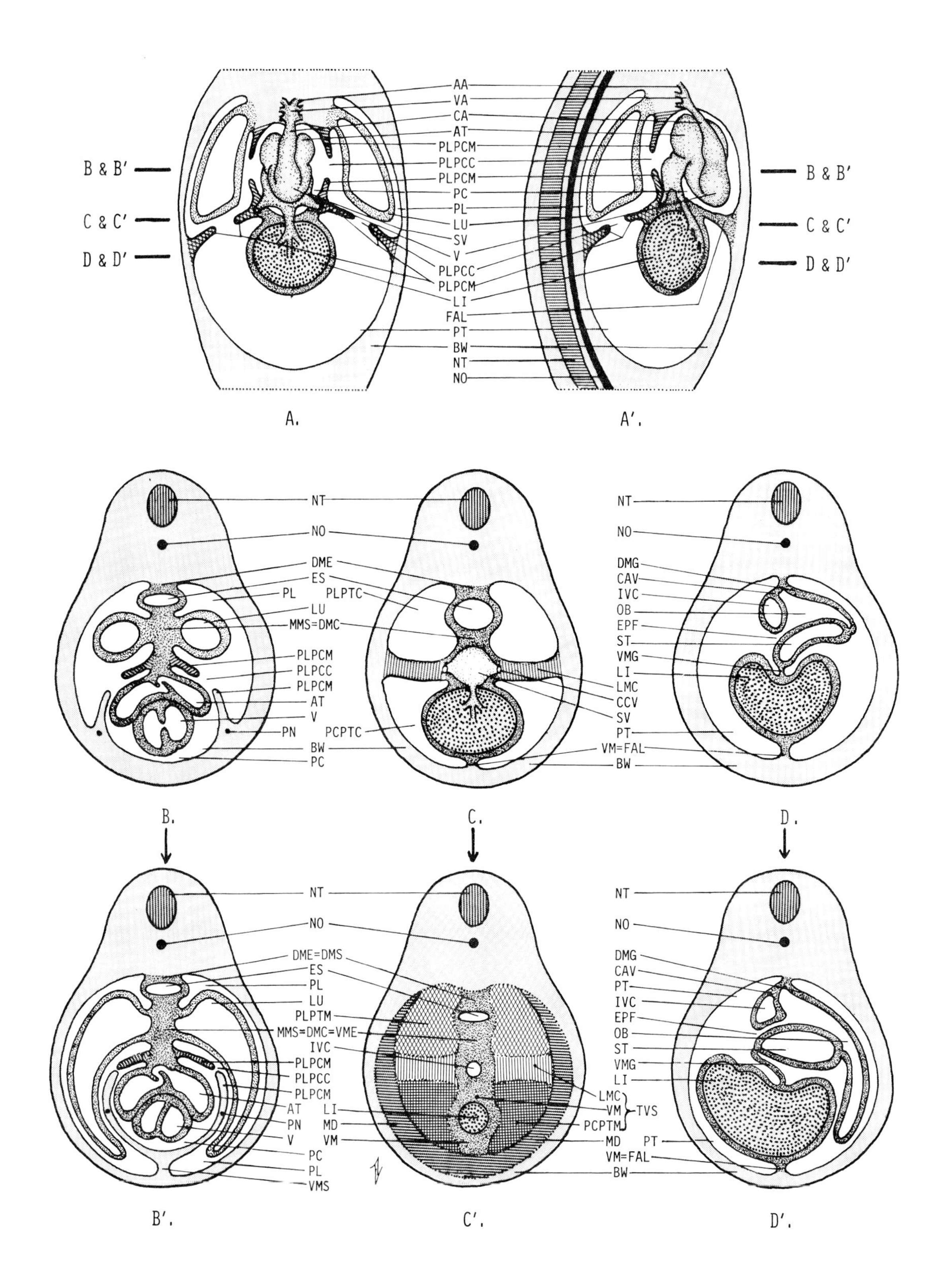

Figure 9-8. Diagrams of partitioning of the vertebrate coelom. 9-8 A & A'. Schematic frontal and parasagittal sections respectively at a primordial stage in mammalian coelom formation. Guide lines show the levels at which Figures 9-8 B–B', C–C', and D–D' are taken. 9-8 B and B'. Cross sections at an early and later stage in mammalian development through the pleuropericardial membranes. 9-8 C and C'. Cross sections at an early and later stage in mammalian development through the lateral mesocardium and diaphragm. 9-8 D & D'. Cross sections at an early and later stage in mammalian development through the liver and stomach. 9-8 E, E' & E". Respectively, frontal, transverse, and sagittal sections of an adult shark. 9-8 F, F' & F". Respectively, frontal, transverse, and sagittal sections of an adult amphibian–reptile–bird. 9-8 G, G' & G". Respectively, frontal, transverse, and sagittal sections of an adult mammal.

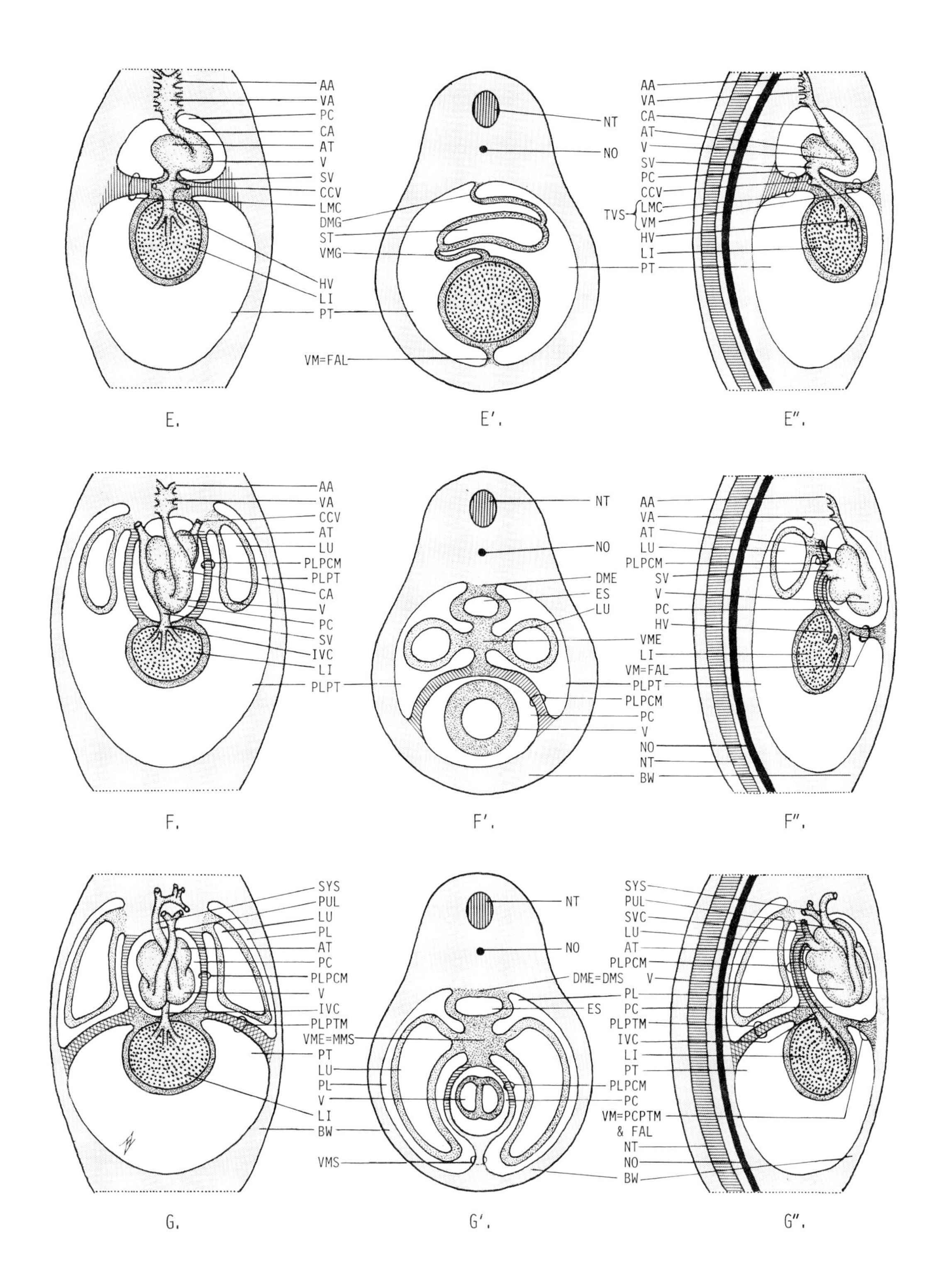

Figure 9-8. Legend for 9-8 A. through G".

AA aortic arches
AT atrium
BW body wall
CA conus arteriosus
CAV caval mesentery
CCV common cardinal vein
DMC dorsal mesocardium
DME dorsal mesoesophagus
DMG dorsal mesogaster, greater omentum
DMS dorsal mediastinal septum
EPF epiploic foramen of omental bursa
ES esophagus
FAL falciform ligament
HV hepatic vein
IVC inferior vena cava
LI liver
LMC lateral mesocardium
LU lung bud, lungs
MD muscular diaphragm
MMS median mediastinal septum
NO notochord
NT neural tube
OB omental bursa, lesser peritoneal cavity
PC pericardial coelom
PCPTC pericardioperitoneal canal
PCPTM pericardioperitoneal membrane
PN phrenic nerve
PL pleural coelom
PLPCC pleuropericardial canal
PLPCM pleuropericardial membrane
PLPT pleuroperitoneal coelom
PLPTC pleuroperitoneal canal
PLPTM pleuroperitoneal membrane
PT peritoneal coelom
PUL pulmonary trunk
ST stomach
SV sinus venosus
SVC superior vena cava
SYS systemic trunk
TVS transverse septum
V ventricle
VA ventral aorta
VM ventral mesentery
VME ventral mesoesophagus
VMG ventral mesogaster, lesser omentum, or gastrohepatic ligament
VMS ventral mediastinal septum

portal veins of adult amphibians and some reptiles but they will degenerate in mammals. They can be traced posteriorly into the tail where they form the **caudal vein** or **veins.**

After finishing the study of serial cross sections, it is quite instructive to examine the parasagittal sections through the mesonephros (Fig. 9-14 C–F). These sections show to special advantage the relationships of the mesonephros to its veins and the relationship of the ducts to the cloaca and urogenital sinus.

c. Lateral plate (hypomere) derivatives. A consideration of the extraembryonic derivatives of the hypomere will be omitted since they do not appear in the material for study, nor do they differ materially from those already described for the 48 and 72 hour chick embryos.

Intraembryonic somatic hypomere. Somatic hypomere derivatives include: the **limb buds;** an indeterminate amount of mesenchyme of the lateral body wall; the general **parietal peritoneum** of the pericardial, pleural, and peritoneal cavities; and **interstitial cells of the genital ridge** on the surface of the mesonephros (discussed in the preceding section; refer to Chapter 2, Figure 2-2 for a summary of vertebrate gonad development). Of these, a word can be said about the developing limb buds (Fig. 9-13 O–R, W–Y). These large masses of mesenchyme will form all elements of the **appendicular skeleton** (see Figure 5-3) and the **intrinsic muscles** of the limb. Intrinsic muscles have their origins and insertions entirely within the limb, whereas limb muscles having their origins outside the limb, in the body wall, are usually considered to be from epaxial or hypaxial myotome derivation. These are the **extrinsic muscles** which grow into the base of the limb from the body wall. The limb bud is covered by a thin surface epidermis in which a thickened **epidermal ridge** can be identified in cross sections through the apex of the limb. The epidermal ridge is induced to form by the limb bud mesenchyme and functions by producing a so-called "mesodermal maintenance factor" which is necessary to maintain limb bud growth essential for forming all parts of the limb distal to the middle of the stylopodium (i.e., the middle of the humerus or femur of the arm or leg respectively). The mesodermal maintenance factor acts more or less as an embryonic hormone responsible for maintaining mitotic activity essential for growth of distal limb elements.

Intraembryonic splanchnic hypomere and coelomic septa. The primary derivatives of the splanchnic lateral mesoderm are the **dorsal** and **ventral mesenteries** of the gut and gut derivatives, and the **connective tissue, visceral muscle,** and **visceral peritoneum** layers that cover these organs. The relationships of the dorsal and ventral mesenteries of the gut posterior to the stomach have been discussed in descriptions of the stomach, liver, pancreas, and omentum. Anterior to the stomach, the mesenteries are highly modified by the development of the pharyngeal pouches, the heart, trachea, and lungs. The splanchnic (visceral) mesoderm of the pharyngeal region of the gut gives rise to the **visceral muscles of the gill arches.** These structures will not be discussed in connection with the study of the 10 mm pig since none of the muscles have differentiated at this time. However, refer to a brief summary of the fate of the gill arch musculature that is given in Table IX-5. The anterior end of the **coelom** extends forward only through the level of the **esophagus.** As was mentioned in previous chapters, the **heart** is formed from the ventral mesentery of the esophagus, and the heart, therefore, is a derivative of splanchnic mesoderm. The heart will be considered separately in the following section. At this time special consideration will be given to the various septa that participate in subdividing the coelom into separate **pericardial, pleural,** and **peritoneal cavities.** Review the description of these parts that is given in the discussion of the 3 day chick. The differentiation of the three coelomic divisions, and the septa which partially divide them, are more clearly delineated in the 10 mm pig. A summary of mammalian coelomic cavities, and the septa that separate them from one another during embryonic development, is given by the diagrams in Figure 9-8. For a more detailed review of this complex anatomical subject, refer to a standard reference in vertebrate anatomy.

Identify the **transverse septum** and note its compound origin from the fusion of **lateral mesocardia** with the **ventral mesentery of the esophagus.** The latter is synonymous with the dorsal mesocardium, or median mediastinal septum. Note the positions of the **liver** and **postcaval vein** (fused vitelline veins) in the transverse septum which separate the **pericardial cavity** ventrally from the remainder of the coelom.

Find a section showing two **lung buds** (Fig. 9-13P) and identify the paired **pleural coeloms** surrounding them. Note that they are separated from the pericardial coelom by the transverse septum. This part of the transverse septum is called the **pleuro-pericardial membrane.**

Trace the pleural coelomic cavities posteriorly until they open into the **peritoneal cavity**

(Fig. 9-13 P). This opening is the **pleuro-peritoneal canal.** The tissue perforated by this opening is the **pleuro-peritoneal membrane.** The latter arises from the fusion of a dorsal outgrowth from the transverse septum and a ventral outgrowth of the parietal peritoneum covering the anterior end of the mesonephros (Fig. 9-13 P, Q).

The mammalian **diaphragm** will eventually be formed by complete fusions between the transverse septum, the paired pleuro-peritoneal membranes, and the mediastinal septum (dorsal mesentery of the esophagus). The diaphragm acquires hypaxial muscles from the lateral body wall and forms a complete separation between the pleural and peritoneal cavities (see Fig. 9-8 C').

d. Circulatory system. The greatest developmental advance to be found in contrasting the structure of the 3 day chick and 10 mm pig is in the circulatory system. This system is truly intermediate between the early embryonic and definitive states. Before attempting the study of this system, thoroughly familiarize yourself with the general circulatory plan of the embryo by reference to the whole mount diagram. Keep in mind that this diagram shows only main arterial and venous channels and that bilateral vessels have been shown on only one side. Review also the circulatory plan of the early embryo and the circulatory path of the definitive mammalian body as shown in Figures 9-9, 9-10, 9-11 and 10-5. Familiarity with the adult structure of the mammalian heart, arterial, and venous plans is essential for a full appreciation of the transitional character of the circulatory system in the 10 mm pig.

(1). Heart (Figs. 9-9, 9-13 K–P). Primary features of interest in the heart of the 10 mm pig concern its division into right and left halves, which anticipates the development of separate **systemic** and **pulmonary circulatory systems.** The early stages in heart development, as illustrated in the 72 hour chick, are essentially similar to those that are in adult fish. The primitive vertebrate heart consists of four chambers, connected in series, which pump blood through the gill arches and then to the dorsal aorta and its branches. In arriving at the adult mammalian condition, the sinus venosus remains undivided but the remaining chambers (namely, the atrium, ventricle, and conus arteriosus) become completely separated by median septa.

In studying the heart in serial sections, begin at the anterior end of the heart and proceed posteriorly from the point at which the **conus arteriosus** is clearly suspended by a **dorsal mesocardium** in the **pericardial cavity** (Fig. 9-13 K–L). This is approximately the point at which the **6th aortic (pulmonary) arch** enters the **ventral aorta.** Note that the lumen of the conus is partially divided into right and left halves by a dorsoventral thickening in the wall of the conus. This thickening is the **bulbar septum.** The right side of the conus is connected anteriorly with the 3rd and 4th aortic arches and constitutes the primordium of the **systemic trunk** or ascending aorta. The left side of the conus, which connects anteriorly with the 6th aortic arch, will form the **pulmonary trunk** and will carry blood to the lungs. Trace the conus posteriorly and pay particular attention to the angle at which the bulbar septum divides the conus lumen. It will be seen that the systemic component will eventually enter the left side of the ventricle and the pulmonary trunk will enter the right side (Figs. 9-9, 9-13 M, N).

The **ventricle** can be identified by the very thick and spongy appearance of the epimyocardial tissue. Note that the outer contour of the heart is distinctly lobed into right and left halves. The two sides are incompletely separated by the **interventricular septum.** The canal connecting the right and left ventricular cavities is the **interventricular foramen.** This canal will eventually be closed when the interventricular septum finally fuses with the **endocardial cushion.** The latter is a thickening in the heart wall which divides the opening between the atrium and ventricle into right and left valves. The **right atrioventricular valve** will form the tricuspid valve of the adult heart, and the **left atrioventricular valve** will form the bicuspid, or mitral valve. In these sections the endocardial cushion appears as a cone of dense cardiac mesenchyme (Fig. 9-13 N).

The **atrium** is also partially divided into right and left halves. This heart region can be identified by its dorsal position, its relatively thin wall, and its extensive irregular lateral expansion. Note that the atrium is divided by the primary **interatrial septum** which extends dorsally from the endocardial cushion to the wall of the atrium. The interatrial septum may appear in some sections to form a complete separation of right and left atria. Progress either forward or backward through the atrium until you find a section in which the primary interatrial septum is incomplete and the right and left atria are continuous through a gap in the membrane. The gap is the **primary foramen ovale.** A second interatrial septum and foramen are in the process of developing in the 10 mm pig. The second septum is to the right of the first and will eventually overlap and form a partial valve covering the primary foramen ovale. At this time the **secondary interatrial septum** can usually be identified as a small fold arising from the dorsal wall of the right atrium (Fig. 9-13 M). Trace it back and note that it becomes continuous with the wall of the **sinuatrial valve** (Fig. 9-13 N). At this time, all blood enters the right atrium through this valve with the exception of a trifling amount that enters the left atrium by way of the newly forming **pulmonary vein.** This vein can be identified as part of the capillary bed in the laryngotracheal mesenchyme that is supplied by the pulmonary arteries. The single pulmonary vein can be traced into the left atrium at about the level where the primary lung buds are given off (Fig. 9-13 N–P). After this time, it can be traced for a short distance posteriorly under the two lung buds. Blood from the **right atrium** can pass directly to the right ventricle through the **right atrioventricular (tricuspid) valve;** alternatively, it can enter the left ventricle by way of the foramen ovale, left atrium, and **left atrioventricular (bicuspid or mitral) valve.**

The **sinus venosus,** although undivided, is perhaps the most modified portion of the heart. It can be identified on the basis of its relationships in early development to the **common cardinal veins** and the **vitelline anastomosis.** The structure of the sinus venosus can best be understood by visualizing the entire heart as being rotated through 90° in a clockwise manner as the heart moves posteriorly from the neck into the thoracic cavity. When this occurred, the common cardinal veins (originally lateral to the heart) shifted in such a manner as to bring the right common cardinal into the **anterior** wall of the sinus venosus, whereas the left common cardinal was forced to open into the sinus **posteriorly.** The fused vitelline veins are in the axis of rotation, and their relative position remained unchanged. Find a section similar to that shown in Figure 9-13 N. The lumen of the sinus venosus is reduced to a small cavity immediately dorsal to the **sinuatrial valve.** The lumen of the sinus is directly continuous with the **right common cardinal vein** (i.e., primordium of

Legend for Figure 9-9. Diagrammatic explanation of cardiac septation and the division of pulmonary and systemic circulatory paths through the mammalian heart.

9-9 A. Normal heart of a 4 to 5 day chick embryo showing the typical S-shaped path of blood through the undivided heart. 9-9 A'. Diagrammatic drawing of how the heart at the same stage would appear if the heart axis were straight and not sigmoid. 9-9 A''. A diagrammatic drawing of a hypothetical "straight heart" such as that shown in 9-9 A' showing the locations of septa that will ultimately divide the conus, ventricle, and atrium in hearts of birds and mammals.

9-9 B. Ventral view of a normal heart of a 12 to 15 mm pig embryo schematically showing the spiral path of the bulbar septum through the conus arteriosus which results in the separation of pulmonary and systemic trunks. 9-9 B'. Dorsal view of a normal heart of the same stage showing the path of the left and right common cardinal veins, the pulmonary veins, and postcaval vein into the heart. 9-9 b''. A diagrammatic optical section through the heart at the same age to show septa dividing the atrium and ventricle. **Legend:**

AA3 3rd aortic arch
AA4 4th aortic arch
AA6 6th aortic arch
AC anterior cardinal vein
ALV allantoic vein
AT atrium
ATL left atrium
ATR right atrium
AVL left atrioventricular (bicuspid, mitral) valve
AVR right atrioventricular (tricuspid) valve
BS bulbar septum
CA conus arteriosus
CC common cardinal vein
CCL left common cardinal
CCR right common cardinal, superior vena cava
CD carotid duct
COCL left common carotid artery
COCR right common carotid artery
DA dorsal aorta
DAL left dorsal aorta
DUA ductus arteriosus
EC external carotid artery
ECC endocardial cushion
IAF' primary interatrial foramen
IAF'' secondary interatrial foramen
IAS' primary interatrial septum
IAS'' secondary interatrial septum
IC internal carotid artery
INA innominate artery
IVF interventricular foramen
IVS interventricular septum
PA pulmonary artery
PAL left pulmonary artery
PAR right pulmonary artery
PCV postcardinal vein
PLV pulmonary vein
PT pulmonary trunk
PV postcaval vein
SCAL left subclavian artery
SCAR right subclavian artery
SV sinus venosus
SY systemic trunk
UV umbilical vein
V ventricle
VA ventral aorta
VL left ventricle
VR right ventricle
VV vitelline veins

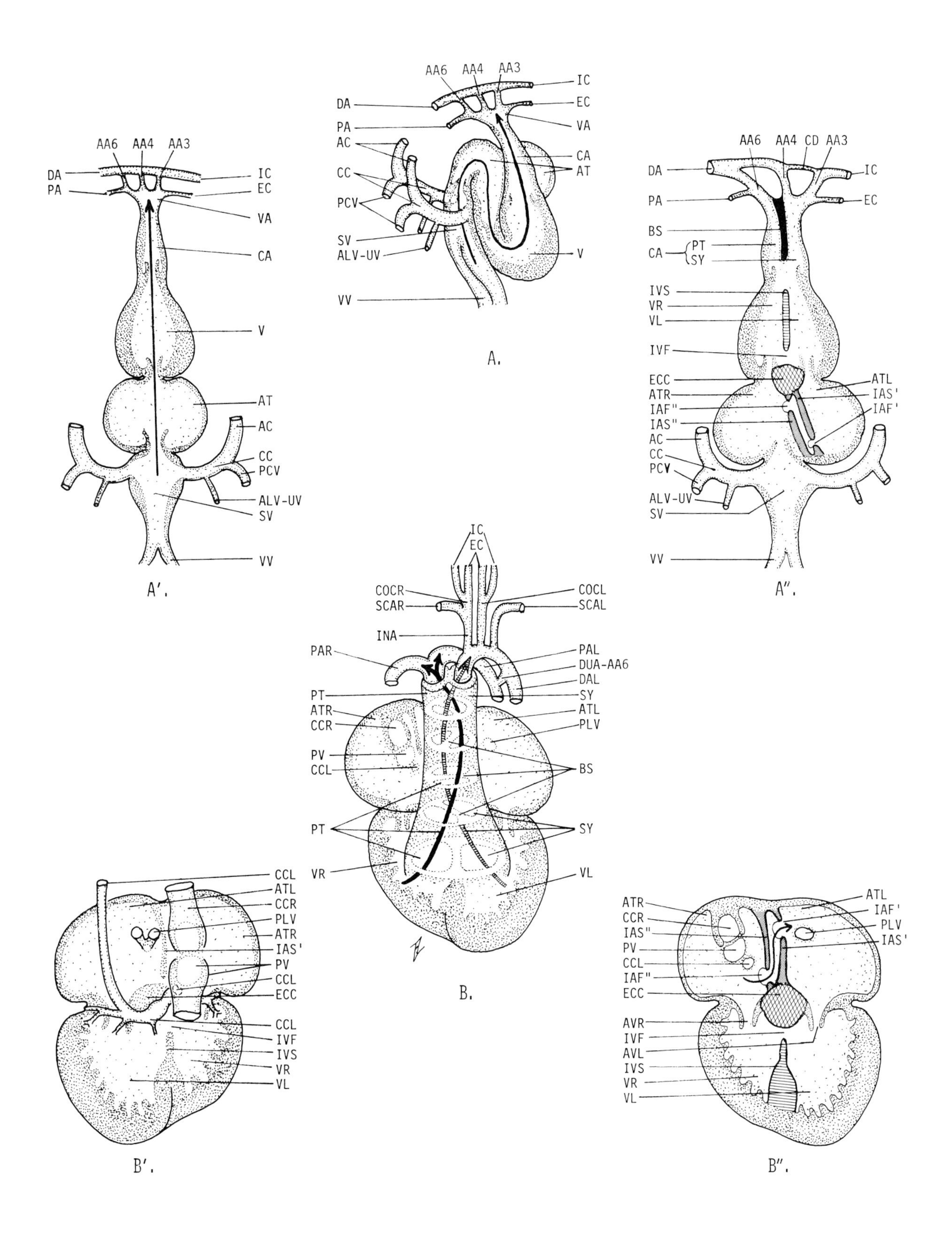

Figure 9-9. Diagrammatic explanation of cardiac septation and the division of the pulmonary and systemic circulatory paths through the mammalian heart. See explanation and legend.

precaval vein). On the left side of the embryo, identify the **left common cardinal vein** and the **left sinus horn** (i.e., primordium of coronary sinus). Trace the connection between these channels and the sinus venosus proper. To do this it will be necessary to go posteriorly through several sections to find the entrance of the left sinus horn into the posterior wall of the sinus venosus. At this same level, the fused vitelline veins (i.e., anterior end of the postcaval vein) also empty into the posterior wall of the sinus venosus (Fig. 9-13 O).

(2). Arterial circulation anterior to the heart (Fig. 9-10). Review the circulatory plan of the 72 hour chick and 10 mm pig as illustrated in the whole mount diagrams (Figs. 8-7, 9-12). Begin the study of the anterior circulation with the conus anteriosus at a level in which the bulbar septum divides the lumen into distinct **systemic** and **pulmonary trunks** (Fig. 9-13 L). For a summary of the fate of the aortic arches in mammals, see Figure 9-10. For simplicity, trace out the arches in the reverse order of their number as suggested in the following directions.

The **6th aortic arches (pulmonary arches)** arise as dorsal outgrowths from the pulmonary trunk. They can be traced through the dorsal mesocardium where the two arches branch around the trachea and esophagus and enter the paired dorsal aortae. If the embryo is sufficiently advanced, a small posterior branch of each 6th arch may be found which runs back toward the lung buds. These vessels, along with the ventral halves of the 6th arches, represent the paired primordia of the **pulmonary arteries** (Fig. 9-13 M, N). The right 6th aortic arch will soon atrophy, but the dorsal half of the left 6th arch will persist until birth as the **ductus arteriosus.** This is an arterial channel which shunts pulmonary blood directly into the dorsal aorta.

The **5th aortic arch** generally will not be seen. It is transitory in its appearance or may not appear at all. It should be looked for at levels of the embryo sectioned between the 4th visceral pouches and the ultimobranchial bodies (i.e., 5th pouch, Fig. 9-13 J).

Legend for Figure 9-10. Schematic representation of the mammalian arterial circulation. See figure.

9-10 A. The left side shows the primitive common plan of the arterial system. 9-10 A'. The right side diagrams a transitional stage in the development of the mammalian arterial plan. 9-10 B. Diagram of the definitive mammalian plan for major arteries. 9-10 C–J. Diagrams of persistent aortic arches in representative adults of vertebrate classes: C. Cyclostome, D. Elasmobranch, E. Teleost, F. Urodele Amphibian, G. Anura Amphibian, H. Reptile, I. Bird, J. Mammal. The conditions shown in E through I occur as rare anomalies in adult mammals including man. **Legend** (a. = arteries):

A 1-7 aortic arches
A3, A4, A6 R & L right & left arches 3, 4, & 6
AL allantoic sac
ALA allantoic a.
AT atrium
ATR right atrium
BA basilar a.
CA conus arteriosus
CCA common carotid a.
CD carotid duct of aorta
CDA caudal a.
CIA common iliac a.
COE coeliac a.
CW circle of Willis
DA dorsal aorta
DAR ductus arteriosus
EC external carotic a.
EIA external iliac a.
FA femoral a.
IC internal carotid a.
IIA internal iliac (hypogastric) a.
IMA internal mammary a.
IME inferior mesenteric a.
INA innominate a.
ISA intersegmental a.
IT intestine
LDA left arch of aorta
LU lung
MNA mesonephric a.
MNE mesonephros
MTN metanephros
OVTE ovary or testis
PA pulmonary a.
PFA profundus femoral a.
PNE pronephros
PT pulmonary trunk
RA renal a.
RDA right arch of aorta
SB swim bladder
SCA subclavian a.
SCAL left subclavian a.
SCAR right subclavian a.
SMA superior mesenteric
ST stomach
SV sinus venosus
SY systemic trunk
V ventricle
VA ventral aorta
VL left ventricle
VR right ventricle
VRTA vertebral artery
VTA vitelline a.
YS yolk sac

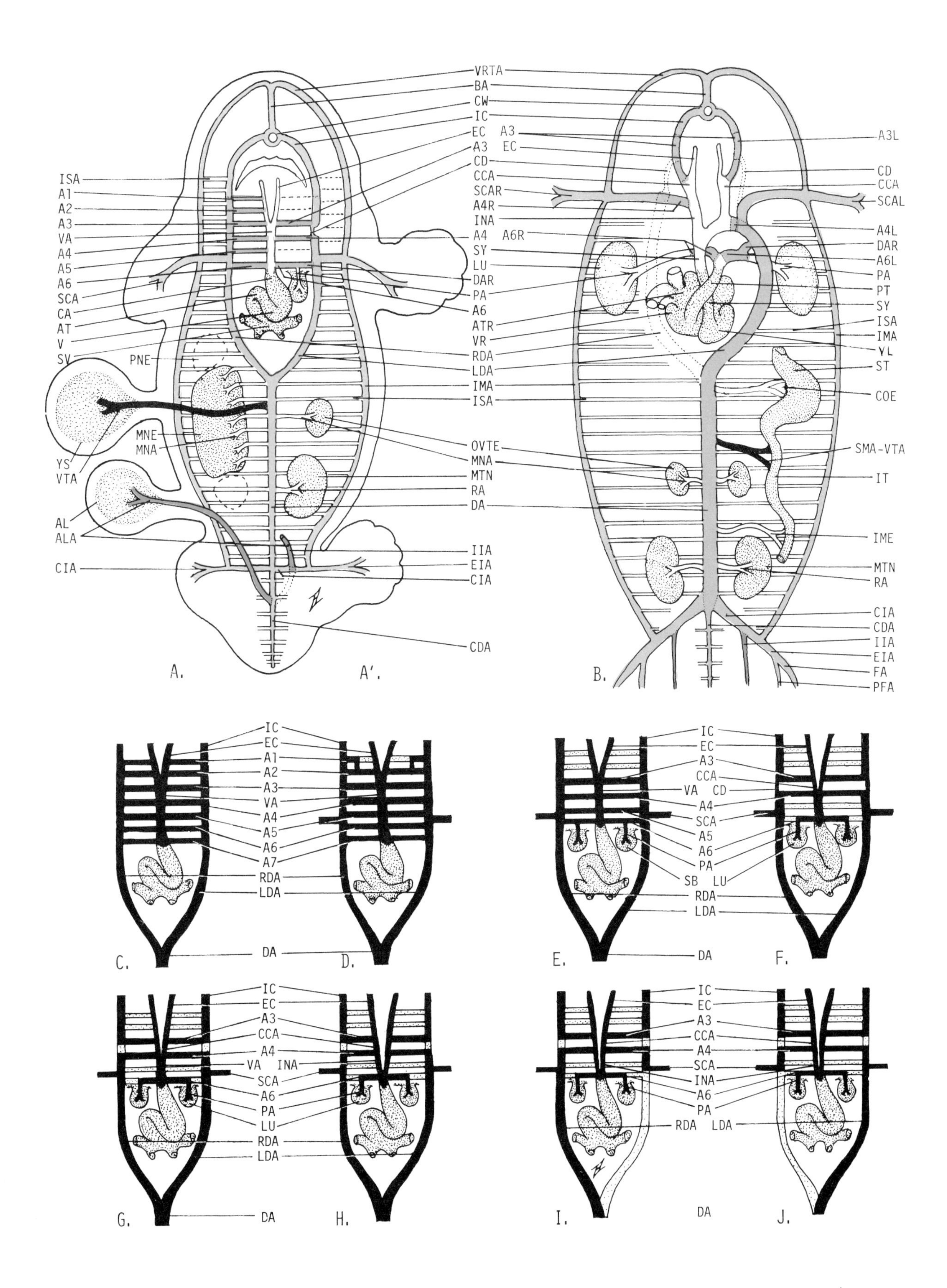

Figure 9-10. Schematic representation of the mammalian arterial circulation. See explanation and legend. Color code: blue — major branches of the dorsal aorta, green — dorsal aorta, red — aortic arches, yellow — ventral aorta, purple — allantoic artery, black — vitelline artery.

The **3rd and 4th aortic arches** arise from the systemic trunk of the conus arteriosus. Trace the systemic trunk forward and note where the arches are given off from the aortic sac (Fig. 9-13 I, J). Follow them through their respective visceral arches around the pharynx to their connections with the dorsal aortae. The dorsal aorta between the 3rd and 4th arches will later degenerate. However, while it persists, this segment of the aorta is called the **carotid duct** (Fig. 9-13 G, H). The fate of the two 4th aortic arches in later development is different. These are the **systemic arches;** however, in mammals only the left 4th arch participates in the formation of the main systemic course of the aorta. The left 4th arch, in the definitive state, will form the connection between the systemic trunk (ascending aorta) and dorsal aorta. The **right 4th arch** will persist as the base of the right subclavian artery. The 3rd aortic arches, after the disappearance of the carotid duct, will form the base of the paired **internal carotid arteries.** For this reason the 3rd aortic arch is named the **carotid arch.**

The **1st and 2nd aortic arches,** which initially supplied the mandibular and hyoid arches respectively, are in the process of degenerating at the 10 mm stage in the development of the pig. Except for minor fragments, they will entirely disappear during the course of development. For example, a small dorsal fragment of the 1st aortic arch is incorporated into each internal carotid artery, and a fragment of the 2nd arch persists as the stapedial artery of the middle ear. Figure 9-10 summarizes these developmental changes in aortic arches.

The **external carotid artery** appears as a pair of anterior branches from the ventral aorta or aortic sac. These replace, in function, the atrophied 1st aortic arches. The paired external carotid arteries are just beginning to form in the 10 mm pig. They can be identified in sections comparable to those shown in Figure 9-13 H–J where the external carotids appear as diffuse capillary extensions of the ventral aorta which branch around the thyroid diverticulum and extend into the mandibular processes.

Trace the **paired internal carotid arteries** anteriorly from their origins in 3rd aortic arches. The posterior part of these arteries is now made up of the dorsal aortae at the levels of the 1st and 2nd aortic arches. Anterior to the 1st arch, the internal carotid arteries pass to the floor of the diencephalon (Fig. 9-13 F–H) along the side of the infundibulum. Posterior to the infundibulum, the two internal carotid arteries fuse into a single vessel.

The **basilar artery** is the median fusion of the right and left internal carotids. Its course is immediately ventral to the midbrain and hindbrain (Fig. 9-13 C–E). At the posterior end of the myelencephalon the basilar artery branches and unites with the paired **vertebral arteries** which lie ventral to the spinal cord and can be traced back to the anterior margin of the fore-limb bud (Fig. 9-13 D–N). There they join the **subclavian arteries** which supply the fore-limb buds.

(3). Arterial circulation posterior to the heart (Fig. 9-10). The arterial supply of blood to the posterior parts of the body are all provided by branches of the **dorsal aorta.** The dorsal aorta itself is paired anterior to the level at which the subclavian arteries are given off (Fig. 9-13 N). Posterior to this point, these vessels fuse into a single median dorsal aorta which passes to the posterior end of the body. The post-sacral extent of the dorsal aorta into the tail region is termed the **caudal artery** (median sacral artery in man).

The **superior mesenteric artery** (Fig. 9-13 U–W) is homologous with the paired vitelline arteries of chick embryos. This vessel in the 10 mm pig embryo supplies the umbilical loop of the intestine and the vestigial **yolk sac.** It is small but can be traced in the dorsal mesentery of the gut into the umbilical cord.

The **renal arteries,** which supply the mesonephroi, have been noted earlier and require no further discussion here.

The **intersegmental arteries** are the remaining branches of the dorsal aorta to be considered. They have been discussed in connection with segmentation of the myotomes and sclerotomes (see Fig. 9-13 Q). However, special consideration will be given to two pairs of intersegmental arteries that supply blood to the limbs. The first of these, the **subclavian arteries,** represent an enlargement of the 7th cervical intersegmental arteries. (Fig. 9-13 N). The paired **vertebral arteries** are of special interest, since they arise from the subclavians by anastomosis of the dorsal ends of the first 6 intersegmental arteries. Their ventral connections to the dorsal aorta atrophy. The relationship of the vertebral arteries to the basilar artery has been discussed.

Throughout the remainder of the trunk, the intersegmental arteries are given off regularly between each pair of somites. They supply blood to the myotomes and other tissues of the body wall but require no particular attention.

The **5th lumbar intersegmental artery** has a special history worthy of emphasis. These arteries will form the **common** and **external iliac arteries** which carry blood to the hind limb bud. These arteries may not be well differentiated, but usually some branches of the dorsal aorta into the hind limb bud can be identified.

The **umbilical or allantoic arteries** are the largest branches of the dorsal aorta. They arise in the lumbar region, pass ventrally into the umbilical cord, and from thence to the chorionic placenta. An anastomosis can occasionally be found between the bases of the umbilical arteries and the 5th lumbar intersegmentals (Fig. 9-13 Z). This is a normal occurrence and in later stages the umbilical arteries will receive all their blood from the 5th lumbars and will appear as mesial branches of these vessels (see Fig. 9-10 A'). When this fusion occurs, the 5th lumbar artery, proximal to the anastomosis, becomes the **common iliac artery.** The 5th lumbar distal to the anastomosis becomes the **external iliac artery.** The base of the umbilical arteries after birth persists as the paired **internal iliac** or **hypogastric arteries** which supply the bladder and pelvic region (Fig. 9-10 B).

(4.) Venous circulation anterior to the heart (Figure 9-11 A, B, C) summarizes the developmental history of mammalian veins. Figure 9-11 A shows the primitive vertebrate common plan; B shows the status of veins in the 10 mm pig embryo, and C gives the definitive condition in adult mammals such as man. A comparison of these figures gives a general account of the way in which the definitive venous condition in mammals is derived from the common plan by addition, deletion, and fusion of veins during development. The anterior veins are relatively unchanged from the condition observed in earlier chick stages. Blood from the external carotid, internal carotid, and basilar arteries passes through capillary beds of the head into the paired **anterior cardinal veins.** The latter are greatly expanded into spaces called **anterior cardinal sinuses** (Fig. 9-13 B–E). Blood in the anterior cardinal veins passes directly posteriorly and flows into the right and left **common cardinal veins** which in turn empty into the **sinus venosus.** Only one branch of the anterior cardinal veins merits special mention. This is the **subclavian vein** which enters each anterior cardinal at the level of the fore-limb buds and returns blood from the capillaries of the limb buds (Fig. 9-13 O). These are the 7th cervical intersegmental veins.

(5). Venous circulation posterior to the heart (Fig. 9-11). The blood channels that return blood from the posterior regions of the body to the heart have undergone considerable modification from those in the 72 hour chick stage. In the 3 day chick the posterior cardinal veins, allantoic veins, and vitelline veins accounted for the return of all embryonic and extraembryonic blood to the heart. The primary modifications are associated with the functional differentiation of the mesonephros and liver which interrupt the earlier channels and are responsible for the development of new pathways, the most important of which is the post caval vein (inferior vena cava of man).

The **posterior cardinal veins** still return most blood from the body wall to the heart. However, the mesonephros, in its growth, interrupts the course of the posterior cardinal vein which is now recognizable as a distinct vessel only at its anterior and posterior extremities (Fig. 9-12). The posterior cardinal veins, posterior to the mesonephroi, receive blood from the hind-limb buds by the external iliac veins (Fig. 9-13 Y). Posterior to this point, the posterior cardinals are termed **internal iliac** or **caudal veins.** The segment of posterior cardinal veins between the external iliacs and the posterior ends of the mesonephroi is homologous with the **renal portal veins** of amphibians (Fig. 9-13 Z). In mammals and most amniotes the "old" renal portal components will disappear with the degeneration of the mesonephroi. However, a short segment of these paired vessels will anastomose to form the posterior end of the **postcaval vein,** and a small amount of the "renal portals" which fail to fuse will form the base of the **common iliac veins.**

In the 10 mm pig embryo, blood from the posterior cardinals (i.e., renal portal) is filtered through the capillary beds of the mesonephroi. In its course through the kidney the blood can be returned to the sinus venosus by either of two routes. It may go through the entire length of the mesonephros, enter the anterior ends of the **subcardinal** or **postcardinal veins,** and from thence to the **common cardinals** and finally the **sinus venosus.** Or, the blood may pass diagonally through the mesonephros to the **subcardinal anastomosis** and be passed to the postcaval vein which enters the sinus venosus (Fig. 9-12). Both pathways should be traced through the serial sections by going forward through the kidney from levels comparable to those in Figure 9-13 N–V.

The return of venous blood from the extraembryonic placenta and yolk sac is accomplished by the **umbilical** and **vitelline veins** respectively. Both veins pass through the tissue of the liver on

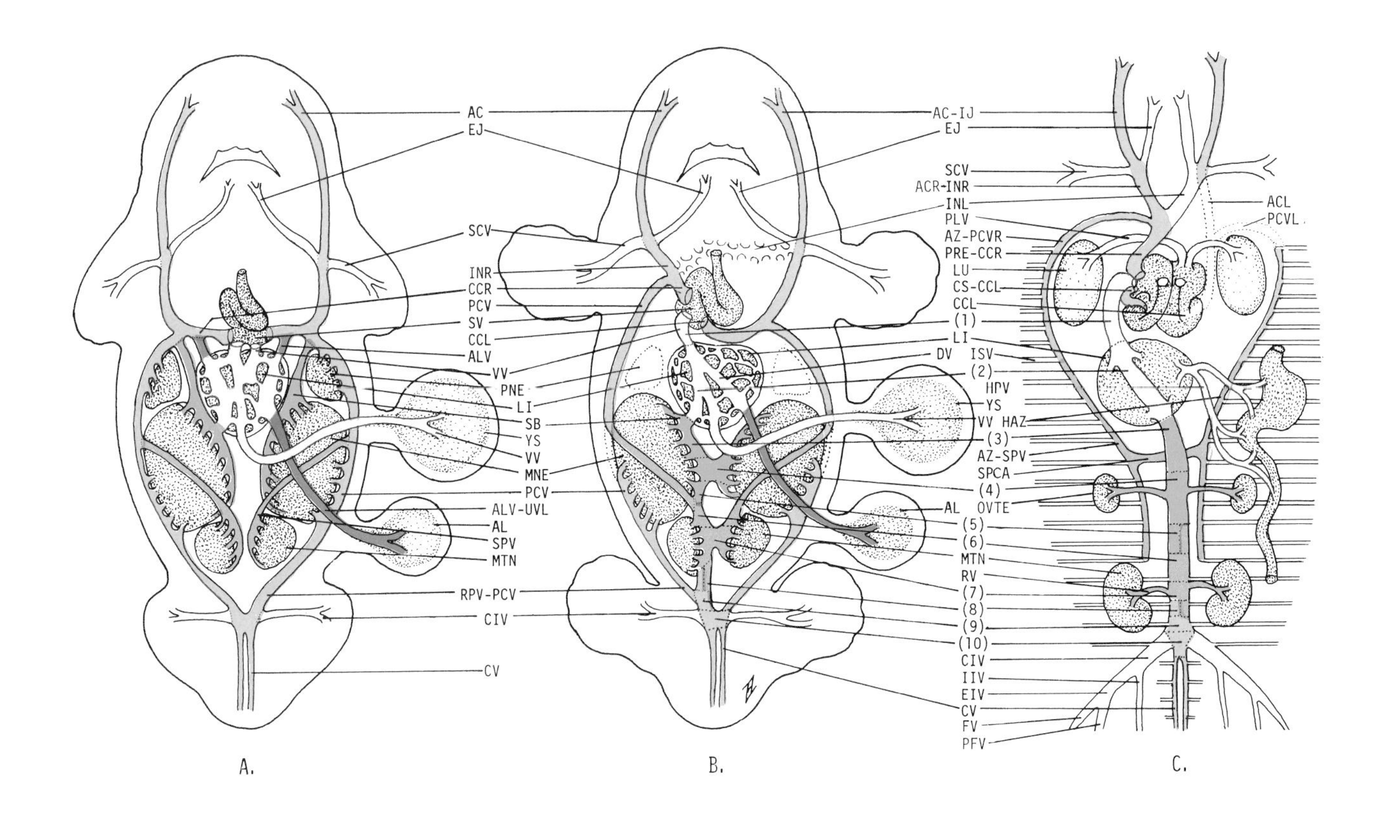

Figure 9-11. Schematic representation of the mammalian venous circulation. 9-11 A. Diagram of the primitive common plan of the venous system. 9-11 B. Diagram of a transitional stage in the development of the mammalian venous plan. 9-11 C. Diagram of the definitive mammalian plan for major veins. **Legend** (v = vein):

Numbers (1) through (10) refer to the following venous components of the inferior vena cava:

(1) right & left vitelline anastomosis
(2) liver sinusoids
(3) right subcardinal v.
(4) right & left subcardinal anastomosis
(5) right sub-supracardinal anastomosis
(6) right supracardinal v.
(7) right & left supracardinal anastomosis
(8) right post-supracardinal anastomosis
(9) right postcardinal v.
(10) right & left postcardinal anastomosis

Color code: green — primary cardinal system (anterior — posterior — common cardinal veins), red — subcardinal veins, blue — supracardinal veins, yellow — vitelline veins, purple — allantoic veins.

AC anterior cardinal v.
ACL left anterior cardinal
ACR right anterior cardinal
AL allantoic sac
ALV allantoic v.
AZ azygous v.
CCL left common cardinal
CCR right common cardinal
CIV common iliac v.
CS coronary sinus
CV caudal v.
DV ductus venosus
EIV external iliac v.
EJ external jugular v
FV femoral v.
HAZ hemiazygous v.
HPV hepatic portal v.
IJ internal jugular v.
IIV internal iliac v.
INL left innominate v.
INR right innominate v.
ISV intersegmental v.
LI liver
LU lung
MNE mesonephros
MTN metanephros
OVTE ovary or testis
PCV postcardinal v.
PCVL left postcardinal v.
PCVR right postcardinal v.
PFV profundus femoral v.
PLV pulmonary v.
PNE pronephros
PRE precava or superior vena cava
RPV renal portal v.
RV renal v.
SBV subcardinal v.
SCV subclavian v.
SPCA supracardinal anastomosis
SPV supracardinal v.
SV sinus venosus
UVL left umbilical v.
VV vitelline v.
YS yolk sac

their way to the sinus venosus, and both of them are interrupted in their course by an invasion of liver tissue (see whole mount, Fig. 9-12).

Trace the **umbilical (allantoic) veins** from the umbilical cord to the liver by going forward through the series from a level comparable to that in Figure 9-13 V. Note that the **left umbilical vein** is larger than the one on the right side. The latter will soon completely degenerate, and the left vein will return all of the blood from the placenta. In the liver note that the **right umbilical vein** breaks up into a spongy capillary network (Fig. 9-13 U), whereas the left umbilical vein can be traced into the liver as a discrete channel. This intra-hepatic continuation to the left umbilical vein is the **ductus venosus** (Fig. 9-13 Q–T). It is a new channel formed from the **liver sinusoids** and does not constitute an anatomical continuation of the primitive allantoic vein. Trace the ductus venosus and note that it empties into the **postcaval vein** deep within the liver (Fig. 9-13 Q). The postcaval vein is also derived at this level from liver sinusoids. Follow the postcaval vein forward until it empties into the posterior wall of the sinus venosus (Fig. 9-13 O, P).

The **vitelline veins** which return blood from the intestine and yolk sac are very small and can best be followed by tracing the postcaval vein from the sinus venosus back into the liver. This segment of the postcaval vein is equivalent to the **fused vitelline vein** of the 72 hour chick. In the liver, the vitelline veins are completely disrupted by the growth of liver cords. Trace the course of the postcaval vein back to the point at which the ductus venosus is given off (Fig. 9-13 R) and then follow the ductus venosus back until it divides into distinct right and left channels (Fig. 9-13 S). The right trunk will be continued posteriorly beyond the liver as the **hepatic portal vein.** It represents an anastomosis of the vitelline veins posterior to the liver. Follow the hepatic portal channel through the liver until it enters the dorsal mesentery of the intestine at the level at which the common bile duct is given off (Fig. 9-13 T). Then trace the hepatic portal vein as it accompanies the superior mesenteric artery and intestine into the umbilical cord (Fig. 9-13 V, W). At different levels along its course, the hepatic portal vein shifts its position from the right to the left sides of the endodermal lining of the intestine. Depending upon the position of the hepatic portal vein at a given level, it represents either the right of left vitelline vein, or an anastomosis between these two embryonic primordia.

The **postcaval vein** (inferior vena cava) has already been discussed in connection with other organs and veins. Here it will be necessary to review only the various components that make up this vein. Proceed backwards from the point at which the postcaval vein enters the sinus venosus. The following are the components of the postcaval vein in the 10 mm pig embryo: **(1). fused vitelline veins** (Fig. 9-13 P); **(2) liver sinusoids** (Fig. 9-13 S); **(3) right subcardinal vein** (Fig. 9-13 T); and **(4) subcardinal anastomosis** (Fig. 9-13 V–X). A total of ten components will finally be involved in the formation of the adult mammalian inferior vena cava. The additional elements listed from anterior to posterior are diagrammatically shown in Figure 9-11; they are: **(5) right sub-supracardinal anastomosis; (6) right supracardinal vein; (7) supracardinal anastomosis; (8) right sub-postcardinal anastomosis; (9) right postcardinal vein;** and **(10) postcardinal anastomosis.** Few if any other major body organs have such a varied and diverse embryonic origin as that shown in the inferior vena cava. It is not surprising that it is one of the most variable of all organs. Virtually every permutation conceivable in the fusion of these originally paired venous elements can occur in different mammalian species and in different human individuals.

In conclusion to this study of the 10 mm pig embryo, it will be informative to return to a consideration of the general chordate characters given in Chapter I, and to review them in the light of your current knowledge of vertebrate development up to this stage in mammalian morphogenesis. It is also recommended that you **go through the charts of germ layer derivatives** given in the Introduction in Tables Int.-5, 6, and 7 for ectodermal, mesodermal, and endodermal derivatives respectively. These tables can be seen to be a graphic outline of increasing specialization predicted by von Baer's second law. These germ layer tables also provide a summary of embryonic differentiation as it has been reviewed in the course up to this point.

After these retrospective reviews on vertebrate origins from both the evolutionary and embryological points of view, one can appreciate the transformations that will convert the embryo into a recognizable mammal or bird. The latter subjects will be considered in the concluding chapter which follows.

Legend for Figure 9-12. Whole mount diagram of a 10 mm pig embryo. Letters in the right and left margins refer to correspondingly lettered cross section diagrams in Figures 9-13 A–Z'.

Figure 9-13 A–Z'. Representative cross section diagrams through the 10 mm pig embryo. The level of each section can be approximated in the whole mount diagram, Figure 9-12, by placing a straight edge across the diagram and connecting correspondingly lettered guide lines. In general, the left side of each cross section diagram is slightly anterior to the right side; accordingly the morphological data in the series in approximately two times as great as the number of diagrams.

Figure 9-14 A–H. Representative sagittal sections through the 10 mm pig embryo.

Legend and checklist of structures to be identified.

Color code: blue — ectoderm, red — mesoderm, yellow — endoderm.

Cranial nerves:

3 III, oculomotor
4 IV, trochlear
5 V, trigeminus
5 G semilunar ganglion of V
5 M mandibular ramus of V
5 O opthalmic ramus of V (profundus)
5 X maxillary ramus of V
6 VI, abducens
7 VII, facialis
7 G geniculate ganglion of VII
8 VIII, acoustic nerve and ganglion
9 IX, glossopharyngeal
9 P petrosal ganglion of IX (ventral)
9 S superior ganglion of IX (dorsal)
10 X, vagus
10 J jugular ganglion of X (dorsal)
10 N nodos ganglion of X (ventral)
11 XI, accessory
12 XII, hypoglossal
12 F Froriep's ganglion (vestigial) (occipital ganglia of XI & XII ?)

General structures:

A 1 remnant of 1st aortic arch
A 2 remnant of 2nd aortic arch
A 3 3rd aortic arch (carotid)
A 4 4th aortic arch (systemic)
A 5 5th aortic arch (vestigial or absent)
A 6 6th aortic arch (pulmonary)
AC anterior cardinal vein (internal jugular vein)
ACS anterior cardinal sinuses
AE apical ectodermal ridge
AG autonomic ganglia
ALS allantoic stalk
AP alar plate (sensory column)
AQ aqueduct of Sylvius (neurocoele of mesencephalon)
AT atrium of heart
ATL left atrium
ATR right atrium
AVL left atrioventricular canal (bicuspid-mitral valve)
AVR right atrioventricular canal (tricuspid valve)
BA basilar artery capillary bed
BD common bile duct
BP basal plate (motor column)
BR brachial plexus of spinal nerves
BS bulbar septum
CA conus arteriosus
CAM caval mesentery
CCL left common cardinal vein (coronary vein of heart)
CCR right common cardinal vein (base of precaval vein)
CD carotid duct (dorsal aorta between 3rd & 4th aortic arches)
CDA caudal artery (post-anal dorsal aorta)
CEF cephalic flexure of mesencephalon
CER cervical flexure
CH choroid fissure of optic cup
CI common iliac arteries (5th lumbar intersegmental arteries)
CL general coelom, peritoneal cavity
CLO cloaca
CM cloacal membrane
CO cornea, conjunctiva
CP 1 1st closing plate (tympanum)
CP 2 2nd closing plate
CP 3 3rd closing plate
CV caudal vein (postcardinal vein in tail)
DA dorsal aorta
DI diencephalon
DM dorsal mesentery of gut
DMC dorsal mesocardium (part of transverse and mediastinal septa)

General structures, *continued*

DMS dorsal mesentery of stomach (dorsal mesogaster, greater omentum)
DMY dermomyotome of somite
DV ductus vensosus (liver sinusoids connecting left umbilical vein to postcaval vein)
EC external carotid artery
ECC endocardial cushion of heart
ED endolymphatic duct
EPF epiploic foramen of omental bursa
EI external iliac vein
FL fore-limb bud
FO 1 foramen ovale I, primary interatrial foramen
FO 2 foramen ovale II, secondary interatrial foramen
FP floor plate of neural tube
GB gall bladder
GL glomerulus
GLO glottis & laryngeal mesenchyme
GR genital ridge
HL hind-limb bud
HP hepatic portal vein (homologue of vitelline veins posterior to liver)
IA 1 primary interatrial septum
IA 2 secondary interatrial septum
IC internal carotid artery (now includes also dorsal aorta anterior to 3rd arch)
IN, INF infundibulum
ISA intersegmental arteries
ISA7 7th cervical intersegmental artery
ISV7 7th cervical intersegmental vein
IT intestine, duodenum
ITA anterior limb of intestinal loop (small intestine)
ITP posterior limb of intestinal loop (large intestine or colon)
IV interventricular septum
IVF interventricular foramen
LB lung bud
LE lens vesicle
LI liver
LMC lateral mesocardium
LS liver sinusoids
LSF lumbo-sacral flexure
LT lamina terminalis
LTG laryngo-tracheal groove
LV left ventricle
MAN mantle layer (gray matter)
MAR marginal layer (white matter)
MB midbrain, mesencephalon
MN mesonephros, mesonephric tubules
MND mesonephric duct (formerly pronephric duct; later Wolffian duct & vas deferens)
MR mammary ridge
MT metencephalon (cerebellum)
MTC metanephric mesomere, renal cortex
MTD metanephric duct
MTM metanephric (renal) medulla
MTP metanephric (renal) pelvis
MY myelencephalon
MYC myocoele of somite
MYO myotome
NCL neurocoele, neural canal
NM neuromere of myelencephalon
NO notochord
NP nasal pit
NT neural tube, spinal cord
OB omental bursa (lesser peritoneal cavity)
OE esophagus
OP optic cup
OPM optic (orbital) muscle mass
OPR optic recess
OPS optic stalk
OT otic vesicle (inner ear)
P pharynx
P 1 1st pharyngeal pouch (middle ear cavity)
P 2 2nd pharyngeal pouch (crypt of tonsil)
P 3 3rd pharyngeal pouch (posterior parathyroid and thymus)
P 4 4th pharyngeal pouch (anterior parathyroid and thymus)
P 5 5th pharyngeal pouch (ultimobranchial body)
PA pulmonary artery
PAN' dorsal pancreas
PAN'' ventral pancreas
PAP palatine process
PC peritoneal cavity
PCP posterior choroid plexus (roof plate of myelencephalon)
PCV posterior cardinal vein
PE pericardial coelom
PF pontine flexure
PL pleural coelom
PLV pulmonary vein
PPC pleuroperitoneal canal
PPM pleuroperitoneal membrane
PR pigmented retina of optic cup
PRE precaval vein, superior vena cava (composed of right common cardinal & right anterior cardinal veins)
PT pulmonary trunk of conus anteriosus
PV postcaval vein, inferior vena cava (composed of vitelline anastomosis, liver sinusoids, right subcardinal, subcardinal anastomosis)
RA renal arteries
RC ramus communicans
REC rectum
RF roof plate of neural tube
RP Rathke's pouch (stomodaeal hypophysis)
RPV renal portal vein (segment of postcardinal between mesonephros and external iliac vein)
RT sensory (nervous) retina of optic cup
RV right ventricle

General structures, *continued*

SAV	sinuatrial valve
SB	subcardinal vein
SBA	subcardinal anastomosis (component of postcaval vein)
SBL	left subcardinal vein
SBR	right subcardinal vein (component of postcaval vein)
SC	sclerotome of somite
SCA	subclavian artery (7th intersegmental artery)
SCC	semicircular canal of inner ear
SCV	subclavian vein
SG	spinal ganglia and nerves
SL	sulcus limitans
SM	somite
SMA	superior mesenteric artery (homologue of vitelline artery)
ST	stomach
STD	stomodaeum (anterior $\frac{1}{2}$ of oral cavity)
SU	sacculus utriculus of inner ear
SVL	left horn of sinus venosus (primordium of coronary sinus)
SVR	right horn of sinus venosus (base of precaval vein)
SY	systemic trunk (ascending aorta) of conus arteriosus
TB	tail bud
TE	telencephalon (cerebrum)
TG	tailgut, postanal gut
TH	thyroid diverticulum
TO	tongue
TR	trachea
TV	transverse septum (capsule of liver composed of ventral mesentery of esophagus + lateral mesocardia)
U	umbilical cord
UA	umbilical (allantoic) artery
UAL	left umbilical (allantoic) artery
UAR	right umbilical (allantoic) artery
UG	urogenital sinus
UR	urorectal septum
UVL	left umbilical vein
UVR	right umbilical vein
V	ventricle of heart
V 1	lateral ventricle of telencephalon
V 3	ventricle of diencephalon
V 4	4th ventricle of myelencephalon
VC	vertebral centrum
VG 1	1st visceral groove (external auditory meatus)
VG 2	2nd visceral groove
VG 3	3rd, 4th, & 5th visceral grooves (cervical sinus)
VMC	ventral mesocardium
VMO	ventral mesentery of esophagus (anterior $\frac{3}{4}$th equivalent to dorsal mesocardium; component of transverse and mediastinal septa)
VMS	ventral mesentery of stomach (gastrohepatic ligament, lesser omentum)
VS 1	1st visceral arch (mandibular)
VS 1'	mandibular process of arch I
VS 1''	maxillary process of arch I
VS 2	2nd visceral arch (hyoid)
VS 3	3rd visceral arch
VS 4	4th visceral arch
VS 5	5th & 6th visceral arch
VTA	vertebral artery (anastomosis of intersegmental arteries 1-6)
VV	vitelline veins
VVA	vitelline anastomosis, fused vitelline veins
VVL	left vitelline vein (hepatic portal vein)
VVR	right vitelline vein (hepatic portal vein)
XCL	extraembryonic coelom
YST	yolk stalk

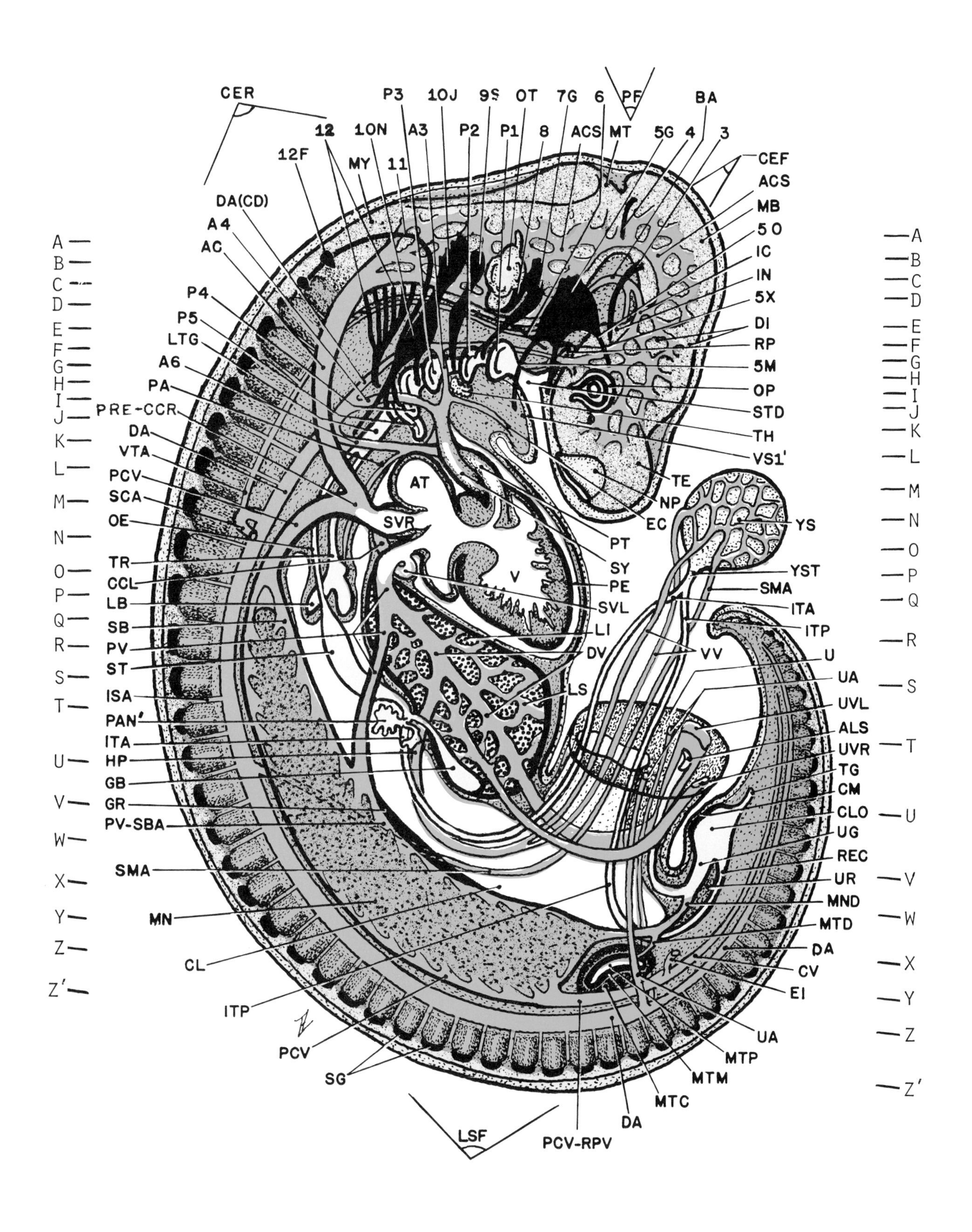

Figure 9-12. Whole mount diagram of a 10 mm embryo. Letters in the right and left margins refer to correspondingly lettered cross section diagrams in 9-13 A–Z'. See legend.

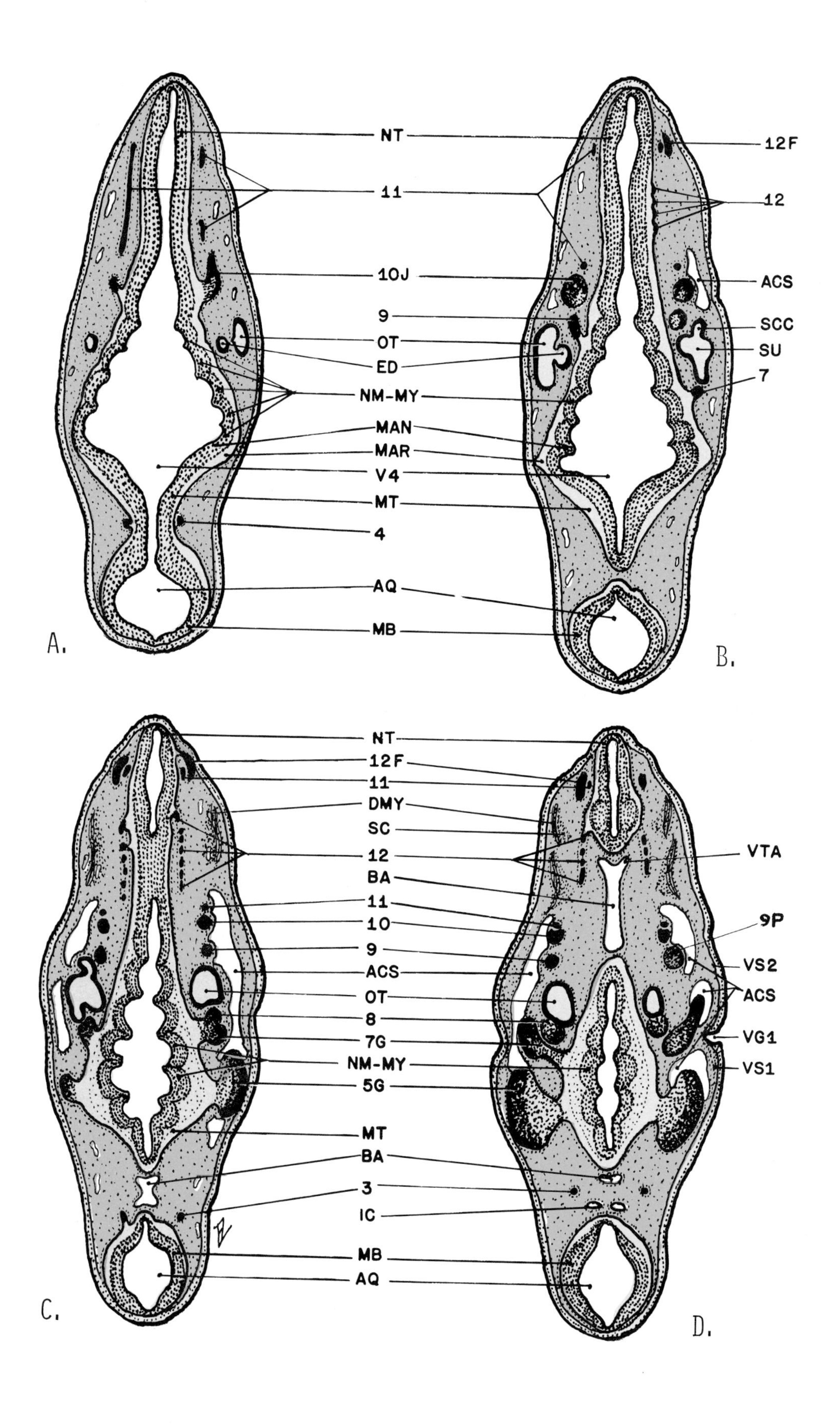

Figure 9-13 A–D. Representative cross sections of the 10 mm pig embryo. Sections at levels shown by correspondingly lettered guide lines in 9-12. Lee legend.

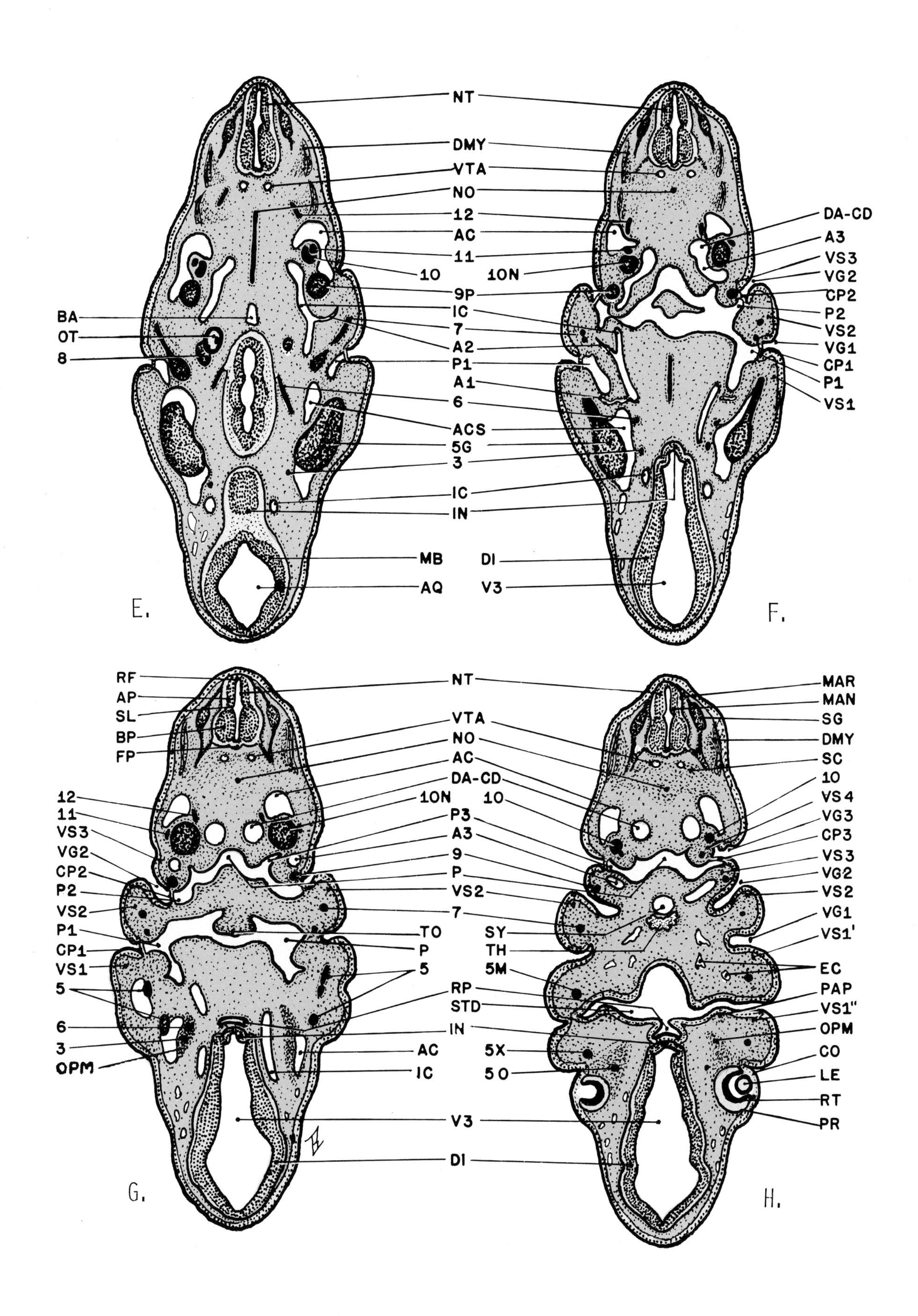

Figure 9-13 E–H. Representative cross sections of the 10 mm pig, continued. Sections at levels shown by correspondingly lettered guide lines in 9-12. See legend.

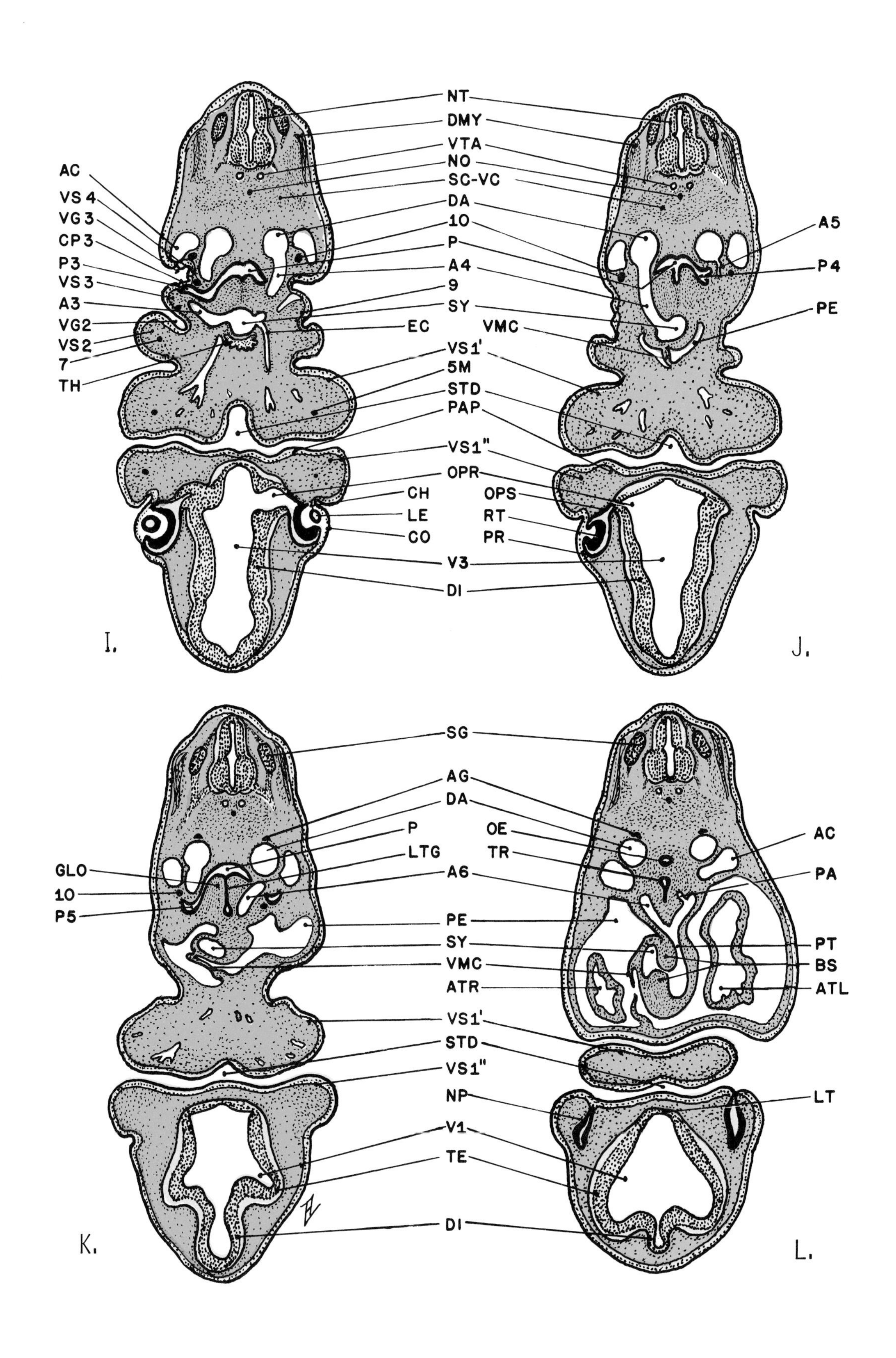

Figure 9-13 I–L. Representative cross sections of the 10 mm pig, continued. Sections at levels shown by correspondingly lettered guide lines in 9-12. See legend.

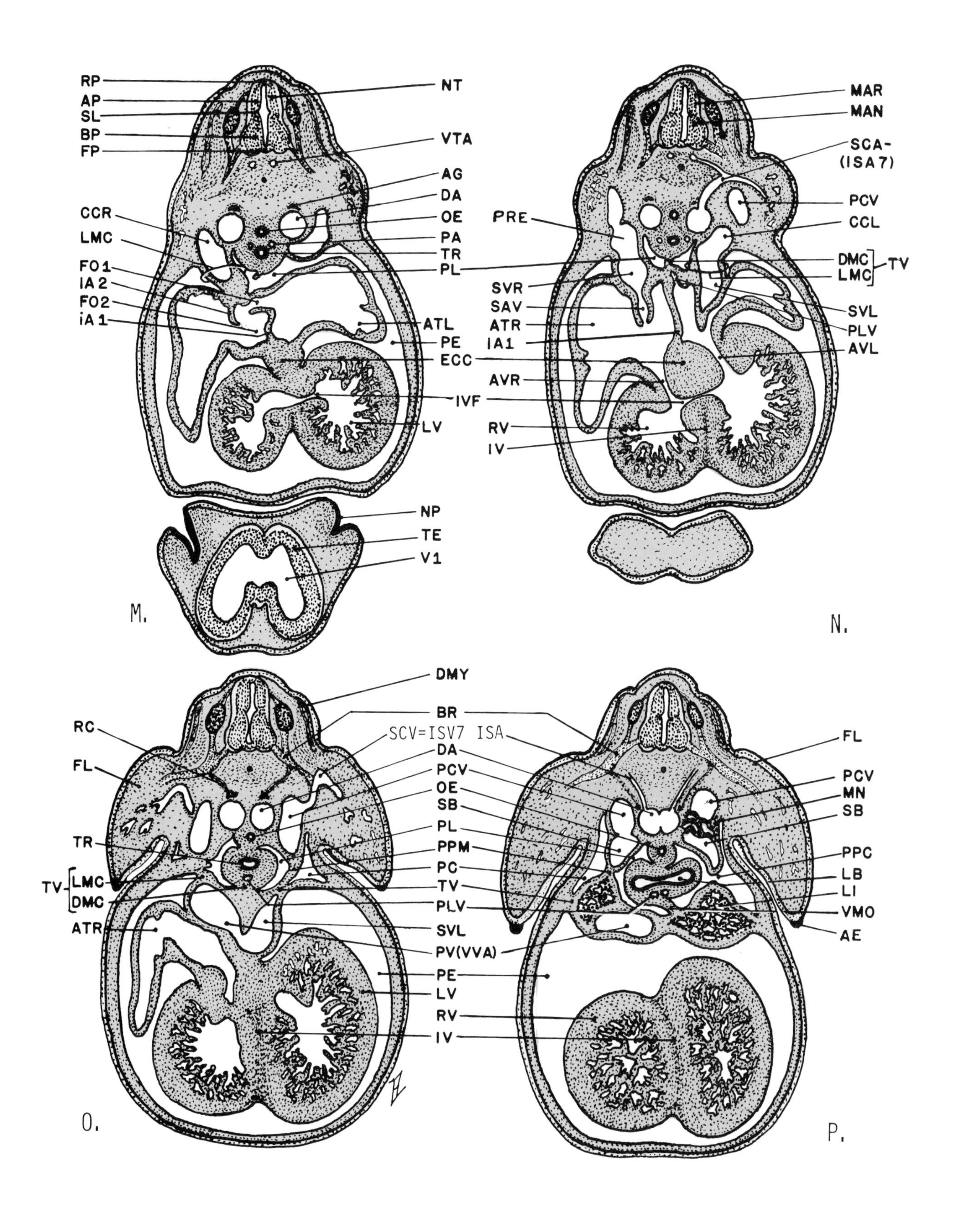

Figure 9-13 M–P. Representative cross sections of the 10 mm pig, continued. Sections at levels shown by correspondingly lettered guide lines in 9-12. See legend.

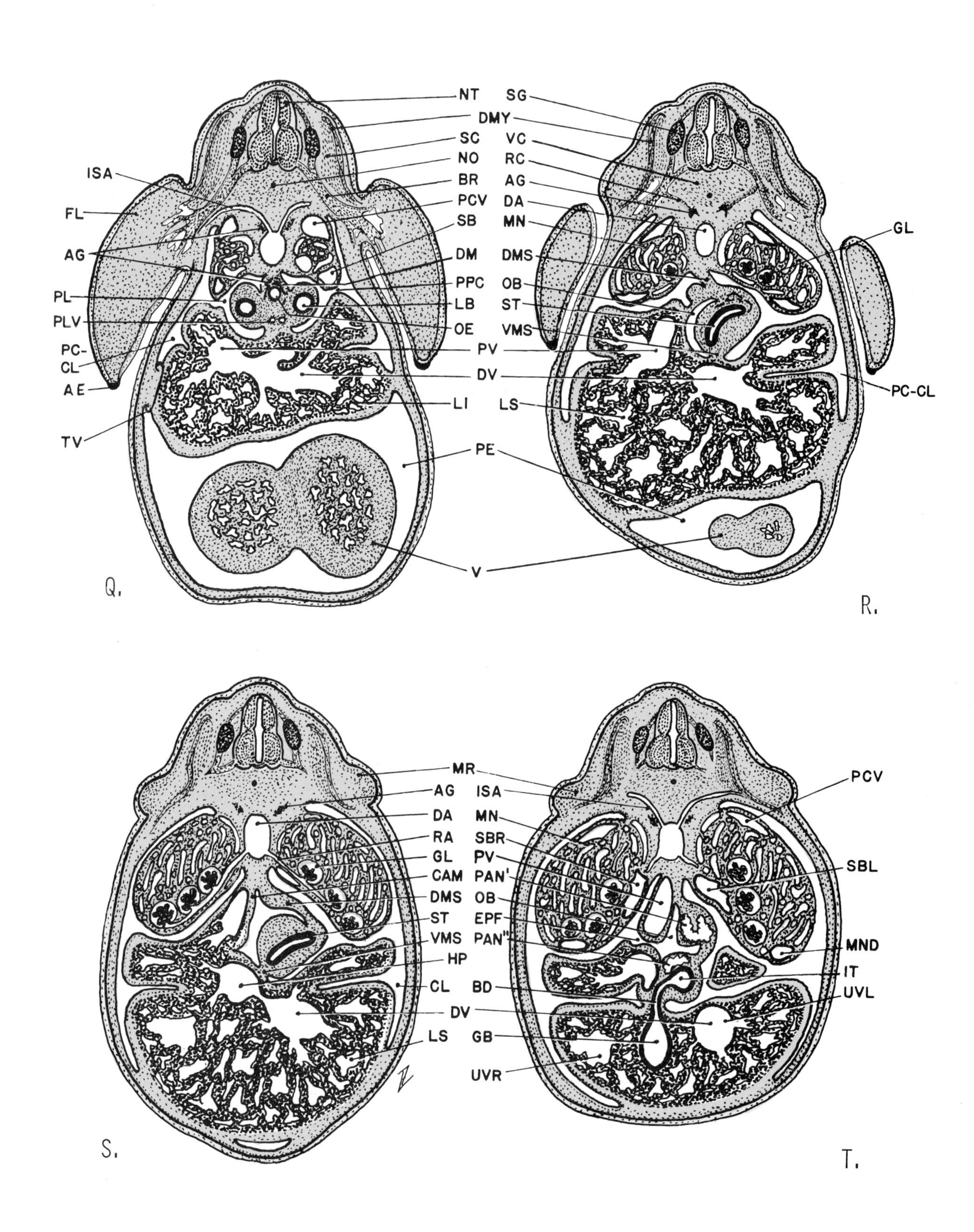

Figure 9-13 Q–T. Representative cross sections of the 10 mm pig, continued. Sections at levels shown by correspondingly lettered guide lines in 9-12. See legend.

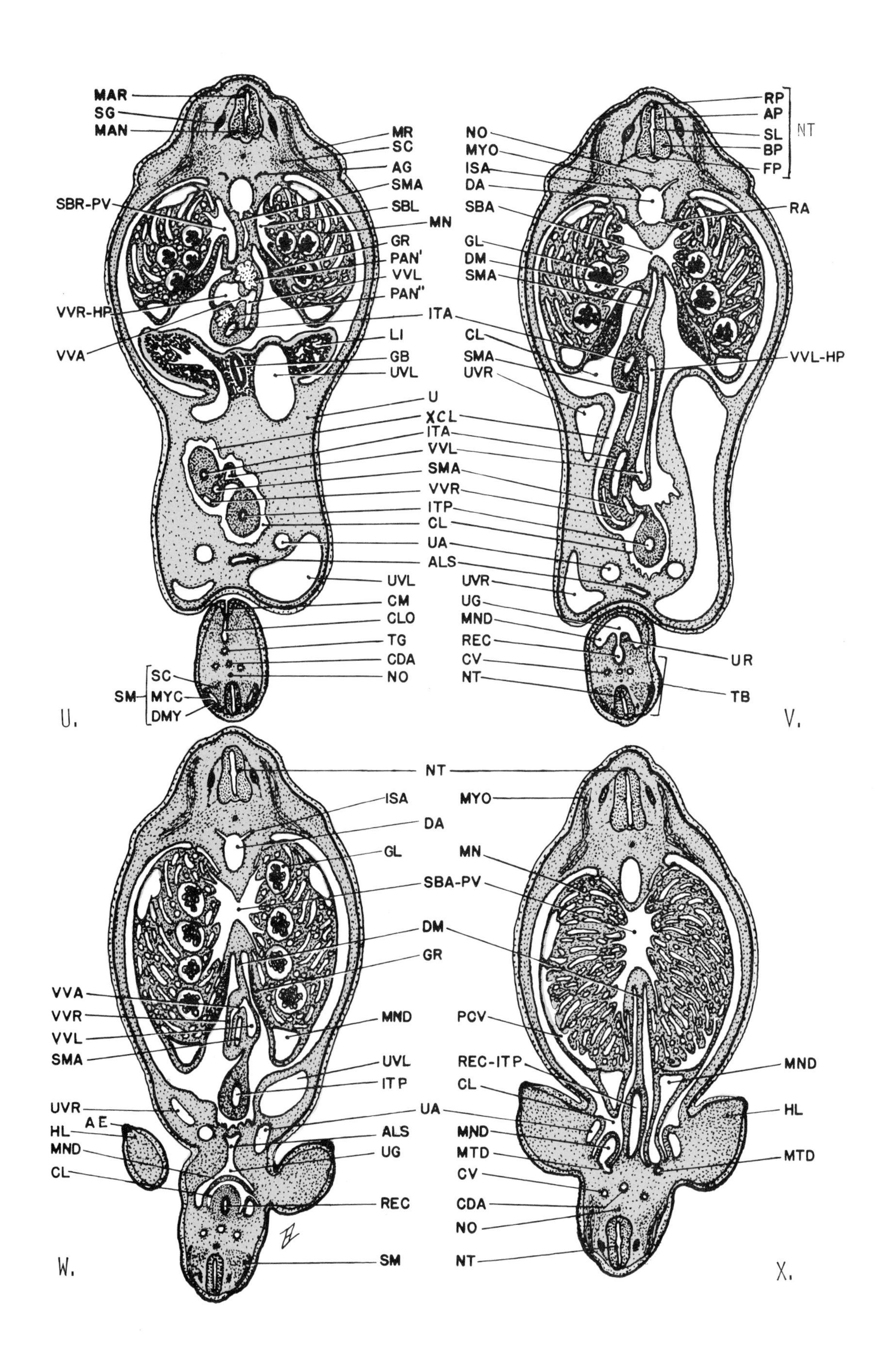

Figure 9-13 U–X. Representative cross sections of the 10 mm pig, continued. Sections at levels shown by correspondingly lettered guide lines in 9-12. See legend.

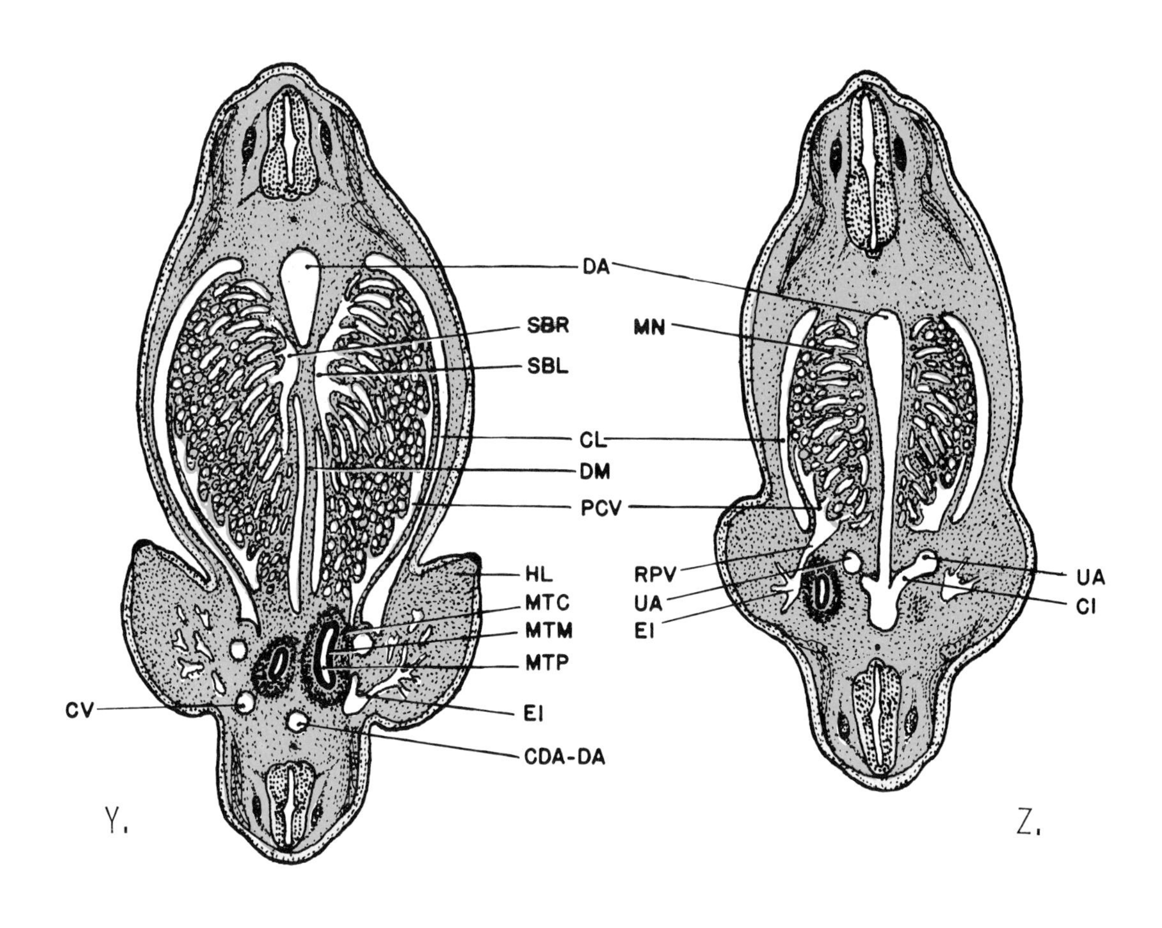

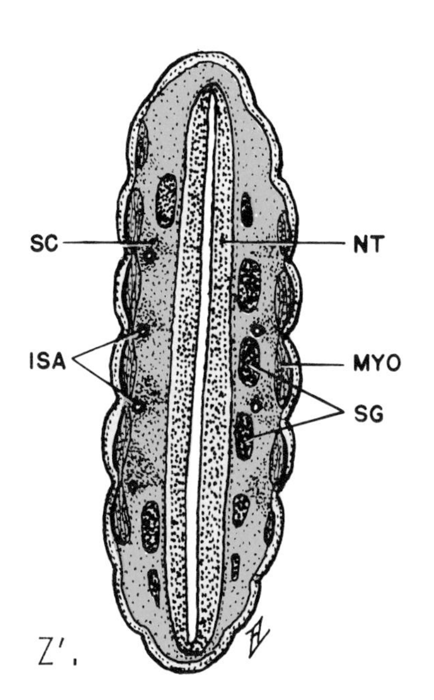

Figure 9-13 Y–Z'. Representative cross sections of the 10 mm pig, continued. Sections at levels shown by correspondingly lettered guide lines in 9-12. See legend.

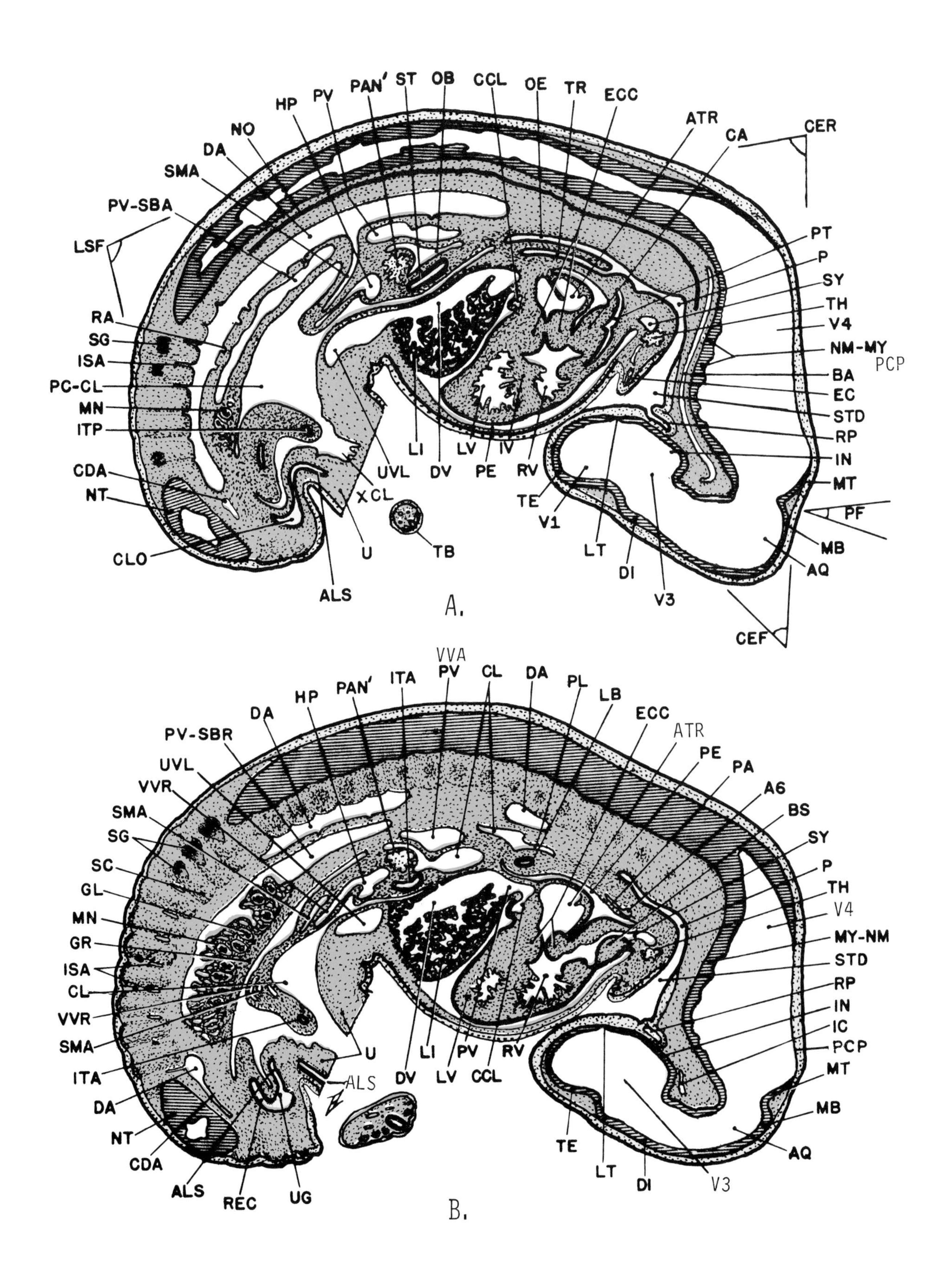

Figure 9-14. Representative sagittal sections of the 10 mm pig. 9-14 A. Median sagittal section. 9-14 B. First left parasaggital section in the series 9-14 B–H. See legend.

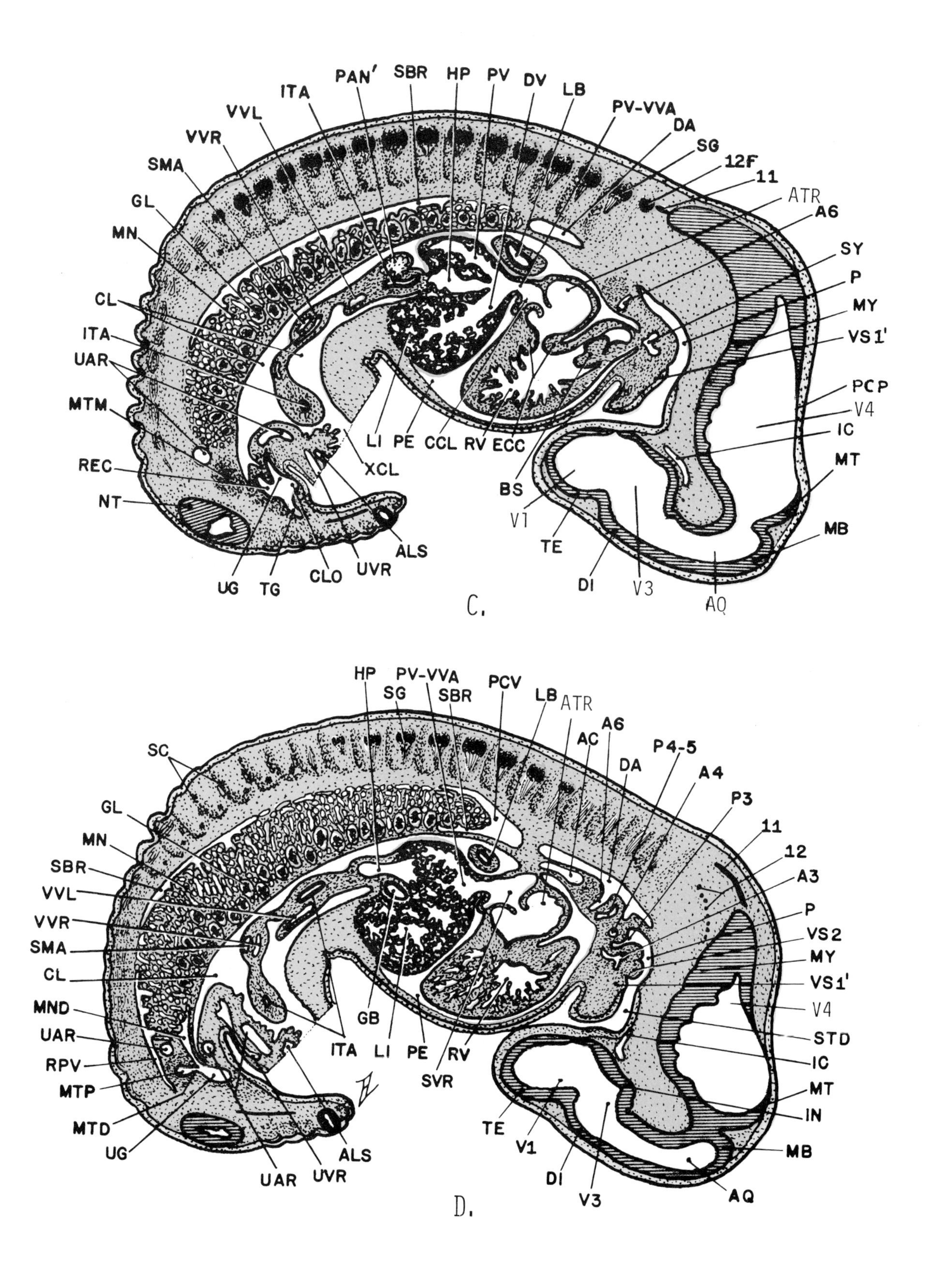

Figure 9-14 C–D. Representative left parasagittal sections of the 10 mm pig, continued. See legend.

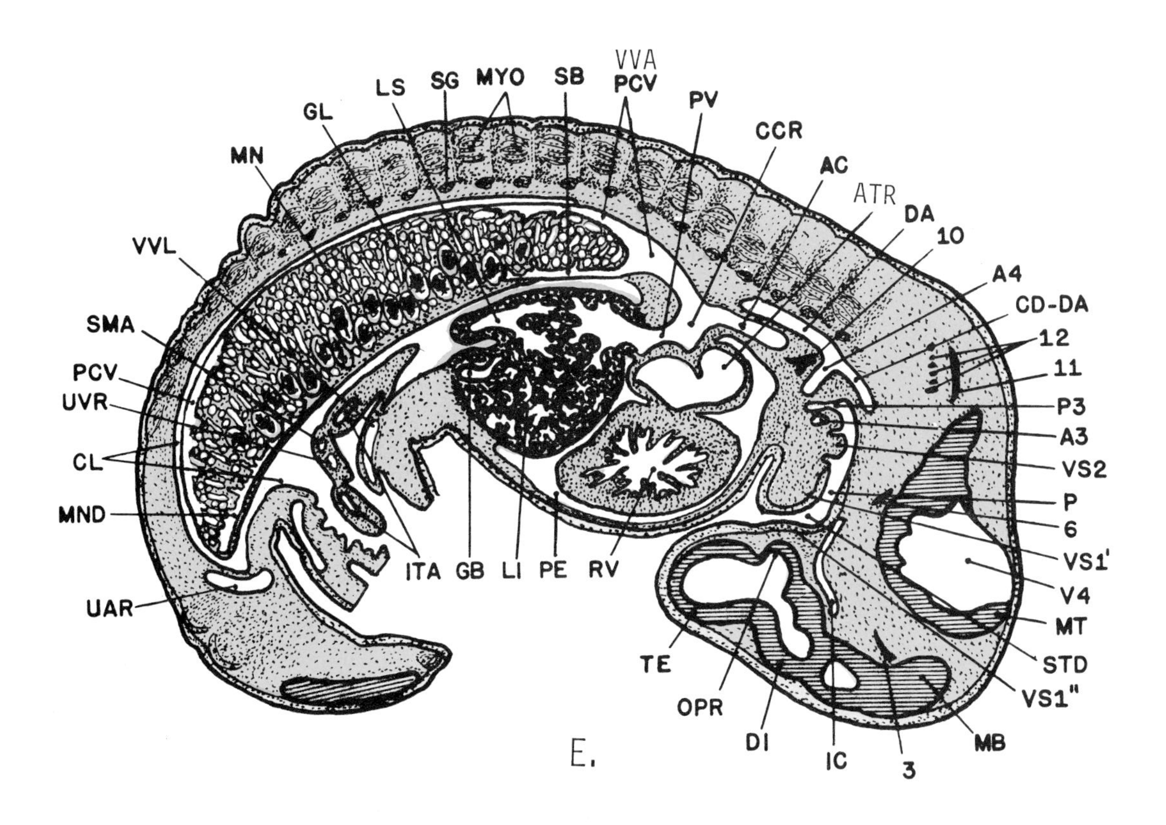

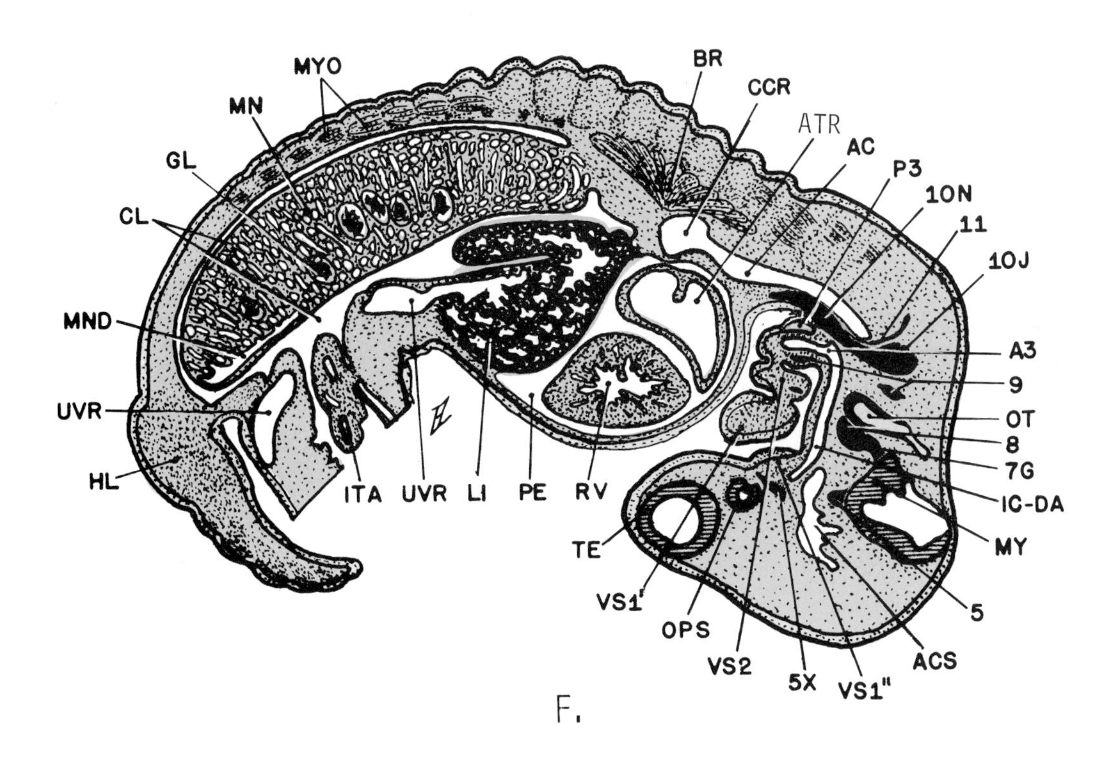

Figure 9-14 E–F. Representative left parasagittal sections of the 10 mm pig, continued. See legend.

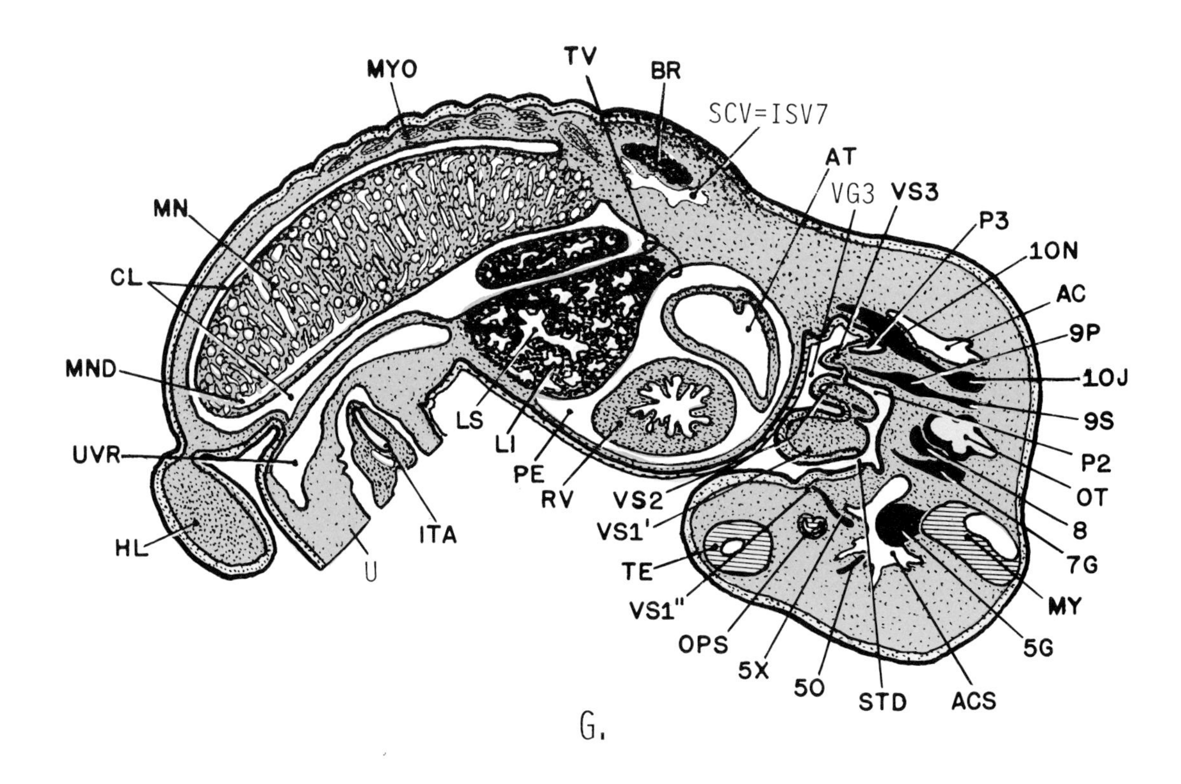

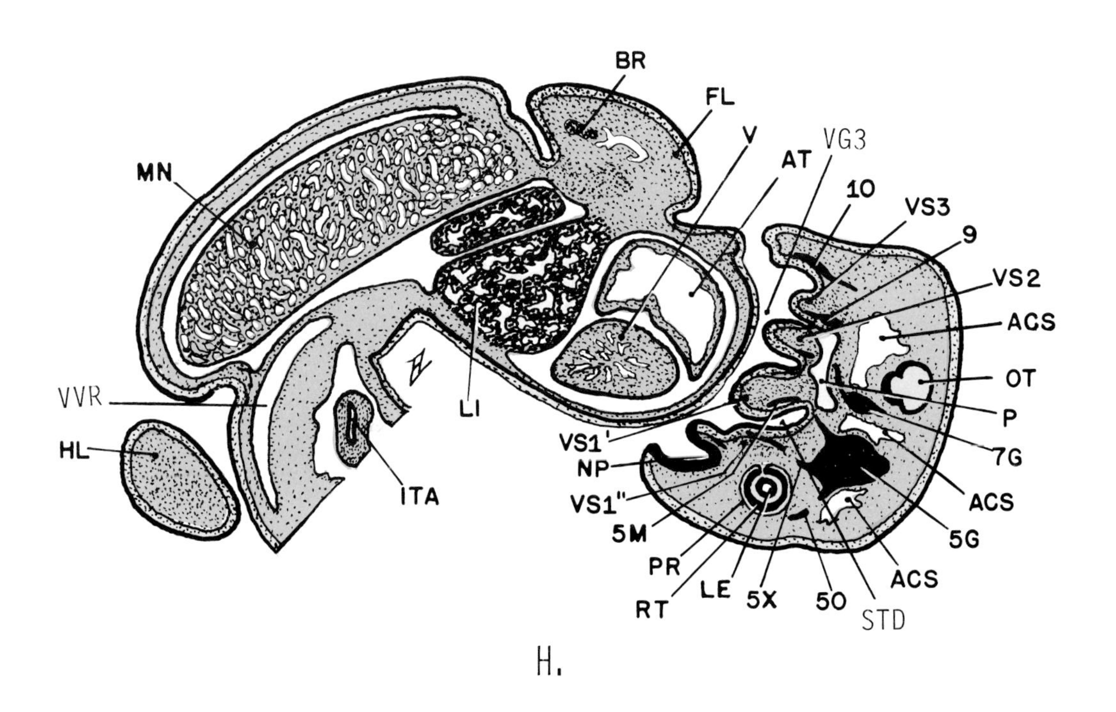

Figure 9-14 G–H. Representative left parasagittal sections of the 10 mm pig, continued. See legend.

CHAPTER X

CONCLUSION

COORDINATION AND INTEGRATION IN THE TRANSITION FROM EMBRYONIC TO FETAL STAGES IN DEVELOPMENT

Illustrated by Advanced Chick, Mouse, and Pig Developmental Stages

TABLE OF CONTENTS

ILLUSTRATIONS

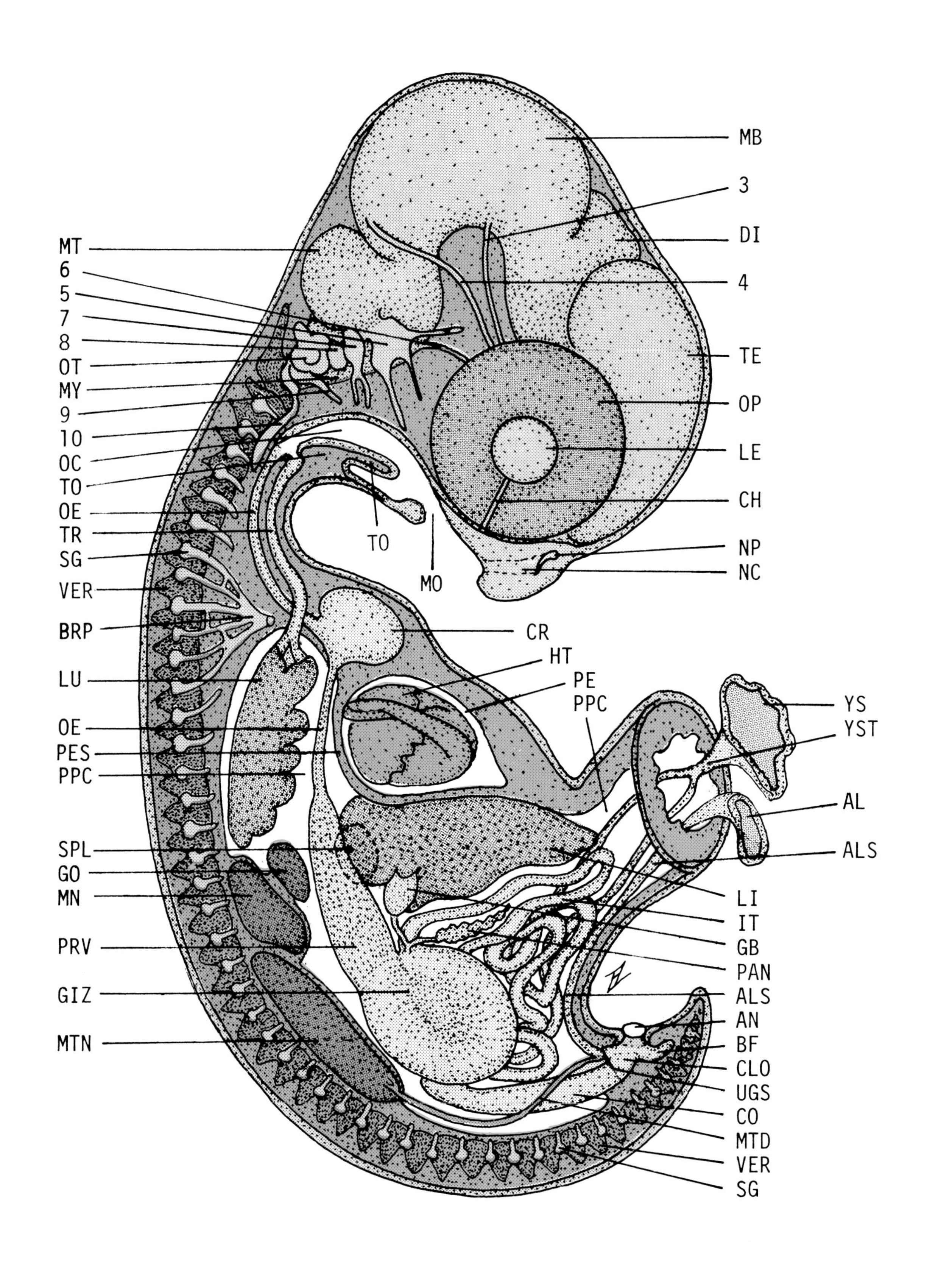

Figure 10-1. Diagram of the general anatomy of a 7 to 10 day chick embryo. See legend.

CHAPTER X

CONCLUSION

COORDINATION AND INTEGRATION IN THE TRANSITION FROM EMBRYONIC TO FETAL STAGES IN DEVELOPMENT

Illustrated by Advanced Chick, Mouse, and Pig Developmental Stages

The **fetal stage** in amniote development is attained when the organization and specialization of body parts are sufficiently advanced to reveal clearly the order and family to which the embryo belongs. The fetus will show diagnostic features permitting us to recognize, for example, that it is a turtle, or a song bird, or a rodent, but its specialization need not be sufficiently detailed to permit recognition of the precise species of the adult. In human development the fetal stage arbitrarily is said to begin at the end of the third month, at which time human characteristics are unmistakable.

In Chapters 7 and 8 the chick embryos at 48 and 72 hours incubation were described as being at the so-called **primordium mosaic stage.** By this it was implied that the embryo was composed of many autonomous, self-differentiating primordia, each of which was largely independent of all other parts in its ability to perform prescribed roles in development when transplanted to atypical positions. This ability to self-differentiate will be expressed irrespective of whether or not the organ carried out any logical function of advantage to the embryo as a whole. For example, an eye self-differentiating as an intracoelomic abdominal graft would have all morphological characteristics of a normal eye but would never develop necessary neuronal connections with optic centers of the brain to carry out visual functions of advantage to the host. At the primordium mosaic stage we can characterize the entire embryo as having a maximum of regional autonomy of all parts and a minimum coordination of growth and function of parts.

Most development following the primordium mosaic stage in early organogenesis is concerned with two parallel activities. **First,** primordia will undergo tissue-specific types of **histogenesis** and cell specialization in each case characteristic for

Legend for Figure 10-1. Diagram of the general anatomy of a 7 to 10 day chick embryo. Legend and checklist of structures to be identified.

Color code: blue — ectoderm, red — mesoderm, yellow — endoderm.

3 oculomotor nerve
4 trochlear nerve
5 trigeminus nerve
6 abducens nerve
7 facial nerve
8 auditory nerve
9 glossopharyngeal nerve
10 vagus nerve
AL allantois
ALS allantoic stalk
AN anus
BF bursa of Fabricius
BRP brachial plexus
CH choroid fissure
CLO cloaca
CO colon, large intestine
CR crop
DI diencephalon
GB gall bladder
GIZ gizzard
GO gonad
HT heart
IT small intestine
LE lens
LI liver
LU lung
MB midbrain
MN mesonephros
MO mouth
MT metencephalon
MTN metanephros
MTD metanephric duct
MY myelencephalon
NC nasal canal
NP nasal pit, nostril
OC oral cavity
OE esophagus
OP optic cup, eye
OT otic vesicle, inner ear
PAN pancreas
PE pericardial coelom
PES pericardial sac
PPC pleuroperitoneal coelom
PRV proventriculus of stomach
SG spinal ganglia
SPL spleen
TE telencephalon
TO tongue
TR trachea
UGS urogenital sinus
VER vertebrae
YS yolk sac
YST yolk stalk

the organ function in question. That is, cells determined as myoblasts will proceed to form functional contractile muscle cells, or neuroblasts will develop axons and dendrites and become capable of conducting impulses. In brief, the simple epithelia and mesenchyme of primordia will specialize as connective tissue, bone, cartilage, glands, etc. **Second,** and of possibly greater importance in achieving organismal status, is the gradual subjugation of all tissues and organs into an integrated whole. **This coordination of regional growth** and **organ function** is achieved primarily by the action of endocrine secretions (hormones) and nerves.

The coordinating and integrating functions of hormones and nerves is very different from the role of induction in early development, although both of them are required in achieving the fully differentiated state. It will be recalled that inductions restrict the capacities for cell specialization and thereby determine cell specificity. Neurohumoral stimuli function to release these capacities for specialization in their so-called target cells. **Inductors are determining stimuli and neurohumors are differentiating stimuli.**

Histogenesis and coordination begin in embryonic stages and continue through fetal, juvenile, and adult stages in a never ending process of growth and learning until senescence reverses the process and ultimately terminates the life cycle in death. Most embryology courses, including this one, are terminated long before an appreciable amount of histogenesis and coordination in structure and function are manifest. As was indicated in Chapter 9, the 10 mm pig embryo carried mammalian development up to a point at which it is possible to anticipate mammalian form as a logical next step in the development of structures. The present chapter is an attempt to provide a more tangible bridge between the mid-stage in primordium development as represented by the 10 mm pig embryo, and the definitive mammalian state. To this end a few dissections of late embryos and early fetuses in chick, mouse, and pig development are suggested along with some simple techniques for manual dissection that are particularly suitable for use on the intermediate size range between the microscopic and usual macroscopic size.

Figure 10-2. Diagram of the general anatomy of the 35 to 45 mm pig embryo. Legend and checklist of structures to be identified.

Color code: blue — ectoderm, red — mesoderm, yellow — endoderm.

3	oculomotor nerve
4	trochlear nerve
5	trigeminus nerve
6	abducens nerve
7	facial nerve
9	glossopharyngeal nerve
10	vagus nerve
12	hypoglossal nerve
AL	allantoic sac
ALS	allantoic stalk
AN	anus
BRP	brachial plexus
CEC	caecum, appendix
CO	colon, large intestine
DI	diencephalon
DPH	diaphragm
GB	gall bladder
GO	gonad
HT	heart
IT	small intestine
LE	lens
LI	liver
LU	lung
MB	midbrain
MN	mesonephros
MND	mesonephric duct
MO	mouth
MT	metencephalon
MTN	metanephros
MTD	metanephric duct
MY	myelencephalon
NC	nasal canal
NP	nasal pit, nostril
OC	oral cavity
OE	esophagus
OP	optic cup, eye
OT	otic vesicle, inner ear
PAL	palate
PAN	pancreas
PC	peritoneal coelom
PE	pericardial coelom
PES	pericardial sac
PIT	pituitary gland
PL	pleural coelom
REC	rectum
SAL	salivary glands
SCP	sciatic plexus
SG	spinal ganglia & nerves
SPL	spleen
SRG	suprarenal gland, adrenal gland
ST	stomach
TE	telencephalon, cerebrum
TH	thyroid gland
THY	thymus gland
TO	tongue
TR	trachea
UGS	urogenital sinus
UGV	urogenital sinus
UGV	urogenital vestibule
VER	vertebrae
YS	yolk sac
YST	yolk stalk

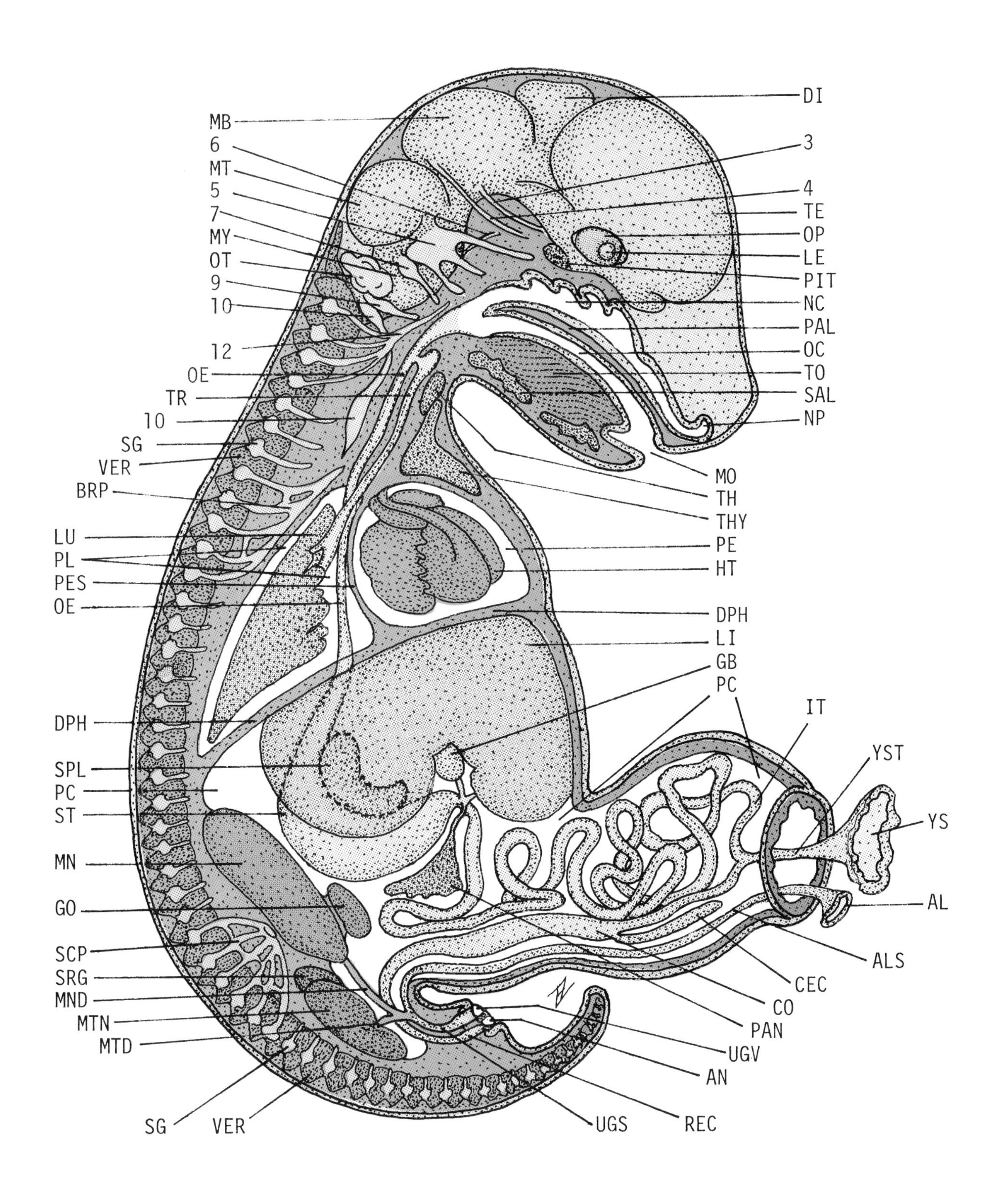

Figure 10-2. Diagram of the general anatomy of the 35 to 45 mm pig embryo. See legend.

A. MICRODISSECTION OF 16 mm TO 50 mm EMBRYOS AND FETUSES

The techniques to be used involve a combination of gross dissection with the unaided eye and microdissection with the aid of a hand lens or low power dissecting microscope. The material will consist of late embryos and early fetuses of 16 mm to 50 mm crown-rump length obtained from the study of living chick embryos in Chapter 8, or from the living mouse in Chapter 9, or small pig fetuses obtained specially for this exercise.

Materials needed: Dissecting microscope or hand lens, focusable micro-illuminator, dissecting pins, standard dissecting instruments (including fine forceps, scissors, dissecting needles, single-edge razor blade, medicine dropper pipettes), 3 wax bottom Petri dissecting dishes (i.e., with about one-quarter inch of paraffin in the bottom).

No attempt will be made to provide a detailed description of bird and mammal adult anatomy. It is assumed that the broad outlines of vertebrate anatomy, and particularly that of mammals, is part of common knowledge acquired by all students in previous exposures to biology. For a quick review, the basic organization of viscera for the bird and mammal is shown in Figures 10-1 and 10-2 respectively.

Place specimens to be dissected in a wax bottom dissecting dish. Rinse the specimen with several changes of tap water and leave the dish awash with about one-eighth inch of water. Adjust the light so that the surface of the embryo or fetus is brightly illuminated.

The following instructions for dissection are of necessity very general since they are intended to apply to any avian or mammalian late embryo or early fetus. It is generally more impressive to move backwards from the most advanced stages to younger ones. In this way you will be alerted to look for structures in early stages that are only beginning to appear and otherwise might be overlooked.

1. External anatomy. On the basis of general knowledge of the general structure of the vertebrate body, identify the following structures.

On the head identify the **nostrils, mouth,** and **eyes,** which may be open or closed depending upon age and development of eyelids. In birds note, in particular, the sclerotic ossicles in bird embryos which appear around the pupil of the eye as a series of 12 to 14 very evenly placed papillae. Note the **ears** with the opening of the external canal representing the first visceral groove. In mammals the pinna of the external ear may be starting to form. The **upper** and **lower jaws** are modified derivatives of the first visceral arch.

On the neck identify the posterior **visceral arches** and **pouches** which will become progressively less distinct with advancing development. In mammals, the laryngeal apparatus is the most conspicuous derivative of these arches.

On the trunk identify the **pectoral appendages** (forelimb or wing), the **pelvic appendages,** and **digits** on both pairs of appendages. In mammalian fetuses the **mammary ridge** and **nipples** form a conspicuous row between the base of the fore-and hind limbs. Examine the cut end of the cord and identify the **yolk stalk, allantoic stalk,** and cut ends of the **vitelline** and **allantoic arteries** and **veins;** the **anus** is located just ventral to the base of the tail.

External genital primordia are shown in Figure 10-3 A. In mammals identify the **external genital primordia,** which lie just anterior to the anus between the hind limbs. They are too imperfectly formed to reveal the final sex of the embryo. The external genital primordia and their fate in male and female mammals are as follows. The **urogenital groove** and **urogenital membrane** lie immediately anterior to the anus; in females they form the urogenital vestibule; in males they form part of the penile urethral canal. The **genital tubercle,** located at the anterior end of the urogenital groove, appears as a prominent papilla; in females it will form the clitoris; in males it forms the glans penis. The **lateral genital folds** form parallel elevations beside the urogenital groove; in females they form the labia minora; in males they form the shaft of the penis. The **lateral genital swellings** are rounded elevations between the base of the hind limbs and the lateral genital folds; in females they form the labia majora; in males they form the scrotal sac which will house the testes after they descend from the body cavity into this region. The general topography of these primordia is given in Figure 10-3 A.

2. Internal anatomy. Embryonic tissues are very soft, and most dissection can be done by needle point rather than scissors and scalpel. Use fine insect pin dissecting needles and finely sharpened forceps for most of the dissection. Use sharp scissors very carefully to cut away parts of the body wall or organs that must be removed to reveal deep structures. **Remember that right and left refer to the embryo's right and left**

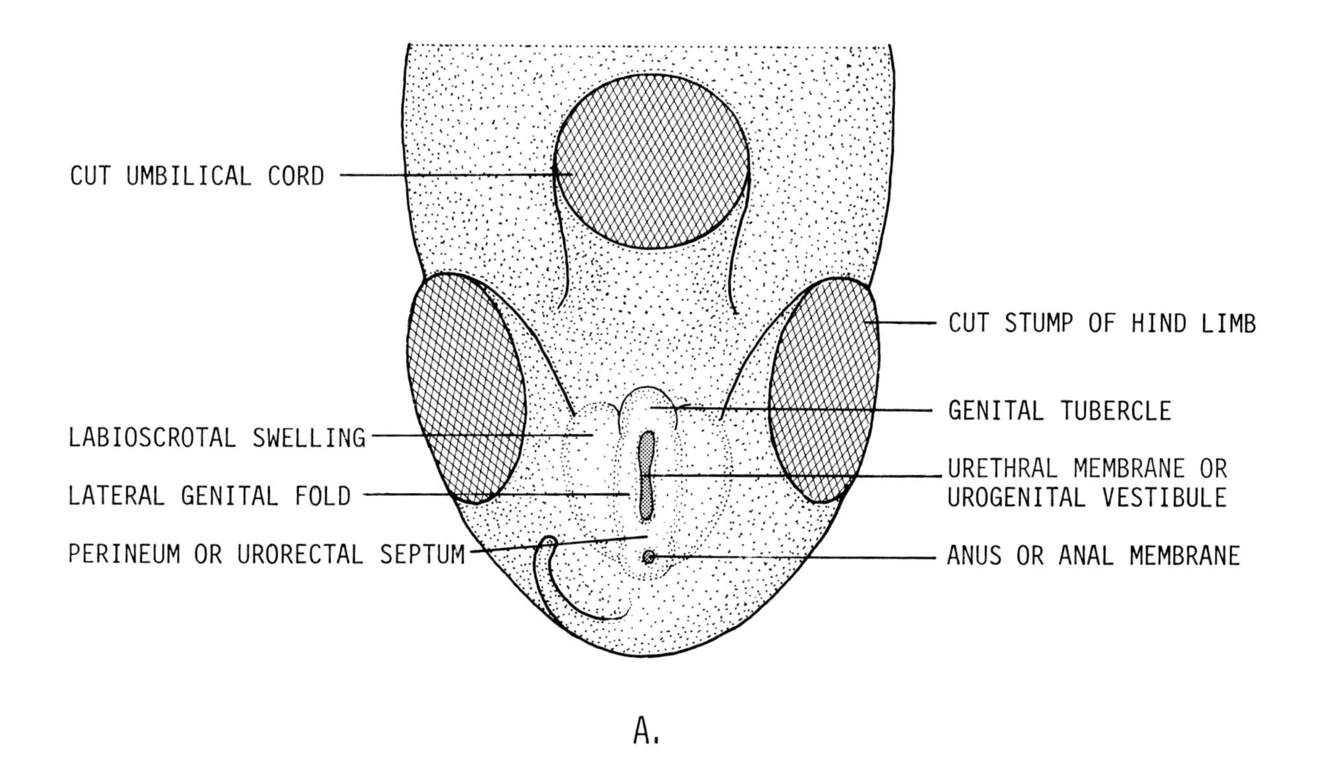

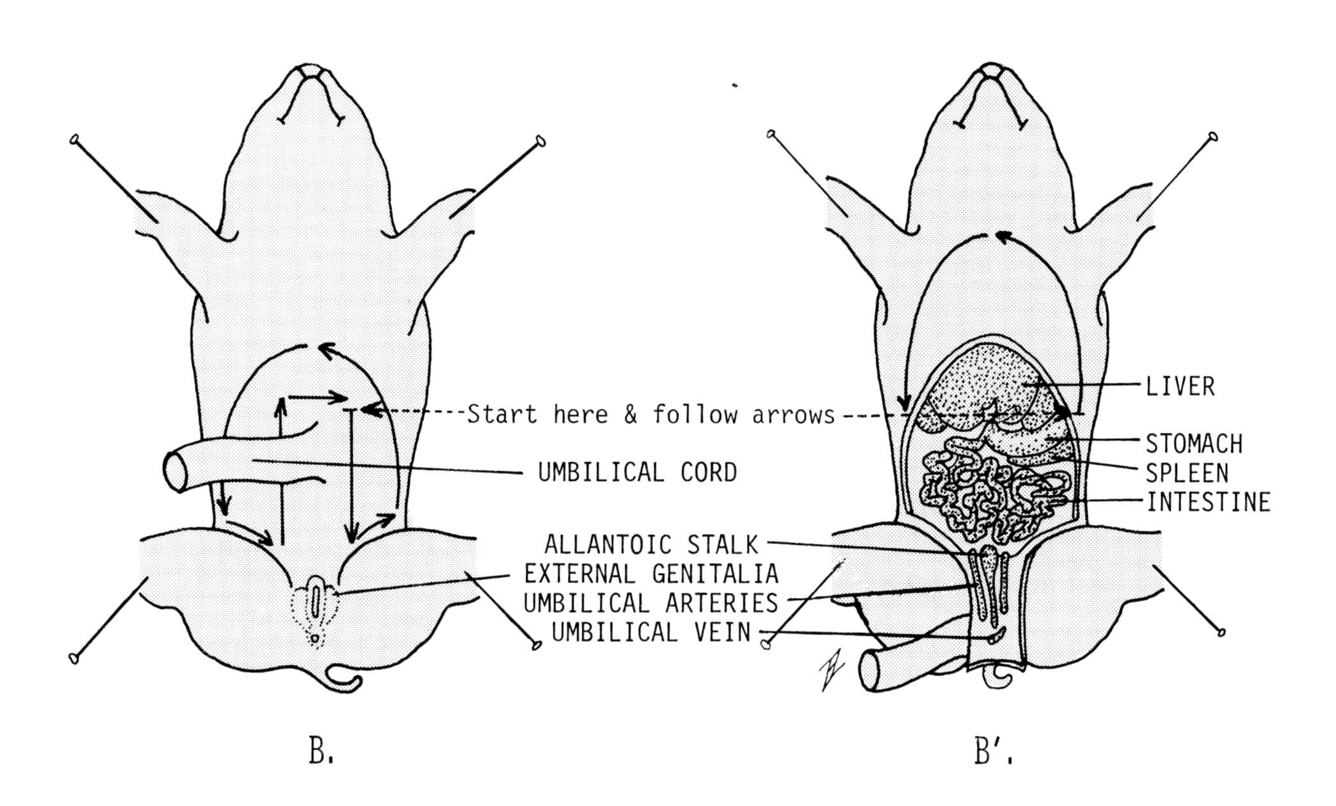

Figure 10-3. Diagram of external genitalia and directions for beginning the dissection of a fetus. See explanation in text.

sides in the following directions; this will be opposite to your own right and left sides.

Place the embryo or fetus on its back and pin it down on the wax with dissecting pins inserted at an acute angle through the appendages to keep them out of the way. Open the peritoneal cavity and expose the abdominal viscera by making **incision Number 1,** as shown in Figure 10-3 B. Cut away and discard the flap of ventral body wall. Identify the diaphragm which in mammals will separate the abdominal and thoracic coelomic cavities. Then proceed to open the thoracic cavity by making **incision Number 2** to expose the heart and lungs.

The visceral organs of the abdominal and thoracic cavities, when exposed, will occupy the relative positions shown in Figure 10-3 B. However, depending upon the age of the specimen, the relative size of different organs may be very different from those shown. In particular, expect great variations to exist in the size of the liver, heart, and kidney which are often comparatively enormous in early stages, whereas lungs, intestine, and reproductive structures are often inordinately small by comparison.

In the thoracic cavities identify the following parts with the aid of Figures 10-1, 10-2, and 10-3 B: the **heart** with right and left atria and ventricles, the **thoracic thymus** which partially covers the anterior end of the heart, the **lungs** which may be very small in young specimens and entirely hidden dorsally behind the heart.

In the abdominal cavity identify the **small intestine** which may form a hernial loop into the umbilical cord, and the **colon,** or **large intestine,** situated toward the left dorsal side of the abdominal cavity and largely hidden by coils of the small intestine. The stomach is located under the diaphragm at the anterior left extremity of the abdominal cavity. The **spleen** is a very large blood-forming gland attached to the left posterior wall of the stomach. The **liver** is located at the right side and fills most of the anterior end of the abdominal cavity. The **pancreas** will be located in the space between the posterior surface of the stomach and the first loop of the small intestine. Cut across the anterior end of the small intestine with a sharp scissors. Pull the small intestine out of the body cavity; cut across the posterior third of the large intestine as it moves to the midline in the body cavity. Discard the intestine and with the aid of Figure 10-4 identify the urogenital structures that are located toward the posterior end deep in the body cavity. The **kidneys** are a pair of bean-shaped organs that are unmistakable; in older embryos they are the metanephric or adult amniote kidney. In younger embryos there may be two pairs of kidneys present. The anterior ones are the mesonephric kidneys and the posterior pair are metanephroi. Eventually the mesonephric kidneys will completely atrophy except for some accessory gonadal structures (see Chapter 2, Fig. 2-2). Attempt to identify the special internal components of **female** and **male reproductive systems** as shown in Figure 10-4. The most conspicuous variations will be associated with the position of the ovaries and testes which originate deep in the body cavity at a level opposite the mesonephroi. Testes move back through the ventral abdominal wall to a final location in the scrotal sacs. Expect, therefore, to find the testes at any point in their path of migration. In birds there is a single left ovary and oviduct (the right elements may persist as vestigial organs) but testes and male tubes are paired in birds. They remain within the body cavity and do not descend into a scrotum as is the case late in mammalian development.

Dissection of the arterial and venous circulatory systems is entirely optional. The specimens have not been injected and the vessels are, for the most part, thin walled and difficult to trace. The general mammalian circulatory plan is given in Figure 10-5 as an aid to the identification of such vessels as you may chance to see. Remove the heart from its anteriodorsal attachment to the superior vena cava and from the inferior vena cava which enters the right atrium through the diaphragm. If the heart is large enough, slice it open in the frontal plane and identify the right and left atria with the foramen ovale in the interatrial septum. In most cases the right and left ventricles will be completely separated by the interventricular septum. Note the connection of the ascending aorta to the left ventricle and the pulmonary trunk to the right ventricle. In mammals the arch of the aorta passes to the left and in birds to the right side of the body on its path around the heart to posterior parts of the body.

Dissection of lungs, trachea, esophagus and **mouth.** With the heart removed from the thoracic cavity, identify the paired *bronchial tubes* attached to the lungs. Trace the bronchial tubes anteriorly in the **trachea.** Dissect open the skin of the throat and trace the trachea forward to the voice box, or **larynx,** which represents specializations of the embryonic visceral arches 4, 5, and 6 (see Fig. 5-5 B). Lateral to the trachea and larynx, in older stages, identify the saddle-like glandular

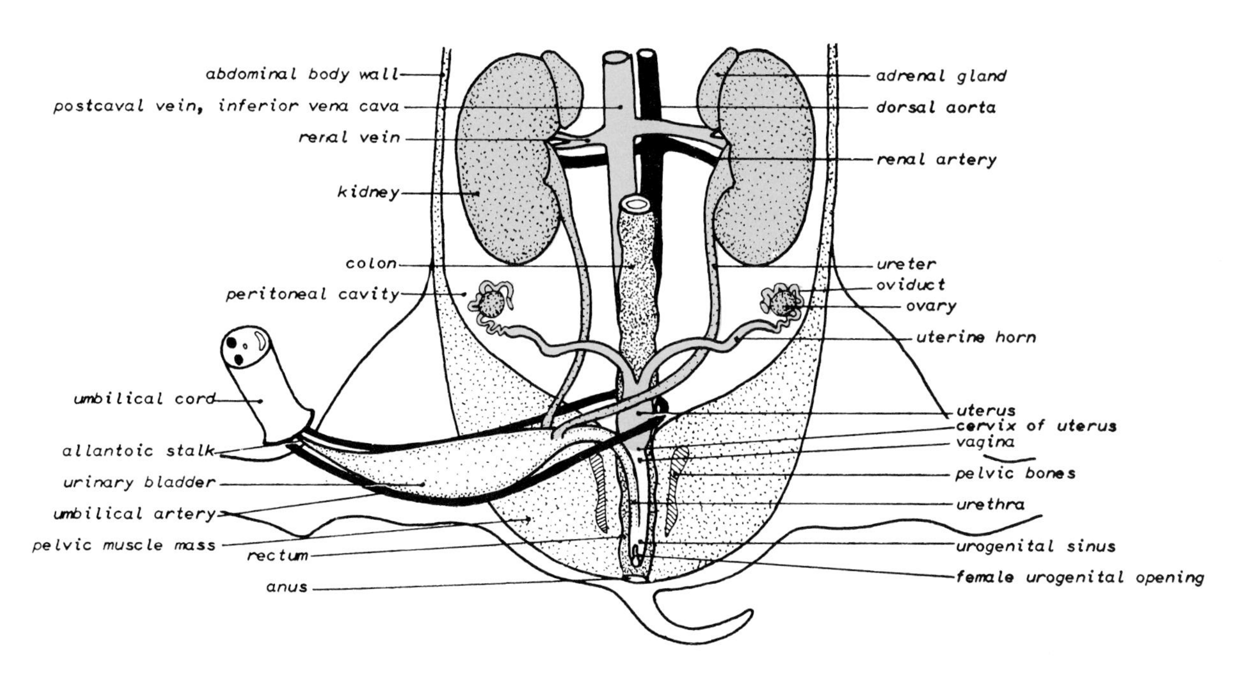

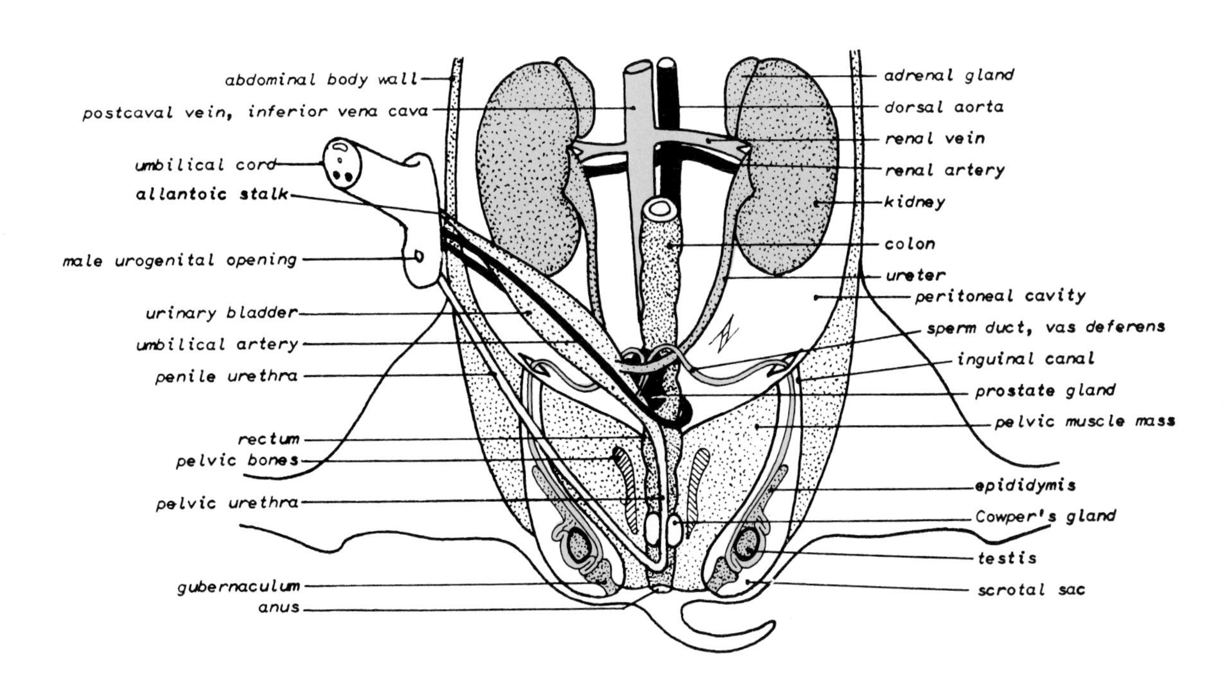

Figure 10-4. Urogenital systems in the fetal pig. 10-4 A. Female urogenital structures. 10-4 B. Male urogenital structures. (From H.E. Lehman, *Laboratory Studies in General Zoology,* Hunter Publishing Co., 1981).

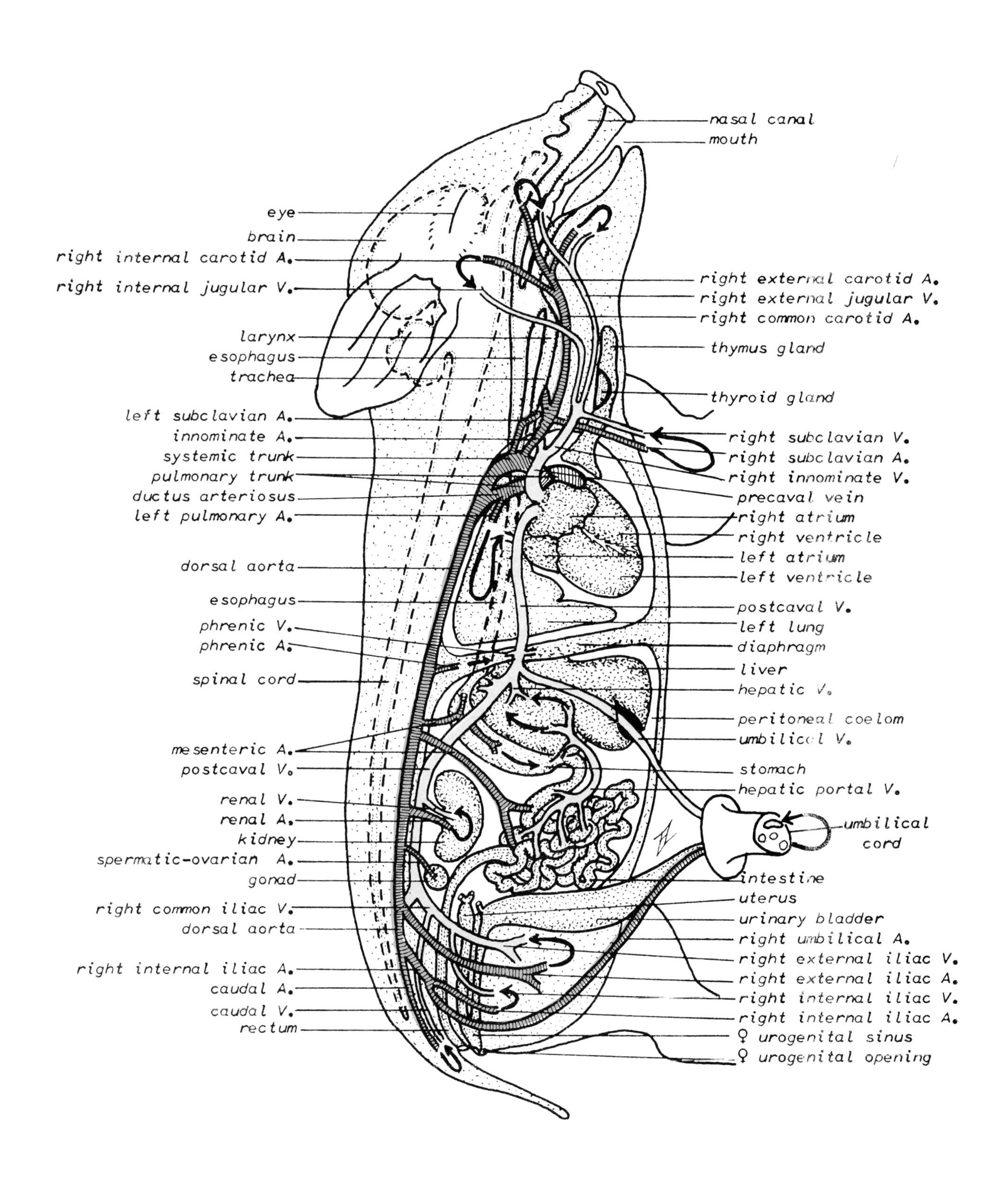

Figure 10-5. Diagram of the circulatory system of the fetal pig. Arteries red and not shaded, veins blue and shaded, and gut and endodermal derivatives yellow. (From H.E. Lehman, *Laboratory Studies in General Zoology,* Hunter Publishing Co., 1981). Color code: blue — veins, red — arteries, yellow — endoderm and gut.

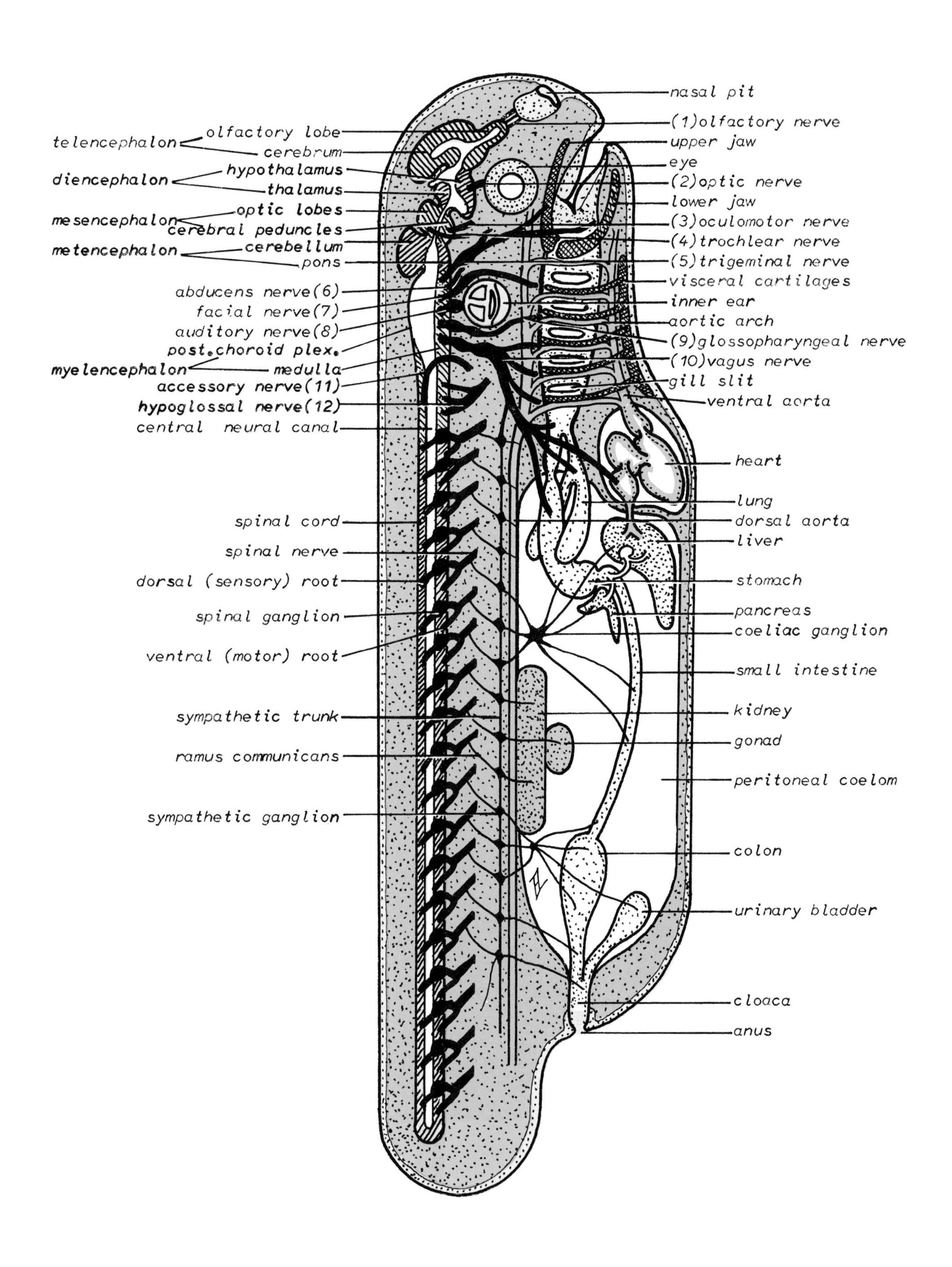

Figure 10-6. Common plan of the vertebrate nervous system. (From H.E. Lehman, *Laboratory Studies in General Zoology*, Hunter Publishing Co., 1981). Color code: blue — ectoderm, red — mesoderm, yellow — endoderm.

masses of **thyroid** and **cervical thymus** tissue which overlie the trachea. Trace the larynx forward into the **mouth cavity.** The **esophagus** will be found immediately under the full length of the trachea. Trace it back through the **diaphragm** into the stomach, and then forward into the mouth cavity. Open the mouth cavity by splitting the lower jaw in the midventral line. Cut the tissue to one side of the **tongue.** Spread the two halves of the lower jaw apart and identify the **palate** which forms the roof of the mouth. Split the palate down the middle and expose the paired **nasal canals** dorsal to it. A **median nasal septum** will separate them. The nasal canals and mouth are derivatives of the embryonic **stomodaeum.** With the point of a dissecting needle, split open the upper or lower jaw ridge at the side of the mouth cavity. In older embryos **tooth buds** will be found like a row of peas inside the skeletal primordia of the jaws. Near the posterior end of the palate, locate the slit-like openings of the **Eustachian tubes,** one of which opens into the posterior end of each nasal canal. At a corresponding level on both sides of the base of the tongue will be found the **crypts** of the **palatine tonsils.** Recall that the Eustachian tubes and crypts of the palatine tonsils are respectively derivatives of the first and second pharyngeal pouches (see Table 9-4).

Dissection of the major sense organs. (1) The **nostrils** and **nasal canals** have already been dissected; the **sensory nasal epithelium** is the lining of the corrugated roof of the nasal canals. **(2)** The **tongue,** as the major sensory organ of taste, has also been described in the dissection of the mouth and needs no additional comment. **(3)** Dissect out an **eye** and slice it open with a sharp razor blade so that the relationships of the cornea, iris, lens, retina, and optic nerve will be seen. **(4)** Dissect open the **middle ear cavity** by approaching it both from the external auditory canal and from the Eustachian tube. **(5)** The **eardrum** will be easily identified as a persistent first gill closing plate. The minute **ear bones** (1 in birds and 3 in mammals) probably will not be visible without considerable magnification. This is also the case with the various components of the **inner ear** including the semicircular canals, cochlea, sacculus, and utriculus. However, they can be looked for in the mesenchyme mesial to the middle ear cavity.

Dissection of the nervous system. Since the skeletal system is still very imperfectly formed, or absent, the dissection of the nervous system is comparatively simple. Skin, connective tissue, and lateral masses of muscles must be removed, but this can be done with a minimum of difficulty. The primary brain regions, the cranial nerves, spinal cord, and spinal nerves and their ganglia can be exposed. It is unlikely that sympathetic ganglia and commissures will be seen in young stages or without considerable magnification. However, all other general associations of the central and peripheral nervous systems can be seen with ease, and it will be both challenging and rewarding to make as careful a dissection of the fetal nervous system as time and skill will permit. A generalized common plan of the vertebrate nervous system is given in Figure 10-6 as an aid for this dissection.

B. SELECTED TECHNIQUES FOR HANDLING AND STUDY OF OLDER EMBRYOS AND SMALL FETUSES

The following methods will probably not be employed in routine laboratory work in embryology classes. They are included as useful methods for the preparation of laboratory demonstration materials by instructors and students who elect to undertake special studies demanding more time. They are particularly useful for biological materials of any type that fall into the difficult size range that is too large for routine serial sectioning and too small for gross dissection; namely, in the size range of from one-half to two inches, or one to five centimeters, in length.

1. Alizarin Red S and Toluidine Blue methods for in toto staining and clearing of skeletons. Variations of these so-called Spaltenholtz-Schultze Alizarin bone techniques go back to the middle of the past century and have been reported with special modifications for different kinds of biological materials including adult fish, amphibia, reptiles, and fetuses of mammals. Refer to the brief accounts provided with short bibliographies in G.L. Humason, *Animal Tissue Techniques,* Freeman, 1962; and R.B. Willey, "Staining shark skeletons with Alizarin Red S," *Turtox Bulletin* **47**:2, 1969. I am also indebted to D.P. Costello for his personal comments and staining schedule for the use of the alizarin method. The following directions will borrow from these sources and from personal experience.

Materials needed:

40% acetone in distilled water (about 50 ml per specimen).

Hydrolyzing solution (1% KOH, about 200 ml per specimen).

Concentrated Alizarin Red S (0.2 g dye per 25 ml 1% KOH).

Concentrated Toluidine Blue (0.2 g dye per 25 ml 95% alcohol).

Hardening and clearing solution (2 ml commercial formalin, 20 ml glycerin, make up to 100 ml with 1% KOH hydrolyzing solution).

Dehydrating solution (75% glycerin, about 100 ml per specimen).

Toluidine Blue destaining solution (70% alcohol, about 200 ml).

Also needed: wide-mouth, low specimen jars, warm stage or desk lamp, ultra-violet lamp or window with direct sunlight, and plastic embryo spoon for handling embryos.

Note: It is desirable to have a small amount of cotton in the bottom of the dish to support the embryo and permit the free diffusion of solutions from all sides; however, new cotton should be added with each change of solution used.

Alkaline alizarin sulfonate will combine with inorganic calcium and form a complex insoluble in alkaline solutions. This is the fundamental feature of alizarin bone staining, although any deposits of calcium in cartilage or teeth will also be nonselectively stained. **Toluidine Blue** staining of cartilage is based on the capacity of this dye to combine with mucoproteins such as cartilaginous matrices in alcoholic solution. The staining technique employs the principle of differential solubility of the dye, which is washed out more rapidly from noncartilaginous tissues and thus leaves the cartilage in relief. **It is not practical to try to stain both cartilage and bone in the same preparation** since the Toluidine Blue, unless carefully controlled, will usually mask alizarin staining.

a. Schedule for staining bone with Alizarin Red S. The times recommended below are for embryos with crown-rump lengths of about two centimeters; for each additional centimeter in length, add an equivalent amount of time. Solutions should equal at least 10 times the volume of tissue being treated.

1. Wash fixatives out of preserved embryos in 2 three-hour changes of tap water.

2. Transfer embryos to 50% acetone for 12 hours or more to remove neutral fats.

3. Wash in 3 one-hour changes of tap water or one hour in running water.

4. Hydrolyze in 1% KOH for 4 to 8 hours, or until the ribs can be clearly seen as white lines through the skin. The embryo will become quite soft and translucent during hydrolysis and should be handled with great care.

5. Pour off the first hydrolyzing solution and replace it with fresh 1% KOH. Add, dropwise, concentrated Alizarin Red S and stir. A strong transparent pink color is desired. Do not tint the solution too strongly.

6. Depending upon the size of the embryo, stain for 6 to 48 hours. If the staining mixture is a light rose color, embryos of any size can be stained for 48 or more hours without overstaining.

7. Pour out staining solution and rinse off the embryo with stock clearing and hardening solution. Cover the jar with Saran Wrap or other thin plastic and place in direct sunlight or under a UV-lamp for decolorizing non-skeletal tissues. The clearing and hardening solution should be changed after two days and thereafter each week until the embryo is reasonably transparent with little stain in any tissues other than those that are calcified.

8. When clearing and destaining is complete, as judged optically, pour off the solution and rinse the embryo with two six-hour changes in 75% glycerin.

9. Dehydrate in glycerin by marking the initial level of fluid and then dehydrate on a warm stage or under a desk lamp until the volume is reduced to three-quarters the starting volume.

10. It is customary to mount the final preparation on a microscope slide with a fine wire or nylon thread that is pushed through the embryo and then tied to the slide. Place the slide in a suitable specimen jar filled with pure glycerin and label for permanent display.

Alternatively use a plastic embedding procedure, in which case follow steps 1 through 7 and then substitute the following method:

8a. From step 7, wash embryo following clearing and destaining in two six-hour changes of tap water to remove hydroxide and glycerin.

9a. Take the embryo through the schedule for acetone dehydration and plastic clearing and mounting outlined at the end of Chapter 7 for plastic embedding.

b. Schedule for staining cartilage with Toluidine Blue

1. Repeat steps 1 through 3 from the preceding Alizarin Red schedule.

4b. From water, transfer the embryo to 70% alcohol and add, dropwise, stock Toluidine Blue until the solution is a clear dark transparent blue. Stain for 2 to 7 days depending upon size.

5b. Destain in successive daily changes of 70% alcohol until the alcohol remains essentially colorless and the ribs, sternum, scapula, pelvis, etc. appear distinctly blue against an essentially colorless background of milky skin and tissues. Care must be taken to avoid excessive destaining.

6b. Proceed from this point to step 7 through 10 in the Alizarin schedule.

It is suggested that parallel stages be prepared from litter mates for bone and cartilage so that the distribution of the two types of skeletal tissue can be seen in embryos or fetuses of the same age.

2. Silver nitrate methods for staining cut surfaces of whole embryos in paraffin blocks. This method combines two older methods in a manner that is, I believe, superior in simplicity and versatility to previously reported related techniques. **The first method** is a special application of the very old technique for depositing black colloidal silver in tissues and is most widely associated with the Golgi, Cajal, or Bodian methods for silver staining of nervous tissues (see G.L. Humason, *Animal Tissue Techniques,* Freeman, 1962, or other standard histochemical references for details). **The second method** entails the staining of tissue surfaces exposed by cutting through embryos embedded in paraffin in the manner first described by E.S. Hegre and A.D. Brashear, *Anatomical record,* **97:** 21-28, 1947.

Materials needed:

50% acetone in distilled water (about 50 ml per specimen jar).

*1% silver nitrate in distilled water (about 50 ml per jar made up immediately before use; a slight milkiness is acceptable).

Standard alcoholic dehydration series (about 100 ml each of 70%, 80%, 95%, 100%, and xylene or toluene for clearing tissues).

Soft paraffin-beeswax (mixture of 1 part 52-54°C melting point paraffin and 1 part white beeswax or, alternatively, white candle wax).

Photographic developer (available in photographic paper developer kits).

Photographic hypo (or 5% sodium hyposulfite in 5% acetic acid, or from photographic paper developer kits at photographic stores).

Also needed: Specimen jars should be new or acid cleaned to remove all traces of soap or alkali; clean plastic embryo spoons (**avoid** all metal dissecting instruments), new single-edge razor blade, alcohol lamp, putty knife, and paraffin oven (one with excellent characteristics can be improvised from a gallon tin can open at one end and with a desk lamp placed over it as a source of heat).

a. Schedule for silver impregnation and paraffin embedding

1. Wash fixative (particularly formalin) out of embryos by 3 twelve-hour changes in tap water, or 6 hours in running water.

2. Transfer embryos to 50% acetone for 24 hours for removal of fats.

3. Wash in 3 one-hour changes in tap water.

4. Transfer to 1% silver nitrate (or alternative silver solutions*) for 3 to 7 days in the dark (i.e., as in a desk drawer or closed box).

5. Rinse in 2 five-minute washes in tap water and carry the embryo through alcoholic dehydration with 2 thirty-minute, or more, changes in

*As alternatives to silver nitrate, ammoniacal silver hydroxide or carbonate can be used. To prepare these dissolve 1 g silver nitrate in 20 ml distilled water, then add 1 ml 1.0 N NaOH (for hydroxide), **or** 0.1 g Na_2CO_3 (for carbonate); allow the dark precipitate to settle; decant off and discard the supernatant. Wash the precipitate 3 times by successive flooding and decanting with 100 ml distilled water; resuspend precipitate in 50 ml distilled water and add, dropwise, concentrated ammonium hydroxide while stirring until the precipitate almost completely dissolves. A water-clear solution should be obtained. Use immediately or refrigerate in a dark bottle for use within a few days; discard brown solutions. Hegre and Brashear use 0.5% lead acetate in 95% alcohol and develop with 5% sodium sulfide.

each of the alcohols and clearing solutions (see above). Tissues can be stopped for overnight storage at any step in the process including xylene or toluene since hardening of tissues is a distinct advantage in this technique.

6. Place second xylene change in paraffin oven to bring tissues up to the temperature of melted paraffin.
7. Replace xylene with melted paraffin; adjust heat so that the temperature is at the just-melting point. Infiltrate in 2 thirty-minute changes of paraffin (longer time is permissible provided that the temperature is regulated to prevent overheating).
8. Use match boxes or fold file card paper to form boxes of a suitable size to hold embryos. Fill the box with melted paraffin and immediately transfer the embryo to the box and orient it in whatever manner is desired (i.e., usually on the left side with a spot placed on the box to mark the anterior end). Use a dissecting needle heated with an alcohol flame to melt the surface film so as to release any trapped air bubbles.
9. When the surface film is hardened and about one-eighth inch thick, submerge the box quickly in ice water and harden the block quickly.
10. After 10 to 20 minutes remove the block from water. If the block is not to be used immediately, it should be labelled with information on the species and embryonic stage of the specimen. Blocks can be stored in a cool place indefinitely for later use.
11. With a sharp razor blade, square the paraffin block by gradually paring down its six sides until the embryo can be seen within the translucent paraffin.
12. Decide on a plane, or planes, of section to be made through the embryo (i.e., median sagittal, parasagittal, transverse, frontal, oblique, cut-out quadrants, etc.); then warm the paraffin block slightly by heating near the desk lamp, and make slow clean cuts with a new sharp razor blade in the predetermined planes of section.
13. Stain the cut surface by adding a drop of photographic developer to the cut surface of the block and note the blackening of the exposed tissue. Development is slow and may take up to an hour or more.
14. Add a drop of photographic hypo to the cut surface, and wipe off after about 30 seconds.
15. Mount the paraffin block, with the stained surface up, on a standard microscope slide. If it is too large, mount it in a plastic box of suitable size. This can be done by heating the slide on an alcohol flame to the melting point of paraffin and quickly pressing the paraffin block to the hot glass slide. If the seal is not strong after cooling, repeat the process with more heat.

b. Preparation of thick paraffin serial sections of large embryos and fetuses. It is obvious that the above method can be extended to the preparation of thick serial sections of embryos that are too large to be handled conveniently in routine histological microslide preparations.

11b. Proceed through steps 1 through 11 as indicated on the preceding page.

12b. Decide on a plane of section and cut the embryo into 1 to 5 mm thick sections. Transfer each section carefully to a clean piece of paper that was previously numbered consecutively in anticipation of the sections that are cut.

13b. After all sections have been cut, add one or more drops of developer as needed to the cut surfaces and develop them as described earlier.

14b. Wipe off the developer, add hypo for 30 or more seconds, and then wipe off the hypo.

15b. Mounting serial sections presents problems of increasing dimensions with increasing size and number of sections to be mounted. It is most easily done on microslides by having no more than one row of sections and no more than about 4 to 6 sections per slide. For larger materials, necessity will dictate appropriate modifications in the following directions. Place microslides on a piece of asbestos and use a thin metal putty knife as a means for transferring heat from an alcohol flame to the underside of slides. Transfer sections one at a time to the cold slide. Then heat the underside of the slide with a hot putty knife and press the section to the heated glass. Cool the slide and see if the section is secure. Repeat if necessary and proceed to the next section until the slide is filled.

3. A method for making 3-dimensional demonstration dissections of de-paraffined embryos and fetuses. This last technique is an extension of the preceding one. However, it can be applied to paraffin embedded embryos with or without silver staining.

Materials needed: These are identical to those required for the preceding series and can use embryos previously carried into paraffin by that procedure. Alternatively, silver impregnation and photographic developing can be omitted from the schedule of treatments and handling. Suitable plastic boxes for mounting specimens are desirable.

Schedule for preparing 3-D demonstration materials

1. Utilize steps 1 through 10 on the preceding schedule, carry embryos or fetuses into paraffin blocks.
2. With a clean, new razor, cut away parts of the embryo in appropriate planes and with whatever stair-step incisions that may be considered appropriate for the purposes in mind. The cut surfaces can be silver stained as directed in steps 11 through 15 on the preceding schedule of procedures.
3. Return the embryo in its paraffin block to the paraffin oven and melt it down.
4. Transfer the dissected embryo into hot xylene for two successive changes to remove all paraffin.
5. Replace the second change of hot xylene with a change of cold xylene.
6. With great care, remove the embryo from the xylene with a plastic spoon; blot off any excess xylene, and then transfer the embryo to a plastic box and quickly orient it in the manner considered most desirable. Most clear plastic boxes are soluble in xylene and the specimen will stick firmly to the box within a few minutes. Alternatively, Duco plastic cement, or Elmers glue, or polymerizing plastic adhesives can be used.
7. Allow the xylene to evaporate. When this takes place, the tissues will become opaque and the cut surfaces of tissues and hollow spaces of natural cavities will become visible in high relief.
8. The specimens are permanent but exceedingly fragile and cannot be touched without damage. It is recommended that these preparations be permanently sealed by adding a drop of xylene to the joint between the cover and bottom of the plastic demonstration box. Appropriate labels can be attached to the cover, and a white or colored card can be cut to fit and be attached to the bottom of the preparation.

APPENDIX I

SELECTED READING ON VERTEBRATE EMBRYOLOGY

A. Textbooks and General References

AREY, L.B. 1974. *Developmental Anatomy,* 7th ed. W.B. Saunders Company.

BALINSKY, B.I. 1975. *An Introduction to Embryology.* W.B. Saunders Company.

BALLARD, W.W. 1964. *Comparative Anatomy and Embryology.* Ronald Press.

BELLAIRS, R. 1971. *Developmental Processes in Higher Vertebrates.* University of Miami Press.

BERRILL, N.J. 1971. *Developmental Biology.* McGraw-Hill Book Company.

BERRILL, N.J. and G. KARP 1976. *Development.* McGraw-Hill Book Company.

BODEMER, C.W. 1968. *Modern Embryology.* Holt, Rinehart and Winston, Inc.

BROWDER, L.W. 1980. *Developmental Biology.* Holt, Rinehart and Winston, Publishers.

CARLSON, B.M. 1981. *Patten's Foundations of Embryology.* McGraw-Hill, Inc.

DAVENPORT, R. 1979. *An Outline of Animal Development.* Addison-Wesley Publishing Company.

DAVIES, J. 1963. *Human Developmental Anatomy.* Ronald Press.

DeHAAN, R.L., and H. URSPRUNG 1965. *Organogenesis.* Holt, Rinehart, and Winston.

DEUCHAR, E.M. 1975. *Cellular Interactions in Animal Development.* Chapman and Hall.

EBERT, J.D., and I.M. SUSSEX 1965. *Interacting Systems in Development.* Holt, Rinehart and Winston.

FULTON, C., and A.O. KLEIN 1976. *Explorations in Developmental Biology.* Harvard University Press.

GILBERT, S.F. 1985. *Developmental Biology.* Sinauer Associates, Inc., Publishers.

GILCHRIST, F.G. 1968. *A Survey of Embryology.* McGraw-Hill Book Company.

GRANT, P. 1978. *Biology of Developing Systems.* Holt, Rinehart and Winston, Publishers.

HAM, R.G. and M.J. VEOMETT. 1980. *Mechanisms of development.* C.V. Mosby Company.

HAMILTON, H.L. 1952. *Lillie's Development of the Chick,* Revised. Henry Holt Company.

HAMILTON, W.J., J.D. BOYD, and H.W. MOSSMAN. 1945. *Human Embryology.* Williams and Wilkins Company.

HOPPER, A. F. and N. H. HART 1985. *Foundations of Animal Development.* Oxford University Press.

HUETTNER, A.F. 1949. *Fundamentals of Comparative Embryology of the Vertebrates.* Macmillan Company.

HUXLEY, J.S., and G.R. DeBEER. 1934. *The Elements of Experimental Embryology.* Cambridge University Press.

KERR, J.G. 1919. *Text-Book of Embryology, Volume 2: Vertebrata with the Exception of the Mammalia.* Macmillan Company.

KUHN, A. 1971. *Lectures on Developmental Physiology* (Translated by R. Milkman). Springer-Verlag.

LANGMAN, J. 1969. *Medical Embryology.* Williams and Wilkins Company.

LASH, J., and J.R. WHITTAKER, Editors. 1974. *Concepts of Development.* Sinauer Associates, Inc., Publishers.

LOVTRUP, S. 1974. *Epigenetics.* John Wiley and Sons.

MARSHALL, A.M. 1893. *Vertebrate Embryology.* G.P. Putnam and Sons.

McEWEN, R.S. 1957. *Vertebrate Embryology.* Henry Holt and Company.

MANNER, H.W. 1964. *Elements of Comparative Embryology.* Macmillan Company.

MONROY A. and A.A. MOSCONA. 1979. *Introductory Concepts in Developmental Biology.* University of Chicago Press.

MOORE, J.A. 1963. *Heredity and Development.* Oxford University Press.
MOORE, K.L. 1973. *The Developing Human: Clinically oriented Embryology.* W.B. Saunders.
MOOG, F. 1949. *Structure and Development of the Vertebrates.* Prentice-Hall, Inc.
NELSON, O.E. 1953. *Comparative Embryology of the Vertebrates.* The Blakiston Company.
OPPENHEIMER, S.B. and LeFEVRE, G. JR. 1984. *Introduction to Embryonic Development,* 2nd ed., Allyn and Bacon Inc.
PATTEN, B.M. 1948. *Embryology of the Pig.* The Blakiston Company.
PATTEN, B.M. 1971. *Early Embryology of the Chick.* McGraw-Hill Book Company.
PHILLIPS, J.B. 1976. *Development of Vertebrate Anatomy.* C.V. Mosby Company.
RUGH, R. 1964. *Vertebrate Embryology.* Harcourt, Brace and World, Inc.
SAUNDERS, J.W., Jr. 1970. *Patterns and Principles of Animal Development.* Macmillan.
SCHUMWAY, W., and F.B. ADAMSTONE. 1942. *Introduction to Vertebrate Embryology.* John Wiley and Sons, Inc.
SPRATT, N.T., Jr. 1971. *Developmental Biology.* Wadsworth Publishing Company.
THOMAS, J.B. 1968. *Introduction to Human Embryology.* Lea and Febiger.
TORREY, T.W. and A. FEDUCCIA. 1979. *Morphogenesis of the Vertebrates.* John Wiley and Sons.
TRINKAUS, J.P. 1984. *Cells into Organs.* Prentice Hall, Inc.
WADDINGTON, C.H. 1957. *Principles of Embryology.* G. Allen and Unwin, Ltd., London.
WALLBOT, V. and H. HOLDER 1987. *Developmental Biology.* Random House.
WEISS, P. 1939. *Principles of Development: A Text in Experimental Embryology.* Henry Holt and Company.
WIEMAN, H.L. 1930. *An Introduction to Vertebrate Embryology.* McGraw-Hill Book Company.
WILLIER, B.H., P.A. WEISS, and V. HAMBURGER. 1955. *Analysis of Development* (Reprinted 1971, Hafner Publishing Company). W.B. Saunders Company.
WITSCHI, E. 1956. *Development of Vertebrates.* W.B. Saunders Company.

B. SOME MAJOR SERIALS AND JOURNALS IN ENGLISH FOR THE PUBLICATION OF ORIGINAL WORK ON VERTEBRATE EMBRYOLOGY

SYMPOSIA OF THE SOCIETY FOR THE STUDY OF DEVELOPMENT AND GROWTH (Annual, beginning in 1939 to the present under several publishers), currently by Academic Press.
ADVANCES IN MORPHOGENESIS (Annual to biannual beginning 1961), Academic Press.
CURRENT TOPICS IN DEVELOPMENTAL BIOLOGY (On occasion beginning 1966), Academic Press.

PERIODICALS AND JOURNALS

THE BIOLOGICAL BULLETIN
THE CELL
CELL DIFFERENTIATION
DEVELOPMENTAL BIOLOGY
DIFFERENTIATION
JOURNAL OF EMBRYOLOGY AND EXPERIMENTAL MORPHOLOGY
JOURNAL OF MORPHOLOGY
JOURNAL OF EXPERIMENTAL ZOOLOGY
NATURE
PROCEEDINGS OF THE NATIONAL ACADEMY OF SCIENCES, USA
QUARTERLY REVIEW OF BIOLOGY
W. ROUX' ARCHIVES OF DEVELOPMENTAL BIOLOGY
SCIENCE:

C. USEFUL REFERENCES FOR LABORATORY METHODS AND TECHNIQUES

General references

BANCROFT, J.D. 1967. *An Introduction to Histochemical Technique.* Appleton-Century-Crofts.
CAMERON, G. 1950. *Tissue Culture Technique.* Academic Press.
GRAY, P. 1952. *Handbook of Basic Microtechnique.* The Blakiston Company.
HUMASON, G.L. 1962. *Animal Tissue Techniques.* W.H. Freeman and Company.
KRUSE, F., and M.K. PATTERSON. 1973. *Tissue Culture.* Academic Press.
MERCHERT, D.J. 1964. *Handbook of Cell and Organ Culture.* Burgess Publishing Company.

NEW, D.A.T. 1966. *The Culture of Vertebrate Embryos.* Logos Press of Academic Press.

PARKER, R.C. 1950. *Methods of Tissue Culture.* P.B. Hoeber, Inc.

PAUL, J. 1961. *Cell and Tissue Culture.* Williams and Wilkins Company.

PREECE, A. 1959. *A Manual for Histologic Technicians.* Little, Brown and Company.

WILT, F.H., and N.K. WESSELLS. 1967. *Methods in Developmental Biology.* T.Y. Crowell Co.

Laboratory handbooks and manuals for descriptive and experimental embryology

ADAMSTONE, F.B., and W. SCHUMWAY. 1947. *A Laboratory Manual of Vertebrate Embryology.* John Wiley and Sons.

DOWNS, L.E. 1968. *Laboratory Embryology of the Frog.* W.C. Brown Company.

EAKIN, R.M. 1964. *Vertebrate Embryology.* University of California Press.

HAMBURGER, V. 1942. *A Manual of Experimental Embryology.* University of Chicago Press.

HARRISON, B.M. 1949. *Embryology of the Chick and Pig.* W.C. Brown Company.

JOHNSON, L.G., and E.P. VOLPE. 1973. *Patterns and Experiments in Developmental Biology.* W.C. Brown Company.

LEHMAN, H.E. 1965. *Laboratory Handbook of Vertebrate Embryology.* Department of Zoology University of North Carolina-Chapel Hill.

MATHEWS, W.W. 1976. *Atlas of Descriptive Embryology.* Macmillan Company.

MOORE, K.L. 1975. *Study Guide and Review Manual of Human Embryology.* W.B. Saunders, Co.

MORGAN, T.H. 1897. *Development of the Frog's Egg: An Introduction to Experimental Embryology.* Macmillan Company.

POSTLETHWAIT, S.N., Director. 1976. *BSCS Minicourse on Animal Growth and Development.* W.B. Saunders Company.

RUGH, R. 1976. *A Laboratory Manual of Vertebrate Embryology.* Burgess Publishing Company.

RUGH, R. 1962. *Experimental Embryology:* A Manual of Techniques and Procedures. Burgess Publishing Company.

RAFFERTY, K.A., Jr. 1970. *Methods in Experimental Embryology of the Mouse.* The Johns Hopkins University Press.

SKOLD, B.H., and E. KÜNZEL. 1972. *Davis' Embryology Laboratory Guide for the 48 Hour Chick and 10 mm Pig.*

TORREY, T.W. 1962. *Laboratory Studies in Developmental Anatomy.* Burgess Publishing Company.

VANABLE, J.W., and J.H. CLARK. 1968. *Developmental Biology: A Laboratory Manual.* Burgess Publishing Company.

WATERMAN, A.J. 1948. *A Laboratory Manual of Comparative Vertebrate Embryology.* Macmillan Company.

WATTERSON, R.L., and R.M. SWEENEY. 1970. *Laboratory Studies of Chick, Pig, and Frog Embryos.* Burgess Publishing Company.

WISCHNITZER, S. 1975. *Atlas and Laboratory Guide for Vertebrate Embryology.* McGraw-Hill Book Company.

APPENDIX II

EDUCATIONAL FILMS ON VERTEBRATE EMBRYOLOGY AND DEVELOPMENT

The following film titles are suggested as supplements (or in the event that living materials are unavailable, in substitution) for observations and experiments described in the laboratory material for individual chapters of this textbook. Multiple titles covering the same general subject matter are often given without an indication of preference on the part of this author. In all cases films are 16 mm unless specifically stated to be otherwise. A semicolon is used to separate the title and producer from the source authorized to sell, rent, or loan the film. Most 16 mm films are single-reel units of 12 to about 30 minutes duration; 8 mm films are in loop-casettes with 2 to 4 minute duration.

FILMS APPROPRIATE FOR CHAPTER I

PATTERNS OF REPRODUCTION, A.I.B.S. Biology Film Series; McGraw-Hill Films, 330 West 42 Street, New York, New York.

TUNICATE REPRODUCTION AND DEVELOPMENT, PARTS I & II, R.D. Allen & M. Allen; Ealing Corporation, Harper and Row Publishers, 2350 Virginia Avenue, Hagerstown, Maryland (8 mm).

OOPLASMIC SEGREGATION DURING ASCIDIAN DEVELOPMENT, J. Lash; BFA Educational Media, 2211 Michigan Avenue, Santa Monica, California.

THE EMBRYONIC DEVELOPMENT OF FISH, J.V. Durden & N.J. Berrill; National Film Board of Canada, 680 Fifth Avenue, New York, New York.

EPIBOLY IN THE KILLIFISH, FUNDULUS HETEROCLITUS, PARTS I, II & III, J.P. Trinkaus; BFA Educational Media, 2211 Michigan Avenue, Santa Monica, California (8 mm).

FILMS APPROPRIATE FOR CHAPTER II

THE SEX CELLS, A.I.B.S. Biology Film Series; McGraw-Hill Films, 330 West 42 Street, New York, New York.

MIGRATION OF THE PRIMORDIAL SEX CELLS INTO THE GENITAL RIDGE, Eaton Laboratories Audio-Visual Corporation; Eaton Medical Film Library, Norwich, New York (Free loan).

OBSERVATION OF LIVING PRIMORDIAL GERM CELLS IN THE MOUSE, R.J. Blandau & R. Hayashi; Washington Film Service, University of Washington Press, Seattle, Washington.

DIFFERENTIATION OF THE GENITAL RIDGE INTO THE OVARY OF THE FULL TERM FETUS, Eaton Laboratories Audio-Visual Corporation; Eaton Medical Film Library, Norwich, New York (Free loan).

EMBRYOLOGY: GROWTH OF THE OOCYTE AND DEVELOPMENT OF THE OVARIAN FOLLICLE, Macmillan Company Ltd, 4 Little Essex Street, London WC2, United Kingdom (8 mm).

MEIOSIS, A.S. Bajer; Harper and Row Publishing Co., 2350 Virginia Avenue, Hagerstown, Maryland (8 mm).

CELL DIVISION: MITOSIS & MEIOSIS, S. Katten; McGraw-Hill Films, 330 West 42 Street, New York, New York.

REMOVING FROG PITUITARY, T.R. Marcus & G.P. Fulton; Ealing Film Loops, Harper and Row Publishers 2350 Virginia Avenue, Hagerstown, Maryland (8 mm).

INDUCING FROG OVULATION, T.R. Marcus & G.P. Fulton; Ealing Film Loops, Harper and Row Publishers, 2350 Virginia Avenue, Hagerstown, Maryland (8 mm).

PHYSIOLOGY OF REPRODUCTION IN THE RAT, R.J. Blandau; Washington Film Service, University of Washington Press, Seattle, Washington.

OVULATION AND EGG TRANSPORT IN THE RAT, R.J. Blandau & R. Hayashi; Washington Film Service, University of Washington Press, Seattle, Washington.

SPERM MATURATION IN THE MALE REPRODUCTIVE TRACT, P. Daddum, R.J. Blandau & R. Hayashi; Washington Film Service, University of Washington Press, Seattle, Washington.

EMBRYOLOGY: OVULATION, Macmillan Company, Ltd., 4 Little Essex Street, London WC2, United Kingdom (8 mm).

FILMS APPROPRIATE FOR CHAPTER III

MITOSIS IN ANIMAL CELLS, A. Owczarzak and J. Mole-Pajer; Harper and Row Publishers, 2350 Virginia Avenue, Hagerstown, Maryland (8 mm).

SAND DOLLAR FERTILIZATION AND CLEAVAGE, R.A. Cloney; Education Development Center, 39 Chapel Street, Newton, Massachusetts.

SAND DOLLAR GASTRULATION, R.A. Cloney; Education Development Center, 39 Chapel Street, Newton, Massachusetts.

STARFISH SPAWNING AND FERTILIZATION, Encyclopaedia Britannica Educational Corporation, 425 North Michigan Avenue, Chicago, Illinois.

FERTILIZATION, A.I.B.S. Biology Film Series; McGraw-Hill Films, 330 W. 42 Street, New York, New York.

AN ANIMAL LIFE CYCLE SEA URCHIN, A.I.B.S. Biology Film Series; McGraw-Hill Films, 330 West 42 Street, New York, New York.

NORMAL DEVELOPMENT OF THE SEA URCHIN, E. Bell; Education Development Center, 39 Chapel Street, Newton, Massachusetts.

GASTRULATION IN THE SEA URCHIN, T. Gustafson; Ealing Film Loops, Harper and Row Publishers, 2350 Virginia Avenue, Hagerstown, Maryland (8 mm).

FERTILIZATION AND CLEAVAGE OF THE SAND DOLLAR, R.D. Allen; Ealing Corporation, Harper and Row Publishers, 2350 Virginia Avenue, Hagerstown, Maryland (8 mm).

EMBRYOLOGY: FERTILIZATION, Macmillan Company, Ltd., 4 Little Essex Street, London WC2, United Kingdom (8 mm).

EMBRYOLOGY: CLEAVAGE AND FORMATION OF THE BLASTOCYST, Macmillan Company, Ltd., 4 Little Essex Street, London WC2, United Kingdom (8 mm).

FILMS APPROPRIATE FOR CHAPTER IV

THEORIES OF DEVELOPMENT (Reproduction and development of the frog), A.I.B.S. Biology Film Series; McGraw-Hill Films, 330 West 42 Street, New York, New York.

FROG DEVELOPMENT: FERTILIZATION TO HATCHING, J.V. Durden & M. Braverman; Education Services, Inc. Box 415, Watertown, Massachusetts.

FROG DEVELOPMENT (5 parts, Pairing and Egg-Laying, First Cell Division to Early Neural Fold, Development of the Body Regions, Continued Development to Hatching, Hatching through Initial Leg Growth, and Metamorphosis, Ealing Film Loops; Harper and Row Publishers, 2350 Virginia Avenue, Hagerstown, Maryland (8 mm).

FERTILIZATION IN RANA PIPIENS, E. Bell; Education Development Center, 39 Chapel Street, Newton, Massachusetts.

CLEAVAGE IN RANA PIPIENS, E. Bell; Education Development Center, 39 Chapel Street, Newton, Massachusetts.

FILMS APPROPRIATE FOR CHAPTER V

TWIN FORMATION IN THE SALAMANDER, W. Luther; Audio-Visual Aids Library, Pennsylvania State University, University Park, Pennsylvania.

MICROSURGERY IN AMPHIBIAN EMBRYOS, I. Brick; Harper and Row Publishers, 2350 Virginia Avenue, Hagerstown, Maryland (8 mm).

NERVE CONTROL OF TRANSPLANTED SALAMANDER LIMBS, P. Weiss; Harper and Row Publishers, 2350 Virginia Avenue, Hagerstown, Maryland (8 mm).

NERVE CELL REGENERATION: AXON GROWTH, Pasadena Foundation of Medical Research; Harper and Row Publishers, 2350 Virginia Avenue, Hagerstown, Maryland (8 mm).

TISSUE CULTURE OF FROG SPLEEN, PARTS I & II (Removal of Spleen and Explanation), J.A. Wagner; Encyclopaedia Britannica Educational Corporation, 425 North Michigan Avenue, Chicago, Illinois (8 mm).

NUCLEAR TRANSFER IN AMPHIBIA, E. Lucey; Research Film Unit, Institute of Animal Genetics, West Mains Road, Edinburg, Scotland, United Kingdom.

RECENT DEVELOPMENTS IN NUCLEAR TRANSPLANTATION, C.L. Markert; Department of Medical Communication, M.D. Anderson Hospital and Tumor Institute, Houston, Texas.

FILMS APPROPRIATE FOR CHAPTER VI

EMBRYONIC DEVELOPMENT OF THE CHICK, E. Lucey; Research Film Unit, Institute of Animal Genetics, West Mains Road, Edinburg, Scotland, United Kingdom.

LIFE BEFORE BIRTH, J.D. Ebert; Carousel Films Inc., 1501 Broadway, New York, New York.

FILMS APPROPRIATE FOR CHAPTER VII

48 HOUR CHICK AND EARLY BLASTODERM, B.M. Patten & T.C. Droner; American Medical Association Film Library, 535 N. Dearborn Avenue, Chicago, Illinois.

FIRST HEARTBEATS AND THE BEGINNING OF THE CIRCULATION OF BLOOD IN THE EMBRYO, T.C. Kramer; American Medical Association Film Library, 535 N. Dearborn Avenue, Chicago, Illinois.

DEVELOPMENT OF THE CARDIO-VASCULAR SYSTEM OF THE CHICK: THE HEART, T.W. Torrey & C.M. Flaten; Audio-Visual Center, Indiana University, Bloomington, Indiana.

CHICK EMBRYO TECHNIQUES, Communicable Disease Center; National Audio-Visual Center, Washington, D.C. (Free loan).

DEVELOPMENT OF ORGANS, A.I.B.S. Biology Film Series; McGraw-Hill Films, 330 West 42 Street, New York, New York.

MYOGENESIS,.E. Konigsberg; BFA Educational Media, 2211 Michigan Avenue, Santa Monica, California.

HEART DEVELOPMENT IN THE CHICK EMBRYO, PARTS I, II & III, J.V. Durden; BFA Educational Media, 2211 Michigan Avenue, Santa Monica, California (8 mm).

OBSERVATIONS OF CULTURED CHICK MYOCARDIAL CELLS, R.J. Blandau & R. Hayashi; Washington Film Service, University of Washington Press, Seattle, Washington.

DEVELOPMENT & DIFFERENTIATION, S. Katten; McGraw-Hill Films, 330 West 42 Street, New York, New York.

FILMS APPROPRIATE FOR CHAPTER VIII

DEVELOPMENT OF THE CHICK EXTRAEMBRYONIC MEMBRANES, T.W. Torrey; Audio-Visual Center, Indiana University, Bloomington, Indiana.

TOPOGRAPHY OF MAJOR DIVISIONS OF THE BRAIN AND THEIR RELATION TO THE EMBRYONIC BRAIN, H.A. Matzke; National Medical Audio-Visual Center, Washington, D.C. (Video tape).

THE EMBRYOLOGY OF THE EYE, American Academy of Opthamology and Otolaryngology and Department of Embryology of the Carnegie Institution of Washington; Sturgis-Grant Productions, Rochester, Minnesota (Sale), or International Eye Film Library, Jersey City, New Jersey (Loan).

THE CHICK EGG-WINDOW TECHNIQUE, T.R. Marcus & G.P. Fulton; Ealing Film Loops, Harper and Row Publishers, 2350 Virginia Avenue, Hagerstown, Maryland (8 mm).

REMOVING YOUNG CHICK EMBRYO: FILTER PAPER RING TECHNIQUE, R. Buchsbaum; Encyclopaedia Britannica Educational Corporation, 425 North Michigan Avenue, Chicago, Illinois (8 mm).

INJECTING EGGS, R. Buchsbaum; Encyclopaedia Britannica Educational Corporation, 425 North Michigan Avenue, Chicago, Illinois (8 mm).

TISSUE CULTURE OF CHICK EMBRYO, PARTS I & II, R. Buchsbaum; Encyclopaedia Britannica Educational Corporation, 425 North Michigan Avenue, Chicago, Illinois (8 mm).

EXPERIMENTS ON THE CHICK EMBRYO, J.W. Saunders; BFA Educational Media, 2211 Michigan Avenue, Santa Monica, California (8 mm).

FILMS APPROPRIATE FOR CHAPTER IX

EMBRYOLOGY ON FILM SERIES (In 7 parts: Growth of the Oocyte, Development of the Ovarian Follicle, Ovulation, Fertilization, Cleavage and Formation of the Blastocyst, and Implantation), W.J. Hamilton, T.W. Glenister, J.D. Boyd and Orriss Animations, Ltd; Macmillan Company, Ltd, 4 Little Essex Street, London WC2, United Kingdom (8 mm).

HUMAN REPRODUCTION AND BIRTH (In 4 parts: Sexual Intercourse, Fertilization and Early Development, Embryo and Fetus, and Human Birth), Ealing Film Loops; Harper and Row Publishers, 2350 Virginia Avenue, Hagerstown, Maryland (8 mm).

REPRODUCTIVE HORMONES, A.I.B.S. Biology Film Series; McGraw-Hill Films, 330 West 42 Street, New York, New York.

DEVELOPMENT AND MAINTENANCE OF THE HUMAN EMBRYO: ORGANOGENESIS OF EMBRYO AND FETUS, Department of Maternal-Child Health, College of Nursing, University of Illinois Medical Center (free loan), or Aldine Publishing Company (sale), Chicago, Illinois.

THE PLACENTA AT TERM, Department of Anatomy, University of Texas Medical School; University of Texas Health Science Center, San Antonio, Texas.

THE FETAL PIG, PART I, T.W. Torrey; Ealing Film Loops, Harper and Row Publishers, 2350 Virginia Avenue, Hagerstown, Maryland (8 mm).

THE ANATOMY OF THE 9 mm PIG AS SEEN IN SERIAL SECTION, E.S. Hegre; Commonwealth Motion Pictures, West Broad Street, Richmond, Virginia.

EMBRYOLOGY OF THE EAR, PARTS I & II (Development of the Inner Ear, and Development of the Middle and External Ear), American Academy of Opthalmology and Otolaryngology; Sturgis-Grant Productions, American Medical Association Film Library, Chicago, Illinois.

DEVELOPMENT OF THE COELOMIC CAVITY, Department of Anatomy, University of Texas Medical School; University of Texas Health Science Center, San Antonio, Texas.

CONGENITAL MALFORMATIONS OF THE HEART, PARTS II & III (Acyanotic Congenital Heart Disease, and Cyanotic Heart Disease), R.F. Rushmer & R.J. Blandau; Washington Film Service, University of Washington Press, Seattle, Washington.

FILMS APPROPRIATE FOR CHAPTER X

DEVELOPMENT OF THE AORTIC ARCH, American College of Surgeons; Sturgis-Grant Productions or Squibb Film Library, Princeton, New Jersey (Free Loan).

ANOMALIES OF THE AORTIC ARCH, American College of Surgeons; Sturgis-Grant Productions or Squibb Film Library, Princeton, New Jersey (Loan free).

EMBRYOLOGY AND PATHOLOGY OF THE INTESTINAL TRACT, L. Chafin & W.H. Snyder; Graphic Films Corporation, Hollywood, California, or National Medical Center, Atlanta, Georgia (Free Loan).

DEVELOPMENT OF THE GASTRO-INTESTINAL TRACT, J.J. McDonald; American Medical Association Film Library, 535 North Dearborn Street, Chicago, Illinois.

DEVELOPMENT OF THE HUMAN GASTROINTESTINAL TRACT, L.G. Audette & E.S. Crelin; National Medical Audio-Visual Center, Atlanta, Georgia (Free Loan).

ANATOMY AND PHYSIOLOGY OF THE HUMAN REPRODUCTIVE ORGANS: EMBRYOLOGY. Department of Maternal-Child Health, College of Nursing, University of Illnois Medical Center; Aldine Publishing Company, Chicago, Illinois (Free Loan).

DEVELOPMENTAL DISTURBANCES (of the mouth and face), Department of Oral Pathology, University of Maryland; Independent Learning Center, University of Maryland, Baltimore, Maryland

DEVELOPMENT OF BEHAVIOR IN THE DUCK EMBRYO, G. Gottlieb & Z.Y. Kuo; Harper and Row, 2350 Virginia Avenue, Hagerstown, Maryland (8 mm).

GENERAL EDUCATIONAL FILM REFERENCES

Educational films and video tapes are available for rental through most state university film libraries and extension services.

A most useful general source on educational films and tapes is *Index to 16 mm Educational Films* (8th Edition, 1984) and *Index to Educational Videotapes* (6th Edition, 1985), available from NICEM, University of Southern California, University Park, Los Angeles, California 90007.

Also recommended as a general information source is *Educational Film and Video Locator,* Consortium of University Film Centers and R. R. Bowker Company, R. R. Bowker Company, New York, New York.

APPENDIX III

SOURCES AND MAINTENANCE OF EMBRYOLOGICAL MATERIALS

Without attempting a complete catalogue of regional and national suppliers, a few general sources from which embryological materials can be obtained are given here as assistance to those who use the foregoing laboratory suggestions.

1. **Preserved whole mounts, prepared slides, and demonstration models of embryological materials.** These can be obtained from the following general laboratory supply companies. The code-word in parenthesis in front of each entry will be used in Part 2, in substitution for a complete citation.

   (Carolina) Carolina Biological Supply Company, Burlington, North Carolina 27215; or Gladstone, Oregon 97027

   (Turtox) Turtox Cambosco, Macmillan Scientific Company, 8200 Hoyne Avenue, Chicago, Illinois 60620

   (Wards) Wards Natural Science Establishment, Inc., P.O. Box 1712, Rochester, New York 14603

   (To my knowledge, Carolina is the only source for developmental stages of *Ascidia nigra* and metamorphosis of *Ecteinascidia turbinata* (Chapter I), or for sectioned salamander larvae (Chapter V). These are not listed in their current catalog but can be obtained by special request).

2. **Living materials.** Living materials suggested in the foregoing laboratory exercises can be obtained from the following sources. The list is given in the numerical sequence of chapters for which animals are recommended. Brief culture or holding methods for maintaining reproductive adults are inserted at relevant points.

   a. Chapter II. **Living sea urchins** (and starfish only in winter or spring). So-called fertilization kits can be obtained from Carolina, Turtox, or Wards; alternatively they can be ordered from the following sources (species in parenthesis).

      The Bermuda Biological Station, St. George's West, Bermuda Islands (Lytechinus)

      Carolina Biological Supply Co., Burlington, North Carolina 27215. (Eucidaris or Lytechinus)

      Florida Marine Biological Specimen Co., Panama City, Florida 32401 (Arbacia or Lytechinus)

      Gulf Specimen Co., P.O. Box 237, Panacea, Florida 32346 (Arbacia or Lytechinus; an excellent brief statement of methods for maintaining marine aquaria is included in their catalog)

      Supply Department, Marine Biological Laboratory, Woods Hole, Massachusetts 02540 (Arbacia)

      Marino's Aquatic Specimens, 301 Evergreen Ave., Daly City, California 94015 (Strongylocentrotus)

   It is possible to set up and maintain a marine aquarium system simply and inexpensively. It is best to have either all glass, all plastic, or stainless steel frame tanks of 7.5 to 10 gallon or more capacity. It is advisable to have about three complete units ready for use to accommodate different species, or males separate from females, or to provide a reserve tank, in the event that one becomes hopelessly fouled. Holding tanks should be set up and ready for use about one week before animals are expected to arrive. First, make up synthetic sea water by dissolving completely 1 lb. synthetic ocean salts per 3 gal. tap water. These salts can be ordered in bulk from Carolina, Turtox, or Wards. Equip each tank (which may be a plastic dishpan or bucket) with an air-lift sub-gravel aquarium filter capable of changing the water completely at the rate of about 3 times per hour. These are available at hobby and pet shops; all submersible parts should be glass or plastic. Cover the filter to a depth of two or three inches with a 1:1 mixture of coarse granite sand and crushed oyster shell, marble chips, or white blackboard chalk chips. The calcium carbonate in the shell or chips will stabilize sea water mixtures to pH 8.2-8.4, which is optimal for marine organisms. Fill the tank and mark the top level so that fresh water can be added later to replace water lost by evaporation. Activate the filter pump and keep it running continuously. Aquaria can be kept at ordinary room temperature (20-22°C, or 68-72°F) if subtropical or Gulf coast animals are used. Northern and Pacific species must be cooled to about

10-15°C, or 50-60°F. The system described here will be self-maintaining over long periods of time, barring massive death from over-crowding. Organic wastes will routinely be trapped in the coarse bottom gravel where bacteria will decompose them without fouling the water above it. A standard charcoal aquarium filter can be inserted into the line of circulating water. This will help to remove many organic pigments and small molecular wastes that will eventually prove harmful if they become too concentrated. A healthy aquarium will grow diatoms and other marine algae that may be unsightly but, nevertheless, algae are evidence of a successful setup.

b. Chapter II. **Living Chironomous larvae.** No commercial source known to me provides living Chironomous larvae. However, Chironomous larvae are common inhabitants of bottom sediments of most natural ponds, slowly flowing streams, or pools. They are particularly abundant in effluent streams from sewage disposal plants, or other enriched water with a high bacterial count and low oxygen concentration. Adult chironomids are ubiquitous in nature and will readily spawn in almost any outside source of standing water having a depth of about one to two feet (including garden pools, barrels, or tanks) and a modest amount of decomposing leaf litter at the bottom. Larvae can be harvested year-round from such artificial sources after an initial summer period of about three months. Larvae form loosely organized soft mucous tubes which will collect particles of detritus that help to conceal the larvae from sight. The tubes and worms can be easily identified if the bottom sludge is transferred to a shallow, white enamel pan. Worms are bright red and can be easily removed from their tubes.

c. Chapter II. **Living Ilyanassa, mud snails.** These marine animals can be kept for long periods in an aquarium system such as that described in Part a, above. Instructions for maintaining, feeding, and harvesting eggs are given in Chapter III, Section A. Animals can be obtained from many muddy-sand marine environments on the East Coast, or from:

Director, Duke Marine Laboratory, Beaufort, North Carolina 28516.

Supply Department, Marine Biological Laboratory, Woods Hole, Massachusetts 02540.

d. Chapters III, IV, & V. **Living amphibians.** Adult and sexually ripe frogs and salamanders can be obtained locally in nature. The season is usually restricted to February, March, and April. Local suppliers can often be found by asking suburban or rural high school teachers for the names of local boys who are known to be prowling the woods and fields in search of frogs and snakes. Such naturalists are invaluable in their knowledge of local resources and are often willing suppliers of eggs or adults. Frog fertilization kits and fertilized eggs can be obtained from Carolina, Turtox, or Wards; they also will supply pituitary powder for injection and induced spawning techniques as described in Chapter III, Section E-3. Other suppliers include the following sources:

Native frogs *Rana pipiens:*

Bay Biological Supply Co., P.O. Box 71, Port Credit, Ontario, Canada.

E.G. Hoffman and Sons, P.O. Box 815, Oshkosh, Wisconsis 54901.

G.W. Nace, Director, Amphibian Facility, Department of Zoology, University of Michigan, Ann Arbor, Michigan 48104

Southwestern Scientific Supply Co., P.O. Box 17222, Tucson, Arizona 85715.

Parco Scientific Co., P.O. Box 595 Vienna, Ohio 44473

(African clawed toad) *Xenopus laevis:*

Jay E. Cook, P.O. Box 87, Cockeysville, Maryland 21031

C.W. Fletcher, Importer and Distributor, P.O. Box 98, Hampstead, Maryland 21074

NASCO, 901 Janesville Ave., Fort Atkinson, Wisconsin 53538, or P.O. Box 3837, Modesto, California 95350

South African Snake Farm, P.O. Box 6, Fish Hoek, Cape Province, South Africa

Living salamanders or salamander eggs:

Axolotl Colony, Department of Zoology, Indiana University, Bloomington, Indiana 47401

(For *Taricha torosa*) Department of Biology, Stanford University, Stanford, California 94301

Holding facilities for adult amphibians can be quite simple. It is best to keep a small number of animals (less than 6) in separate plastic containers such as dishpans or breadboxes. They should be covered with half-inch mesh, galvanized, wire cloth crimped around the edges to prevent escape. About one liter of water should be added to each pan. One side of the pan should be elevated so that animals can seek out wet or dry places as they choose. Water should be changed every day and the pan and animals should be squirted to wash out shed skin and feces. This can be done very quickly without removing the wire cover or handling the animals. Minor fungus infections (red leg) can be checked by routinely adding 2 grams NaCl per liter of pan water, and/or by adding daily 1 ml 0.1% potassium permanganate per liter of pan water. When the permanganate changes from pink to brown it no longer has any fungicidal or bactericidal value. Late winter frogs are often very emaciated. Their health can be greatly improved by feeding them earthworms or meal worms, or by daily injections with 1 or 2 ml 0.1% glucose into the dorsal lymph space under the skin.

For additional information on amphibian care, see the bibliography for major references, including Wilt and Wessels, or the following:

For maintenance of frogs and toads, see: G.W. Nace, 1974, *Amphibians, guidelines for the breeding, care, and management of laboratory animals,* Printing and Publishing Office, National Academy of Sciences, 2101 Constitution Avenue, Washington, D.C. 20418.

For maintenance and induced spawning of Xenopus, see: A.L. Etheridge and S.M.A. Richter, 1975 *Xenopus laevis, rearing and breeding the African clawed frog.* NASCO, 901 Janesville Avenue, Fort Atkinson, Wisconsin. 53538.

For care of axolotls, see: G.M. Malacinski and A.J. Brothers, *The Axolotl News Letter,* Axolotl Colony, Department of Zoology, Indiana University, Bloomington, Indiana 47401.

e. Chapters VI, VII, and VIII. **Living fertile chicken eggs.** Fertile chicken eggs must be obtained from local hatcheries. They can be identified in telephone business directories. Methods for holding and incubating eggs are given in Chapter VI, Section E.

f. Chapters IX and X. **Living albino mice.** Living mice can be obtained from Wards or from local pet shops. If a medical school, medical research institute, or pharmaceutical testing center is in the vicinity, arrangements can often be made to obtain pregnant mice without the necessity of maintaining a colony. However, it is generally necessary to maintain a small colony for class use. One adult pregnant mouse is needed for each group of four students in a class. It will take approximately two to three months to complete a life cycle from birth to sexual maturity. In general, it will take from four to six months to build up a colony from which six to twelve pregnant females can be taken on a given date.

3. **Sources of special materials needed for experimental procedures.** Sources and suppliers of many special materials are included in the text. Additional general agents include the following.

a. **General sources for chemicals and equipment.** Most inorganic, organic, and dye compounds, general laboratory equipment, and dissecting supplies can be obtained from many sources; the following are reliable:

Fisher Scientific Supply Co., Seaboard Boulevard, 3600 Broad Street, Richmond, Virginia 23230 (and other regional offices)

Scientific Products, Inc., 1430 Waukegan Road, McGraw Park, Illinois 60085

A.H. Thomas, Co., P.O. Box 779, Philadelphia, Pennsylvania 19105 (and other regional offices)

b. **Special sources for organic compounds.** These include amino acids, enzymes, metabolic intermediates, and inhibitors. The following are among the more widely known national distributors.

Nutritional Biochemical Co., 26201 Miles Road, Cleveland, Ohio 44128

Sigma Biochemical Corp., P.O. Box 14508, St. Louis, Missouri 63178

Worthington Biochemical Corp., Freehold, New Jersey 07728

(For frog pituitary powder) Connecticut Valley Supply Co., South Hampton, Massachusetts 01073

(For amphibian anaesthetic, MS222, or Finquel) Ayerst Laboratories, New York City, New York

c. **Sources of laboratory plastic supplies, culture dishes, and equipment.** Catalogs of Carolina, Turtox, Wards, Fisher, and Thomas named earlier, all list various kinds of plastic supplies.

d. **Sources of high quality biological dyes for vital staining and histology.** Biological dyes and stains often are highly variable and it is unwise to economize on them; this is particularly true for vital stains. The following firms are listed in the order of my own preference which is probably subjective.

Chroma Dyes, distributed in the USA by Roboz Surgical Instrument Co., Inc., 801 18th St., NW Washington, D.C. 20006.

George T. Gurr, Ltd., 136 New King's Road, London SW 6, England (strongly to be recommended for the combination of high quality and low cost).

Harleco Dyes, Hartman-Leddon Co., Philadelphia, Pennsylvania 19116

National Aniline Dyes, Allied Chemical and Dye Corp., 40 Rector Street, New York City, New York 10006

GLOSSARY OF SELECTED TERMS OF SPECIAL USAGE IN EMBRYOLOGY

The terms included in this list are part of the common vocabulary of embryology. The definitions provided are planned primarily to identify, rather than to give complete meanings of terms. The latter are provided by the text in association with the discussion of relevant aspects of development. Some terms are included that are not used in the text but are considered to be part of the working knowledge of general embryology. Abbreviations are given in parentheses for most terms for which common symbols are used in figures throughout much of the accompanying text.

ABDUCENS NERVE (6 or VI): the sixth cranial nerve (pure somatic motor) to the external rectus eye muscles.

ABSCISSION: the active separation of a single sheet of cells into two parts along a line.

ACCESSORY or ACCESSORIUS NERVE (11 or XI): the eleventh cranial nerve (pure visceral motor) to muscles of post-sixth visceral arches.

ACOUSTIC or ACOUSTICUS NERVE or GANGLION (8 or VIII): the eighth cranial nerve (pure special somatic sensory) of the ear; derived from the otic vesicle epithelium.

ACROSOME: a special area of sperm cytoplasm, usually opposite the flagellum; functionally associated with the attachment of the sperm to the egg surface.

ACROSOME FILAMENT: in some species the sperm acrosome forms a slender filament that serves to penetrate surface coats and make contact with the egg surface; it may also be directly involved in mechanisms for sperm entry at fertilization.

AFFERENT: a term meaning to carry toward an object or structure of interest or concern.

AFFERENT BRANCHIAL ARTERIES (AB 3-6): arteries that carry blood from the ventral aorta to the gill capillaries, i.e., the ventral halves of embryonic aortic arches.

ALAR PLATE (AP): the dorsal sensory column of the brain and spinal cord.

ALBUMEN SAC: an extension of the extraembryonic serosal membrane that surrounds most of the egg albumen late in chick development.

ALLANTOIC ARTERIES (ALA): paired arteries that carry blood from the dorsal aorta to the allantoic sac, chorion, and chorionic placenta; also called the umbilical arteries.

ALLANTOIC CAVITY (ALC): the cavity of the allantoic sac; the chorionic vesicle of nonprimate amniotes; it is derived from the urogenital sinus of the cloaca.

ALLANTOIC MEMBRANE or SAC (AL): a midventral evagination from the urogenital sinus and formed of splanchnopleure; it fuses with the serosa to form the chorion and chorionic placenta; homologous with the amphibian urinary bladder.

ALANTOIC STALK (ALS): the connecting tube between the allantoic stalk and urogenital sinus.

ALLANTOIC VEINS (ALV): paired veins that drain blood from the allantoic sac, chorion, and placenta to the ductus venosus and heart; also called umbilical veins.

AMMOCETES: the characteristic larval stage of cyclostomes.

AMNIOTE: vertebrates with an extraembryonic amniotic membrane and cavity surrounding the embryo; i.e., reptiles, birds, and mammals.

AMNIOTIC CAVITY (AMC): the cavity surrounded by the amniotic membrane and filled with fluid that supports and protects the embryo in its own small lake of amniotic fluid.

AMNIOTIC FOLDS, HEAD FOLD (AMH), LATERAL FOLDS (AML), or TAIL FOLD (AMT): double layered folds of extraembryonic somatopleure that overgrow the embryo and enclose it in an amniotic cavity; the outer layer is the serosa; the inner layer is the amnion.

AMNIOTIC MEMBRANE or SAC (AM): the primary extraembryonic membrane of amniotes; the membrane is composed of somatopleure and immediately surrounds the developing embryo.

ANALOGY or ANALOGOUS STRUCTURES: structures in different species that share a common function; they may be convergent adaptations or may have homologous origins.

ANAMNIOTE: vertebrates in which the embryo lacks an amniotic sac; i.e., fish and amphibia.

ANAPHASE: the third period in mitosis during which diplotene chromosomes are separated and their hoplotene halves move to opposite poles of the spindle.

ANASTOMOSIS: the fusion of two or more originally separate parts.

ANIMAL POLE (AN): the point on the surface of an egg at which polar bodies are given off.

ANIMAL-VEGETAL AXIS (AN-VEG): an imaginary line that passes from the point of polar body attachment through the center of the egg; the animal pole coincides with the polar body attachment, and the vegetal pole is opposite it.

ANISOGAMETES or ANISOGAMY: sexual reproduction by the union of gametes that are morphologically differentiated only by size into micro- and macrogametes, both of which are motile and contain stored nutrients; characteristic of some lower algae.

ANLAGE (*an'-lag-e;* plural, *anlagen):* an embryonic primordium determined to form a particular organ or part of an organ; an organ rudiment.

ANTERIOR CARDINAL VEIN (AC): a vessel that carries blood from the interior of the skull and brain to the common cardinal or precaval veins; the primordium of the internal jugular vein of adult mammals.

ANTERIOR CHOROID PLEXUS (ACP): the thin roof plate of the diencephalon; it functions in the secretion of the fluid of the neurocoele.

ANTERIOR INTESTINAL PORTAL (AIP): the opening of the foregut into the open midgut in early amniote development.

ANTERIOR NEUROPORE (ANP): an opening at the anterior end of the neural tube.

ANTERIOR POLE (A): the head end of the body.

ANUS (AN): the terminal opening of the gut tube.

AORTIC ARCHES (A 1-6 or AA 1-6): paired arteries connecting the ventral aorta to the dorsal aorta around the lateral walls of the pharynx; in cyclostomes there are 7 or more pairs, in elasmobranchs 7 pairs, in higher vertebrates 6 pairs of arches.

APICAL ECTODERMAL RIDGE (AE): a ridge of epidermal ectoderm on the outer border of the limb bud that provides a maintenance factor essential for limb growth and development.

APPENDICULAR SKELETON: bones of the fore and hind limbs and their girdles, derived from lateral plate, hypomere, mesoderm.

APPENDIX TESTIS: a small remnant of the distal end of the Müllerian (female) duct that is attached to the testis of the male mammal.

AQUEDUCT OF SYLVIUS (AQ): the neurocoele of the mesencephalon; also called the *iter*.

AQUEOUS HUMOR (AH): the fluid-filled cavity of the eye between the lens, iris, and cornea.

ARCHAETYPE: the concept of an hypothetical ancestral species from which modern forms of related organisms are descended; it is a little used concept today.

ARCHENTERON (AR): the primitive gut cavity formed during gastrulation, and its cellular wall which may be composed of endoderm or of endoderm and mesoderm.

ARCHINEPHRIC DUCT (PND): the Wolffian or pronehpric duct of the kidney.

AREA OPACA (AO): regions of the chick blastoderm that surround the area pellucida and embryonal area; it has a heavy staining capacity and includes the areas vasculosa and vitellina, and the zone of junction.

AREA PELLUCIDA (AP or APL): the relatively thin and transparent part of the chick blastoderm immediately surrounding the embryo and bordered distally by the area vitellina.

AREA VASCULOSA (VAS): the part of the area opaca with blood islands and/or vessles, located between the areas vitellina and pellucida.

AREA VITELLINA (AV): the part of the distal extraembryonic blastoderm lacking invaginated mesoderm; the blastoderm between the zone of junction and area vasculosa.

ASSOCIATION NEURON: nerve cells entirely within the central nervous system that link nervous stimuli from peripheral sensory neurons to peripheral motor neurons.

ATRIAL CAVITY or SAC: an ectodermally-lined respiratory chamber into which the gill slits of protochordates open.

ATRIOPORE: the opening of an atrial sac to the exterior in protochordates.

ATRIOVENTRICULAR VALVES (AV or AV' & AV"): openings between the right and left atria and ventricles of the heart.

ATRIUM (AT): the heart chamber between the sinus venosus and ventricle.

ATTACHMENT PAPILLAE: anterior epidermal organs of larval tunicates that function during metamorphosis in attachment of larvae to the substrate.

AUTONOMIC GANGLIA (AG): the ganglia of the sympathetic trunk, derived from neural crest.

AUTOPODIUM: the distal segment of the paired appendages corresponding with the levels of the wrist–hand and/or ankle–foot.

AUTOSTYLIC JAW SUSPENSION: a type of joint by which the lower jaw moves against a socket formed from another skeletal element of the first visceral arch; i.e., the articulation of Meckel's cartilage against the quadrate cartilage in salamanders.

AXES OF SYMMETRY: linear patterns for the organization of parts in the body:
Dorsoventral (DV) axis: the axis from back to belly.
Anteroposterior (AP) axis: the axis from head to tail.
Mediolateral (ML) axis: the axis from midline to right or left side.

AXIAL ELONGATION: the general movements of chordamesoderm and somite mesoderm cells after they move to the interior; movements that orient chordamesoderm and somites in the anteroposterior axis of the body.

AXIAL SKELETON: derivatives of the sclerotome including vertebrae, ribs, sternum, and neurocranium of the skull; the latter are bones of the brain case.

AZYGOUS VEIN (AZ): a single vessel of mammals that drains intersegmental blood from the right side of the body wall into the precaval vein; a homologue of the anterior end of the right posterior cardinal vein and right supracardinal vein.

BALANCER (BAL): small paired epidermal evaginations from the side of the head of salamander larvae that assist in balance until the forelimbs develop; they then degenerate.

BALBIANI RING or PUFF: the expanded areas of polytene chromosomes that are sites of major gene activity.

BASAL PLATE (BP): the ventral motor column of the neural tube.

BASIBRANCHIAL CARTILAGE (BC): the single midventral cartilage bar under the pharynx to which the hypobranchial visceral cartilages of the gill arches attach.

BASILAR ARTERY (BA): the artery under the midbrain connecting the paired internal carotid arteries to the paired vertebral arteries.

BASISCLEROTOME (BSC): the posterior half of a segmental sclerotome, but the anterior half of a definitive vertebra.

BICUSPID or MITRAL VALVE (AVL): the left atrioventricular canal or valve.

BIOLOGICAL DISCIPLINE: any of the generally recognized subject areas of biological science.

BIPINNARIA LARVA: the characteristic larva of starfish.

BIVALENTS: synapsed homologous chromosomes of primary gonocytes; i.e., precursors of tetrads.

BLASTEMA: a group of cells that will develop into an organ or new individual.

BLASTOCOELE (BC): the cavity in a blastula stage embryo.

BLASTOCYST: the name usually given to a mammalian blastula stage embryo.

BLASTODERM (BD): the entire cellular area of an early embryo, including embryonic and extraembryonic regions.

BLASTOMERE: an embryonic cleavage cell.

BLASTOPORE (BP): the opening to the embryonic archenteron; in chordates it corresponds to the anus or the location where the proctodaeum and anus will later form.

BLASTOPORE LIPS (BP), DORSAL LIP (DL), LATERAL LIP (LL), VENTRAL LIP (VL): the borders of the blastopore lumen; chordamesoderm forms dorsal lip, lateral mesoderm forms lateral and ventral lips of the blastopore.

BLASTULA: the embryonic stage following cleavage when the embryo consists of a single cell layer; it may be a hollow coeloblastula, solid stereoblastula, or flat discoblastula.

BLOOD ISLAND (BI): the primordium of blood cells and much of the endothelial tissue that lines a closed circulatory system, including capillary walls.

BODY FOLDS, HEAD (HF), TAIL (TF), and LATERAL (LBF): undercutting folds of the blastoderm in the area pellucida that separate the embryo from extraembryonic regions except in the belly region where the folds meet and help to form the umbilical cord.

BODY STALK or UMBILICAL CORD (U): the general connection of an embryo or fetus to its extraembryonic membranes and/or its placenta.

BOTTLE CELLS (FC): See Flask Cells.

BOWMAN'S CAPSULE: the distal double-walled expansion of mesonephric and metanephric kidney tubules that surround an internal glomerulus; the capsule of a renal corpuscle.

BRACHIAL: referring to the arm or shoulder.

BRACHIAL PLEXUS (BR): the spinal nerve network leading to the forelimbs.

BRAIN PAN: See Chondrocranium.

BRANCHIAL: referring to gills or gill respiration.

BRANCHIAL GROOVES: See Visceral Grooves.

BRANCHIAL MUSCLES (VSM 1-6): hypomere muscles of the pharynx and gill arches; a synonym for visceral muscles of the gill arches.

BULBOURETHRAL or COWPER'S GLAND: an accessory male sex gland derived from endoderm of the urethra; it secretes an alkaline fluid into the penile urethra.

BURSA OF FABRICIUS: An embryonic specialization of the tailgut endoderm which in birds functions as a reservoir for immune system cells before the spleen and other lymphoid organs form; it is not found in mammals in which the thymus glands serve this function.

CAENOMORPHOSIS or CAENOMORPHIC CHARACTERS: aspects of development that have been modified by evolutionary change and do not exhibit ancestral ways of forming organs; usually they are adaptive to new kinds of environmental or developmental conditions, i.e., the formation of the amnion and other extraembryonic membranes by early reptiles.

CALVARIUM: bones forming the roof of the skull (i.e., frontals, parietals & supraoccipital) of neural crest origin.

CAMERA LUCIDA: a drawing apparatus designed for use with the microscope.

CARDIAC MESODERM (HT): See Heart Primordium.

CAROTID ARCH (A 3 or AA 3): the third pair of aortic arches; they form the base of the adult internal carotid arteries.

CAROTID DUCT (CD): the dorsal aorta between the third and fourth aortic arches; it usually atrophies completely but rarely may persist in adult tetrapods.

CAUDAL: a term referring to the postanal tail of chordates.

CAUDAL ARTERY (CDA): the median artery of the tail derived from the dorsal aorta posterior to the attachments of the paired common iliac arteries to the aorta.

CAUDAL VEINS (CV): single or paired veins of the tail derived from the embryonic posterior cardinal veins posterior to the junction of the common iliac veins with the postcava.

CAUDALIZATION: the differentiation of the tail blastema into a typical chordate tail with appropriately organized axial organs including a neural tube, notochord, myotomes, and often fins in aquatic vertebrates.

CAVAL MESENTERY (CAV or CAM): a part of the dorsal mesentery of the stomach that carries the postcaval vein.

CAVITATION: morphogenetic cell movements that result in hollowing out of a solid cell mass as in the formation of a myocoele in a somite.

CELL CYCLE: the life history of cells from interphase through mitotic cell division.

CELL DIFFERENTIATION: the production of special gene products that are characteristic for the special structure or function in different tissues.

CEMENT GLANDS (CG): paired mucous glands that develop on the ventral surface of the head of the frog hatching larva.

CENTRAL NERVOUS SYSTEM (CNS): the vertebrate brain and spinal cord.

CENTRIOLE or DIVISION CENTER: the cell organelle that organizes the formation of the spindle–aster system of dividing animal cells; it is absent in most plant cells.

CENTROLECITHAL: eggs with yolk unequally concentrated at the center of the ovum; it is characteristic of eggs of insects and other arthropods.

CEPHALIC FLEXURE (CEF): the strong ventral bend in the brain stem at the level of the mesencephalon.

CEREBRAL GANGLION: the central ganglion of metamorphosed ascidians which represents the collapsed adult derivative of the larval cerebral vesicle; brain centers of many invertebrates.

CEREBRAL VESICLE: the brain-like enlargement at the anterior end of the neural tube of protochordates.

CERVICAL: pertaining to the neck region, as in cervical vertebrae or cervical spinal nerves.

CERVICAL FLEXURE (CER): the strong ventral bend in the brain stem at the occipital and cervical body levels.

CERVICAL SINUS (VC 4-5): an alternative name for the fourth and fifth visceral grooves which, in mammals, are united into a single epidermal depression on the neck.

CHONDROCRANIUM or BRAIN PAN: the ventral parts of the neurocranium of the skull bordering the neural canal that are first formed from cartilage; i.e., parts of the skull derived from the embryonic sensory capsules, trabeculae, parachordals, and occipital sclerotomes.

CHORDAMESODERM (CM): the embryonic primordium of the notochord or chorda dorsalis.

CHORION: the term means *membrane;* it can refer to simple fertilization membranes around eggs, or to very complex extraembryonic respiratory membranes of bird and reptile eggs; it is also used as a synonym for the embryonic serosa, but this source of confusion should be avoided.

CHORION CELLS: in tunicate eggs the chorionic membrane is secreted by follicle cells that remain with the egg after it is shed; these chorion cells are discarded when embryos hatch.

CHORIONIC GONADOTROPIN: the hormone produced by the placenta that has essentially the same function as leuteinizing hormone of the pituitary in maintaining corpus luteum.

CHORIONIC PLACENTA: the typical placenta characteristic of advanced stages in fetal development; a placenta utilizing allantoic circulation as the embryonic source of fetal blood to the placenta.

CHORIONIC VESICLE: a term usually used in mammals to refer to the allantoic sac; however, in primates, the allantoic sac is vestigial and therefore (unfortunately) in man and most primates the chorionic vesicle is homologous with the extraembryonic coelom of lower mammals.

CHOROID FISSURE (CH): the medioventral defect in the wall of the optic cup through which optic nerve fibers will leave the retina and grow to the brain floor.

CHROMOMERE: a single transverse band with strong staining affinity in polytene chromosomes; i.e., somewhat equivalent to a single gene locus on a chromosome.

CIRCLE OF WILLIS (CW): an arterial loop around the infundibulum that connects the paired internal carotid arteries with the median basal artery under the brain.

CLASSIFICATION: the naming and taxonomic organization of species; see Tables Int.-3 and Int.-4 for a survey of classification with particular reference to chordates.

CLEAVAGE: early mitotic stages following fertilization that result in a cluster of a few dozen cells.

CLEAVAGE SYMMETRY: patterns of cleavage that are related to adult axiate organization:

- Asymmetric cleavage: cleavage planes bear no constant relation to the adult axes.
- Radial cleavage: cleavage planes are parallel or transverse to only one adult axis.
- Bilateral cleavage: cleavage planes are parallel or transverse to AP, DV, or ML axes.
- Spiral or oblique cleavage: with cleavage planes bearing a constant relation to, but oblique to, AP, DV, or ML axes of adult symmetry.

CLOACA (CLO): the terminal portion of the endodermal hindgut of chordates; it is the general receptacle for ducts and products of the digestive, excretory, and reproductive systems.

CLOACAL MEMBRANE (CM): the temporary closing plate that separates the hindgut from the proctodaeum; a double membrane composed of hindgut endoderm and proctodaeal ectoderm.

CLOSING PLATE (CP): double membranes of epidermal ectoderm and endoderm that will ordinarily degenerate and create openings of the mouth, anus, or gill slits.

CLUB SHAPED GLAND: the region in Amphioxus larvae that will be associated with the wheel organ in metamorphosed juveniles and adults.

COELIAC ARTERY (COE): the major unpaired artery to the stomach.

COELOM or COELOMIC CAVITY (CL): any cavity formed in the hypomere mesoderm and lined by peritoneal epithelium.

COELOMIC VESICLE (CL): an evagination from the archenteron of echinoderm gastrulae that will give rise to the larval mesoderm and coelomic cavity.

COMMON BILE DUCT (BD): the duct from the gall bladder and pancreas leading to the small intestine.

COMMON CARDINAL VEINS (CC): the primary embryonic veins connecting the anterior and posterior cardinal veins to the sinus venosus; the ducts of Cuvier.

COMMON CAROTID ARTERIES (COCR and COCL): paired branches of the systemic trunk of the aorta that carry blood to the external and internal carotid arteries; they are derived from the embryonic ventral aorta or aortic sac between the third and fourth aortic arches.

COMMON ILIAC ARTERIES (CI): the fifth pair of lumbar intersegmental arteries which will carry most of the blood to the hind limbs.

COMPETENCE: a term referring to all capacities that an embryonic tissue or primordium has at a given time to form various kinds of structures including, and in addition to, the part it would form in normal development; it is usually tested by isolating, transplanting, or removing parts of an embryo in experimental tests for regulation and self-differentiation.

CONCEPTUS: a single fetal implantation in the mammalian uterus.

CONSTRICTION: a morphodynamic movement in which embryonic parts become separated by pinching off; i.e., as in separation of anterior Amphioxus somites from the gut wall.

CONUS ARTERIOSUS (CA): the most anterior of the four primary heart chambers connecting the ventricle to the ventral aorta.

CONVERGENCE: the movement of chordamesoderm and somite mesoderm toward the dorsal lip of the blastopore prior to involution at gastrulation.

CORACOSCAPULAR CARTILAGE (CS): a cartilage bar in the shoulder region that is the primordium of the amphibian pectoral girdle.

CORNEA and CONJUNCTIVA (CO): the modified transparent skin over the eye; the surface epidermal layer is the conjuctiva; the deeper connective tissue layer is of neural crest origin.

CORONARY VEIN and SINUS (CCL): the rudiment of the embryonic left common cardinal vein which in adult mammals drains blood from the heart wall into the sinuatrial valve.

CORPUS LUTEUM: the modified follicle cells that produce the hormone progesterone that is responsible in part for maintenance of mammalian embryos in the uterus.

CORTICAL ACTIVATION GRANULES: special granules near the surface of many ripe eggs that participate in the formation of the fertilization, zygote, and hyaline membranes.

CORTICAL CYTOPLASM: the surface layer of structured cytoplasm and cell membrane in eggs or cells; usually it is more viscous than deeper layers of cytoplasm; it is often associated with the determination of cell competence.

CORTICAL INTERSTITIAL CELLS: cells, probably derived from the theca externa, that form follicle cells of the vertebrate ovary.

CORTICAL ISLANDS: clusters of germ cells in the embryonic ovotestis that are primordia of oogonia in the adult ovary.

COWPER'S GLAND: See Bulbourethral Gland.

CRANIATE: chordates with a distinct head; i.e., an alternative name for vertebrates and distinct from the acraniates, or protochordates.

CUMULUS OOPHORUS: a column of follicle cells that elevates the primary oocyte into the antrum of a mammalian Graafian follicle.

CYTOKINESIS: the cytoplasmic division of cells which occurs normally at telophase of mitosis.

DEEP (SENSORY) EPIDERMAL LAYER (EED): the layer of epidermis of frog embryos that forms sensory placodes.

DEEP GANGLION LAYER (RTG'): a layer of nerve cell bodies adjacent to the rod and cone layer of the sensory retina.

DERMAL SCALE PLATES: bony supporting tissue formed from the dermis and underlying some kinds of vertebrate scales of the skin and teeth; they are the evolutionary origin of dermal bones of the axial skeleton and other special regions of the body.

DERMATOCRANIUM: the parts of the skull formed from dermal scale plates during evolution.

DERMOMYOTOME (DMY): the part of the mesodermal somite located dorsolateral to the myocoele; it is the primordium of the dermis and segmental skeletal muscles, or myotome.

DETERMINATE CLEAVAGE: cleavage divisions that result in sister cells with different developmental competence unequal separation of cytoplasmic determinants for tissue or organ differentiation; i.e., cleavage that results in ooplasmic segregation.

DEUTEROSTOMES: the phyla of animals, including echinoderms and chordates, in which the embryonic blastopore marks the position of the adult anus; the mouth is a new opening.

DEVELOPMENTAL COMMON PLAN: a generalized morphological diagram, devoid of special characters, that is diagnostic for the structure of embryonic stages or taxonomic groups.

DIAPHRAGM: a unique mammalian structure which separates the pleural and peritoneal cavities and aids in breathing movements with muscles derived from the 3-5th cervical myotomes.

DIENCEPHALON (DI): the second of five major brain regions, located between the telencephalon and mesencephalon; the thalmic region of the adult brain.

DIPLOID: the number of chromosomes characteristic for the fertilized egg and all cells derived from it by ordinary mitosis; i.e., double the haploid number.

DIPLOTENE: chromosomes with two chromatids or chromatin threads with two chromonemata; i.e., chromatin of the G^2 phase in the cell cycle.

DISTAL: toward the end of a structure farthest from the body midline.

DORSAL AORTA (DA): the major systemic artery efferent to the embryonic aortic arches.

DORSAL LIP (DLB): the dorsal border of the blastopore; in chordates it is composed of chordamesoderm cells. See also Blastopore Lips.

DORSAL MEDIASTINAL SEPTUM (DMS): a special part of the dorsal mesoesophagus that is part of the support for the pericardial sac and a component of the mammalian diaphragm.

DORSAL MESENTERY (DM): a double fold of splanchnic mesoderm that suspends the gut in the coelomic cavity; subdivisions of it are named for the organs they support, i.e., dorsal mesoesophagus, dorsal mesogaster, dorsal mesocolon, etc.

DORSAL MESOCARDIUM (DMC): the segment of ventral mesentery of the gut at the level of the esophagus that supports the heart in its pericardial cavity; in regions of the ventricle and atrium it degenerates soon after forming.

DORSAL POLE (D): the point or line on an egg or embryo that coincides with the back side of the body.

DORSAL ROOT GANGLION: the enlargement in the dorsal roots of spinal nerves containing the cell bodies of somatic and visceral sensory nerve cells.

DORSAL ROOT OF SPINAL NERVES: the bundles of somatic sensory and visceral sensory nerve fibers from a spinal nerve that pass into the alar plate of the central nervous system.

DUCTS OF CUVIER (CC): an alternative name for the common cardinal veins.

DUCTUS ARTERIOSUS (DAR): the fetal dorsal half of the left sixth aortic arch which connects the left pulmonary artery to the left arch of the aorta; in mammals it usually closes shortly after birth.

DUCTUS VENOSUS (DV): a major fetal venous channel through the liver which connects the left umbilical (allantoic) vein with the postcaval vein; it degenerates after birth but its walls remain as the ligamentum venosum of the adult liver.

DUODENUM or DUODENAL INTESTINE: the anterior end of the small intestine at the attachment to the stomach and including the connection of the common bile duct to the gut.

DYAD CHROMOSOME: maternal or paternal diplotene chromosomes remaining in secondary gonocytes following the first meiotic reductional division.

ECTODERM (EC): the outer of the three primary germ layers.

ECTOMESENCHYME: skeletal and other supporting tissues of neural crest origin; a synonym for mesectoderm.

ECTOPLASM (ECP): a cytoplasmic region of the egg or early cleavage cell that is determined to differentiate as ectodermal derivatives; also a term used as a synonym for cortical cytoplasm.

EFFECTOR CELL: a motor cell; usually a muscle or gland cell.

EFFERENT: a term meaning, *to carry away* from a region of interest or concern.

EFFERENT BRANCHIAL ARTERIES (EG 3-6): arteries that carry blood from functional gill capillaries to the dorsal aorta; i.e., the dorsal halves of embryonic aortic arches.

EMBOLY: the active invagination of prospective endoderm and mesoderm during gastrulation by a sudden collapse of vegetal regions to the interior as in protochordate eggs.

EMBRYO: an early developmental stage of higher animals; usually reserved for stages between gastrulation and fetal or larval periods.

EMBRYO CELLS (EMC): cells of early mammalian cleavage stages that will form the embryo and some extraembryonic parts, as distinct from trophoblast cells.

EMBRYONAL AREA (EMB): the part of the chick blastoderm that will form the embryo, as distinct from surrounding extraembryonic parts of the blastoderm.

EMBRYONIC DISC (EMD): a thick placode of cells in the early mammalian blastocyst derived from embryo cells and representing the early embryonal area.

EMBRYONIC INDUCTION: any process whereby one type of embryonic cell (the inductor) determines another cell type (the reactor) to embark on a path of self-differentiation that it was previously incapable of carrying out.

EMBRYONIC SHIELD: the region of delaminated endoderm in the blastoderm of the very early chick embryo; i.e., the region of the hypoblast in the unincubated blastoderm.

ENDOCARDIAL CUSHION (ECC): a thick ingrowth from the heart wall that separates the heart lumen into right and left atrioventricular valves.

ENDOCARDIUM (END): the endothelial lining of the early heart tube; it is the inductor of the myocardial layer of the heart in some vertebrates.

ENDODERM (EN): the inner of the three primary germ layers.

ENDOLYMPHATIC DUCT (ED): the persistent remnant of the connection between the epidermis and the otic vesicle of the inner ear.

ENDOPLASM (ENP): a cytoplasmic region of the egg or early cleavage cell that is determined to differentiate as endodermal derivatives; also a term sometimes used as a synonym for the central cytoplasm of cells filled with nutrient or secretion droplets and granules.

ENDOSTYLE: a deep, midventral, ciliated band of epithelium in the floor of the pharynx of protochordates that aids in the transport of trapped food to the stomach; homologous with the vertebrate thyroid gland.

ENDOTHELIUM: the thin epithelial lining of capillaries, veins, arteries, and heart chambers in closed circulatory systems.

ENTEROCOELIC MYOCOELE: a myocoele cavity of somites formed by outpocketing from the wall of the archenteron, as in the anterior myocoeles of Amphioxus and cyclostome embryos.

EPAXIAL MYOTOME (EM): the major dorsal subdivision of segmental myotomes above the horizontal septum; the primordium of dorsal back muscles and extrinsic levators of the limbs.

EPIBLAST (EPB): the superficial layer of cells in the early gastrula at the two layer stage; the part of the blastoderm that will form the ectoderm and also may form mesoderm; the latter is the case in most amniotes.

EPIBOLY: a morphodynamic movement that results in a sheet of cells spreading to cover other regions by radial expansion; i.e., the expansion of ectoderm during gastrulation.

EPIBRANCHIAL GROOVE: a ciliated middorsal band of epithelium in the pharynx of Amphioxus that aids in the transport of mucous strands with trapped food to the stomach.

EPIBRANCHIAL PLACODES (LL): the thickened epidermal primordia of the lateral line system of sensory canals of amphibians and other aquatic lower vertebrates.

EPIBRANCHIAL VISCERAL CARTILAGES (EBCL 1-6): the dorsal half of the paired visceral cartilages of the gill arches.

EPICARDIAC DIVERTICULUM: an asexual reproductive diverticulum from the floor of the tunicate pharynx that will form asexual reproductive stolons that bud off secondary zooids.

EPIDERMIS or EPIDERMAL ECTODERM (E or EE): the outer covering of the body and its glandular or sensory derivatives.

EPIDIDYMUS TESTIS: the part of the mammalian testis with vasa efferential tubules; it is a vestige of the embryonic mesonephros attached to the testis.

EPIGENETICS: all factors that operate from outside a cell which tend to regulate the expression of its genes; in development these are usually referred to as determining or inductive stimuli.

EPIMERE MESODERM (SM): somite or segmental mesoderm.

EPIMYOCARDIUM or MYOCARDIUM (EMY or MC): the heavy musuclar wall of the heart derived from the ventral mesentery of the esophagus and pharynx.

EPIPHYSIS or PINEAL BODY (EP): a glandular evagination from the roof plate of the diencephalon with probable endocrine functions relating to photoregulation of reproductive cycles in adult vertebrates.

EPIPLOIC FORAMEN (EPF): the opening from the omental bursa, or lesser peritoneal sac, into the general peritoneal cavity.

EPITHELIUM: pavement tissues with few intercellular spaces; these are tissues which form all glands and line all body cavities and surfaces.

EPOOPHORON and PAROOPHORON: vestigial mesonephric tubules in female mammals.

ESOPHAGUS or OESOPHAGUS (OE): the posterior foregut back of the pharynx and dorsal to the laryngotracheal groove.

ESTRONE or FEMALE SEX HORMONE: the hormone produced by follicle cells of the ovary that elicit responses of femaleness.

ETHMOID PLATE CARTILAGE (ET): the primordium of the ethmoid bone; derived by fusion from the anterior ends of the trabecular and nasal capsule cartilages.

EUSTACHIAN TUBES (P 1): paired canals connecting the mouth and middle ear cavities; derived from the base of the first pharyngeal pouches.

EXTERNAL AUDITORY CANAL or MEATUS (VG 1): the outer ear canal; derived from the first embryonic visceral groove.

EXTERNAL CAROTID ARTERIES (EC): the paired anterior extensions of the ventral aorta which carry blood to the jaws and surface of the head.

EXTERNAL GILLS (EG): external outgrowths from the visceral arches that contain capillary loops from the aortic arches in larvae of amphibians.

EXTERNAL ILIAC ARTERIES (EIA): the major branches of the common iliac arteries that carry blood from the aorta to the hind limbs.

EXTERNAL ILIAC VEINS (EIV): the major branches of the common iliac veins that return blood from the hind limbs to the common iliac veins and postcaval vein.

EXTERNAL JUGULAR VEINS (EJ): paired veins that drain blood from the surface of the head to the precaval veins (or innominate veins in mammals).

EXTIRPATION: the experimental removal of a primordium, tissue, or organ from an embryo.

EXTRAEMBRYONIC COELOM (XCL): an extension of the embryonic coelom through the body stalk or umbilical cord to the extraembryonic areas of the blastoderm.

FACIAL or FACIALIS NERVE (7 or VII): the seventh cranial nerve (mixed visceral sensory and motor) to the second visceral arch.

FALCIFORM LIGAMENT (FAL): the segment of ventral mesentery of the stomach connecting the capsule of the liver to the ventral body wall.

FATE MAP: a drawing of an early embryonic stage showing the regions that will give rise in normal development to specific organs and tissues of the adult.

FENESTRA (HF): See Hypophyseal Foramen.

FERTILIZATION: the union of gametes with activation of the zygote to undergo development.

FERTILIZATION MEMBRANE (FM): in many species a membrane is freed from the egg surface as part of the fertilization reaction; there are many differences in the structure and manner of elevation of these membranes in different species.

FETUS or FOETUS: late embryonic stages after the basic characters have appeared that permit it to be recognized as the young of its taxonomic group; in man it is the embryo after three months to full term in gestation.

FIBROUS LAYER OF RETINA (RTF): an intermediate layer of nerve fibers between the ganglionic layers of the retina.

FIN, DORSAL and VENTRAL (DF or VF): mesenchyme-filled elevations of the dorsal and ventral epidermis of the trunk and/or tail of aquatic vertebrate embryos and larvae.

FINQUEL, MS-222, or TRICAIN METHANOSULFONATE: an anaesthetic for coldblooded vertebrates.

FIRST MATURATION or MEIOTIC DIVISION: a cell division in which genetic segregation of homologous maternal and paternal chromosomes occurs as the first step in genetic reduction of chromosome number in germ cells.

FLASK CELLS or BOTTLE CELLS (FC): special cells that initiate invagination of the blastopore lips in amphibian gastrulation.

FLOOR PLATE (FP): the thin, non-nervous, ventral wall of the developing neural tube.

FOLLICLE CELLS: the so-called nurse cells of the vertebrate ovary which surround developing primary oocytes and contribute estrone (female sex hormone) to the hormonal feedback system that regulates periodic cycles of oocyte growth.

FORAMEN OF MUNRO (FM): the opening from the lateral ventricles of the telencephalon into the third ventricle of the diencephalon neurocoele.

FORAMEN OVALE, PRIMARY AND SECONDARY (FO 1 & 2): the openings in the primary and secondary interatrial septa which permit free flow of blood between the right and left atria during mammalian fetal stages; they will normally close after birth.

FOREBRAIN or PROSENCEPHALON (FB): the anterior primary brain vesicle; it will form the telencephalon and diencephalon.

FOREGUT (FG): the anterior part of the endodermal gut tube including the pharynx and esophagus–trachea regions.

FORE LIMB BUD (FLB): the primordium of arms or wings.

FOURTH VENTRICLE (V 4): the neurocoele of the myelencephalon.

FRONTAL PLANE OF SYMMETRY: the horizontal plane of the body, parallel with the AP and ML body axes.

FRORIEP'S GANGLIA (12 F or XII F): the vestigial dorsal root ganglia of occipital nerves; especially the 4th.

G^1 PHASE OF THE CELL CYCLE: the first period of interphase when chromatin is haplotene.

G^2 PHASE OF THE CELL CYCLE: the last period of interphase when chromatin is diplotene.

GALL BLADDER (GB): a ventral endodermal outgrowth from the midgut at the junction of the small intestine and stomach; it is associated with storage of liver bile.

GAMETOGENESIS: the period in the life cycle that deals with the origin, growth, and specialization of eggs and sperm cells in parent bodies.

GAMONES: chemical products of gametes that aid either in gamete attachment, in coding for species specificity, in mechanisms for assuring monospermy, or in activation of the egg during fertilization; they are so-called "gamete hormones."

GANGLION: a cluster of peripheral nerve cell bodies located outside the brain or spinal cord.

GANGLIONIC CREST (GC): the part of the neural crest that will form ganglia of cranial nerves, and spinal, and/or sympathetic ganglia of the peripheral nervous system.

GARTNER'S CANAL: a vestigial remnant of the mesonephric (Wolffian or archinephric) duct in female mammals.

GASSERIAN GANGLION (5g or Vg): See Seminlunar Ganglion.

GASTROCOELE (GC): the cavity of the archenteron or primitive gut.

GASTROHEPATIC DIVERTICULUM: a large endodermal evagination from the anterior midgut of Amphioxus; with glandular and digestive function, possibly homologous with the vertebrate liver.

GASTRULATION: the embryonic period in which the primitive gut or archenteron is formed, when the primary germ layers are separated, and embryonic axial symmetries are fixed.

GENETIC REDUCTION: meiotic mechanisms responsible for reducing the chromosome number from diploid to haploid in the process of gamete production.

GENETIC SEGREGATION: the separation of maternal from paternal homologous chromosomes, which usually occurs at the first maturation division of meiosis.

GENICULATE GANGLION (7g or VIII g): the ganglion of the seventh cranial nerve.

GENITAL FOLD or LATERAL GENITAL FOLD: lateral ridges beside the urethral groove; the embryonic primordia of the body of the penis in male, or of the labia minora in female mammals.

GENITAL RIDGE (GR): the paired primordia of the reproductive germinal epithelium of the gonad; it is a thickening composed of peritoneum, primordial germ cells, and mesenchyme near the mesonephric kidneys.

GENITAL SWELLINGS or LABIOSCROTAL SWELLINGS: lateral elevations beside the genital folds; the embryonic primordia for the scrotal sac of male, or of the labia majora of female mammals.

GENOME: the total genetic makeup of an individual organism.

GERM LAYERS or PRIMARY GERM LAYERS: ectoderm, mesoderm, and endoderm of the gastrula and post-gastrula stages in development.

GERMINAL EPITHELIUM: any mitotically active epithelial layer that produces cells; i.e., the theca externa of the ovary in some vertebrates, etc.

GERMINAL VESICLE: the enlarged nucleus of primary oocytes.

GESTATION: the period of intrauterine development of mammalian embryos and fetuses.

GLOMERULUS or GLOMUS (GL): capillary knots that filter blood fluids from renal arterioles into Bowman's capsules of mesonephric and metanephric kidney tubules; a glomus serves the same function for pronephric tubules but excretes fluids into the coelom.

GLOSSOPHARYNGEAL NERVE (9 or IX): the ninth cranial nerve (mixed visceral sensory and motor) to the third visceral arch.

GLOTTIS (GLO): the anterior end of the trachea or laryngotracheal groove at its union with the esophagus.

GONAD: the nonspecific term for a reproductive gland including ovaries and testes.

GONIAL CELLS, OOGONIA, and SPERMATOGONIA: germ cells in a gonad that are still capable of mitotic replication and the production of more germ cells.

GONOTID, OOTID, SPERMATID: meiotically reduced haploid germ cells which may, but need not, be physiologically ripe gametes ready for fertilization.

GRAAFIAN OVARIAN FOLLICLE: a mature follicle with a single very large antrum (cavity) in the follicle layer surrounding the primary oocyte.

GRAY CRESCENT (GC): a light gray region on the dorsal side of fertilized amphibian eggs that marks the location of the dorsal surface and prospective chordamesoderm.

GREATER OMENTUM (DMS): the dorsal mesogaster of mammals that forms the wall of a large sac, the lumen of which is the lesser peritoneal cavity or omental bursa; the wall of this cavity is the greater omentum.

GULAR CAVITY: See Opercular Cavity.

GULAR FOLD (GF): the operculum-like covering of the external gill chamber in amphibian larvae.

HAPLOID: the reduced chromosome number characteristic for gametes of a species; half the normal diploid number of somatic cells.

HAPLOTENE: chromosomes with a single chromatid, or chromatin threads with a single chromonemata; i.e., chromatin of the G^1 phase of the cell cycle.

HATSCHEK'S PIT: an opening to the exterior in the roof of the oral hood of Amphioxus from the left first myocoele, a suggested (but improbable) homologue of the hypophysis of vertebrates.

HEAD COELOM (MYC'): the right first myocoele of Amphioxus.

HEAD FOLD (AMH): see Amniotic Folds.

HEAD FOLD (HF): See Body Folds.

HEAD MESODERM, HEAD MESENCHYME, or PRECHORDAL PLATE (HM): a small region of mesoderm that invaginates at the dorsal lip of the blastopore in advance of chordamesoderm; it gives rise to diverse head structures including eye muscles and parts of the cranium.

HEART PRIMORDIUM or CARDIAC MESODERM (HT): splanchnic lateral plate mesoderm of the ventral mesentery at levels of the esophagus and pharynx that gives rise to the heart.

HEAD PROCESS (HP): the anterior end of the embryonic chordamesoderm of the chick after posterior regression of Hensen's node begins at late primitive streak stages.

HEMATOBLASTS (HEM): embryonic primordial blood cells of the blood island and other tissues.

HEMIAZYGOUS VEIN (HAZ): in adult mammals this vein drains blood from the left side of the body wall to the azygous vein on the right side; it is derived from the embryonic left supracardinal vein of the metanephric kidney.

HENSEN'S NODE or PRIMITIVE KNOT (HN): the most anterior end of the primitive streak; the part corresponding with the prospective chordamesoderm region and homologous with the dorsal lip of the amphibian blastopore.

HEPATIC PORTAL VEIN (HP): the vein that carries blood from gut capillaries to liver capillaries, derived from the embryonic vitelline veins posterior to the liver.

HEPATIC VEINS (HV): paired vessels that drain blood from the liver capillary bed into the postcaval vein of amniotes, or directly into the sinus venosus in anamniotes; they are vessels derived from the embryonic vitelline veins anterior to the liver.

HERMAPHRODITE or HERMAPHRODITIC: the presence of a single individual of male and female reproductive competence.

HETEROGAMETES or HETEROGAMY: sexual reproduction by the union of haploid gametes that are morphologically differentiated into motile sperm and yolk-rich nonmotile eggs; the characteristic gametes of all multicellular animals and most higher plants.

HETEROGENIC INDUCTION: embryonic induction in which the inductor tissue causes a different kind of tissue to be formed by a reactor than the tissue type of the inductor itself.

HETEROPLASTIC: transplantations of embryonic tissues or organs between hosts and donors of different species of the same genus.

HETEROTOPIC: the transplantation of organs or tissues to abnormal locations in host embryos.

HIND BRAIN or RHOMBENCEPHALON (HB): the posterior primary brain vesicle; it will form the metencephalon and myelencephalon.

HIND LIMB BUD (HL): the primordium of the posterior paired appendages.

HINDGUT (HG): the region of the endodermal gut posterior to the colon; it is usually considered to be synonomous with the cloacal regions of the gut tube.

HOLOBLASTIC CLEAVAGE: cleavage with complete total separation of sister cells.

HOMOGENIC INDUCTION: embryonic inductions in which the inductor tissue causes more of its own kind of tissue to be formed by a reactor.

HOMOLOGY or HOMOLOGOUS STRUCTURES: structures in different species derived from the same embryonic primordium, exhibiting similarity in fundamental adult morphological plan, and derived paleontologically from the same ancestral beginnings; it is not essential for such structures to have similar functions.

HOMOPLASTIC: the transplantation of organs or tissues into normal positions in host embryos.

HORIZONTAL SEPTUM: the connective tissue sheath that separates epaxial and hypaxial components of myotomes in the frontal plane of the body axis.

HORNS OF THE SINUS VENOSUS, RIGHT and LEFT HORNS (SVR or SVL): the lateral extensions of the sinus venosis where they unite with the paired common cardinal and vitelline veins.

HUMERUS (HU): the skeletal element of the upper fore limb

HYALINE CYTOPLASM: transparent cytoplasm near the cell surface of eggs that is recruited to the sperm entrance point; it forms the fertilization cone and surrounds the sperm pronucleus after sperm entry and is incorporated into the first cleavage spindle.

HYALINE FUNNEL: an extension of hyaline cytoplasm beneath the sperm entry point into which the sperm pronucleus will migrate.

HYALINE PLASMA MEMBRANE: a transparent glycoprotein membrane formed around echinoderm eggs at fertilization that functions in holding blastomeres together until blastula stages.

HYOID ARCH (VS2): the second visceral arch.

HYOMANDIBULAR CARTILAGE (VC 2): the epibranchial or dorsal end of the second visceral cartilage; the homologue of the stapes and columella of vertebrate middle ears.

HYPAXIAL MYOTOME (HY): the major ventral subdivision of segmental myotomes below the horizontal septum, the primordium of striated muscles of the belly wall and the extrinsic depressor muscles of the limbs.

HYPOBLAST (HPB): the deep layer of cells in the early gastrula at the two layer stage; the part of the blastoderm that will form endoderm and may also form mesoderm, as is the case with protochordates, the frog, and many other anamniotes.

HYPOBRANCHIAL VISCERAL CARTILAGE (HBCL 1-6): the ventral half of the paired visceral cartilages of the gill arches.

HYPOGASTRIC ARTERIES (IIA or ALA): See Internal Iliac Arteries.

HYPOGASTRIC VEINS: See Internal Iliac Veins.

HYPOGLOSSAL MUSCLES (OC or OCM): the occipital myotomes that form tongue muscles and are innervated by the twelfth cranial nerve.

HYPOGLOSSAL NERVE (12 or XII); the twelfth cranial nerve (pure somatic motor) to tongue muscles derived from occipital myotomes.

HYPOMERE MESODERM (LM): a synonym for the lateral plate mesoderm, i.e., the somatic and splanchnic mesodermal linings of the coelomic cavities.

HYPOPHYSEAL ECTODERM (HY): See Hypophysis.

HYPOPHYSEAL FORAMEN or FENESTRA (HF): the space between the paired trabecular cartilages into which the infundibulum of the diencephalon projects ventrally.

HYPOPHYSIS or HYOPOPHYSEAL ECTODERM (HY): an evagination from the roof of the stomodaeum that underlies the infundibulum and forms the anterior and intermediate lobes of the pituitary gland (adenohypophysis); a derivative of Rathke's pocket.

IMPLANTATION: the attachment of a mammalian embryo to the maternal uterine wall with the development of functional placental associations between fetal and maternal bloods.

INDUCTION SERIES: chains of inductions that follow one after another, in which the reactor of one stage becomes the inductor of the next; in this way complex organs of multiple germ layer derivation can be programmed.

INFERIOR MESENTERIC ARTERY (IMA): the major unpaired artery to the colon.

INFERIOR VENA CAVA or POSTCAVAL VEIN (PV): the major posterior vein of tetrapod vertebrates for the return of blood from the trunk and tail to the heart.

INFUNDIBULUM (IN): the ventral evagination from the floor of the diencephalon that will contribute the posterior lobe of the pituitary gland.

INNER CELL MASS: this name is usually given to the homologue of the embryonic disc in primate development; however, it forms more than the embryonic discs of lower mammals.

INNOMINATE ARTERY (INA): an unpaired artery from the systemic arch of the aorta that transports blood to the right subclavian and right common carotid arteries; it is homologous with the right side of the ventral aorta between the right fourth and sixth aortic arches.

INNOMINATE VEINS (INR & INL): in mammals the paired veins carry blood from the right and left subclavians, and internal and external jugular veins to the precaval vein; the right innominate is derived from the embryonic right anterior cardinal vein; the left innominate is a newly formed vein.

INTEGRATION AND COORDINATION: the period in late embryonic and fetal development when hormones and nervous controls regulate the growth and final functional specialization of tissues and organs.

INTERATRIAL SEPTUM, PRIMARY and SECONDARY (IA 1 & 2): ingrowths of the heart wall that partially subdivide the atrium into right and left atrial chambers.

INTERMEDIATE MESODERM (MM): a synonym for mesomere or nephrogenic cord; the mesodermal primordium of the pro-, meso-, and metanephric kidneys and their ducts.

INTERNAL CAROTID ARTERIES (IC): the primary embryonic sources of blood to the brain and interior of the skull; the anterior continuations of the paired dorsal aortae.

INTERNAL ILIAC or HYPOGASTRIC ARTERIES (IIA or ALA): paired branches of the common iliac arteries to the urinary bladder and genital region; they are remnants in the adult mammal of embryonic allantoic (umbilical) arteries.

INTERNAL ILIAC or HYPOGASTRIC VEINS: paired branches of the common iliac veins from the bladder and genital region.

INTERNAL JUGULAR VEIN (AC): paired veins that drain blood from the brain to the innominate veins; derived from the embryonic anterior cardinal veins.

INTERNAL MAMMARY ARTERIES (IMA): paired arteries of the lateral body wall that provide an alternative route than the aorta for blood going to intersegmental arteries of the thoracic and lumbar regions.

INTERNAL SPREADING: the radial migration of invaginated hypomere mesoderm and endoderm cells away from the blastopore or primitive streak to more distant regions of the embryo.

INTERSCLEROTOME (ISC): the anterior half of a segmental sclerotome, but the posterior half of a definitive vertebra.

INTERSEGMENTAL VESSELS, ARTERIES (ISA) or VEINS (ISV): the metameric branches of the dorsal aorta or anterior and posterior cardinal veins; they pass between somites or myotomes.

INTERVENTRICULAR FORAMEN (IVF): the embryonic opening between right and left ventricular cavities; in birds and mammals it is normally closed completely by the interventricular septum.

INTERVENTRICULAR SEPTUM (IV): the ingrowth of the heart wall that divides the lumen into right and left ventricles.

INTESTINE (IT): the midgut region of the gut tube exclusive of the stomach.

INVAGINATION: a morphodynamic cell movement that results in cells moving to the interior as a continuous layer.

INVERSION: a morphodynamic cell movement in frog gastrulation in which surface endoderm moves toward the dorsal lip of the blastopore and then reverses its direction as it goes to the interior and forms the floor of the archenteron cavity.

IN VITRO: the cultivation of embryonic tissues or primordia in nonliving media.

IN VIVO: the cultivation of embryonic tissues or primordia in living embryos or adults.

INVOLUTION: a morphodynamic cell movement of mesoderm around the blastopore lips as a belt passing over a pulley.

IRIS (IR): the circular pigmented membrane of the eye in front of the lens; it is derived from the wall of the optic cup.

ISOGAMETES or ISOGAMY: sexual reproduction by the union of gametes that are not morphologically differentiated into eggs and sperm cells, usually they are motile and distinguishable only as physiological mating types; found in some lower algae.

ISOLECITHAL: eggs with yolk uniformly distributed in the cytoplasm of the ovum.

ISTHMUS (IS): a narrow region of the brain stem between the metencephalon and mesencephalon near the region of the pontine flexure.

JUGULAR GANGLION (10j or Xj): the dorsal ganglion of the tenth cranial nerve.

KARYOGAMY: the fusion of haploid egg and sperm pronuclei during fertilization.

LABILE: the ability of cells or embryonic tissues to regulate or form structures appropriate to new regions to which they are experimentally transferred; in contrast with the quality of self-differentiation.

LABILE DETERMINATION: this describes the nature of a cell at some midpoint in its determination, when there is some reduction in general competence to regulate but the cell or tissue is not yet completely determined or able to self-differentiate as a specific single type of tissue or organ.

LABIOSCROTAL SWELLINGS: See Genital Swellings.

LAMINA TERMINALIS (LT): a thin plate of non-nervous tissue in the anterior median wall of the diencephalon; it represents the closing scar of the anterior neuropore.

LAMPBRUSH CHROMOSOMES: chromosomes of primary gonocytes with visible loops projecting laterally from synapsed (or chiasmata linked) chromosomes; the lateral loops are interpreted as local regions of gene activity.

LARGE INTESTINE (ITP): the major part of the intestine posterior to the attachment of the embryonic yolk stalk; the colon.

LARVA or LARVAL STAGES: a phase in the life cycle during which the individual is an independent, freeliving or parasitic organism, but it is fundamentally different from the adult in structure and/or in ecological life habit, i.e., amphibian tadpoles.

LARYNGOTRACHEAL GROOVE (LTG): a ventral trough-like evagination from the foregut floor posterior to the pharynx; it is the primordium of the trachea and lung buds.

LATERAL (R or L): toward the right or left side of the body.

LATERAL ABDOMINAL VEINS (LA): paired veins of the lateral body wall of many anamniotes and homologous with the allantoic veins of amniotes.

LATERAL FOLDS (AML) OR LATERAL BODY FOLDS (LBF): See Amniotic Folds, Body Folds.

LATERAL LINE PLACODES or LATERALIS PLACODES (LL): thickenings in the epidermis that are primordia of the lateral line organs of fish and amphibians.

LATERAL LIP OF THE BLASTOPORE (LLB): the right and left sides of the blastopore; in chordates they are usually represented by cells of somite and/or lateral plate mesoderm. See also Blastopore Lips.

LATERAL MESOCARDIUM (LMC): the contact and fusion points between the lateral body wall and lateral walls of the sinus venosus through which the common cardinal veins enter the sinus venosus; major component of the transverse septum and mammalian diaphragm.

LATERAL PLATE MESODERM (LM): a synonym for hypomere mesoderm; the primordium for the somatic and splanchnic mesodermal linings of the coelom, gut, and body walls.

LATERAL VENTRICLES (V 1): the neurocoeles of the right and left lobes of the telencephalon; the cavities of the cerebral hemispheres.

LEFT ATRIUM (LA): the pulmonary side of the atrium that receives blood from the lungs after the embryonic atrium is divided by interatrial septa.

LEFT VENTRICLE (LV): the systemic side of the ventricle that pumps blood to the general body by way of the systemic trunk and aorta after the embryonic ventricle is divided by the interventricular septum.

LEIDIG CELLS: testosterone-secreting interstitial cells of the testis.

LENS PLACODE or VESICLE (LE): the paired epidermal primordia of the lens of the eyes.

LUTEINIZING HORMONE: a pituitary hormone that stimulates ovulation of eggs, the conversion of follicle cells into corpus luteum in the ovary, and the release of spermatozoa from Sertoli cells of the testis.

LIVER (LI): an endodermal evagination of the gut wall into the ventral mesogaster beneath the stomach.

LIVER SINUSOIDS (LS): the anastomosing blood channels in the liver which are derived from the embryonic vitelline veins.

LUMBAR: a term referring to the lower back or abdominal levels of the body between the thoracic and sacral regions.

LUMBOSACRAL FLEXURE (LSF): a ventral bend in the posterior body axis at the level of the hind limb buds.

LUMEN: the cavity of a hollow organ or body structure.

LUNG BUDS (LB): paired posterior outgrowths from the laryngotracheal groove; the primordia of the lungs and homologous with swim bladders of fish.

MACROMERE (MAC): large cells produced by unequal cleavage divisions of cells.

MADREPORIC PORE: an opening to the exterior from the coelomic vesicles of early echinoderm embryos.

MAMMARY RIDGE (MR): paired elongated epidermal placodes along the lateral body wall between the fore and hind limb buds; the primordia of mammary glands.

MANDIBULAR ARCH (VS 1): the first visceral arch; its maxillary and mandibular processes form the upper and lower jaws respectively.

MANDIBULAR RAMUS (5m or Vm): the branch of the fifth cranial nerve to the lower jaw.

MANTLE CYTOPLASM (MC): RNA-rich cortical cytoplasm distributed in the animal half of the amphibian egg prior to fertilization; after fertilization it moves to the regions of the egg that coincide with the gray crescent and prospective mesoderm.

MANTLE LAYER (MAN): the gray matter layer of the spinal cord and brain wall possessing nucleated cell bodies of developing neuroblasts of the central nervous system.

MARGIN OF THE FREE HEAD (MFH): the edge of the head of the early chick embryo after the head fold has undercut it from the area pellucida and proamnion.

MARGINAL LAYER (MAR): the white matter layer of the brain and spinal cord composed primarily of fibrous parts of developing neuroblasts of the central nervous system.

MASSETER MUSCLES (MAS): the major epibranchial visceral muscles of the first visceral arch; the major muscles of the lower jaw.

MAXILLARY RAMUS (5x or Vx): the branch of the fifth cranial nerve to the upper jaw.

MEATUS (VG 1): See External Auditory Canal.

MECKEL'S CARTILAGE (VC 1): the hypobranchial component of the first visceral cartilage, the primary cartilage of the lower jaw; the articular end of this cartilage is the primordium of the malleus of the middle ear in mammals.

MECKEL'S DIVERTICULUM: a vestigial remnant of the embryonic yolk stalk that abnormally may remain as a permanent appendage of the intestine in a small percentage of mammals.

MEDIAN or MEDIAL: a term meaning toward or near the middle of the body.

MEDIAN MEDIASTINAL SEPTUM (MMS): the embryonic ventral mesoesophagus at the level of the dorsal mesocardium, which supports the pericardial sac between the two pleural cavities.

MEDIAN SAGITTAL PLANE: the plane of bilateral symmetry, parallel with the AP and DV axes.

MEDULLARY CORDS: primordia of seminiferous tubules of the testis in the embryonic genital ridge and ovotestis.

MEDULLARY INTERSTITIAL CELLS OF THE TESTIS: cells, probably derived from the mesonephros, between the seminiferous tubules; they give rise to the testosterone-secreting endocrine cells of Leidig in the adult testis.

MEGALECITHAL and MESOLECITHAL EGGS: eggs with relatively large amounts of stored yolk as in bird or reptile and frog eggs respectively.

MEIOTIC DIVISION: See First Maturation, Second Maturation divisions.

MEROBLASTIC CLEAVAGE: cleavage with incomplete separation of sister cells so that parts of the blastoderm is syncytial for a time as is the case in eggs of birds.

MEROGONY: the development of egg fragments; usually this refers to experimental treatments that remove the egg nucleus and then inseminate or activate the remaining cytoplasm.

MESECTODERM: skeletal and other supporting tissues of neural crest origin; a synonym for ectomesenchyme.

MESENCEPHALON (MB): See Midbrain.

MESENCHYME: wandering amoeboid cells possessing either phagocytic, connective tissue, or undifferentiated cell characters.

MESENDODERM: general mesoderm derived from cells that invaginate at the blastopore or primitive streak, or move to the interior by other means during chordate gastrulation; typical general mesoderm as distinct from mesectoderm.

MESOBLAST (MES): an alternative name for mesendoderm, only rarely used in current embryology.

MESOCOELE: the primitive cavity formed in the mesomere or intermediate mesoderm; the lumen for primary pronephric and mesonephric kidney tubules; a synonym for primary nephrocoeles including their nephrostomal openings into the coelom; this is also a term used as a synonym for the aqueduct of Sylvius or lumen of the mesencephalon.

MESODERM (MES): the middle of the three primary germ layers.

MESOLECITHAL EGGS: See Megalecithal Eggs.

MESOMERE (MM): the nephrotome, intermediate mesoderm, and primordium of the kidneys.

MESONEPHRIC ARTERIES (MNA): branches of intersegmental arteries to the mesonephros; they are primordia of the spermatic and ovarian arteries of adult mammals.

MESONEPHRIC DUCT (MND): the duct of the mesonephros, derived from the pronephric duct, Wolffian, or archinephric duct.

MESONEPHROS and MESONEPHRIC TUBULES (MN): the second vertebrate kidney to appear in the development of vertebrates; the tubules have a Bowman's capsule, internal glomerulus, and often a nephrostome opening into the coelom.

MESOPLASM (MEP): the cytoplasmic region of an egg or early cleavage cell that is determined to form embryonic mesoderm.

METAMERISM or METAMERIC STRUCTURES: these are similar structures in the same individual that are repeated at periodic intervals along one or more of the body axes; metameric structures are said to be serially homologous, i.e., somites, vertebrae, etc.

METAMORPHOSIS: the termination of larval or juvenile stages in the life cycle by hormonally induced major transformations in a previous morphological form.

METANEPHRIC DUCT (MTD): the ureter or duct from the metanephric kidney of adult amniotes to the urinary bladder; derived as an outgrowth from the pronephric or archinephric duct.

METANEPHRIC MESOMERE (MTC): the primordium of the renal cortex of adult amniotes; it is derived from the posterior segments of the nephrogenic cord.

METANEPHRIC RENAL MEDULLA (MTM): the collecting ducts and renal pelvis of the adult amniote kidney; it is derived from the ureteric bud of the pronephric duct.

METAPHASE: the second period in mitosis during which diplotene chromosomes move to the equator of the spindle prior to division.

METAPLASTIC or METAPLASIA: the transformation of a differentiated cell into another kind of specialized tissue type.

METAPLEURAL FOLDS: ventrolateral extensions of the body wall in Amphioxus.

METENCEPHALON (MT): the fourth of five major brain regions; located between the mesencephalon and myelencephalon; the cerebellar region of the adult brain.

MICROMERE (MIC): small cells produced by unequal cleavage cell divisions.

MIDBRAIN or MESENCEPHALON (MB): the third of five major brain regions; it is located between the diencephalon and metencephalon; it is the region of the optic tectum and tegmentum of the adult brain.

MIDDLE EAR CAVITY (P 1): the cavity of the middle ear across which the auditory bone or bones are suspended between the ear drum and inner ear; the cavity is derived from the first embryonic pharyngeal pouch.

MIDGUT or MIDGUT LUMEN (MG): the part of the endodermal gut including the stomach, small intestine, and colon and their endodermal derivatives.

MITOSIS: normal cell division in which the chromosomes of the mother cell are duplicated and then equally separated to two daughter cells.

MITRAL VALVE (AVL): See Bicuspid Valve.

MONAD: a fully reduced haplotene chromosome of a germ cell at the gonotid or gamete stage.

MORPHODYNAMIC or MORPHOGENETIC MOVEMENTS: cell movements during development that are responsible for the molding of organ primordia during embryonic development.

MORPHOGENESIS: the processes in embryonic development that are responsible for the formation and differentiation of body structures.

MORPHOLOGY: the scientific study of structure and the relationships among parts of organisms.

MORULA: late cleavage stages of 32 or more cells; a little used term of no great value.

MOSAIC CLEAVAGE: types of cleavage divisions in which many different cytoplasmic determinants are isolated into specific blastomeres, each of which will be determined to form a particular kind of tissue or organ and no other; i.e., determinate cleavage.

MOTOR NUCLEUS, SOMATIC or VISCERAL MOTOR NUCLEI: aggregations of somatic and visceral preganglionic nerve cell bodies in the lateral walls of the basal plate of the neural tube; the only neurons of the peripheral nervous system that are derived from the neural tube (as opposed to the neural crest) and with cell bodies in the neural tube.

MOUTH (MO): the external opening to the stomodaeum or oral cavity.

MUCOSA: any glandular epithelium that secretes mucous, as in many regions of the gut tube.

MÜLLERIAN DUCT: the female reproductive tube, the embryonic primordium for the oviducts and uterus; it is induced by the pronephric duct from the parietal peritoneum.

MUSCLE–MESENCHYME MESOPLASM: the segregated region in tunicate and Amphioxus eggs that is determined and will give rise to all mesoderm except chordamesoderm.

MUSCULAR DIAPHRAGM (MD): the outer edge of the mammalian diaphragm that contains cervical myotomes used in breathing; the part of the diaphragm distinct from the central tendinous diaphragm that lacks muscle.

MYELENCEPHALON (MY): the fifth and most posterior of five major brain regions; the part located between the metencephalon and anterior spinal cord; the medulla region of the adult brain.

MYOCARDIUM or EPIMYOCARDIUM (MC or EMY): the heavy muscular wall of the heart derived from the ventral mesentery of the esophagus which surrounds the endocardial heart tubes.

MYOCOELE (MYC): the lumen of mesodermal somites; primitively they are derived from the archenteron lumen but in most vertebrates the cavity arises directly by splitting of the somite mass; the cavity will degenerate completely shortly after forming.

MYOTOME (MYO): the derivative of somite mesoderm that will give rise to the axial segmental muscles of the body wall.

MYOTUBE: very primitive chordate muscle cells in which the nucleus is located centrally and the myofibrils form a ring under the cell surface.

NARIS, EXTERNAL and INTERNAL (EN & INA): openings of the nasal canals to the exterior or into the mouth cavity; the nostrils; the plural of naris is nares.

NASAL CANAL (NAC): the channel connecting the internal and external nares or nostrils; the canal is a derivative of the embryonic nasal placode and nasal pit.

NASAL CAPSULE (NCC): the cartilage of neural crest origin that encloses much of the nasal canals.

NASAL PLACODE and NASAL PIT (NP): epidermal invaginations at the anterior end of the head which will form the lining of the nasal canals and the olfactory nerves.

NASOHYPOPHYSEAL PIT: a single median dorsal invagination of epidermis in the Ammocetes larva that is probably homologous with both the nasal pits and hypophysis (adenohypophyseal glands) of higher vertebrates.

NEOPLASM or NEOPLASIA: abnormal growth or differentiation of normal or abnormal cell types; or unregulated cell growth usually related to tumors or cancer.

NEOTENIC or NEOTENY: the presence of sexual maturity in individuals that have not undergone metamorphosis from a typical larval or juvenile body form; an example is the presence of sexually mature salamanders that have not metamorphosed and lost their external gills.

NEPHROSTOME (NES): the opening of a pronephric or mesonephric tubule into the coelom; in functional tubules the opening is strongly ciliated.

NEURAL ARCH (NA): a dorsal outgrowth of sclerotome that grows over the neural tube; it will unite with the neural spine and vertebral centrum to form the vertebral neural canal.

NEURAL CANAL: the cavity in a vertebra bounded by the vertebral centrum, neural arches, and neural spine; a term also used as a synonym for the neurocoele of the neural tube.

NEURAL CREST (NC): the cells derived from the neural folds; they give rise to the ganglionic crest and mesectoderm.

NEURAL ECTODERM (NE): the embryonic primordium of the neural tube and all its derivatives.

NEURAL FOLDS (NF): the cells that make up the bordering edge of the neural plate; they will form the embryonic neural crest.

NEURAL KEEL: a solid, delaminated band of dorsal ectoderm from which the neural tube and neural crest will later be formed in the development of cyclostomes and bony fish; it is homologous with the neural plate of other vertebrates.

NEURAL PLATE (NP): the thickened dorsal placode of ectoderm that will form the neural tube.

NEURAL TUBE (NT): the primordium of the dorsal tubular brain and spinal cord of chordates.

NEURENTERIC CANAL (NEC): an opening that briefly connects the lumen of the neural tube with the cavity of the archenteron following neurulation in some protochordates and chordates; it normally closes without a trace during later development of the tail bud.

NEUROCOELE (NCL): the lumen of the dorsal tubular nerve cord at all levels of the brain and spinal cord.

NEUROMERE (NM): metameric swellings in the wall of the embryonic brain; they are most characteristic of the myelencephalon floor.

NODOS GANGLION (10n or Xn): the ventral ganglion of the tenth cranial nerve.

NOTOCHORD (NO): the chorda dorsalis; a supporting rod in the AP body axis between the neural tube and gut which is the diagnostic character for the Phylum Chordata; it is derived from chordamesoderm.

OCCIPITAL (OC): a term that refers to the posterior cranial region of the skull that is derived from the first four (occipital) somites; it also refers to associated segmental nerves, myotomes, and blood vessels of the region.

OCELLUS: unicellular or simple multicellular eye spots of invertebrates and protochordates.

OCULOMOTOR NERVE (3 or III): the third cranial nerve (pure somatic motor) to the inferior oblique, superior-, inferior-, and internal rectus eye muscles.

OLFACTORY NERVE (1 or I): the first cranial nerve (special somatic sensory) an epidermal nerve derived from the nasal epithelium.

OLIGOLECITHAL: eggs with very little yolk, as with echinoderm and mammal eggs.

OMENTAL BURSA (OB): the lesser peritoneal cavity; it is enclosed by the greater omentum, the caval mesentery, liver capsule, and wall of the liver.

ONTOGENY or ONTOGENIC CYCLE: the life history of individuals of a species from fertilization through sexual maturity, senescence, and natural death.

OOGENESIS: the development of eggs in the ovary.

OOGONIA: See Gonial Cells.

OOPLASMIC SEGREGATION: the localization in a particular part of the egg, or cleavage cell, of cytoplasmic determinants for specific tissue or organ differentiation; i.e., it occurs with cleavages that result in the embryonic determination of particular blastomeres for special cell fates; it is the underlying basis for determinate cleavage.

OOTID: See Gonotid.

OPEN MIDGUT (MG): the part of the embryonic gut that has not yet acquired a ventral floor.

OPERCULAR or GULAR CAVITY (OPC): the cavity lined by epidermal ectoderm formed by the overgrowth of external gills by opercular or gular folds in fish and amphibia in tadpoles.

OPTHALMIC or PROFUNDUS RAMUS (5o or Vo): a sensory branch of the fifth cranial nerve to the eye socket and snout.

OPTIC or SCLEROTIC CAPSULE: the embryonic primordium for the sclera of the eye; it is of complex derivation from neural crest and head mesoderm.

OPTIC CUP (OP): the primordia of the retina; they are paired evaginations from the lateral wall of the fore brain.

OPTIC FORAMEN (OPF): an opening in the trabecular cartilages through which the optic nerve passes from the eye to the brain floor.

OPTIC LOBES (OPT): See Optic Tectum.

OPTIC or ORBITAL MUSCLE MASS (OPM): the primordium of the six external muscles of the eye; although they are derived from head mesoderm in tetrapods, they are homologous with the preotic myotomes of primitive fish.

OPTIC NERVE (2 or II): the second cranial nerve (special somatic sensory) derived from neural ectoderm of the sensory retina of the optic cup.

OPTIC RECESS (OPR): an expansion of the lumen of the diencephalon that, in the adult, marks the point at which the optic vesicles evaginated from the brain wall.

OPTIC STALK (OPS): the narrow connection between the optic cup and brain wall; it will later contribute to the sheath of the optic nerve.

OPTIC TECTUM or OPTIC LOBES (OPT): the roof of the mesencephalon; the visual centers of the brain.

OPTIC VESICLES (OP): paired lateral evaginations from the fore brain that are primordia of the optic cups and retinal parts of the eye.

ORAL CAVITY (OR): the mouth cavity; in mammals it is the ventral half of the ectodermally lined stomodaeum after the latter is divided into oral and nasal cavities by the palate.

ORAL CIRRI: small sensory filaments surrounding the oral cavity of Amphioxus.

ORAL FILAMENTS or TENTACLES: rings of filtering tentacles at the entrance to the pharynx of tunicates, Amphioxus, and the Ammocetes larva.

ORAL HOOD: a muscular extension of the sides of the head over the mouth cavity in Amphioxus and the Ammocetes larva.

ORAL MEMBRANE (OM): the embryonic closing plate between the stomodaeum and foregut; it consists of a layer of stomodaeal ectoderm applied to a layer of pharyngeal endoderm.

ORGANOGENESIS: the embryonic period during which organ primordia become determined and transformed into recognizable organs.

ORTHOTOPIC: the transplantation of an embryonic tissue or organ into its normal location on a host embryo.

OTIC CAPSULE (OTC): the cartilage or bone that surrounds the inner ear.

OTIC PLACODE, OTIC VESICLE, or OTOCYST (OT): the epidermal invaginations that are the embryonic primordia of the inner ear.

OVARY: the female sex gland; in chordates it represents the cortical parts of the ovotestis.

OVOTESTIS: the vertebrate gonad during early development possesses both male and female structural elements at the same time; i.e., a period in development that exhibits hermaphroditic bipotentiality for sexual differentiation.

OVULATION: the release of fully developed eggs from ovarian follicles; in vertebrates it is stimulated by the pituitary leuteinizing hormone, LH.

OVUM: a morphologically and physiologically ripe female sex cell.

PAEDOGENESIS: asexual reproduction by sexually immature larval or juvenile stages in the life cycle of a species; it is a common occurrence in many parasitic species.

PAEDOMORPHOSIS or PAEDOMORPHIC CHARACTERS: the aspects of embryonic development in which typical embryonic or juvenile conditions are retained into adult stages; i.e., the retention of atavistic and/or larval conditions into sexual maturity.

PALATINE PROCESS (PAP): the paired internal evaginations from the upper jaws which will divide the stomodaeum of mammals into dorsal nasal (paired) and ventral oral cavities.

PALEOMORPHOSIS or PALEOMORPHIC STRUCTURE: any structural evidence in embryonic development revealing the primitive ancestral manner of development; i.e., development of aortic arches or pronephric nephrostomes etc.

PANCREAS, DORSAL and VENTRAL (PAN): endodermal digestive glands derived from a ventral and dorsal evagination from the anterior end of the small intestine; they use the common bile duct to carry secretions to the gut lumen.

PARABIOSIS: the natural or experimental production of two individuals that are fused together and share a common circulatory system; i.e., Siamese twinning.

PARACHORDAL CARTILAGE (PAR): paired cartilage bars under the midbrain and hindbrain that lie parallel with, and lateral to, the anterior end of the notochord.

POLYTENE CHROMOSOMES: chromosomes composed of many times the normal diploid number of chromonemata (chromatin threads); i.e., the giant salivary gland chromosomes of Chironomous or Drosophila.

POSTANAL GUT (TG): See Tailbud.

POSTERIOR POLE (P): the tail end of the body.

POSTGANGLIONIC VISCERAL MOTOR NEURON: a cell of the autonomic nervous system with its cell body in sympathetic ganglia or in ganglionic centers of visceral endorgans.

PONTINE FLEXURE (PF): the dorsal bend in the brain stem at the level of the metencephalon between the cephalic and cervical flexures of the brain stem.

PORTAL VEINS: veins that originate and terminate in capillary networks; i.e., the hepatic portal and renal portal veins, etc.

POSTCAVAL VEIN or INFERIOR VENA CAVA (PV): the major vein of tetrapod vertebrates for the return of blood to the heart from posterior body levels; it is derived from many different embryonic venous channels.

POSTERIOR CARDINAL VEINS (PCV): the major paired systemic veins of the early embryo posterior to the heart; they carry intersegmental blood from body tissues and the pronephric kidney to the common cardinal veins.

POSTERIOR CHOROID PLEXUS (PCP): the thin roof plate of the myelencephalon.

PRECAVA or SUPERIOR VENA CAVA (CCR): in mammals the unpaired precava carries blood from the paired innominate veins to the sinuatrial valve of the right atrium; it is derived from the embryonic right common cardinal vein; when two precaval veins are present in other tetrapods, they represent right and left common cardinals.

PRECHORDAL PLATE (HM): a synonym for head mesoderm or head mesenchyme.

PREGANGLIONIC VISCERAL MOTOR NEURON: a cell of the autonomic nervous system with its cell body in the visceral motor nucleus of the spinal cord.

PREORAL GUT or SEESSEL'S POCKET: a small anterior diverticulum of the foregut anterior to the oral membrane and mouth; it has no special significance and degenerates completely.

PRIMARY EGG POLARITY: this is associated with qualitative and quantitative differences in the cytoplasm of the egg along the animal–vegetal axis.

PRIMARY GONOCYTE, PRIMARY OOCYTE, PRIMARY SPERMATOCYTE: germ cells in the growth phase preceding meiotic reduction of egg and sperm cells.

PRIMARY ORGANIZER: the chordamesoderm or roof of the archenteron in amphibia; the primary inductor of the embryonic axis and central nervous system.

PRIMARY OVARIAN FOLLICLE: a follicle with a single layer of follicle cells surrounding the primary oocyte.

PRIMITIVE FOLD (PF): the parallel lateral edges of the primitive streak; they are homologous with the lateral lips of the amphibian blastopore.

PRIMITIVE GROOVE (PG): the slight depression in the midline of the primitive streak between the paired primitive folds; it is homologous with the blastopore of amphibia.

PRIMITIVE KNOT (PN): the most anterior end of the primitive streak in mammals; it corresponds with the prospective chordamesoderm region of the blastoderm and is homologous with the dorsal lip of the amphibian blastopore and with Hensen's node of bird gastrulae.

PRIMITIVE PIT (PP): a slight depression (in the chick) or an actual opening to the interior archenteron cavity (in mammals) at the anterior end of the primitive groove; the pit is homologous with the blastopore and/or the neurenteric canal of lower chordates.

PRIMITIVE STREAK (PS): a linear structure in the early blastoderm of amniotes that is a center for involution and proliferation of mesoderm and endoderm; it is homologous with the blastopore and blastopore lips of amphibian gastrulae.

PRIMORDIAL GERM CELLS (GP): early embryonic stem cells with embryonic competence to form cells of the reproductive germ line in the adult gonads.

PRIMORDIAL GERM PLASM (GP): determined cytoplasm in some eggs that determines the cells that acquire it to form primordial germ cells.

PRIMORDIUM: an embryonic region that is determined to form a particular organ or part of an organ; an organ rudiment or *anlage.*

PRISM LARVA: a pre-feeding stage in the development of the sea urchin between the gastrula and pluteus stages of development.

PROAMNION (PA): the part of the chick blastoderm anterior to the embryonal area that lacks invaginated mesoderm; i.e., a part of the blastoderm containing only extraembryonic ectoderm and endoderm.

PROCTODAEUM (PD): a short epidermal invagination in the region of the closing blastopore that will open into the cloacal region of the hindgut and form the anus.

PROGESTERONE: the hormone produced by the corpus luteum of the ovary that stimulates the uterine wall to final readiness for implantation of the embryo, and for maintenance of placental connections with the uterine wall during gestation.

PROLIFERATION: the rapid mitotic replication of cells, as in blood islands or the primitive streak of early embryos.

PRONEPHRIC DUCT (PND): the collecting duct of the pronephric tubules formed by posterior fusion of pronephric tubules and the growth of the duct back to the cloaca, synonomous with the archinephric and Wolffian ducts.

PRONEPHRIC TUBULES (PNT): primitive kidney tubules with nephrostomes opening into the coelom but lacking a Bowman's capsule and internal glomerulus; they are functional in embryos or larvae of fish and amphibia, but degenerate in all adult vertebrates.

PRONUCLEUS (PNN): a reduced maternal or paternal haploid nucleus in a fertilized egg before karyogamy unites them into a diploid zygote nucleus.

PROPHASE: the first period of mitosis, which involves the condensation of diffuse chromatin threads into diplotene chromosomes.

PROSENCEPHALON (FB or FG): the forebrain from which the telencephalon and diencephalon are derived. See also Forebrain.

PROSTATE GLAND: an accessory male sex gland derived from endoderm of the urogenital sinus; it secretes most of the alkaline sperm-activating fluid of semen.

PROSTATIC UTRICLE: a small remnant of the base of the Müllerian (female) duct imbedded in the prostate gland of the male; also called the utriculus masculinum.

PROTONEPHRIDIA: excretory organs of Amphioxus consisting of clusters of single celled flame cells called solenocytes; not to be confused with pronephric tubules which are multicellular complex organs.

PROXIMAL: a term referring to the end of a structure nearest the body midline.

PSEUDOREDUCTION: the period during primary gonocyte development in which homologous chromosomes are synapsed and therefore superficially appear to be haploid; i.e., the zygotene period in the first meiotic prophase.

PSEUDOSEGMENTS: simple metameric clusters of myotube muscle cells in the tunicate tadpole tail.

PULMONARY ARCHES (A 6 or AA 6): the paired sixth aortic arches; they are partly included in the base of the adult pulmonary arteries.

PULMONARY ARTERIES (PA): paired arteries arising from the sixth aortic arches and going to the lung buds; they are homologous with the pneumatic arteries to swim bladders in fish.

PULMONARY TRUNK (PT): the arterial trunk that carries blood from the right ventricle to the pulmonary arteries; it is derived by splitting from the embryonic conus arteriosus.

PULMONARY VEINS (PLV): vessels carrying blood from the lungs to the left atrium; they are derived by direct outgrowth from the atrium wall into the median mediastinal septum and lungs.

QUADRATE CARTILAGE (VC 1'): the articular end of the epibranchial cartilage of the first visceral arch; the articular part of the upper jaw skeleton and homologous with the incus of the middle ear bones in mammals.

RAMUS COMMUNICANS (RC): the connecting trunks of pre- and postganglionic visceral motor nerve fibers that connect autonomic ganglia with the ventral roots of spinal nerves.

RATHKE'S POUCH or POCKET (RP): a dorsal evagination from the stomodaeal ectoderm that will form the anterior and intermediate lobes of the pituitary gland; the hypophysis or hypophyseal duct.

RECAPITULATION THEORY: the concept by Ernst Haeckel suggesting that the embryonic stages of animals correspond with adult stages passed through in the evolution of the species; i.e., ontogeny recapitulates phylogeny; a theory mainly of historical value.

RECEPTOR CELL: a sensory receptor cell of any sense organ.

RECTUM (REC): the dorsal half of the embryonic cloaca cut off by the urorectal septum in mammals; the terminal end of the gut tube posterior to the coelom in mammals.

REGENERATION: the replacement of differentiated organs that are lost or removed by accident or experimentation; a more extensive replacement of lost parts than simple wound healing.

REGULATION or REGULATORY: the ability of embryonic parts to form more than the structures that they would form in normal development; it is usually tested by isolating, transplanting, or removing parts of an embryo.

REICHERT'S MEMBRANE: a special extraembryonic membrane in rodents consisting of fusion between serosa and yolk sac without yolk sac placenta formation.

RENAL ARTERIES (RA): branches of intersegmental arteries to the kidneys; usually the metanephric kidney is implied.

RENAL CORPUSCLE: the filtration unit of mesonephric and metanephric kidney tubules including a Bowman's capsule and its internal glomerulus.

RENAL PORTAL VEINS (RPV): the short segment of the embryonic posterior cardinal veins between the common iliac veins and the posterior end of the mesonephros; they carry blood from the hind limb and tail capillaries to capillaries of the mesonephros in amniote embryos and anamniote adults.

RENAL VEINS (RV): veins draining blood from the metanephros into the postcaval vein; they are derived from the embryonic supracardinal veins.

RETEOVARIAL SAC: a cavity between the developing genital ridge and the mesonephric kidney; it will form the ovarial sac of the adult ovary and the canal system of the rete testis of the adult testis.

RHOMBENCEPHALON (HV): the hind brain from which the metencephalon and myelencephaon will develop. See also Hind Brain.

RIGHT ATRIUM (RA): the systemic side of the atrium; it receives blood from precaval, postcaval, and coronary veins.

RIGHT VENTRICLE (RA): the systemic side of the ventricle; it pumps blood out through the systemic trunk to the aorta.

ROD & CONE LAYER OF THE RETINA (RTS): See Sensory Epithelium.

ROOF PLATE (RF): the thin non-nervous wall of the neural tube; the precursor of the anterior and posterior choroid plexes of the brain and the dorsal sulcus of the spinal cord.

S PHASE OF THE CELL CYCLE: the period after the middle of interphase when DNA replication occurs and chromatin threads go from haplotene to diplotene.

SACCULUS and UTRICULUS (SU); the major cavities of the inner ear derived from the embryonic otic vesicles.

SACRAL: a term referring to the hip region, the trunk level for the attachment of hind limbs.

SCHIZOCOELIC MYOCOELE: a myocoele cavity that forms directly by splitting within the somite mesoderm; i.e., as in the posterior myotomes of Amphioxus and the Ammocetes, and in all somites of higher vertebrates.

SCLEROTIC OSSICLES: small bones that develop as a ring in the sclerotic capsule of the eye in reptiles, birds, and some anamniotes.

SCLEROTOME (SC): the median ventral part of a somite that will give rise to the major components of the metameric axial skeleton including vertebrae, ribs, and sternum.

SECOND MATURATION or MEIOTIC DIVISION: the reduction of diplotene dyad chromosomes to haplotene as the second and final step in the reduction of chromosome number in germ cells.

SECONDARY GONOCYTE, SECONDARY OOCYTE, SECONDARY SPERMATOCYTE: a germ cell following the first maturation or meiotic division and possessing dyad chromosomes revealing primary meiotic reduction.

SECONDARY OVARIAN FOLLICLE: a follicle with two or more layers of follicle cells around a primary oocyte, but lacking antra (spaces) in the follicle layer.

SEESSEL'S POCKET: See Preoral Gut.

SEGMENTAL PLATE or UNSEGMENTED SOMITE MESODERM (USM): the part of the prospective somite mesoderm that has not become visibly subdivided into somite blocks.

SEMICIRCULAR CANALS (SCC): anterior, horizontal, and posterior canals attached to the sacculus of the inner ear; the sensory receptors of motion.

SEMILUNAR or GASSERIAN GANGLION (5g or Vg): the ganglion of the fifth cranial nerve.

SEMINAL VESICLE: an expanded portion of the vas deferens (Wolffian duct) of males that functions in the temporary storage and ripening of sperm.

SENSORY CRISTA (SCR): patches of sensory hair-cells in the semicircular canals of the inner ear that are the sensory receptors for the sense of motion.

SENSORY EPITHELIUM or ROD & CONE LAYER OF THE RETINA (RTS): the layer of light-sensitive cells of the retina; the layer in contact with the pigmented retina.

SENSORY MACULA (SM): patches of sensory hair-cells in the walls of the sacculus and utriculus of the inner ear that are the sensory receptors for equilibrium and balance.

SENSORY RETINA (RT): the inner layer of the optic cup including the layer of rod and cone cells, and the ganglionic nervous layers.

SERIAL SECTIONS: the preparation of embryos in such a manner as to yield an unbroken sequence of slices that are then stained and mounted for microscopic study.

SERIALLY HOMOLOGOUS STRUCTURES: structures from the same individual that are metameric and structurally and developmentally similar in their manner of embryonic origin but different in adult structure; i.e., vertebrae, gill arches, or spinal nerves, etc.

SEROAMNIOTIC CAVITY (SA): the part of the extraembryonic coelom between the serosa and amniotic membranes.

SEROAMNIOTIC RAPHE (SAR): a remnant of the point of fusion of right and left lateral amniotic folds after they overgrow and fuse over the embryo.

SEROSA or SEROSAL MEMBRANE (SR): the outermost of the extraembryonic membranes; it is composed of extraembryonic somatopleure and is the outer component of the amniotic folds that overgrow the embryo.

SERTOLI NURSE CELLS or SUSTENTACULAR CELLS OF SERTOLI: cells of the seminiferous tubules in which spermatids mature into spermatozoa.

SHEARING: a morphodynamic movement of gastrulation in Amphibia that results in the separation of lateral and ventral mesoderm from the endoderm; this is the consequence of mesoderm and endoderm cells moving in conflictingly different directions during gastrulation.

SINUATRIAL VALVE (SAV): the opening into the right atrium for the precaval and postcaval veins; it is homologous with the embryonic sinus venosus.

SINUS RHOMBOIDALIS: the open neural plate at the posterior end of the chick embryo during the first and second days of incubation.

SINUS VENOSUS (SV): the most posterior of the four primary heart chambers; it receives blood from the paired common cardinal and vitelline veins of the early embryo.

SITUS INVERSUS: the abnormal reversal of asymmetries in body organization, as, for example, the retention of the right (instead of the left) arch of the aorta in mammals.

SMALL INTESTINE (ITA): the part of the midgut between the stomach and colon; it is for the most part anterior to the attachment of the yolk stalk.

SOMATIC LATERAL PLATE MESODERM (SO): hypomere mesoderm lining the body wall surface of the coelomic cavities.

SOMATIC MOTOR NUCLEI: See Motor Nucleus.

SOMATOPLEURE (SOP): the double layer consisting of ectoderm and somatic hypomere mesoderm; the primordium of the serosal and amniotic membranes and body wall of the embryo.

SOMITE (SM): the metameric blocks of mesoderm lateral to the notochord; it is also called epimere and is the basis for primary metamerism in the vertebrate body.

SPERM or SPERMATOZOON: a morphologically and physiologically ripe male sex cell.

SPERMATID: See Gonotid.

SPERMATOGENESIS: the development of sperm in the testis.

SPERMATOGONIA: See Gonial Cells.

SPINAL GANGLIA (SG): the dorsal root ganglia of spinal nerves; the location of nerve cell bodies for somatic and visceral sensory neurons of the peripheral nervous system.

SPINAL NERVES (SN): segmental bundles of nerve fibers arising by the union of dorsal (sensory) and ventral (motor) roots of spinal nerves.

SPLANCHNIC MESODERM (SP): hypomere mesoderm lining the gut wall and dorsal and ventral mesenteries of the gut; i.e., the visceral lining of the coelomic cavities.

SPLANCHNOCRANIUM: the part of the skull that is functionally associated with the jaws, and nasal and oral cavities; it is embryonically derived primarily from cranial neural crest (mesectoderm); it is the so-called facial part of the skull.

SPLANCHNOPLEURE (SPP): the double layer consisting of endoderm and splanchnic hypomere mesoderm; the embryonic gut wall and the primordium of the yolk sac and allantoic membranes; these are the only extraembryonic membranes capable of autonomous development of blood vessels; they contribute vessels to the yolk sac placenta and chorionic membrane and placenta.

SPLEEN (SP): a lymphoid organ that develops in the splanchnic mesoderm of the stomach; it is a major source of phagocytic lymphocytes in the blood.

STOMACH (ST): the anterior component of the midgut.

STOMODAEUM (STD): the epidermal invagination that will form the oral cavity of most chordates; in mammals it will form the oral and nasal cavities.

STYLOPODIUM: the upper segment of paired appendages corresponding with the levels of the humerus and/or femur.

SUBCARDINAL ANASTOMOSIS (SBA): the segment of the postcaval vein derived from the fusion of right and left subcardinal veins where veins from the gonads enter the postcava.

SUBCARDINAL VEINS (SB): paired veins that develop as primary drainage for the mesonephric kidneys; they are major contributors to the developing postcaval vein.

SUBCEPHALIC POCKET (SCP): the space between the free head and the proamnion.

SUBCLAVIAN ARTERY (SCA): the major artery to the forelimb; it is homologous with the seventh cervical intersegmental artery.

SUBINTESTINAL VEIN: a vein of the gut in Amphioxus and the Ammocetes larva that is homologous with the hepatic portal vein of higher vertebrates.

SULCUS LIMITANS (SL): a lateral groove in the neurocoele of the spinal cord and brain that marks the boundary between the ventral basal plate (motor) and dorsal alar plate (sensory); it terminates in the mesencephalon.

SUPERFICIAL EPIDERMAL LAYER (EES): the surface layer of epidermal cells in frog embryos.

SUPERIOR GANGLION (9s or IXs): the dorsal ganglion of the ninth cranial nerve.

SUPERIOR MESENTERIC ARTERY (SMA): the vessel in adult vertebrates that supplies blood to the anterior end of the intestine; it is homologous with an anastomosis of embryonic vitelline arteries to the yolk sac.

SUPERIOR VENA CAVA or PRECAVA (SVC or CCR): the vein or veins that return blood to the sinus venosus or sinuatrial valve of the heart from body levels anterior to the heart; it is homologous with the common cardinal veins or right common cardinal vein in mammals.

SUPRACARDINAL VEINS (SPC): embryonic veins that develop as the major drainage of the metanephric kidneys; they give rise to the renal veins of adult amniotes.

SUSTENTACULAR CELLS OF SERTOLI: See Sertoli Nurse Cells.

SWIM BLADDER (SB): the flotation organ of bony fish; it is the homologue of tetrapod lungs.

SYMMETRIZATION: all embryonic activities that lead to embryonic determination of adult planes of body symmetry.

SYMMETRY: any patterned organization of parts in any of the body axes.

SYMPATHETIC GANGLIA: segmental ganglia of the autonomic nervous system.

SYMPATHETIC NERVES: nerves that leave the sympathetic ganglia and go to visceral endorgans.

SYMPATHETIC NERVOUS SYSTEM: autonomic (visceral) neurons located between cervical and thoracolumbar levels of the spinal cord with antagonistic action to parasympathetic nerves to the same endorgans.

SYMPATHETIC TRUNK: nerve fibers that connect segmental ganglia of the autonomic nervous system into an anteroposterior chain of ganglia.

SYNAPSIS: the point by point fusion of allelic gene loci of homologous chromosomes during first meiotic prophase.

SYNCYTIUM: a multinucleate cytoplasmic mass.

SYNGAMY: the fusion of male and female gametes in fertilization.

SYSTEMIC ARCHES (A4 or AA4): the fourth aortic arches; in mammals the left systemic arch is incorporated into the great arch of the aorta; the right systemic arch is the base of the right subclavian artery.

SYSTEMIC TRUNK (SY): the ascending aorta; the systemic component of the embryonic conus arteriosus after it is split by the bulbar septum; the systemic trunk receives blood from the left ventricle and carries it to the aorta.

TACHYMORPHOSIS or TACHYMORPHIC CHARACTERS: the aspects of embryonic development that result from the elimination of primitive steps in organ formation; i.e., the embryo may form advanced structures directly without repeating ancestral paleomorphic steps.

TAILBUD (TB): the mass of undifferentiated cells in the tail of early embryos which will differentiate into neural, epidermal, and mesodermal tissues of the tail; it is synonomous with the tail blastema.

TAILGUT or POSTANAL GUT (TG): a vestigial extension of the hindgut that extends into the tail; it normally degenerates completely.

TAIL FOLD (AMT) or TAIL FOLDS (TF): See Amniotic Folds, Body Folds.

TAIL MESODERM (TM): the cells that will form the tail blastema as shown in a blastula map of prospective regions.

TAXONOMY: the biological discipline that deals with the classification of animals in keeping with their evolutionary relationships.

TEETH (T): complex organs that evolved from placoid scales of fish; they have an outer shell of enamel, a middle bony layer of dentine, and a central pulp cavity.

TELENCEPHALON (TE): the first and most anterior of five major brain divisions; it is the primordium of the cerebral hemispheres and olfactory lobes of the brain.

TELOLECITHAL: eggs with yolk unequally concentrated at the vegetal pole as in eggs of amphibians and birds.

TELOPHASE: the last period in mitosis during which the nuclear envelope is reformed and the cytoplasm divides into two daughter cells; the latter activity is also called cytokinesis.

TENDINOUS DIAPHRAGM (TEN): the central part of the mammalian diaphragm that is not muscular.

TERATOMA or TERATOLOGY: the abnormal growth of differentiated structures in atypical body locations;

they are suspected of being the result of differentiation by primordial germ cells that failed to find the genital ridge and were lost in their wanderings about the body.

TERTIARY FOLLICLE: an ovarian follicle with one or more antra (spaces) in the follicle cell layer surrounding the developing primary oocyte.

TEST CELLS: in tunicates some maternal follicle cells of the ovary remain with the embryo after hatching; these cells secrete the larval test or tunic.

TESTIS: the male sex gland, representing the medullary elements of the embryonic ovotestis.

TESTOSTERONE: the hormone produced by the interstitial cells of Leidig in the testis; the hormone that elicits responses of maleness.

TETRAD CHROMOSOMES: bivalent, or chiasmata-linked homologous chromosomes that are characteristic of chromatin in later stages of primary gonocyte development preceding the first meiotic division.

TETRAPODS: vertebrates with two pairs of limbs suitable for terrestrial locomotion, or with appendages derived evolutionarily from such kinds of limbs; i.e., amphibians, repitles, birds, and mammals.

THECA EXTERNA: the peritoneum of the vertebrate ovary; it may also function as the germinal epithelium of the ovary in instances in which primoridal germ cells are not the source of all adult gametes.

THECA INTERNA: the inner lining of the vertebrate ovary; it is derived from the wall of the embryonic reteovarial sac of the ovotestis.

THIRD VENTRICLE (V 3): the neurocoele of the diencephalon.

THORACIC: a term referring to the chest region of the body with true ribs.

THORACIC TORSION (TT): a lateral twisting of the chick embryo in the region of the thorax; a temporary modification to permit the embryo in the region of the open midgut to lie with its ventral surface against the yolk and, at the same time, to permit the anterior parts of the embryo to lie on its left side.

THYMUS (P3 or P4): glands of the neck and upper chest that are derived from the third and fourth pharyngeal pouches; they are glands associated with lymphoid functions.

THYROGLOSSAL DUCT: a normal temporary connection between the floor of the pharynx at the level of the second visceral arch (base of the adult tongue) and the thyroid gland; it may abnormally persist and mark the path of posterior migration of the thyroid.

THYROID DIVERTICULUM (TH): the primordium of the thyroid gland; it is a midventral evagination of pharyngeal endoderm at the level of the second visceral arch; it is homologous with the endostyle of protochordates.

TONGUE (TO): a muscular organ in the floor of the mouth that carries the visceral sensory epithelium for taste; it evolved as an accessory organ for water circulation through the gill chambers; in tetrapods its muscles are a combination of visceral muscles of the first four visceral arches and occipital hypaxial myotomes.

TOTIPOTENCY: the state of an embryonic part having the ability, or competence, to form a complete embryo; i.e., the zygote of all species, or the prospective ectodermal region of the amphibian blastula, or asexual reproductive buds and blastemas, etc.

TRABECULAE (TRA): (singular, trabeculum) a pair of cartilage rods that extend under the brain anterior to the notochord; they are primary anterior primordia for the chondrocranium; anterior to the optic foramen they are of neural crest origin; posterior to the optic foramen they are derived from head mesenchyme.

TRACHEA (TR): the wind pipe; a connecting tube that is lined by foregut endoderm and connects the lungs to the back of the mouth cavity and larynx.

TRANSVERSE PLANE OF SYMMETRY: the cross sectional plane of the body; it is parallel with the DV and ML axes.

TRANSVERSE PROCESS: paired lateral extensions of the vertebral centrum at the level of the horizontal septum of the body wall; they are primordia of true ribs.

TRANSVERSE SEPTUM (TV): the posterior membrane that separates the pericardial cavity and heart from the peritoneal cavity; the ventral mesoesophagus and lateral mesocardia contribute to its formation; it is a posterior part of the pericardial sac of tetrapods.

TRICUSPID VALVE (AVR): the right atrioventricular valve between the right atrium and right ventricle.

TRIGEMINUS NERVE (5 or V): the fifth cranial nerve (mixed somatic sensory and visceral sensory and motor) to the first visceral arch.

TRIGONE OF THE URINARY BLADDER: a region in the dorsal wall of the urinary bladder that is derived from the pronephric mesoderm of the Wolffian duct.

TROCHLEAR NERVE (4 or IV): the fourth cranial nerve (pure somatic motor) to the superior oblique eye muscles.

TROPHOBLAST: the cells in the mammalian blastocyst exclusive of the embryonic disc or inner cell mass; these cells will form the extraembryonic ectoderm in most mammals; in primates; including man, the trophoblast also forms most extraembryonic mesoderm.

TUBULATION: the gastrulation movement whereby germ layers round up and close off forming tubes; i.e., the closing of endoderm to form the gut tube, or the closing of the neural plate to form the neural tube.

TUNIC: the outer coat of ascidians consisting of a cellulose-like polysaccharide called tunicin.

TUNICA ALBUGINEA: a connective tissue layer in the developing gonad that separates the female (cortical) and male (medullary) parts of the ovotestis; it persists as a vestige in the interior of the adult ovary, and as the connective tissue capsule of the adult testis.

TYMPANUM or TYMPANIC MEMBRANE (CP1): the ear drum; it is derived from the closing plate of the first gill slit, or is re-formed at the site of this closing plate.

ULTIMOBRANCHIAL BODY (P5): a glandular organ that produces the hormone thyrocalcitonin; the gland is derived from the fifth pharyngeal pouch.

UMIBILICAL or ALLANTOIC ARTERIES (UA or ALA): paired arteries carrying blood from the dorsal aorta to the allantoic stalk, chorion, and chorionic placenta.

UMBILICAL or ALLANTOIC VEINS (UV or ALV): paired veins that carry blood from the allantois, chorion, or chorionic placenta back to the common cardinal veins, or in later stages to the ductus venosus and postcaval veins; homologous with lateral abdominal veins of anamniotes.

UMBILICAL CORD or BODY STALK (U): the general connection of an embryo or fetus to its extraembryonic membranes and/or placenta.

UNSEGMENTED SOMITE MESODERM (USM): See Segmental Plate.

URACHUS: a vestige of the allantoic stalk that occasionally persists in adult mammals and connects the urinary bladder to the umbilicus.

URETER (MTD): the metanephric duct which carries urine from the metanephric renál pelvis to the urinary bladder; it is derived from the ureteric bud of the pronephric (Wolffian) duct.

URETHRA: the tube that carries urine from the urinary bladder to the exterior; in female mammals it is derived from the ventral half of the urogenital sinus by a splitting of this cavity by the urethrovaginal septum; in male mammals the urethra consists of two parts: the pelvic urethra (the urogenital sinus between the urinary bladder and urogenital vestibule), and the penile urethra (the urogenital vestibule and urethral groove).

URETHRAL GROOVE: the depression between the lateral genital folds; it is derived from the anteroventral part of the proctodaeum (exclusive of the anal part separated by the urorectal septum), and will contribute to the urogenital vestibule of female mammals and the penile urethra of male mammals.

URETHROVAGINAL SEPTUM: a partition that develops in female mammals and divides the urogenital sinus into a dorsal vagina and ventral urethra, both of which open into the urogenital vestibule.

URINARY BLADDER: a midventral endodermal evagination from the floor of the cloaca or from its ventral subdivision, the urogenital sinus; the urinary bladder of amphibians is homologous with the allantoic sac of amniotes; the urinary bladder of mammals is derived from the base of the allantoic stalk.

UROGENITAL SINUS (UG): the ventral subdivision of the embryonic cloaca that is produced when the cloaca is split by the urorectal septum.

UROGENITAL VESTIBULE: the common opening for the urethral (excretory) and vaginal (reproductive) tubes to the exterior in female mammals; it is derived from the embryonic urogenital sinus and urethral groove.

URORECTAL SEPTUM (UR): the membrane that divides the cloaca into a dorsal rectum and ventral urogenital sinus in mammals.

UTRICULUS (SU): See Sacculus.

UTRICULUS MASCULINUM: a small remnant of the posterior end of the Müllerian (female) duct that is embedded in the prostate gland of the male mammal; it is also called the prostatic utricle.

VAGUS NERVE (10 or X): the tenth cranial nerve (mixed visceral sensory and motor) to the fourth, fifth, and sixth visceral arches, and the major parasympathetic trunks to the thoracic and abdominal visceral organs.

VAS DEFERENS: the sperm duct in males; it is embryonically derived from the mesonephric (Wolffian or archinephric) duct.

VASA EFFERENTIA: modified mesonephric tubules that participate in the transport of sperm from the rete testis to the vas deferens.

VEGETAL POLE (VEG): the point on the surface of an egg opposite the animal pole; it will usually coincide with either the posterior or ventral surface of the future embryo.

VELUM: a ring of small filtering tentacles at the entrance to the pharynx of Amphioxus; in the Ammocetes larva the velum is a muscular valve lacking tentacles.

VENTRAL AORTA (VA): the primary embryonic artery that connects the conus arteriosus to the aortic arches; in adult mammals it contributes to the formation of the innominate and common carotid arteries.

VENTRAL LIP OF THE BLASTOPORE (VLB): the ventral lining of the blastopore; in chordates it is composed of lateral plate mesoderm of the blood island and ventral mesentery. See also Blastopore Lips.

VENTRAL MEDIASTINAL SEPTUM (VMS): the attachment of the pericardial sac to the ventral midline between the two pleural cavities; although it appears to be a ventral mesentery, it is derived from parietal peritoneum of the thoracic body wall.

VENTRAL MESENTERY (VM): the double layer of splanchnic mesoderm that primitively connects the gut to the ventral body wall; the ventral mesoesophagus will form the heart, the ventral mesogaster will form the capsule of the liver and parts of the transverse septum, and the ventral mesocolon will support the urinary bladder; other parts of the ventral mesentery of the gut degenerate.

VENTRAL MESOCARDIUM (VMC): the temporary mesentery for the heart tube which soon disappears except at the anterior end of the conus arteriosus and posterior end of the sinus venosus; it will contribute to the pericardial sac and median mediastinal septum.

VENTRAL MESOESOPHAGUS (VMO): the ventral mesentery of the esophagus; it will form the heart, parts of the median mediastinal septum, and ventral parts of the transverse septum.

VENTRAL MESOGASTER (VMS): the ventral mesentery of the stomach through which the common bile duct passes from the liver to the stomach; it is also called the gastrohepatic ligament.

VENTRAL POLE (V): the belly side of the body.

VENTRAL ROOT OF SPINAL NERVES: bundles of somatic and visceral motor nerve fibers that leave the basal plate of the spinal cord before they unite with the dorsal root of a spinal nerve.

VENTRICLE OF THE HEART (V): one of the primary heart chambers; it is located between the atrium and conus arteriosus.

VERTEBRAL ARTERY (VTA): paired vessels formed by anastomosis from the first six pairs of intersegmental arteries; they are united anteriorly with the basilar artery and posteriorly with the subclavian arteries.

VERTEBRAL CENTRUM (VC): cylindrical blocs of sclerotome that surround the notochord; neural arches, haemal arches, and transverse processes attach to the centrum to complete a typical vertebra.

VERTEBRAL NEURAL SPINE: the most dorsal part of a vertebra; it unites the lateral neural arches and thereby closes the vertebral neural canal dorsally; in evolution (and possibly also embryologically) neural spines are derived in part from dermal scale plates and therefore may be a dermatome (instead of sclerotome) derivative.

VISCERAL ARCH MESENCHYME (VAM): general mesenchyme of visceral arches early in development, including primordia of visceral arch muscle, cartilage, and blood vessels.

VISCERAL ARCHES (VS 1-6): the paired bars of tissue that separate successive pharyngeal pouches; each arch contains a visceral cartilage, aortic arch, cranial nerve, and visceral muscles.

VISCERAL or BRANCHIAL GROOVES (VG 1-5): the epidermal invaginations from the neck region that end in closing plates of gill pouches; the ectodermal component of a gill slit.

VISCERAL MOTOR NUCLEI: See Motor Nucleus.

VISCERAL MUSCLE (VSM 1-6): muscle derived from the splanchnic hypomere covering the gut wall; usually the muscles are smooth but in the pharyngeal gill region they are striated.

VISCERAL SKELETON (VC): cartilages and bones derived from skeletal supports of the gill arches and jaws; all parts are of neural crest origin.

VITAL DYE: any of a number of dyes of low toxicity that are able to pass into living cells and bind with membranes or intracellular organelles without killing cells; these are used to make the distribution of structures or macromolecules visible and to make the mitotic products of embryonic cells visible in living embryos.

VITELLINE ANASTOMOSIS (VVA): the fusion of right and left vitelline veins; it forms the most anterior part of the postcaval vein, the sinusoids of the liver, and parts of the hepatic portal venous system.

VITELLINE ARTERIES (VTA): the paired vessels that carry blood from the dorsal aorta to the yolk sac. In adult mammals they persist as the superior mesenteric artery.

VITELLINE MEMBRANE (VIT): a surface layer closely applied to the living plasma membrane of eggs of many animals; it often becomes part of a fertilization membrane.

VITELLINE VEINS (VV): paired vessels that carry blood from the yolk sac back to the sinus venosus; in the chick there are four vitelline veins; single dorsal and ventral vitelline veins, and paired lateral vitelline veins; the lateral veins will contribute to the formation of the postcaval and hepatic veins of adult amniotes.

VITREUS HUMOR (VH): the fluid-filled cavity of the optic cup behind the lens and iris.

VON BAER'S LAWS or COROLLARIES: the primary generalizations for comparative embryology:

I. The law of convergent development: development proceeds from the general to the specific.

II. The law of increasing specialization: general characters gradually give rise to specific ones.

III. The law of progressive divergence: embryos of different species become more different as development proceeds.

IV. The law of corresponding stages: embryos of advanced taxa pass through early stages resembling early embryos of primitiva taxa, but late embryos of advanced and primitive taxa do not resemble each other; therefore, homologies can only be revealed by comparisons of corresponding stages.

WAVE OF ACTIVATION: the physiological excitation that passes from the sperm entry point around the egg surface and initiates fertilization in the egg.

WHEEL ORGAN: the ciliary organ in the anterior pharyngeal region of Amphioxus that moves water through the gill slits.

WHOLE MOUNT: the preservation, staining, and preparation of total embryos for observation, usually for use with the microscope.

WOLFFIAN BODY: a synonym for the mesonephric kidney, named for the pioneer embryologist, Caspar Friedrich Wolff.

WOLFFIAN DUCT (PND): a synonym for the pronephric, mesonephric, and archinephric duct; the embryonic primordium for the vas deferens of male amniotes.

XENOPLASTIC: the transplantation of tissues or organs between donor and host embryos that belong to different genera or higher taxonomic levels; i.e., the graft of a frog organ onto a salamander embryo.

YOLK PLUG (YP): the endoderm cells that fill the blastopore lumen before invagination of endoderm is complete in amphibian eggs.

YOLK SAC CAVITY (YCS): the space enclosed by the yolk sac; in birds it is filled with yolk; in mammals it is usually vestigial and filled with mucous secretions.

YOLK SAC MEMBRANE (YS): the extraembryonic membrane of reptiles and birds that surrounds the yolk mass; it is composed of splanchnopleure and is vestigial in most mammals.

YOLK SAC PLACENTA (YSP): the primitive type of placenta that is characteristic of marsupials and of early embryonic stages in most placental mammals; it utilizes the yolk sac circulation to establish a close vascular association with the maternal uterine capillary system for metabolic exchanges between fetal and maternal circulations.

YOLK STALK (YST): the connection of the yolk sac to the midgut; it will normally degenerate entirely; however, if it persists, it is called a Meckel's diverticulum from the gut.

ZEUGOPODIUM: the middle segment of the paired appendages which correspond with the levels of the radius–ulna and/or the tibia–fibula.

ZONE OF JUNCTION (ZJ): the outermost edge of the chick blastoderm where cells are in open syncytial contact with the yolk mass.

ZOOID or INTERNAL ZOOID: the living organism inside the tunic of an ascidian.

ZYGOTE: a fertilized egg cell after fusion of the egg and sperm pronuclei.

ZYGOTE MEMBRANE: the living cell membrane of a fertilized egg.

INDEX